Calculus in a Real and Complex World,
Preliminary Edition

Frank Wattenberg
University of Massachusetts

PWS-KENT PUBLISHING COMPANY
Boston

© 1993 by PWS-KENT Publishing Company.
© 1992 by Wadsworth, Inc.

All rights reserved. No part of this book may be reproduced, stored in a retrieval system, or transcribed, in any form or by any means—electronic, mechanical, photocopying, recording, or otherwise—without prior written permission of PWS-KENT Publishing Company.

PWS-KENT Publishing Company is a division of Wadsworth, Inc.

2 3 4 5 6 7 8 9 10—97 96 95 94 93

ISBN 0-534-18733-1

Contents

I Approximation, Limits and Derivatives   **1**
- §1. Population Models 1
- §2. Equilibrium Points and Limits: Experimentation 11
- §3. Limits: Theory 24
- §4. The Bisection Method 38
- §5. The Intermediate Value Theorem and Continuity 48
- §6. When Does an Equilibrium Attract?—Experiment 60
- §7. When Does an Equilibrium Attract?—Theory 70
  - The Economics of Supply and Demand 76
- §8. The Tangent to a Curve 83
- §9. What is the Derivative? 100
  - The Derivative and the Microscope 100
  - The Derivative and Velocity 104
  - Rates of Change 106
  - Sensitivity 108
  - Notation and Terminology 110
- §10. Summary[+] 110
  - The context of calculus 110
  - Estimation 114
  - Limits 117
  - The Derivative 118

II Differentiation   **121**
- §1. Polynomials 121
  - Logistic Models Revisited 130
- §2. Products and Quotients 133
  - The Power Rule Revisited 143
- §3. The Trigonometric Functions 147

CONTENTS

Trigonometry Review	148
Derivatives of the trigonometric functions	158
§4. Optimization, I	162
An Optimization Recipe:	168
§5. Curve-Sketching and Optimization	183
§6. Optimization, II	191
§7. Inverse Functions	217
Another Interpretation of the Derivative	223
The Inverse Trigonometric Functions	227
§8. The Chain Rule	236
Composing Functions	236
The Chain Rule	238
2-Cycles	246
§9. Optics	249
Reflections in a flat mirror	250
Light Caustics	255
Refraction	261
§10. The Hare: Newton's Method	274
§11. When Does an Equilibrium Attract?	280
§12. Summary[+]	291
The Differentiation Formulas	291
Curve Sketching	295
Optimization	297
Stable vs. Unstable Equilibrium Points	298
Newton's Method	300
Geometric Optics	300
Advanced Topics	300

III Differential Equations   303

§1. Acceleration—the Second Derivative	304
§2. Gravity	317
Falling Bodies	317
Motion in Two Dimensions	324
A Theorem	328
§3. Visualizing Differential Equations	330
Some Terminology	330
Visualizing a Differential Equation	334
§4. Euler's Method	345
Euler's Method: An Example	347

CONTENTS

		Euler's Method 351
		Variations on a Theme by Euler 354
	§5.	The Exponential Function 360
		An Interesting Example 361
		The Exponential Function 365
		Properties of the Exponential Function 367
	§6.	The Natural Logarithm 372
		The Functions $y = a^x$ 375
	§7.	Vertical Asymptotes 383
	§8.	L'Hôpital's Rule 394
		How Big is Infinity? 394
		L'Hôpital's Rule 395
	§9.	Summary$^+$ 401

IV Integration   403

- §1. The Definite Integral 403
  - Riemann Sums 409
  - Summation Notation 410
  - The Definite Integral 411
- §2. Properties of the Definite Integral 418
- §3. The Fundamental Theorem of Calculus 425
- §4. Antidifferentiation 436
- §5. Integration by Substitution 445
  - Area by Cross Sections 455
- §6. Numerical Integration, I 459
  - The Midpoint Method 460
  - The Trapezoid Method 460
- §7. Numerical Integration II: Simpson's Rule . 463
- §8. Computer-Based Antidifferentiation 466
- §9. Summary$^+$ 474

V Methods and Applications of Integration   477

- §1. Integration by Parts 477
- §2. Solids of Revolution 487
- §3. Black Holes and Escape Velocity 494
  - Mass, Work, Energy and Force 495
  - Escape Velocity 500
  - Improper Integrals 505
  - A Hint About "Black Holes" 506

CONTENTS

	Improper Integrals, An Encore	508
§4.	Implicit Differentiation	509
§5.	Separation of Variables	519
§6.	The Logistic Equation and Partial Fractions	525
	Continuous Exponential Models	525
	Continuous Logistic Models	528
	The Method of Partial Fractions	531
	Integrating the Simple Terms	537
	Mathematica and Computer Algebra Systems	544
§7.	Arclength	546
	The Arclength of a Curve Given by $y = f(x)$	546
	Arclength of a Curve Given by $x(t), y(t)$	549
§8.	Integration Miscellany	552
	Integrals of the Form $\int \sin^n \theta \cos^m \theta \, d\theta$	553
	Trigonometric Substitution	556
	Rational Functions	562
§9.	Summary†	568
	Techniques of Integration	568
	Applications of Integration	574

VI Real and Complex Numbers.   585

§1.	Polar Coordinates	585
	Area in Polar Coordinates	594
§2.	Complex Numbers	596
§3.	Lotteries and Geometric Series	603
	Geometric Series	605
§4.	Infinite Series	609
	The Harmonic Series	613
	The One-Way Convergence Test	615
	The Integral Test for Convergence	618
	The Comparison Test	621
	More Theorems	623
	The Ratio Test	626
§5.	Some Technical Tools	628
	Factorials	628
	Notation for Higher Derivatives	629
	Extended Initial Value Tables	629
§6.	Power Series	635
	The Taylor Series for a Function	645

CONTENTS

- §7. Taylor's Theorem 655
- §8. Spring-and-Mass Systems 671
  - Differential Equations Revisited 671
  - Spring-and-Mass Systems 675
  - The Differential Equation $ax'' + bx' + cx = 0$ . . . 681
- §9. Linearity 690
  - The Derivative Revisited 697
  - Newton's Method Revisited 699
  - Stability of an Equilibrium 703
- §10. Summary$^+$ 705

VII Dynamical Systems   711
- §1. One-Dimensional Dynamical Systems 711
- §2. Two-Dimensional Dynamical Systems 727
- §3. Energy 745
- §4. Classification of Equilibria, I 755
  - One-Dimensional Autonomous Systems 755
  - Two-Dimensional Autonomous Systems 758
- §5. Optimization 776
- §6. Peaks, Valleys, and Saddle Points 789
- §7. Nonlinear Dynamical Systems 797
- §8. Summary$^+$ 809
  - The SIR Model 811

A   Selected Solutions   817

Introduction

Calculus is an intensely concrete subject that can help us understand and improve the world in which we live. Its name comes from the Latin word for pebble[1] by way of the use of pebbles as counters in games and counting boards like the abacus. Although an abacus with its smooth counters and polished wood can be a beautiful object in and of itself, the real interest of the abacus goes beyond any physical beauty. The abacus allows us to represent important quantities in a way that allows us to reason about them. We can learn much more about acres and elephants by manipulating the small pebbles that represent them than we possibly could by manipulating the originals. Calculus is both a powerful language that allows us to express a great deal about our world and a way of using that language to reason about the world.

The ways in which we work with the ideas of calculus are not dry rules handed down from generation to generation, but come from the real things that these ideas represent. We know a great deal about the real world, phenomena like velocity and acceleration, and this knowledge can help us understand mathematical ideas like the derivative and the second derivative because the mathematical ideas correspond to the real phenomena. The relationship between calculus and the real world is a two-way relationship—calculus helps us to understand the real world and the real world helps us to understand calculus. One cannot truly understand calculus without knowing its roots in the real world. One cannot truly understand the real world without the intellectual power of calculus.

This book is motivated by the idea that calculus should be taught and learned[2] in the context of the real and complex world that it studies. By doing so we will learn more both about calculus and about the real world. Moreover, by studying calculus as it is used, we will be better able to use calculus outside the calculus classroom.

---

[1] *Calculus* is the diminutive of *calx* the Latin word for stone.

[2] It is no accident that young children often confuse the words *teach* and *learn* and that the French verb *apprendre* is sometimes translated *to learn* and sometimes *to teach*.

This book is first and foremost a calculus book. Its purpose is the study of calculus and the ways in which calculus can be used to solve real and complex problems. In large part, however, this book is made possible by the widespread availability of calculators and computers. Without such tools we could not study real problems in a reasonable amount of time. Calculus, calculators, and computers form a powerful team. Computers and calculators provide the computational power, the brute force, needed to solve often very complex real problems. Calculus provides the intelligence and finesse that enables us to use the power of computers and calculators to maximum advantage.

Ideally, you should use this book together with both a good calculator and a computer. Each has its advantages. A calculator is always available. Modern programmable scientific graphics calculators like the HP-48 and TI-85 provide almost all the computational power needed for this course. Computers, however, are faster and more powerful. Because of their speed and displays they can produce far better graphics than even the best programmable calculators and they are much easier to program.

Your instructor will describe the computer facitlities available for your class. In addition, you should have a programmable scientific calculator for use in connection with this and other classes. Ideally, it should be able to do arithmetic with complex numbers, and have built in procedures for numerical integration and numerical solution of equations of the form $f(x) = 0$. If you do not have access to a personal computer then a calculator with graphics capability will be very useful.

Because of the variety of computers, software, and calculators that are suitable for use in connection with this course, specific instructions are not included in this volume. There are lab manuals available for use with particular hardware and software.

We will use calculators and computers in three ways.

- Many of the exercises in this book depend on the computational power of a calculator or computer for their solution. Calculators or computers, like paper and pencil, should be a regular part of your mathematical tool kit. Whenever necessary you should use either a computer or a calculator to solve a particular problem *without any particular instructions to do so.*

- Mathematics is a laboratory science. You learn mathematics by doing mathematics. You should think of computers and calculator as tools like microscopes that you can use to do experiments on your own, to explore beyond the specific assignments given in class.

- The lab manuals for this course contain specific laboratory projects.

Because we deal with the real world, writing is a more important part of this course than is common in more traditional math courses. Exercises do not involve mere numbers divorced from meaning, but numbers that represent real dollars, real weight, or real velocity. We will investigate, for example, what happens to the selling price of a particular product when its manufacturing cost rises. The answer should not be an abstract number, but a contribution to our understanding of inflation, the way it spreads throughout an economy and its impact on both producers and consumers. The answer in real dollars may surprise you in ways that a mere number cannot. You should always test your answers against your intuition. Sometimes an answer will surprise you. Sometimes that surprise will reveal a mathematical error, but sometimes that surprise will reveal a flaw in your intuition. Never hesitate to write a few sentences. Words can express different things than numbers. When dealing with the real world we have need for both.

Finally, and most importantly, approach this course with an experimental spirit. Calculus is a participatory sport. Do the exercises as you read the book. This book deliberately has fewer exercises than many calculus books and they are sprinkled throughout the text rather than being gathered at the end of each section. You will not find 85 routine problems at the end of each section, but you will find more interesting problems. The problems are an essential part of the course, which is another way of saying that you must be an active participant. The more you do, the more you will learn. The most important exercises in this course are not in this book. They are the exercises that you discover on your own, the questions you ask yourself, your friends, and your instructor.

Developing a new calculus course is an enormous enterprise. I would like to thank everybody who contributed to this enterprise but it is impossible because the contributers are so many and their contributions so great. The most important contributions are often the most difficult to pin down. Above all, I would like to thank the students who took the first two versions of this course at the University of Massachusetts, not only for their patience and good nature in working from the rough first drafts of this book, but because their enthusiasm made the whole effort worthwhile. This book has gained enormously from their comments. Six of the students who took the first version of this course went on to become undergraduate assistants for the second version. They deserve special thanks—Vincent Chen, Kim DuPrie, Olaf Johnson, Jennifer Jordan, Keelee MacPhee, and Shuka Shwartz.

I would also like to thank the instructors who taught early versions of this course—Virginia Bastable, Fran Caporello, Richard Ellis, Dave Foulis, Elena Galaktionova, Michael Houser, Mitchell Kotler, Patrick Miller, Peter Norman, Jon Sicks, and Bruce Turkington. All of them have made enormous contributions to the course. Peter deserves special thanks. Not only did he teach the very first and roughest version of this course but he was involved in its prehistory. This course owes much to supplementary materials that Peter and I and others developed for an earlier calculus course. Brian Morris taught an early version of this course using an early version of the *Mathematica* based computer lab. He made numerous valuable suggestions for improvements.

This project was funded by the National Science Foundation as part of the Five College Calculus in Context Project. I would like to thank the NSF both for their financial support and, more signficantly, for the recognition that calculus reform is an important and worthy effort.

This book reflects many of the ideas and long discussions among the participants in the Calculus in Context Project. I would particularly like to thank Jim Callahan, David Cox, Jim Henle, Ken Hoffman, Donal O'Shea, Harriet Pollatsek, and Lester Senechal.

Finally, I'd like to thank Julie, Marty, and Alina whose enjoyment of learning has stimulated mine in teaching.

Chapter I

Approximation, Limits and Derivatives

§1. Population Models

We share the earth's finite resources with bumblebees and elephants, with turnips and our fellow human beings. From day to day and year to year, our numbers and the numbers of our earthly neighbors vary. This section is about mathematical models of population change.

We begin by studying models with one species in an enclosed habitat. Later this year we will look at models involving more than one species. Even single species models can be surprisingly complex. We will see, for example, a single species model that varies as shown in Figure I.1. You might guess that this kind of cyclical variation would necessarily be caused by some external cyclical factor, perhaps sunspots or harvesting by farmers, but we shall see that it can occur without any outside interference.

We begin with some extremely simple models of population growth—models that are almost completely unrealistic. This is typical of mathematical modeling. The first models are frequently not very good, but we can learn from their shortcomings.

One of the most important themes of this course is the use of mathematics to solve *real* problems. Real problems are often extremely complex and cannot be solved completely in a single day. Many will not even be solved completely in our lifetime. Usually one starts with a simple approximation to the real problem, an approximation that in-

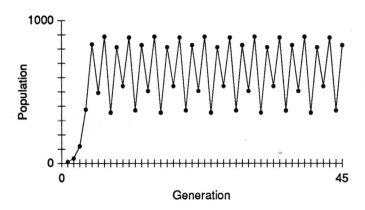

Figure I.1: An example of the population growth of a single species

cludes only the most important factors. Part of the art of mathematical modeling is identifying these factors so that one can build a model that is simple enough to be tractable and yet close enough to reality to give some insight into the original problem. For example, suppose one wanted to study a falling object. If the object were a heavy brick, a model including just the force of gravity would be reasonable. If, however, the object were a feather then air resistance could not be ignored. One often builds a series of models beginning with a simple model including only a few of the most important factors and then adding more factors to get more complicated and better models.

Building mathematical models is very much like other everyday model building activities—building verbal models (describing things in words) or building visual models (drawing pictures). The very act of describing something verbally or drawing a picture forces one to think carefully about reality and to identify its most important features.

We begin by studying species that have distinct generations. Temperate zone insects, for example, have a life cycle that is tied to the four seasons. A new generation hatches each spring, reaches maturity during the summer, and lays eggs in the fall. A graph of the population of such a species might look something like Figure I.2. Our first simplification will be to measure the population in each year by a single number. For example, one might choose to count the number of individuals alive on one particular day, say June 13, in each year. The method used is based on the life cycle of the species being studied.

§1. Population Models

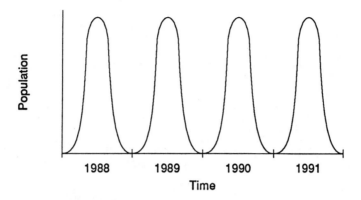

Figure I.2: Population changes for a temperate zone insect

We describe the population of such a species by a sequence of numbers

$$p_1, p_2, p_3, \ldots p_n, \ldots$$

The first term, p_1, denotes the population of the first generation; the second term, p_2, denotes the population of the second generation; and, in general, the nth term, p_n, denotes the population of the nth generation.

We begin with a very simple example. The following table gives the population figures for 1990 and 1991 for a particular species in a particular habitat.

1990	500
1991	600

If these two years were the first two in which we were interested we could write this

$$p_1 = 500$$
$$p_2 = 600.$$

Notice that the net population growth from the first year to the second is 20%. If this same pattern (20% growth each year) were to continue we would see

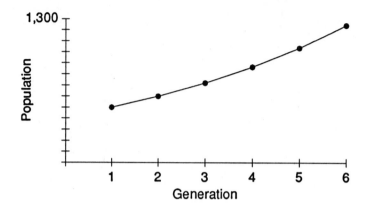

Figure I.3: $p_{n+1} = 1.2p_n$, $p_1 = 500$

1990	p_1	500
1991	p_2	600
1992	p_3	720
1993	p_4	864
⋮	⋮	⋮

Figure I.3 shows the population for the first six years. Each year's population is indicated by a single dot. In order to make our graphs easier to read we connect the dots by straight lines. This makes it look as if the population increases steadily during each year although in reality the population usually varies in a more complicated way.

A model like this one in which the population grows by the same rate each year is called an **exponential model** and is described by an equation of the form

$$p_{n+1} = Rp_n$$

where R is a constant called the **population multiplier.** In the example above $R = 1.2$.

Example:

The population growth of a particular species is described by an exponential model. In 1989 its population was 1500 and in 1990 it was 1650. What will its population be in 1993?

Answer:

§1. Population Models

The ratio of the 1990 population to the 1989 population is

$$\frac{1650}{1500} = 1.1.$$

In an exponential model this ratio will be the same for every year. Thus in 1991 the population will be

$$(1.1)(1990 \text{ population}) = (1.1)(1650) = 1815,$$

in 1992 it will be

$$(1.1)(1815) = 1996.5,$$

and in 1993 it will be

$$(1.1)(1996.5) = 2196.15.$$

For mathematical simplicity we have allowed fractional individuals. Of course, it really doesn't make any sense to talk about 2196.15 individuals. In actual practice, biologists frequently (especially with insect populations) work with the total weight, or **biomass**, of the population rather than the number of individuals. In this case fractions do make sense.

The following table shows population figures for ten years for each of three different exponential models—with $R = 0.9$, $R = 1.0$ and $R = 1.2$. All three models start with $p_1 = 1000$. Figure I.4 shows this same information graphically.

Year	$R = 0.9$	$R = 1.0$	$R = 1.2$
1	1,000.00	1,000.00	1,000.00
2	900.00	1,000.00	1,200.00
3	810.00	1,000.00	1,440.00
4	729.00	1,000.00	1,728.00
5	656.10	1,000.00	2,073.60
6	590.49	1,000.00	2,488.32
7	531.44	1,000.00	2,985.98
8	478.30	1,000.00	3,583.18
9	430.47	1,000.00	4,299.82
10	387.42	1,000.00	5,159.78

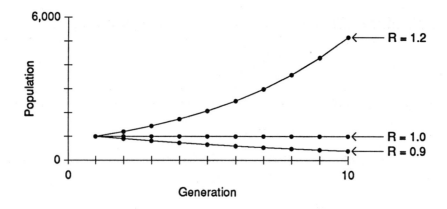

Figure I.4: Comparison of three exponential models

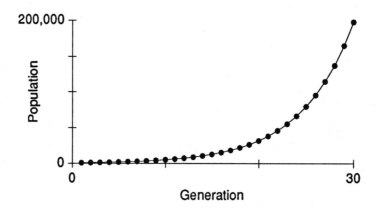

Figure I.5: Thirty years of exponential growth

Notice that when $R = 1$ the population remains constant from year to year, when $R < 1$ the population declines, and when $R > 1$ the population increases. Figure I.5 shows the $R = 1.2$ model for 30 years. After 30 years the population has skyrocketed to about 200,000. Such a model might be reasonable for a few years after a new species is introduced into a habitat that has plenty of food and plenty of room, but clearly it cannot go on forever. At some point the habitat will begin to get crowded; there won't be enough food; there won't be enough shelter. The population cannot continue to rise without any bound.

Exponential models are simple, but not terribly interesting. When $R < 1$ the population dies out. When $R > 1$ the model eventually becomes unrealistic. If $R = 1$ the population remains unchanged. It is, however, unlikely that R would be exactly 1.

§1. Population Models

Exercises

These exercises all involve a species in a very unfavorable habitat. Left alone such a species might behave like an exponential model with $R < 1$, for example,
$$p_{n+1} = 0.8p_n.$$
Suppose, however, that each year a certain number of new individuals migrate to the habitat. Thus we might have a model like
$$p_{n+1} = 0.8p_n + 100$$
in which 100 new individuals migrate to the habitat each year. Thus if the population in 1989 was 1000 the population in 1990 would be
$$(0.8)(1000) + 100 = 800 + 100 = 900$$
and the population in 1991 would be
$$(0.8)(900) + 100 = 720 + 100 = 820.$$
This kind of model is called an **exponential model with immigration**. The general formula describing such a model is
$$p_{n+1} = Rp_n + m.$$
The constant R determines the way in which the population would change in the absence of immigration. In our example above $R = 0.8$. The constant m represents the number of new individuals who migrate to the habitat each year. In the example above m is 100.

Using a computer or calculator[1] calculate $p_2, p_3, \ldots p_{20}$ for each of the following models. Write a few lines describing what you see in each of these exercises.

*Exercise I.1.1: $p_1 = 100$ $p_{n+1} = 0.8p_n + 100$

Exercise I.1.2: $p_1 = 100$ $p_{n+1} = 0.8p_n + 200$

Exercise I.1.3: $p_1 = 100$ $p_{n+1} = 0.8p_n + 50$

Exercise I.1.4: $p_1 = 100$ $p_{n+1} = 0.9p_n + 100$

---

[1]From now on you will be expected to use computers or calculators as appropriate without specific instructions to do so.

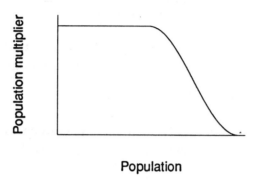

Figure I.6: One possible population multiplier

In exponential models the population multiplier remains constant. In practice, however, it is not constant. Many factors can affect the population multiplier, for example, weather or competition with another species. The single most important factor affecting the population multiplier is the size of the population. If the population is very small, only a few individuals, then there will be plenty of food, water and shelter available. The animals will be healthy, and most will survive long enough to reproduce. On the other hand, if the population is too large for its habitat, there won't be enough food, water, and shelter to go around, the animals will not be well-fed and healthy, and many of them will die before reproducing. The exponential model

$$P_{n+1} = Rp_n$$

with its constant R fails to reflect the variation in the birth and death rates caused by the variation in the size of the population. To get a more realistic model we must replace the constant R by a function $R(p)$ that depends on the population. Such a function usually looks roughly like the one shown in Figure I.6. When the population p is low, $R(p)$ is high because there is an abundance of food, water and shelter. When p is high, $R(p)$ is low because shortages of food, water, and shelter result in less healthy animals, more deaths, and fewer births.

The function $R(p)$ can, however, be quite different than Figure I.6. For a species that hunts in packs, for example, the population multiplier might look like Figure I.7 because if the population were very low the

§1. POPULATION MODELS

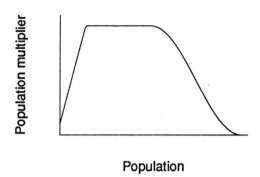

Figure I.7: The population multiplier for a species that hunts in packs

animals would be unable to form efficient hunting packs and, thus, would have difficulty catching their usual prey.

Exercises

Exercise I.1.5: Graph both of the following functions on the same piece of graph paper.

$$R_1(p) = 2.3 - .0001p$$
$$R_2(p) = 2.3 - .00005p$$

These two functions describe the population multiplier for the same species in two different habitats. Which of the two habitats is generally more favorable? That is, which of the two habitats has more food, water shelter, and a better climate? Explain your answer.

Exercise I.1.6: Graph each of the following functions.

$$R_1(p) = \begin{cases} 2.4 & \text{if } p \leq 5000; \\ 2.9 - .0001p & \text{if } 5000 < p. \end{cases}$$

$$R_2(p) = \begin{cases} .00048p & \text{if } p \leq 5000; \\ 2.9 - .0001p & \text{if } 5000 < p. \end{cases}$$

Which of these two functions describes the population multiplier for a species that usually hunts in packs? Explain your answer.

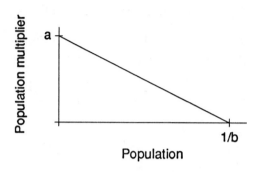

Figure I.8: A linear population multiplier

In the final exercises for this section we work with some particularly simple population multipliers—linear functions of the form

$$R(p) = a(1 - bp)$$

where a and b are positive constants. See Figure I.8. With this function describing the population multiplier, the population will change from year to year according to the formula

$$p_{n+1} = a(1 - bp_n)p_n.$$

This kind of model is called a **logistic model**.

The constants a and b reflect the underlying biology. The constant a determines the value of the population multiplier when the population is very close to zero and there is very little competition for resources. Thus, the value of this constant reflects the biology of the species in a habitat of abundance. The constant b determines how quickly the population multiplier falls as the population rises. The value of this constant reflects how much food, water and shelter are found in the habitat.

Exercises

For each of the following logistic models calculate p_2, p_3, ... p_{20}. Draw graphs similar to Figure I.1 showing your results. After you have done the exercises below try some experiments of your own using different values of p_1, different values of a, and different values of b to see what happens. Discuss your results with other members of the class.

*Exercise I.1.7: $p_1 = 10$, $p_{n+1} = 1.8(1 - .001p_n)p_n$
Exercise I.1.8: $p_1 = 10$, $p_{n+1} = 2.8(1 - .001p_n)p_n$
*Exercise I.1.9: $p_1 = 10$, $p_{n+1} = 3.4(1 - .001p_n)p_n$
Exercise I.1.10: $p_1 = 10$, $p_{n+1} = 3.52(1 - .001p_n)p_n$

§2. Equilibrium Points and Limits: Experimentation

The interplay between experiment and theory in mathematics is one of the most important themes of this course. In this section we continue the experimentation begun in section 1. We turn to theory in the next section. Consider the logistic model

$$p_1 = 10, \quad p_{n+1} = 2.7(1 - .001p_n)p_n.$$

This model's predictions are shown below and in Figure I.9.

Year	Population	Year	Population	Year	Population
1	10.0000	16	630.0342	31	629.6277
2	26.7300	17	629.3460	32	629.6310
3	70.2419	18	629.8279	33	629.6287
4	176.3315	19	629.4907	34	629.6303
5	392.1444	20	629.7268	35	629.6292
6	643.5914	21	629.5616	36	629.6300
7	619.3301	22	629.6773	37	629.6294
8	636.5529	23	629.5963	38	629.6298
9	624.6539	24	629.6530	39	629.6295
10	633.0458	25	629.6133	40	629.6297
11	627.2068	26	629.6411	41	629.6296
12	631.3097	27	629.6216	42	629.6297
13	628.4459	28	629.6352	43	629.6296
14	630.4544	29	629.6257	44	629.6296
15	629.0504	30	629.6324	45	629.6296

Notice that the population seems to be settling down to the number 629.6296. These results raise some questions, for example, would the same end result occur with a different initial population? To explore this question you might try the same basic model

$$p_{n+1} = 2.7(1 - .001p_n)p_n$$

with several different initial populations. The six columns of figures in the table below show the results of six such experiments. The initial population is the top figure in each column. The same results are shown in Figure I.10.

Year	Pop.	Pop.	Pop.	Pop.	Pop.	Pop.
1	25.0000	100.0000	200.0000	400.0000	600.0000	800.0000
2	65.8125	243.0000	432.0000	648.0000	648.0000	432.0000
3	165.9993	496.6677	662.5152	615.8592	615.8592	662.5152
4	373.7975	674.9700	603.6898	638.7569	638.7569	603.6898
5	631.9969	592.3408	645.9708	623.0156	623.0156	645.9708
6	627.9574	651.9776	617.4699	634.1414	634.1414	617.4699
7	630.7926	612.6376	637.7423	626.4165	626.4165	637.7423
8	628.8119	640.7445	623.7731	631.8510	631.8510	623.7731
9	630.2003	621.5157	633.6366	628.0614	628.0614	633.6366
10	629.2293	635.1316	626.7814	630.7208	630.7208	626.7814
11	629.9094	625.6965	631.6015	628.8626	628.8626	631.6015
12	629.4336	632.3411	628.2388	630.1650	630.1650	628.2388
13	629.7668	627.7118	630.5980	629.2541	629.2541	630.5980
14	629.5336	630.9622	628.9493	629.8921	629.8921	628.9493
15	629.6968	628.6920	630.1046	629.4457	629.4457	630.1046
16	629.5826	630.2836	629.2965	629.7583	629.7583	629.2965
17	629.6626	629.1707	629.8625	629.5395	629.5395	629.8625
18	629.6066	629.9503	629.4665	629.6927	629.6927	629.4665
19	629.6458	629.4049	629.7438	629.5855	629.5855	629.7438
20	629.6183	629.7868	629.5497	629.6605	629.6605	629.5497
21	629.6375	629.5195	629.6856	629.6080	629.6080	629.6856
22	629.6241	629.7067	629.5905	629.6448	629.6448	629.5905
23	629.6335	629.5757	629.6570	629.6190	629.6190	629.6570
24	629.6269	629.6674	629.6104	629.6370	629.6370	629.6104
25	629.6315	629.6032	629.6431	629.6244	629.6244	629.6431
26	629.6283	629.6481	629.6202	629.6333	629.6333	629.6202
27	629.6306	629.6167	629.6362	629.6271	629.6271	629.6362
28	629.6290	629.6387	629.6250	629.6314	629.6314	629.6250
29	629.6301	629.6233	629.6329	629.6284	629.6284	629.6329
30	629.6293	629.6341	629.6274	629.6305	629.6305	629.6274

Notice that in the long run all of these experiments look similar. The population bounces around a bit but then after many years seems

§2. Equilibrium Points and Limits: Experimentation

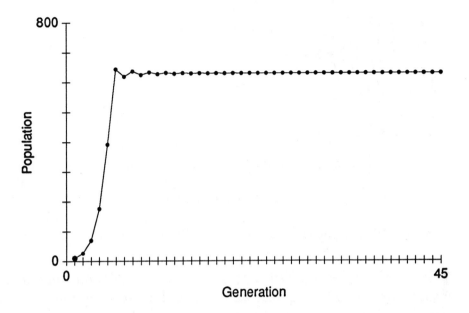

Figure I.9: $p_{n+1} = 2.7 p_n (1 - .001 p_n)$ $p_1 = 10$

to settle down to around 629.63. Based on these experiments one might speculate that this particular model describes a situation in which the habitat can support around 629.63 individuals.

We can explore this idea further by looking more closely at our model. We can rewrite this model

$$p_{n+1} = 2.7(1 - .001 p_n) p_n$$

as

$$p_{n+1} = f(p_n)$$

where $f(p)$ is the function

$$f(p) = 2.7(1 - .001p)p.$$

If we think of p_n as being the *current* population and p_{n+1} as being *next year's* population, then this equation says

Next year's population will be f(the current population).

Thus, next year's population will be greater than the current population if $f(p_n) > p_n$, and next year's population will be less than the current

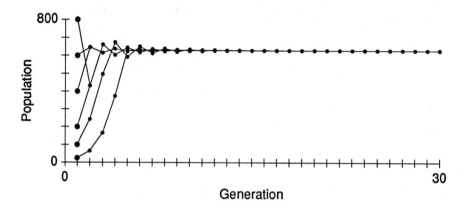

Figure I.10: Six logistic models with different initial populations

population if $f(p_n) < p_n$. When $f(p_n) = p_n$ the population will remain the same for the following year and, hence, for every year after that. Figure I.11 shows the function $y = f(p)$ and the diagonal line $y = p$.

In view of the preceding remarks we are interested in when $f(p) < p$, $f(p) > p$, and $f(p) = p$. We determine when $f(p) = p$ as follows

$$\begin{aligned} f(p) &= p \\ 2.7(1 - .001p)p &= p \\ 2.7p - .0027p^2 &= p \\ 1.7p - .0027p^2 &= 0 \\ p(1.7 - .0027p) &= 0 \end{aligned}$$

Thus $f(p) = p$ when $p = 0$ or when

$$\begin{aligned} 1.7 - .0027p &= 0 \\ p &= 1.7/.0027 \\ &\approx 629.6296 \end{aligned}$$

These two points are called **equilibrium points**. If p_n is an equilibrium point then $p_{n+1}, p_{n+2}, \ldots$ will all have the same value. The first equilibrium point, $p = 0$, corresponds to the fact that if the population starts at zero then in the absence of immigration it will remain there. The second equilibrium point, $p = 1.7/.0027$, is more interesting. This is the level at which the population is in balance with its habitat. There

§2. Equilibrium Points and Limits: Experimentation

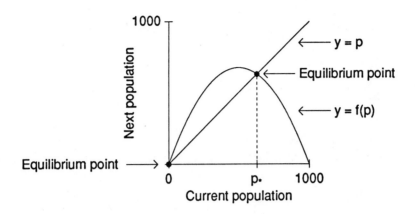

Figure I.11: $y = f(p)$ and $y = p$

is enough food, water and shelter so that the birth and death rates are in equilibrium. We use the notation p_* for this equilibrium point.

Looking at Figure I.11 we see that if the current population p_n is below p_* then the population will rise the following year because the graph of $y = f(p)$ is above the graph of $y = p$. If the current population is above p_* then the population will fall the following year because the graph of $y = f(p)$ is below the graph of $y = p$.

Life is not always quite so simple. For example, consider a species that usually hunts for food in packs. When the population is very low the population multiplier will also be very low because individuals hunting alone or in small groups will be very inefficient. As the population rises, the population multiplier will rise until there are enough individuals to form efficient hunting packs. Then, as the population rises even further, the population multiplier will begin to fall because of shortages of food, water and shelter. An example of this kind of population multiplier is shown in Figure I.12.

The formula for this population multiplier is

$$R(p) = \begin{cases} .0243p & \text{if } p \leq 100; \\ 2.7(1 - .001p) & \text{if } 100 < p \leq 1000. \end{cases}$$

Notice that this formula has two clauses. The first clause describes the computation of $R(p)$ if $p \leq 100$. The second clause describes the computation of $R(p)$ if $p > 100$.

This population multiplier leads to the following model for the pop-

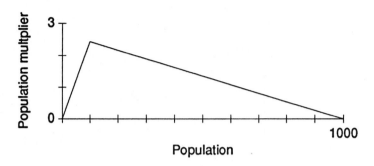

Figure I.12: Population multiplier for pack-hunters

ulation growth for this species.

$$p_{n+1} = R(p_n)p_n$$
$$= f(p_n)$$

where $f(p)$ is the function

$$f(p) = \begin{cases} .0243p^2 & \text{if } p \leq 100; \\ 2.7(1 - .001p)p & \text{if } 100 < p \leq 1000. \end{cases}$$

A graph of the function $f(p)$ is shown in Figure I.13. Notice that this model has three equilibrium points, marked by dots in Figure I.13.

To find the three equilibrium points we must work with each clause separately. The first clause gives us

$$.0243p^2 = p$$
$$p - .0243p^2 = 0$$
$$p(1 - .0243p) = 0$$

which has two solutions $p = 0$ and $p = 1/.0243 \approx 41.1523$.

The second clause gives us

$$2.7(1 - .001p)p = p$$
$$2.7p - .0027p^2 = p$$
$$1.7p - .0027p^2 = 0$$

which has two solutions $p = 0$ and $p = 1.7/.0027 \approx 629.6296$. The first of these solutions, $p = 0$ is spurious since this clause of the definition of $f(p)$ only applies when $100 < p$.

§2. Equilibrium Points and Limits: Experimentation

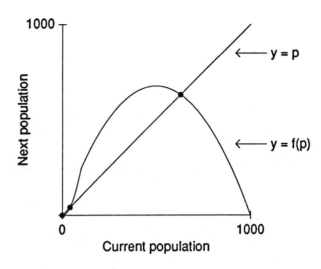

Figure I.13: $f(p)$ for pack-hunters

Putting the two clauses together we obtain three equilibrium points $p = 0$ and $p = 1/.0243$ from the first clause and $p = 1.7/.0027$ from the second clause.

Next we try this dynamical system with six different initial populations. The results are shown in the table below and in Figure I.14.

Year	Pop.	Pop.	Pop.	Pop.	Pop.	Pop.
1	20.0000	40.0000	60.0000	100.0000	200.0000	800.0000
2	9.7200	38.8800	87.4800	243.0000	432.0000	432.0000
3	2.2958	36.7332	185.9618	496.6677	662.5152	662.5152
4	.1281	32.7887	408.7261	674.9700	603.6898	603.6898
5	.0004	26.1249	652.5065	592.3408	645.9708	645.9708
6	.0000	16.5850	612.2028	651.9776	617.4699	617.4699
7	.0000	6.6840	641.0084	612.6376	637.7423	637.7423
8	.0000	1.0856	621.3149	640.7445	623.7731	623.7731
9	.0000	.0286	635.2633	621.5157	633.6366	633.6366
10	.0000	.0000	625.6004	635.1316	626.7814	626.7814
11	.0000	.0000	632.4063	625.6965	631.6015	631.6015
12	.0000	.0000	627.6652	632.3411	628.2388	628.2388
13	.0000	.0000	630.9943	627.7118	630.5980	630.5980
14	.0000	.0000	628.6693	630.9622	628.9493	628.9493
15	.0000	.0000	630.2994	628.6920	630.1046	630.1046

Notice that when the population starts out below the equilibrium point $1/.0243$ it drops because the population is too low to form ef-

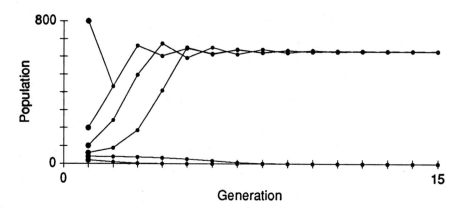

Figure I.14: Six pack-hunter models

ficient hunting packs but when the population starts out above this equilibrium point it goes toward the equilibrium point 1.7/.0027.

Next we introduce some terminology to describe some of the phenomena that we have observed in our experimentation.

Definition:

> A **discrete dynamical system** is a rule
>
> $$p_{n+1} = f(p_n)$$
>
> that can be used to generate each term of a sequence from the preceding term.

Definition:

> An **equilibrium point** for a discrete dynamical system is a solution of the equation
>
> $$p = f(p),$$
>
> that is, a point at which the two curves $y = f(p)$ and $y = p$ intersect as shown in Figures I.11 and I.13.

Frequently our nonmathematical knowledge gives us mathematical information about a model. For example, the sequence

$$100, 85, 70, 55, 40, 25, 10, -5, -20, \ldots.$$

§2. Equilibrium Points and Limits: Experimentation

cannot be a population record because negative population doesn't make any sense.

Example:

Books, tables, chairs and other objects are made up of many molecules that are usually in constant random motion. This motion produces the sensation we call **heat**. The temperature of an object indicates the speed of this random motion. If the molecules are absolutely still then the temperature is -459.6 degrees Fahrenheit or -273.1 degrees Celsius. This temperature is called **absolute zero** because there is no way that an object can have a lower temperature. Thus the sequence
$$-200, -250, -275, -287.5, \ldots$$
could not be a daily record of the temperature of an object in degrees Celsius because no object can have a temperature below -273.1 degrees Celsius.

This leads us to the following definitions.

Definition:

Suppose that $p_1, p_2, p_3, \ldots$ is a sequence.

We say a number L is a **lower bound** for this sequence if every term p_n is greater than or equal to L. We sometimes say that the sequence is **bounded below by** L.

We say a number U is an **upper bound** for this sequence if every term p_n is less than or equal to U. We sometimes say that the sequence is **bounded above by** U.

A sequence is said to be **bounded** if it is bounded below and bounded above.

Example:

Suppose that a sequence is the record of the daily temperature (in degrees Fahrenheit) of the water in a particular lake. This sequence will be bounded below by 32 since water freezes at 32 degrees Fahrenheit and bounded above by 212 since water boils at 212 degrees Fahrenheit.

Example:

Consider the sequence produced by the model

$$p_{n+1} = 1.5p_n, \quad p_1 = 100.$$

This sequence begins

$$100, \ 150, \ 225, \ 337.5, \ 506.25, \ 759.375, \ldots.$$

This sequence continues to get larger and larger with no upper bound. For example, $p_{20} = 221,684$ and $p_{30} = 12,783,404$.

Because this sequence is not bounded above, this model cannot possibly be a model of the population of any species in a finite habitat.

Example:

Consider the sequence given by the logistic model

$$p_{n+1} = 2.7p_n(1 - .001p_n), \quad p_1 = 100.$$

Figure I.15 shows the first 45 terms of this sequence. The dotted line indicates an upper bound. Zero is a lower bound. Notice that whenever a sequence is bounded above it has many different upper bounds. For example, 1000 is an upper bound for this sequence and 345,678 is another upper bound for the same sequence. Similarly, a sequence that is bounded below has many different lower bounds.

In ordinary English we often use the word "limit" instead of the word "bound." However, we will be using the word "limit" in a different way, one that will be explained below. Be careful not to confuse these two words. They both have very precise and different mathematical definitions even though their less precise English definitions overlap.

If you look back at the first model in this section you will observe that after a large number of years the population is very close to $1.7/.0027$. In this situation we say that after a large number of years the population **approaches** $1.7/.0027$. We will express this notion with the formal notation described below. This definition is an intuitive one. In the next section we will give a more precise definition.

§2. Equilibrium Points and Limits: Experimentation

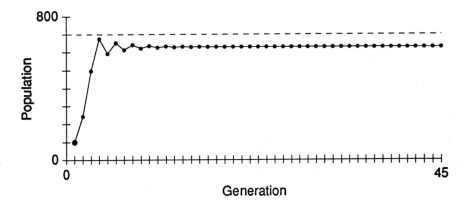

Figure I.15: A bounded sequence

Definition:

We say that a sequence $p_1, p_2, p_3, \ldots$ **approaches the limit** L or **converges to the limit** L and write

$$\lim_{n \to \infty} p_n = L$$

if for every very large n, p_n is very close to L.

If the sequence $p_1, p_2, p_3, \ldots$ comes from a population model then

$$\lim_{n \to \infty} p_n = L$$

means that after many years the population will be close to L.

Do the following exercises using the definitions given in this section. In particular, answer questions about limits based on the intuitive definition given in this section.

Exercises

Questions: *(For each exercise)*

- Is the sequence bounded above? If so, give an upper bound (any upper bound will do).

- Does the dynamical system have any equilibrium points? If so, what are they?

- Does the sequence have a limit? If so, what is it?

*Exercise I.2.1: $p_{n+1} = 0.7p_n + 100$, $p_1 = 100$.

Exercise I.2.2: $p_{n+1} = 0.7p_n + 100$, $p_1 = 500$.

Exercise I.2.3: $p_{n+1} = 0.5p_n + 100$, $p_1 = 100$.

Exercise I.2.4: $p_{n+1} = 0.5p_n + 100$, $p_1 = 500$.

*Exercise I.2.5: $p_{n+1} = 3.4(1 - .001p_n)p_n$, $p_1 = 100$.

Exercise I.2.6: $p_{n+1} = 3.4(1 - .001p_n)p_n$, $p_1 = 300$.

Exercise I.2.7: $p_{n+1} = 3.6(1 - .001p_n)p_n$, $p_1 = 100$.

Exercise I.2.8: $p_{n+1} = 3.6(1 - .001p_n)p_n$, $p_1 = 300$.

*Exercise I.2.9: $p_{n+1} = 3.96(1 - .001p_n)p_n$, $p_1 = 100$.

Exercise I.2.10: $p_{n+1} = 3.96(1 - .001p_n)p_n$, $p_1 = 300$.

There are many other experiments that you might perform on your own. The first thing you might do is to experiment with other examples of logistic models

$$p_{n+1} = a(1 - bp_n)p_n$$

using different values for the constants a and b. You might try to find some pattern in the behavior of these models for different values of a and b.

If you do a number of experiments with these models you will see some very surprising results. You might wonder whether these surprising results are a peculiar property of these particular models. To explore this question you might look at a family of models that is scientifically very much like the family of logistic models. Logistic models are based on the equation

$$p_{n+1} = R(p_n)p_n$$

where the population multiplier, $R(p)$, is given by a formula like

$$R(p) = a(1 - bp)$$

whose graph looks like Figure I.16

Figure I.17 shows another family of very natural population multipliers. This family of functions is described by

§2. Equilibrium Points and Limits: Experimentation

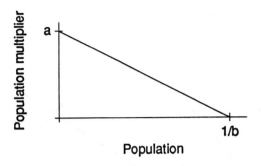

Figure I.16: Linear population multiplier

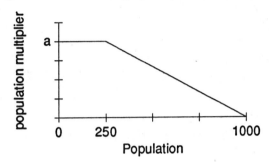

Figure I.17: Another family of natural population multipliers

$$R(p) = \begin{cases} a & \text{if } 0 \leq p \leq 250; \\ a - a(\frac{p-250}{750}) & \text{if } 250 \leq p \leq 1000. \end{cases}$$

*Exercise I.2.11: Experiment with some population models of the form

$$p_{n+1} = R(p_n)p_n, \quad p_1 = 50$$

built around the family of population multipliers described above. Try different values of a and different values of p_1. See how your results with these models compare with some of the results for logistic models.

Computers and Calculators

Many computers and scientific calculators have two capabilities that can be very helpful for investigating equilibrium points. They can draw graphs and they can solve equations of the form $f(x) = 0$ either numerically or symbolically. The following *Mathematica* session, for example,

24 CHAPTER I. APPROXIMATION, LIMITS AND DERIVATIVES

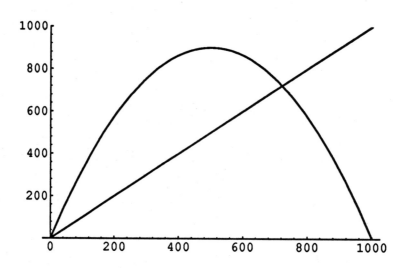

Figure I.18: *Mathematica graph*

examines the equilibrium points of the dynamical system

$$p_{n+1} = 3.6(1 - .001p_n)p_n$$

two ways—first by drawing a graph showing the two curves $y = p$ and $y = 3.6(1 - .001p)p$ (see Figure I.18) and then by solving the equation $p = 3.6(1 - .001p)p$ symbolically.

```
Plot[{3.6 (1 - .001 p) p, p},
    {p,0,1000},
    PlotRange -> {0, 1000}]

Solve[3.6 (1 - .001 p) p == p,p]

{{p -> 0.}, {p -> 722.222}}
```

§3. Limits: Theory

In the last section we looked at the following intuitive definition for one of the most important ideas in calculus, the idea of a limit.

§3. Limits: Theory

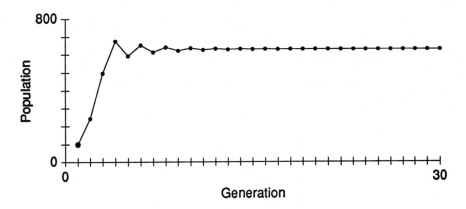

Figure I.19: $p_{n+1} = 2.7p_n(1 - .001p_n)$ $p_1 = 100$

Definition:

Suppose that $p_1, p_2, p_3, \ldots$ is a sequence. We say that the number L is the **limit** of this sequence, and write

$$\lim_{n \to \infty} p_n = L.$$

if for every very large n, p_n is very close to L.

This definition gives a good *intuitive* description of the behavior of a model like the one shown in Figure I.19 and described by

$$p_{n+1} = 2.7p_n(1 - .001p_n), \quad p_1 = 100.$$

In this model after many years the population is very close to $1.7/.0027$. In other words,

$$\lim_{n \to \infty} p_n = 1.7/.0027.$$

This intuitive definition is, however, almost completely useless for many practical problems. Consider the following example.

Example:

Suppose that a particular lake was polluted by a massive dumping of toxic wastes in 1987. As a result, in 1987 the lake had a concentration of 10 ppm (parts per million) of a toxic chemical called Agent QZ. The lake empties into a river and receives plenty of rainfall. To make this problem mathematically tractable we will assume that the lake is constantly and thoroughly stirred up, so that Agent QZ is uniformly distributed throughout the lake. State environmental protection officers have made the following measurements of the concentration of Agent QZ in parts per million.

Year	Concentration
1987	10.00
1988	8.00
1989	6.40
1990	5.12

Based on these measurements they predict that the concentration of Agent QZ in the lake will be given by the model

$$p_1 = 10.00 \text{ ppm}, \quad p_{n+1} = 0.8 p_n$$

where p_1 is the concentration in 1987.

This model depends on a number of assumptions about the lake. The most important is that the general pattern of evaporation, of waterflow from the lake into the river, and of rainfall into the lake will stay constant. If there were several years of drought the water level in the lake would drop, less Agent QZ would flow out of the lake, and less fresh water would flow into the lake. Thus the concentration of Agent QZ would not drop as rapidly as it did in 1987-90. During an extreme drought the *concentration* of Agent QZ might even increase because water was still being lost by evaporation while very little Agent QZ was flowing out of the lake. Our assumption that the water in the lake was thoroughly stirred is also extremely important. In practice, pollutants often sink to the bottom and collect in troughs undisturbed by the surface water flow.

Figure I.20 shows the result of using this model to make predictions through the year 2007. The large dots indicate actual measurements. The smaller dots indicate predictions. Notice that after many years the pollution level is very close to zero or in other words that

$$\lim_{n \to \infty} p_n = 0.$$

This information while mathematically interesting is almost completely useless. We really want to know *when we can go swimming* and *when we can safely drink the water.*

§3. LIMITS: THEORY

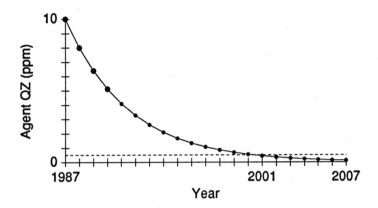

Figure I.20: Model of a polluted lake

In order to answer these questions we need to know what we mean by "safe for swimming" or "safe for drinking." Specifying what is meant by "safe for swimming" or "safe for drinking" is difficult and extremely controversial. As you can imagine, the company responsible for the pollution might claim that swimmers could tolerate a lot more pollution than the swimmers themselves might believe was safe.

Suppose that it has somehow been established that 0.5 ppm is a safe level of Agent QZ for swimming. This level is indicated by a dotted line in Figure I.20. Looking at this graph we can see that it will be safe to swim in the lake in the year 2001. We can read the same result from the following table.

Term	Year	Pollution	Term	Year	Pollution
p_1	1987	10.00	p_{11}	1997	1.07
p_2	1988	8.00	p_{12}	1998	0.86
p_3	1989	6.40	p_{13}	1999	0.69
p_4	1990	5.12	p_{14}	2000	0.55
p_5	1991	4.10	p_{15}	2001	0.44
p_6	1992	3.28	p_{16}	2002	0.35
p_7	1993	2.62	p_{17}	2003	0.28
p_8	1994	2.10	p_{18}	2004	0.23
p_9	1995	1.68	p_{19}	2005	0.18
p_{10}	1996	1.34	p_{20}	2006	0.14

Notice that with this particular model the level of pollution is steadily decreasing. As we mentioned earlier this is not always true.

Because the pollution level in this model is steadily decreasing we can determine when it will be safe to swim simply by using the model to compute the predicted level of pollution until that level falls below the safe level. Since we know that the pollution level will continue decreasing we do not have to worry about the possibility that the lake will become unsafe for swimming at a later date.

Now suppose that the company responsible for polluting the lake claims that it is perfectly safe to swim in the lake even if the concentration of Agent QZ is as high as 1.5 ppm. Looking at the preceding table we see that if we accept the company's claim then it will be safe to swim in the year 1996. The length of time we must wait before it is safe to swim in the lake depends on the definition of "safe for swimming." The more tolerant we are of Agent QZ, the sooner we will be able to swim in the lake. The more cautious we are, the longer we will have to wait.

Exercises

*Exercise I.3.1: Suppose that according to federal regulations the maximum allowable concentration of Agent QZ in water used for drinking is 0.25 ppm. When will it be safe to use the water in this lake for drinking?

Exercise I.3.2: Suppose that according to state regulations the maximum allowable concentration of Agent QZ in water used for drinking is 0.15 ppm. When will it be safe to use the water in this lake for drinking?

This example motivates a much more useful definition of the notion of limit. The basic idea is that for very large n, p_n is very close to the limit, L. The practical meaning of "very large" depends on how fussy we are about "very close." In this example if "very close" means 0.5 then "very large n" means n must be at least 15; if "very close" means 1.5 then "very large n" means n must be at least 10.

We can summarize this with the following table that includes the results from our example above and the exercises.

§3. Limits: Theory

"Very close"	"Very large"
0.15	$n \geq 20$
0.25	$n \geq 18$
0.50	$n \geq 15$
1.50	$n \geq 10$

The number in the "very close" column is usually called the **tolerance**. Thus we say that according to state regulations the tolerance for Agent QZ in water used for drinking is 0.15 ppm. Now we are ready to give a useful definition of limit.

Definition:

Suppose that $p_1, p_2, p_3, \ldots$ is a sequence of numbers. We say that this sequence **approaches the limit** L if for every tolerance $\epsilon > 0$ there is some N such that for every $n \geq N$

$$|p_n - L| < \epsilon.$$

We write this
$$\lim_{n \to \infty} p_n = L.$$

Notice that the quantity

$$|p_n - L|$$

is the distance between the term p_n and the limit L. Thus this definition captures the idea that for very large n, p_n is very close to L. This new definition rephrases our original definition in a way that is more useful for practical problems. Using the notation in this definition we can rephrase our earlier results as follows.

Tolerance	N
0.15	20
0.25	18
0.50	15
1.50	10

Exercises

*Exercise I.3.3:

Suppose that a lake has been polluted by Agent VR and that state environmental protection officers have built the following model describing the level of pollution in the lake.

$$p_{n+1} = 0.75 p_n, \quad p_1 = 8 \text{ ppm}.$$

Notice that
$$\lim_{n \to \infty} p_n = 0.$$

Fill in the blanks in the following table.

Tolerance	N
1.0	
0.5	
0.1	

Exercise I.3.4:

Suppose that a lake has been polluted by Agent LZ and that state environmental protection officers have built the following model describing the level of pollution in the lake.

$$p_{n+1} = 0.85 p_n, \quad p_1 = 5 \text{ ppm}.$$

Notice that
$$\lim_{n \to \infty} p_n = 0.$$

Fill in the blanks in the following table.

Tolerance	N
1.0	
0.5	
0.1	

The idea of limit is extremely important and occurs in many different situations. The following example shows this idea in another context.

§3. Limits: Theory

Example:

Consider the population model

$$p_{n+1} = 0.8p_n + 100, \quad p_1 = 20.$$

This is an example of an exponential model with immigration. The first step in investigating any population model is to look for equilibrium points. In this case we solve the equation

$$\begin{aligned} p &= 0.8p + 100 \\ 0.2p &= 100 \\ p &= 100/0.2 \\ p &= 500 \end{aligned}$$

If we simulate this model for thirty years we obtain the results shown in the following table. The same results are shown graphically in Figure I.21. From these results it appears that

$$\lim_{n \to \infty} p_n = 500.$$

Term	Population	Term	Population	Term	Population
p_1	20.00	p_{11}	448.46	p_{21}	494.47
p_2	116.00	p_{12}	458.77	p_{22}	495.57
p_3	192.80	p_{13}	467.01	p_{23}	496.46
p_4	254.24	p_{14}	473.61	p_{24}	497.17
p_5	303.39	p_{15}	478.89	p_{25}	497.73
p_6	342.71	p_{16}	483.11	p_{26}	498.19
p_7	374.17	p_{17}	486.49	p_{27}	498.55
p_8	399.34	p_{18}	489.19	p_{28}	498.84
p_9	419.47	p_{19}	491.35	p_{29}	499.07
p_{10}	435.58	p_{20}	493.08	p_{30}	499.26

In order to investigate this matter more closely we can add an additional column to the preceding table. This new column will give the distance $|p_n - 500|$ between each term of the sequence and the possible limit 500. This new column is shown in the table below.

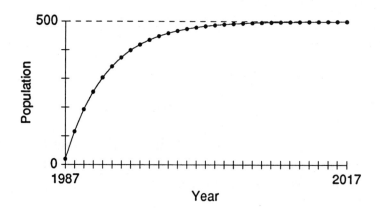

Figure I.21: $p_{n+1} = 0.8p_n + 100$, $p_1 = 20$

Term	Pop.	Dist.	Term	Pop.	Dist.	Term	Pop.	Dist.
p_1	20.00	480.00	p_{11}	448.46	51.54	p_{21}	494.47	5.53
p_2	116.00	384.00	p_{12}	458.77	41.23	p_{22}	495.57	4.43
p_3	192.80	307.20	p_{13}	467.01	32.99	p_{23}	496.46	3.54
p_4	254.24	245.76	p_{14}	473.61	26.39	p_{24}	497.17	2.83
p_5	303.39	196.61	p_{15}	478.89	21.11	p_{25}	497.73	2.27
p_6	342.71	157.29	p_{16}	483.11	16.89	p_{26}	498.19	1.81
p_7	374.17	125.83	p_{17}	486.49	13.51	p_{27}	498.55	1.45
p_8	399.34	100.66	p_{18}	489.19	10.81	p_{28}	498.84	1.16
p_9	419.47	80.53	p_{19}	491.35	8.65	p_{29}	499.07	0.93
p_{10}	435.58	64.42	p_{20}	493.08	6.92	p_{30}	499.26	0.74

Notice that the distance between p_n and 500 is steadily decreasing as n gets larger. As in the previous example this makes it particularly easy to determine a value of N corresponding to each possible tolerance, ϵ.

Exercises

For each of the following exercises do the following.

- Guess what the limit of the sequence would be if it had a limit by finding the equilibrium point of the dynamical system.

- Calculate $p_1, p_2, p_3, \ldots, p_{20}$ to see if the sequence seems to be approaching this equilibrium point as a limit.

- Determine a value of N for the indicated tolerance, ϵ.

§3. Limits: Theory

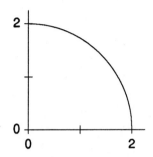

Figure I.22: One-fourth of a circle of radius 2

*Exercise I.3.5: $p_{n+1} = 0.5p_n + 200,$ $p_1 = 100,$ $\epsilon = 1.0.$

Exercise I.3.6: $p_{n+1} = 0.5p_n + 200,$ $p_1 = 100,$ $\epsilon = 0.1.$

Exercise I.3.7: $p_{n+1} = 0.7p_n + 200,$ $p_1 = 100,$ $\epsilon = 1.0.$

Exercise I.3.8: $p_{n+1} = 0.7p_n + 200,$ $p_1 = 100,$ $\epsilon = 0.1.$

*Exercise I.3.9: $p_{n+1} = 1.2p_n + 200,$ $p_1 = 100,$ $\epsilon = 1.0.$

Exercise I.3.10: $p_{n+1} = 1.2p_n + 200,$ $p_1 = 100,$ $\epsilon = 0.1.$

The idea of a limit is closely related to the idea of estimation or approximation. We illustrate this relationship by showing how we can estimate the number π. As you know, the formula for the area A of a circle of radius r is

$$A = \pi r^2.$$

Thus, for example, a circle of radius 2 will have an area of 4π. One-fourth of a circle of radius 2 will have an area of π. See Figure I.22. The curved edge of the quarter circle in Figure I.22 is described by the function

$$f(x) = \sqrt{4 - x^2}.$$

We can estimate π by estimating the area of this quarter circle. We can find such an estimate by dividing the quarter circle up into thin strips. For example, Figure I.23 shows this region divided into five thin strips of equal width. Each of these five strips has width 0.4. The following table describes the strips in more detail.

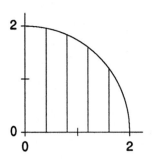

Figure I.23: Five strips

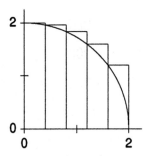

Figure I.24: Approximating the strips by rectangles

Strip	Left edge	Right edge	Width
1	0.0	0.4	0.4
2	0.4	0.8	0.4
3	0.8	1.2	0.4
4	1.2	1.6	0.4
5	1.6	2.0	0.4

We estimate the area of each strip by enclosing it in a rectangle as shown in Figure I.24. We know how to find the area of a rectangle given its base and height. So all we need to do in order to find the area of each of these rectangles is to find its height. We can do this by using the fact that the curved edge of the quarter circle shown in these figures is described by the function

$$f(x) = \sqrt{4 - x^2}.$$

§3. LIMITS: THEORY

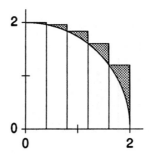

Figure I.25: Estimation error

Thus we can calculate the height and area of each rectangle as shown in the following table.

Rectangle	Left edge	Height	Width	Area
1	0.0	2.000	0.4	0.800
2	0.4	1.960	0.4	0.784
3	0.8	1.833	0.4	0.733
4	1.2	1.600	0.4	0.640
5	1.6	1.200	0.4	0.480
Total				3.437

This gives us an estimate, 3.437, for π. Of course, you already know the value of π is $3.14159\ldots$. Our estimate is a rather rough overestimate. You can see why our estimate is an overestimate by looking at Figure I.25. Each of our rectangles overestimates the area of the corresponding strip. The amount by which our estimate overestimates the true value of π is the area of the shaded regions.

We can find out how large the total overestimate might be by noticing that the overestimates shown in Figure I.25 can be slid over to the right so that they stack up on top of each other. See Figure I.26. Notice that the total overestimate fits into a large rectangle that has a width of 0.4 and a height of 2. Thus the worst possible error for our estimate is the area of this rectangle, $(0.4)(2) = 0.8$. The actual error is less than half of this worst possible error.

There is nothing special about using five rectangles. We could use ten rectangles or twenty rectangles or even one million rectangles. If we use n rectangles we get an estimate, x_n for the number π. This gives

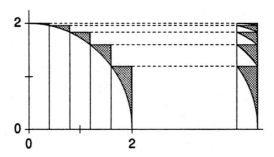

Figure I.26: Estimation error

us a sequence $x_1, x_2, x_3, \ldots$ of estimates for π. We claim that we can get as good an estimate for π as might be necessary for any particular purpose by using this procedure. In other words we claim that

$$\lim_{n \to \infty} x_n = \pi$$

We can prove this as follows.

Proof:

Given a tolerance ϵ we must find an N so that for every $n \geq N$, $|x_n - \pi| < \epsilon$. In other words, given some tolerance ϵ we must determine how many rectangles should be used to obtain an estimate within ϵ of the true value of π.

Notice that if we use n rectangles, then we get a picture very similar to Figure I.26 except that the width of each rectangle is $2/n$. Thus all the errors fit into a rectangle that has width $2/n$ and height 2. Hence the worst possible error is $(2)(2/n) = 4/n$. Now to insure that $4/n < \epsilon$ we make the following calculation

$$\frac{4}{n} < \epsilon$$

$$4 < n\epsilon$$

$$\frac{4}{\epsilon} < n$$

§3. Limits: Theory

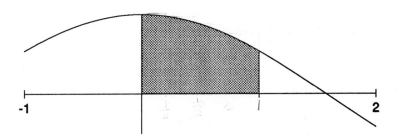

Figure I.27: Area under the curve $y = \cos x$ between $x = 0$ and $x = 1$

For example, to obtain an estimate with an error less than .001 we need to use more than $4/.001 = 4000$ rectangles. That is, we should use at least 4001 rectangles.

Notice that we have actually accomplished two things. We have given a rigorous mathematical proof that

$$\lim_{n \to \infty} x_n = \pi$$

and, as part of that proof, a method that allows us to estimate as precisely as might be necessary for any particular purpose the number π.

Exercises

*Exercise I.3.11: Estimate π using the method of this section with ten rectangles.

*Exercise I.3.12: How many rectangles would be necessary using the method above to obtain an estimate of π that is within .01 of its real value?

*Exercise I.3.13: Consider the region that is bounded by the x-axis, the y-axis, the line $x = 1$, and the curve[2] $y = \cos x$ as shown in Figure I.27. Obtain an estimate for the area of this region that is within .1 of its true value.

---

[2]Warning: We always measure angles in radians. Be sure that your calculator is set to radians.

§4. The Bisection Method

In the last section we studied a method for estimating the area of a region under a curve based on approximating that region by rectangles. In this section we discuss another incarnation of the idea of estimation. We want to develop a very general method of solving equations of the form

$$f(x) = 0.$$

You already know some specific methods of solving some equations of this form.

Theorem 1 *Any linear equation*

$$mx + b = 0$$

where m is not zero has exactly one solution

$$x = -b/m.$$

Proof:

$$\begin{aligned} mx + b &= 0 \\ mx &= -b \\ x &= -b/m. \end{aligned}$$

Linear equations are particularly nice. If we look at a graph of the lefthand side of this equation

$$f(x) = mx + b$$

we see that it is a straight line. The solution of the equation

$$mx + b = 0$$

is just the point at which this straight line intersects the x-axis. See Figure I.28. The equation has exactly one solution because two straight lines (the line $y = mx + b$ and the x-axis) intersect in a single point. There is an exception. If the two lines are parallel then they may not intersect at all or they may lie on top of each other. This is the reason for the clause "where m is not zero" in the statement of the theorem.

§4. The Bisection Method

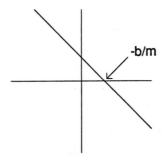

Figure I.28: $y = mx + b$

For a more complicated equation of the form

$$f(x) = 0$$

the situation may be more complicated. They may be no solutions or several solutions and they may be very difficult to find. You probably learned several methods for solving particular examples of such equations, for example, the following theorem.

Theorem 2 *(The Quadratic Formula)*
Consider the equation

$$ax^2 + bx + c = 0$$

where a is not zero.

This equation may have zero, one, or two solutions depending on the quantity $b^2 - 4ac$.

- *If $b^2 - 4ac$ is negative then the equation has no[3] solutions.*

- *If $b^2 - 4ac$ is zero then the equation has one solution, $x = -b/2a$.*

- *If $b^2 - 4ac$ is positive then the equation has two solutions,*

$$x = \frac{-b \pm \sqrt{b^2 - 4ac}}{2a}.$$

---

[3]Although the equation has no real solutions it does have two complex solutions. Complex numbers are discussed in Chapter VI.

CHAPTER I. APPROXIMATION, LIMITS AND DERIVATIVES

Figure I.29: Three quadratic functions

Proof:

$$ax^2 + bx + c = 0$$

$$x^2 + \left(\frac{b}{a}\right)x + \left(\frac{c}{a}\right) = 0$$

$$x^2 + \left(\frac{b}{a}\right)x = -\left(\frac{c}{a}\right)$$

$$x^2 + \left(\frac{b}{a}\right)x + \left(\frac{b}{2a}\right)^2 = -\left(\frac{c}{a}\right) + \left(\frac{b}{2a}\right)^2$$

$$\left(x + \frac{b}{2a}\right)^2 = -\left(\frac{c}{a}\right) + \left(\frac{b^2}{4a^2}\right)$$

$$\left(x + \frac{b}{2a}\right)^2 = \frac{b^2 - 4ac}{4a^2}$$

$$x + \frac{b}{2a} = \pm\frac{\sqrt{b^2 - 4ac}}{2a}$$

$$x = \frac{-b \pm \sqrt{b^2 - 4ac}}{2a}$$

The graph of the function

$$f(x) = ax^2 + bx + c$$

is U-shaped as shown in Figure I.29. Because it is U-shaped it can intersect the x-axis in zero, one or two places. This is the geometric reason behind the three clauses in the preceding theorem.

§4. The Bisection Method

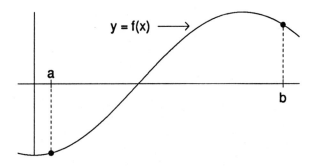

Figure I.30: $y = f(x)$

There are some other particular equations for which specific methods can find solutions. There is, however, no general algebraic method for finding the exact solution of an arbitrary equation of the form

$$f(x) = 0.$$

The main purpose of this section is to find a general method for *estimating* a solution of an arbitrary equation of the form

$$f(x) = 0.$$

This method is called the **Bisection Method** and applies to (almost) any equation of the form

$$f(x) = 0$$

provided we a rough idea about where a solution is located. More specifically, we need to know two points a and b at one of which f is positive and at the other of which f is negative. The Bisection Method is based on Figure I.30. We are looking for a place at which the curve $y = f(x)$ crosses the x-axis and we know one point at which the curve is above the x-axis and another point at which the curve is below the x-axis. These two points are marked by dots in Figure I.30.

The heart of the Bisection Method gives it its name. We chop the original interval $[a, b]$ into two pieces (or bisect it) and then determine which of the two smaller intervals contains the point we're looking for. Let c be the point halfway between a and b.

$$c = \frac{a + b}{2}$$

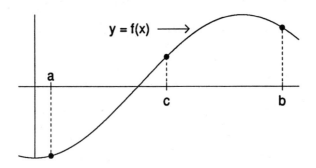

Figure I.31: $f(c)$ and $f(b)$ have the same sign.

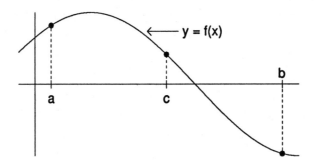

Figure I.32: $f(c)$ and $f(a)$ have the same sign.

Next, calculate the value of the function f at the point c. There are three possibilities

- $f(c)$ and $f(b)$ have the same sign. In this case the point we're looking for is inside the interval $[a, c]$. This situation is shown in Figure I.31

- $f(c)$ and $f(a)$ have the same sign. In this case the point we're looking for is inside the interval $[c, b]$. This situation is shown in Figure I.32.

- $f(c) = 0$. This is an unusual stroke of luck. In this case we've found an exact solution, $x = c$.

§4. The Bisection Method

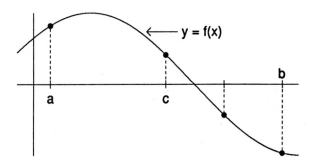

Figure I.33: Bisecting a second time

In either of the first two cases a solution to the equation $f(x) = 0$ is trapped inside an interval half the size of the original interval. Next we repeat the same procedure. Look at the midpoint of the new interval, evaluate the function at this new midpoint, and based on the result trap the solution inside a new interval one-fourth the size of the original interval. This second step is shown in Figure I.33.

This procedure can be repeated as often as necessary. Each time the size of the interval enclosing the solution is divided by two. The following theorem summarizes the Bisection Method.

Theorem 3 *(The Bisection Method)*

This method is designed to find a sequence of estimates for a solution of an equation of the form

$$f(x) = 0$$

given two points a and b at one of which f is positive and at the other of which f is negative.

Suppose that f is a nice[4] function and that either

$$f(a) < 0 < f(b)$$

or

$$f(a) > 0 > f(b)$$

as shown in Figure I.34.

---
[4]We will discuss this qualifying adjective in section 5.

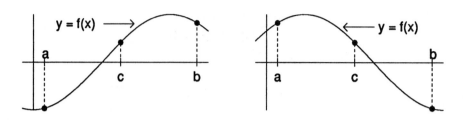

Figure I.34: $y = f(x)$

We define a sequence of intervals

$$[a_1, b_1], [a_2, b_2], [a_3, b_3] \ldots$$

as follows.

$$[a_1, b_1] = [a, b]$$

Given the interval $[a_n, b_n]$ define the next interval $[a_{n+1}, b_{n+1}]$ by evaluating the function f at the midpoint

$$c_n = \frac{a_n + b_n}{2}.$$

- If $f(c_n)$ has the same sign as $f(b_n)$ then

$$[a_{n+1}, b_{n+1}] = [a_n, c_n].$$

- If $f(c_n)$ has the same sign as $f(a_n)$ then

$$[a_{n+1}, b_{n+1}] = [c_n, b_n].$$

- If $f(c_n) = 0$ then by a fluke we have found an exact solution of the original equation. Stop here.

The limit of the sequence $c_1, c_2, c_3, \ldots$ is a solution of the original equation.

Proof:
Given a tolerance $\epsilon > 0$ we need to show that if we repeat this procedure enough times then c_n will be within ϵ of a solution. Notice

§4. THE BISECTION METHOD

that each time the procedure is repeated the new interval is one-half the length of the preceding interval. That is,

$$b_2 - a_2 = \frac{b_1 - a_1}{2}$$
$$b_3 - a_3 = \frac{b_2 - a_2}{2}$$
$$= \frac{b_1 - a_1}{4}$$

and, in general,

$$b_n - a_n = \frac{b_1 - a_1}{2^{n-1}}.$$

Now, since the solution x is trapped in the interval $[a_n, b_n]$ and c_n is the midpoint of this interval,

$$|x - c_n| \leq \frac{(b_n - a_n)}{2} = \frac{b_1 - a_1}{2^n}$$

and all we need to do is choose n large enough so that

$$\frac{b_1 - a_1}{2^n} < \epsilon.$$

Example:

Estimate a solution of the equation[5]

$$\cos x - \frac{x}{2} = 0.$$

Let

$$f(x) = \cos x - \frac{x}{2}.$$

Answer:

Before using the Bisection Method we need a rough idea of the location of a solution. One way to get such a rough idea is by using a graphics calculator or a computer to sketch the function

$$f(x) = \cos x - \frac{x}{2}.$$

The following *Mathematica* procedure does just that. See Figure I.35.

---

[5]Warning: We always measure angles in radians. Be sure that your calculator is set to radians.

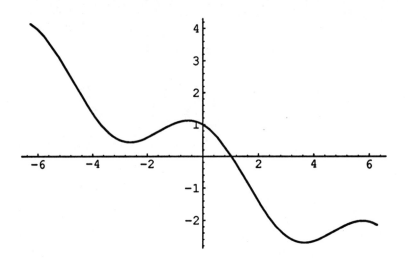

Figure I.35: *Mathematica* graph of $f(x) = \cos x - x/2$

`Plot[Cos[x] - x/2, {x, -2 Pi, 2 Pi}]`

By trial-and-error or by looking at Figure I.35 we see that there is a solution someplace in the interval $[0, \pi/2]$. Notice that $f(0) = 1$ and $f(\pi/2) = -\pi/4$. Thus the first interval is

$$[a_1, b_1] = [0, \pi/2].$$

Looking at the midpoint of this interval, $c_1 = \pi/4$, we see that

$$f(\pi/4) = 0.3144$$

has the same sign as $f(a_1)$, so the second interval is

$$[a_2, b_2] = [\pi/4, \pi/2].$$

Looking at the midpoint of this interval, $c_2 = 3\pi/8$, we see that

$$f(3\pi/8) = -0.2064$$

has the same sign as $f(b_2)$, so the third interval is

$$[a_3, b_3] = [\pi/4, 3\pi/8].$$

The following table summarizes the results thus far

§4. The Bisection Method

n	a_n	b_n	$f(a_n)$	$f(b_n)$	c_n	$f(c_n)$
1	0.0000	1.5708	1.0000	-0.7854	0.7854	0.3144
2	0.7854	1.5708	0.3144	-0.7854	1.1781	-0.2064
3	0.7854	1.1781				

Notice that at this point we have trapped a solution to the equation inside an interval of width $\pi/8 = 0.3927$. This procedure can repeated as many times as necessary to trap the solution inside an interval as small as one might like. This particular problem asked for an estimate within 0.1 of the exact value of a solution. So we need to repeat this procedure once more. This gives us the following results and an estimate, $x = 1.0799$ that is within 0.1 of an exact solution.

n	a_n	b_n	$f(a_n)$	$f(b_n)$	c_n	$f(c_n)$
1	0.0000	1.5708	1.0000	-0.7854	0.7854	0.3144
2	0.7854	1.5708	0.3144	-0.7854	1.1781	-0.2064
3	0.7854	1.1781	0.3144	-0.2064	0.9817	0.0647
4	0.9817	1.1781			1.0799	

We developed the Bisection Method to solve an equation of the form

$$f(x) = 0$$

with zero on the righthand side. It can also be used to solve an equation like

$$\sin x = x/2$$

by simply rewriting it as

$$\sin x - x/2 = 0.$$

Exercises

*Exercise I.4.1:   Find an estimate for a solution of the equation

$$x^4 + x - 1 = 0$$

that is within 0.1 of the exact answer. Notice that if $f(x) = x^4 + x - 1$ then $f(0) = -1$ and $f(1) = 1$.

*Exercise I.4.2: Find an estimate for a solution of the equation

$$x^3 - 3x + 1 = 0$$

that is within 0.1 of an exact solution. Find an initial interval containing a solution using computer- or calculator-based graphics or by trial-and-error.

Exercise I.4.3: Find an estimate for a solution of the equation

$$x^3 - 3x + 1 = 4$$

that is within 0.1 of an exact solution. Find an initial interval containing a solution using computer- or calculator-based graphics or by trial-and-error.

Exercise I.4.4: The equation

$$\sin x = x/2$$

has three solutions, one of which is $x = 0$. Find an estimate that is within 0.1 of one of the other two solutions.

---

§5. The Intermediate Value Theorem and Continuity

Our formulation of the Bisection Method in section 4 had a qualifying adjective

Suppose that f is a *nice* function and ...

We need this qualifying adjective because there are functions that are not "*nice*" for which the Bisection Method does not work. Consider the following example.

§5. THE INTERMEDIATE VALUE THEOREM AND CONTINUITY

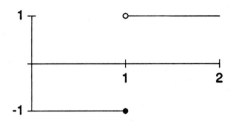

Figure I.36: The "jump" function $y = f(x)$

Example:

Consider the "jump" function

$$f(x) = \begin{cases} -1.00, & \text{if } x \leq 1; \\ 1.00, & \text{if } 1 < x \end{cases}$$

shown in Figure I.36. Notice that $f(1)$ is -1 not 1. This fact is indicated in Figure I.36 by a solid dot at the point $(1,-1)$ and a hollow dot at the point $(1,1)$. Notice that the equation

$$f(x) = 0$$

has no solution. Even though $f(0)$ is negative and $f(2)$ is positive, there is no point x between 0 and 2 where f is zero. The problem with the function f is that very tiny changes in x can cause large changes in the value of $f(x)$. Notice $f(1) = -1$ but if x is just the slightest bit larger than 1 then $f(x) = 1$. The graph of the function $f(x)$ jumps right over the x-axis.

This function seems artificial and it is. Functions like this do, however, sometimes occur in the real world. For example, the price p of mailing a package whose weight is w pounds is often given by a function like

$$p = \begin{cases} \$1.00, & \text{if } w \leq 1; \\ \$2.00, & \text{if } 1 < w \leq 2. \end{cases}$$

The qualifying adjective "nice" in our formulation of the Bisection Method was meant to avoid exactly this kind of nasty behavior. If we picture a function $f(x)$ as a machine that takes some input x and

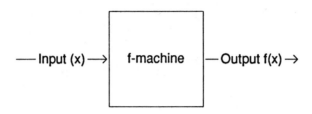

Figure I.37: A function as a machine

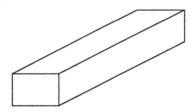

Figure I.38: A block of cheese

produces some output $f(x)$ as shown in Figure I.37 then when a "nice" function is given two different inputs that are very close together the corresponding outputs will also be very close. Some of the same ideas are involved here as in our earlier definition of limit. Consider the following example.

Example:

Suppose that you work in a fancy cheese store. The store displays huge blocks of many different kinds of cheese. Customers can order as much or as little of each cheese as they would like. For example, a customer might enter the store and order three pounds of cheddar. Your job as a salesperson is to cut a piece of cheese whose weight is 3.0 lb. Of course, it is impossible to cut a piece whose weight is absolutely, exactly 3.0 lb. The best you can do is come close. Suppose that a block of cheese looks like the one shown in Figure I.38. That is, the cheese is a rectangular block with nice flat sides meeting in right angles. The store has a cutting frame like the one shown in Figure I.39.

§5. The Intermediate Value Theorem and Continuity

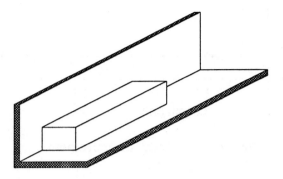

Figure I.39: Cutting frame

You can place the block of cheese in this frame and carefully slide the cheese back and forth in the frame so that a particular length of cheese is extending out from the front of the frame. Then, by sliding a sharp knife along the front edge of the frame, you can make a clean and precise cut. Suppose that you know that the entire block of cheese weighs ten lbs or 160 ounces and that the entire block of cheese is 24 inches long. Thus the cheese weighs $160/24 = 20/3$ ounces per inch and the weight of a piece of cheese is related to its length by the formula

$$w = \left(\frac{20}{3}\right) x$$

where w is weight in ounces and x is length in inches. Thus to obtain a piece of cheese whose weight is 48 ounces we solve the equation

$$\left(\frac{20}{3}\right) x = 48$$
$$x = 48 \left(\frac{3}{20}\right)$$
$$= 7.2$$

as shown in Figure I.40. In order to give your customer a piece of cheese that weighs exactly 48 ounces you must cut a piece that is exactly 7.2 inches long.

Fortunately, your customer is very nice and realizes that you cannot cut the cheese so that it weighs exactly 3.0 lbs. He says, "I'd like three

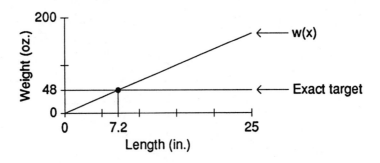

Figure I.40: Exact measurements

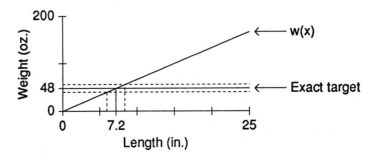

Figure I.41: Nice customer

pounds of cheddar give or take half a pound." Thus he will be happy as long as the piece of cheese that you cut is in the region between the dotted lines in Figure I.41, that is, as long as the piece of cheese that you cut weighs between 40 ounces and 56 ounces. If we do some calculations we see that we want

$$40 < w < 56$$
$$40 < \left(\frac{20}{3}\right)x < 56$$
$$40\left(\frac{3}{20}\right) < x < 56\left(\frac{3}{20}\right)$$
$$6 < x < 8.4$$

So the customer will be satisfied if you cut a piece of cheese whose length is 7 ± 1.2 in. We say that the **input tolerance** 1.2 in. will ensure an **output tolerance** of 8 oz.

Now suppose that a second customer comes in who also wants 3.0 lbs of the cheddar cheese. He is much more fussy, however, than your

§5. The Intermediate Value Theorem and Continuity

first customer. He growls and says, "Last time I was in here I ordered 3.0 lbs of cheese and you gave me 3.4 lbs, more cheese than I really wanted. I only want 3.0 lbs. Of course, I'm a reasonable man and realize that you can't be completely accurate. So, I will accept a piece of cheese that is within one ounce of 3.0 pounds."

You will have to be much more careful to satisfy this customer. He requires a piece of cheese whose weight is between 47 and 49 ounces. Doing some calculations

$$47 < w < 49$$
$$47 < \left(\frac{20}{3}\right)x < 49$$
$$47\left(\frac{3}{20}\right) < x < 49\left(\frac{3}{20}\right)$$
$$7.05 < x < 7.35$$

we see that you will satisfy this customer provided you measure a piece of cheese whose length is 7.2 ± 0.15 in. In this case we say that an input tolerance of 0.15 in. will ensure an output tolerance of 1 oz. The following table summarizes the results of this example.

Exact output	Exact input	Output tolerance	Input tolerance
48	7.2	8	1.20
48	7.2	1	0.15

The function in the example above

$$w = \frac{20x}{3}$$

is an example of a "nice" function. As long as the length x is very close to 7.2 inches, the weight w will be very close to 48 ounces. This function is completely different from the function in our first example. With that function we could not guarantee that $f(x)$ would be very close to $f(1) = -1$ no matter how close x was to 1 because if x was just the slightest bit greater than 1 then $f(x)$ was 1. These examples lead to the following definition.

Definition:

Suppose that $y = f(x)$ is a function and that a is a point. We say that f is **continuous at the point** a if given any **output tolerance** $\epsilon > 0$ there is an **input tolerance** $\delta > 0$ such that for any x within δ of a, $f(x)$ will be within ϵ of $f(a)$. Symbolically,

$$\text{if } |x - a| < \delta \text{ then } |f(x) - f(a)| < \epsilon.$$

We say that f is **continuous** if it is continuous at every point.

Example:

Suppose that after two years of employment in the same cheese store you have been promoted to the exotic cheese department. One of the exotic cheeses is made in triangular blocks like the one shown in figure I.42. The blocks have a uniform thickness of one inch and each block weighs a total of ten pounds. The first piece cut from this block is cut as shown in Figure I.43. Because the block has uniform thickness we can determine the weight of this piece by comparing its area, $x^2/2$, to the area of the entire block, $10^2/2 = 50$, to get

$$w(x) = \left(\frac{\frac{x^2}{2}}{50}\right)(10)$$
$$= x^2/10.$$

A graph of this function is shown in Figure I.44.

Suppose that a customer wants 3.0 pounds of this cheese and that her output tolerance is four ounces (0.25 pounds). What is the input tolerance? First notice that the exact length of a piece of cheese whose weight is 3.0 pounds can be found as follows.

$$x^2/10 = 3.0$$
$$x^2 = 30$$
$$x = \sqrt{30}$$

§5. The Intermediate Value Theorem and Continuity

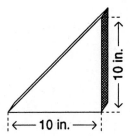

Figure I.42: A triangular block of cheese

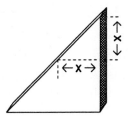

Figure I.43: A piece of triangular cheese

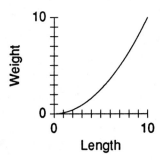

Figure I.44: The weight of a piece of triangular cheese

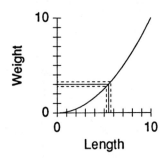

Figure I.45: Cutting a piece whose weight is 3.0 ± 0.25 lbs.

Thus a fictional perfect salesperson would cut a piece of cheese whose length was $\sqrt{30} = 5.4772$ inches. Our customer is willing to accept a piece of cheese whose weight is between 2.75 and 3.25 pounds. So, we compute

$$\begin{aligned} 2.75 &< x^2/10 < 3.25 \\ 27.50 &< x^2 < 32.5 \\ \sqrt{27.5} &< x \quad \sqrt{32.5} \\ 5.2440 &< x \quad < 5.7009 \end{aligned}$$

Thus a real salesperson can cut a piece of cheese whose length is as much as 0.2237 inches too long or 0.2332 inches too short. See Figure I.45. Taking the smaller of these two numbers we see that an input tolerance of 0.2237 inches will ensure an output tolerance of 0.25 pounds.

Example:

Consider the function

$$p(w) = \begin{cases} 1, & \text{if } w \leq 1; \\ 2, & \text{if } 1 < w \leq 2; \\ 3, & \text{if } 2 < w \leq 3; \\ 4, & \text{if } 3 < w \leq 4; \\ 5, & \text{if } 4 < w \leq 5. \end{cases}$$

shown in Figure I.46.
Is this function continuous at the point $w = 3.75$? Is it continuous at the point $w = 4$?

§5. The Intermediate Value Theorem and Continuity

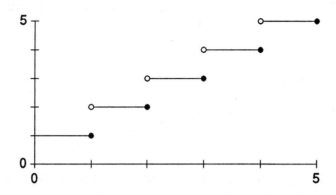

Figure I.46: The function $p(w)$

Answer:

The function is continuous at the point $w = 3.75$ since $p(3.75) = 4$ and if w is close to 3.75 (within 0.25 of 3.75) then $p(w)$ will actually be equal to 4.

The function is not continuous at the point $w = 4$ since $p(4) = 4$ and if w is the least tiny bit above 4 then $p(w)$ will jump up to 5. Thus there is no input tolerance that will insure an output tolerance of, for example, 0.50.

Exercises

*Exercise I.5.1: Suppose that you have been working in the rectangular cheese section of the fancy cheese store for several weeks. You seem to have a large number of customers buying various amounts of cheddar cheese. Some of your customers are quite fussy and others are more laid back. You have been doing a lot of arithmetic to help cut cheese and finally decide that it would be nice to make a chart to help you cut cheese more quickly without stopping to do the arithmetic. Fill in the blanks in the chart below. Some of the entries were computed earlier.

Exact output	Exact input	Output tolerance	Input tolerance
48	7.2	8	1.20
48	7.2	6	
48	7.2	4	
48	7.2	1	0.15
32		8	
32		6	
32		4	
32		2	

*Exercise I.5.2: Suppose that you have been working in the exotic cheese section of the fancy cheese store for several weeks. You seem to have a large number of customers buying various amounts of triangular cheese. Some of your customers are quite fussy and others are more laid back. You have been doing a lot of arithmetic to help cut cheese and finally decide that it would be nice to make a chart to help you cut cheese more quickly without stopping to do the arithmetic. Fill in the blanks in the chart below. Some of the entries were computed earlier.

Exact output	Exact input	Output tolerance	Input tolerance
3.0	5.4772	0.25	0.2237
3.0	5.4772	0.20	
3.0	5.4772	0.15	
4.0		0.25	
4.0		0.20	
4.0		0.15	

*Exercise I.5.3:
Sketch a graph of the function

$$f(x) = \begin{cases} x, & \text{if } x \leq 2; \\ 5 - x, & \text{if } 2 < x. \end{cases}$$

At what points is this function continuous? At what points is it not continuous?

Exercise I.5.4:
Sketch a graph of the function

$$f(x) = \begin{cases} x, & \text{if } x \leq 2; \\ x + 2, & \text{if } 2 < x. \end{cases}$$

§5. The Intermediate Value Theorem and Continuity

At what points is this function continuous? At what points is it not continuous?

*Exercise I.5.5:
Sketch a graph of the function
$$f(x) = \begin{cases} x, & \text{if } x \leq 2; \\ 4-x, & \text{if } 2 < x. \end{cases}$$

At what points is this function continuous? At what points is it not continuous?

*Exercise I.5.6: Suppose that the function $f(x)$ is continuous at the point a. Prove that the function $h(x) = 3f(x)$ is also continuous at the point a.

*Exercise I.5.7: Suppose that the functions $f(x)$ and $g(x)$ are both continuous at the point a. Prove that the function $h(x) = f(x) + g(x)$ is also continuous at the point a.

*Exercise I.5.8: Suppose that the functions $f(x)$ and $g(x)$ are both continuous at the point a. Prove that the function $h(x) = f(x)g(x)$ is also continuous at the point a.

Now that we have a good definition of "nice" we can state a very useful theorem.

Theorem 4 *(Intermediate Value Theorem)*
Suppose that $f(x)$ is a continuous function and that either
$$f(a) < 0 < f(b)$$
or
$$f(a) > 0 > f(b).$$
Then there is a point x between a and b such that
$$f(x) = 0.$$

The Intermediate Value Theorem and the Bisection Method go hand in hand. The Intermediate Value Theorem is a theoretical theorem. It tells us that a particular thing exists, a point x at which the function f is zero. The Bisection Method is a procedure for approximating or estimating that point.

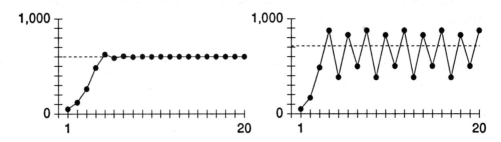

Figure I.47: Two logistic models

§6. When Does an Equilibrium Attract?— Experiment

In section 2 we looked at logistic models of population growth like the two models below.

$$p_{n+1} = 2.5p_n(1 - .001p_n), \qquad p_1 = 50$$
$$p_{n+1} = 3.5p_n(1 - .001p_n), \qquad p_1 = 50$$

Each of these models has two equilibrium points, one of which is zero. The nonzero equilibrium point for the first model is $1.5/.0025 = 600$ and for the second model is $2.5/.0035 \approx 714.29$. Figure I.47 shows the first 20 generations for each of these two models with the nonzero equilibrium point marked by a dotted line. The first model is on the left and the second model is on the right.

For the first model,
$$\lim_{n \to \infty} p_n = 600$$
and we might say that the equilibrium point 600 is "attracting" the population. For the second model, however, we see that there is no limit.

Based on this experimental evidence we might make one conjecture and one observation.

> **Conjecture:** A limit of a dynamical system is an equilibrium point.

> **Observation:** An equilibrium point of a dynamical system does not necessarily "attract" the sequence $p_1, p_2, \ldots$.

§6. When Does an Equilibrium Attract?—Experiment

Our conjecture is correct. More specifically, we have the following theorem.

Theorem 5 *Consider a model given by an initial point p_1 and a dynamical system*
$$p_{n+1} = f(p_n)$$
where f is a continuous function. Suppose that
$$\lim_{n \to \infty} p_n = L.$$

Then L is an equilibrium point.

Proof:

The proof of this theorem is based on the following lemma which is itself interesting.

Lemma 1 *Suppose that f is a continuous function and that $p_1, p_2, \ldots$ is a sequence such that*
$$\lim_{n \to \infty} p_n = L.$$
Then
$$\lim_{n \to \infty} f(p_n) = f(L).$$

Proof:

Suppose that $\epsilon > 0$. We must find an N such that for all $n \geq N$, $|f(p_n) - f(L)| < \epsilon$. Since f is continuous, there is a $\delta > 0$ such that whenever $|x - L| < \delta$ then $|f(x) - f(L)| < \epsilon$. Since
$$\lim_{n \to \infty} p_n = L$$
there is an N such that for all $n \geq N$, $|p_n - L| < \delta$ and, hence, $|f(p_n) - f(L)| < \epsilon$. This proves the lemma.

Now the proof of the theorem follows from the observation that
$$\lim_{n \to \infty} f(p_n) = f(L).$$

But the sequence $f(p_n)$ is just the sequence $p_2, p_3, \ldots$, that is, the same as the original sequence except the terms are numbered differently. Thus,
$$\lim_{n \to \infty} f(p_n) = \lim_{n \to \infty} p_{n+1} = L$$

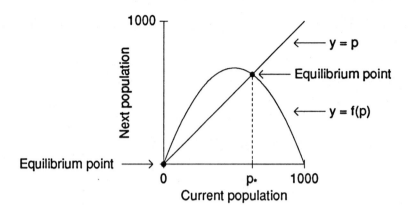

Figure I.48: Equilibrium points

and $f(L) = L$, as we wanted to prove.

Based on this theorem and our observations, we see that a limit is an equilibrium point but that an equilibrium point is not necessarily attracting. We have seen many examples of equilibrium points that are attracting and many examples of equilibrium points that are not attracting.

- For the dynamical system $p_{n+1} = 2.5p_n(1 - .001p_n)$, the equilibrium point 600 attracts the sequence p_n for any initial p_1 such that $0 < p_1 < 1000$. If $p_1 = 0$ or $p_1 = 1000$ then $\lim_{n\to\infty} p_n = 0$.

- For the dynamical system $p_{n+1} = 3.5p_n(1 - .001p_n)$, unless p_1 is 0, 1000, or $2.5/.0035$, the sequence $p_1, p_2, \ldots$ has no limit.

Before reading on, you may want to review other experiments involving equilibrium points and limits, for example, models of population growth for a species that hunts in packs. We want to investigate the question of when an equilibrium point is attracting and when it is not. We begin this investigation by developing a graphical technique for visualizing a dynamical system. In section 2 we investigated a dynamical system, $p_{n+1} = f(p_n)$, by drawing a graph of the function $y = f(p)$ like the one shown in Figure I.48. This graph shows the next generation's population p_{n+1} as a function $f(p_n)$ of the current population p_n. Recall that an equilibrium point is a point where $f(p) = p$, in other words, where the two curves $y = f(p)$ and $y = p$ intersect.

§6. When Does an Equilibrium Attract?—Experiment

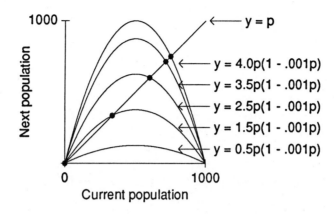

Figure I.49: Comparing several dynamical systems

We can compare several different models by looking at their graphs simultaneously. For example, Figure I.49 shows graphs of five dynamical systems

$$p_{n+1} = 4.0p_n(1 - .001p_n)$$
$$p_{n+1} = 3.5p_n(1 - .001p_n)$$
$$p_{n+1} = 2.5p_n(1 - .001p_n)$$
$$p_{n+1} = 1.5p_n(1 - .001p_n)$$
$$p_{n+1} = 0.5p_n(1 - .001p_n)$$

Looking at this graph, we can see that the dynamical system $p_{n+1} = 0.5p_n(1 - .001p_n)$ does not have a nonzero equilibrium point and that the other four dynamical systems do have nonzero equilibrium points. We want to consider the family of dynamical systems described by

$$p_{n+1} = ap_n(1 - .001p_n)$$

Figure I.49 shows five examples of this family of dynamical systems ($a = 0.5, 1.5, 2.5, 3.5,$ and 4.0). Notice that as a increases, the nonzero equilibrium point seems to be shifting to the right. We can confirm this visual impression by computing the nonzero equilibrium point as follows.

$$p = ap(1 - .001p)$$
$$p = ap - .001ap^2$$
$$0 = (a-1)p - .001ap^2$$

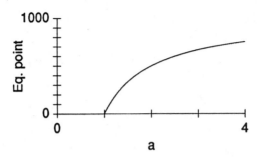

Figure I.50: The dependence of the nonzero equilibrium point on a

$$0 = (a-1) - .001ap$$
$$.001ap = a - 1$$
$$p = (a-1)/(.001a)$$

Notice that this equilibrium point is positive only if $a > 1$. Figure I.50 shows how the nonzero equilibrium point is related to a.

Now we want to use graphical methods to give us some insight into the *dynamics* of these models—the way in which the population *changes* from one generation to the next generation. Mathematically, we have a function $y = f(p)$. We start with an initial population p_1 and compute the population for subsequent generations by repeating the same procedure; $p_2 = f(p_1)$, then $p_3 = f(p_2)$, then $p_4 = f(p_3)$, etc. We can picture this sequence of events as follows.

Step 1:

To compute $p_2 = f(p_1)$, start at the point p_1 on the p-axis. Then draw a vertical line up to the curve $y = f(p)$ as shown in Figure I.51. Now we are at the point (p_1, p_2).

Step 2:

To compute $p_3 = f(p_2)$, draw a line horizontally from the point (p_1, p_2) on the curve $y = f(p)$ to the point (p_2, p_2) on the diagonal line $y = p$ as shown in Figure I.52. Then draw a line vertically from this point to the point (p_2, p_3) on the curve $y = f(p)$. This works because $p_3 = f(p_2)$.

Now we can continue this procedure—going horizontally to the diagonal line $y = f(p)$, then vertically to the curve $y = f(p)$—as often

§6. WHEN DOES AN EQUILIBRIUM ATTRACT?—EXPERIMENT 65

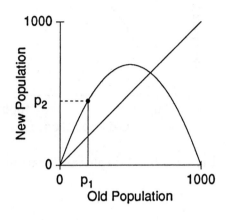

Figure I.51: Graphical iteration—step 1

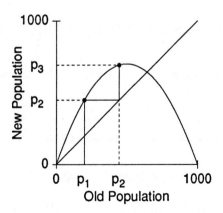

Figure I.52: Graphical iteration—step 2

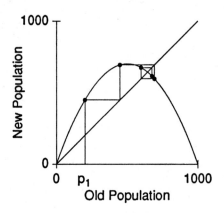

Figure I.53: Graphical iteration

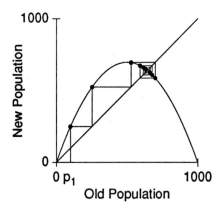

Figure I.54: $p_{n+1} = 2.8p_n(1 - .001p_n)$, $p_1 = 100$

as we would like to get a picture of how the population continues to change. See Figure I.53.

The next two figures show similar graphical iterations for the dynamical system,
$$p_{n+1} = 2.8p_n(1 - .001p_n),$$
with different starting populations; $p_1 = 100$ (Figure I.54), and $p_1 = 900$ (Figure I.55).

By looking at these figures we can see that whatever the initial population p_1 we always get:

$$\lim_{n \to \infty} p_n = 1.8/.0028 \approx 642.86.$$

§6. WHEN DOES AN EQUILIBRIUM ATTRACT?—EXPERIMENT

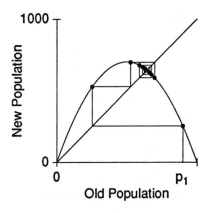

Figure I.55: $p_{n+1} = 2.8p_n(1 - .001p_n)$, $p_1 = 900$

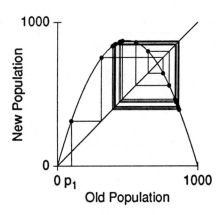

Figure I.56: $p_{n+1} = 3.5p_n(1 - .001p_n)$, $p_1 = 100$

If the initial population is low, then we see the sequence $p_1, p_2, p_3, \ldots$, "climbing stairs" as in Figure I.54 until it gets close to the nonzero equilibrium population. Then it "spirals in" towards the equilibrium. If the initial population is very high, it first drops below the equilibrium population, as in Figure I.55 and then continues very much like Figure I.55. Whatever the initial population, the results are the same—after many years the population will be very close to the nonzero equilibrium population.

Figure I.56 shows one result of using the same graphical technique to investigate the dynamical system, $p_{n+1} = 3.5p_n(1 - .001p_n)$. Notice that in this case we have a much more complex pattern than in our earlier figures. In particular, there is no limit!

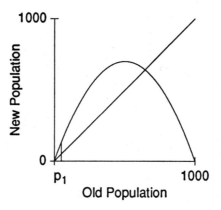

Figure I.57: $p_{n+1} = 2.8p_n(1 - .001p_n)$, $p_1 = 50$

Exercises

The first two exercises below look at logistic models like those studied in this section.

*Exercise I.6.1: Consider the dynamical system

$$p_{n+1} = 2.8p_n(1 - .001p_n), \quad p_1 = 50.$$

First compute p_2, p_3, p_4, and p_5 with a calculator. Then compute this same sequence graphically on Figure I.57.

*Exercise I.6.2: Consider the dynamical system

$$p_{n+1} = 3.2p_n(1 - .001p_n), \quad p_1 = 50.$$

First compute p_2, p_3, p_4, and p_5 with a calculator. Then compute this same sequence graphically on Figure I.58.

The next five exercises look at particularly simple dynamical systems of the form

$$p_{n+1} = mp_n + b.$$

Since the graph of a function of the form $y = mx + b$ is a straight line, you will be able to do these problems without a computer-drawn graph.

§6. When Does an Equilibrium Attract?—Experiment

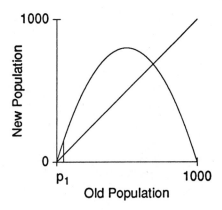

Figure I.58: $p_{n+1} = 3.2p_n(1 - .001p_n)$, $p_1 = 50$

*Exercise I.6.3: Consider the dynamical system

$$p_{n+1} = 0.8p_n + 100, \quad p_1 = 50.$$

First compute $p_2, p_3, p_4, \ldots p_{10}$ with a calculator. Then compute this same sequence graphically.

*Exercise I.6.4: Consider the dynamical system

$$p_{n+1} = 1.2p_n - 100, \quad p_1 = 50.$$

First compute $p_2, p_3, p_4, \ldots p_{10}$ with a calculator. Then compute this same sequence graphically.

*Exercise I.6.5: Consider the dynamical system

$$p_{n+1} = 100 - 0.8p_n, \quad p_1 = 50.$$

First compute $p_2, p_3, p_4, \ldots p_{10}$ with a calculator. Then compute this same sequence graphically.

Exercise I.6.6: Consider the dynamical system

$$p_{n+1} = 100 - 1.2p_n, \quad p_1 = 50.$$

First compute $p_2, p_3, p_4, \ldots p_{10}$ with a calculator. Then compute this same sequence graphically.

Exercise I.6.7: Consider the dynamical system

$$p_{n+1} = 100 - p_n, \quad p_1 = 50.$$

First compute $p_2, p_3, p_4, \ldots p_{10}$ with a calculator. Then compute this same sequence graphically.

Exercise I.6.8: Based on your experience with the exercises above, what appears to determine whether or not an equilibrium point will be attracting? Answer this question first for dynamical systems of the form

$$p_{n+1} = mp_n + b$$

and then for more general dynamical systems like logistic models.

§7. When Does an Equilibrium Attract?—Theory

We are interested in determining when a equilibrium point for a dynamical system is attracting. We have already done a number of experiments and now it is time to step back, look at the experiments that we have done, and see if we can see a pattern. This is the theoretical side of mathematics, the discovery and proof of general principles.

We begin with some particularly simple dynamical systems called **linear dynamical systems**. Linear dynamical systems are of the form

$$p_{n+1} = f(p_n) = mp_n + b.$$

The constant m is called the **slope** of the function $f(p) = mp + b$ since it describes how steeply the graph of the function $f(p) = mp + b$ is rising (if m is positive) or falling (if m is negative). The constant b is called the **y-intercept** of $f(p)$ since it describes the point at which this graph crosses the y-axis.

The exponential models with immigration for population growth that we studied in section 1 were examples of linear dynamical systems although in that section we used notation that was more suggestive of the biology,

$$p_{n+1} = Rp_n + m,$$

using the letter m for the y-intercept because the value of m represented the rate of *immigration*, and the letter R for the slope since this letter is traditionally used to denote the population multiplier for an exponential model. Exercises I.6.3–I.6.7 of section 6 were also linear dynamical systems.

§7. When Does an Equilibrium Attract?—Theory

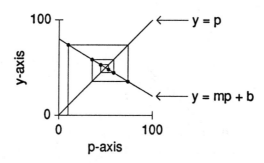

Figure I.59: A linear dynamical system

A linear dynamical system normally has exactly one equilibrium point determined as follows.

$$\begin{aligned} p &= mp + b \\ p - mp &= b \\ (1-m)p &= b \\ p &= \frac{b}{1-m} \end{aligned}$$

There are some exceptions to the general rule that every linear dynamical system has exactly one equilibrium. If $m = 1$ then the two lines, $y = p$ and $y = p + b$, are parallel. In this case if $b = 0$ then $p_{n+1} = p_n$ and every point is an equilibrium point, and if $b \neq 0$ then there are no equilibrium points because the two lines $y = p$ and $y = p + b$ do not intersect.

Figures I.59–I.62 show graphical computations of four different linear dynamical systems. We have deliberately omitted the formulas for these dynamical systems because we want to concentrate on the geometry rather than on the calculations.

Based on these four graphs and your earlier work in section 6, you can probably make a good guess for a general principle that determines whether the equilibrium point of a linear dynamical system is attracting or not.

The equilibrium points for the dynamical systems shown in Figures I.59 and I.60 are attracting. The equilibrium points for the dynamical systems shown in Figures I.61 and I.62 are not attracting. If you compare these four graphs you will see that the crucial difference appears

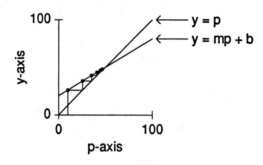

Figure I.60: A linear dynamical system

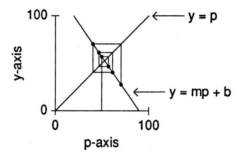

Figure I.61: A linear dynamical system

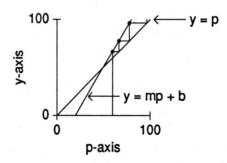

Figure I.62: A linear dynamical system

§7. When Does an Equilibrium Attract?—Theory

to be the slope of the line $y = mp + b$. In both Figures I.59 and I.60 the line $y = mp + b$ is not very steep and the equilibrium point is attracting. In both Figures I.61 and I.62 the line is much steeper and the equilibrium point is not attracting. If you examine these four graphs you might observe that if the slope of line $y = mp + b$ is between -1 and $+1$ then the equilibrium point is attracting. Otherwise it is not attracting. These observations lead us to the following conjecture.

Conjecture:

For a linear dynamical system,

$$p_{n+1} = mp_n + b,$$

the equilibrium point is attracting if $|m| < 1$ and not attracting if $|m| > 1$.

Do you agree with this conjecture? Can you see any geometric reasons why it might be true.

Having made this conjecture, there are several things you might do. The first thing is to test the conjecture by doing several additional experiments. The following series of exercises asks you to do just that. This is the essence of the "Scientific Method." We do a number of experiments. Based on the results of those experiments we try to formulate some general principles. Then we test those conjectured principles with additional experiments. Mathematicians then go one step further. If they think the conjecture is probably true then they attempt to prove it. If the attempt is successful then the conjecture is elevated to the status of a theorem.

Exercises

*Exercise I.7.1: Investigate the linear dynamical system

$$p_{n+1} = 0.8p_n + 100.$$

Based on our conjecture what should happen? Do the results of your experiments support or contradict the conjecture?

*Exercise I.7.2: Investigate the linear dynamical system

$$p_{n+1} = -0.8p_n + 100.$$

Based on our conjecture what should happen? Do the results of your experiments support or contradict the conjecture?

Exercise I.7.3: Investigate the linear dynamical system

$$p_{n+1} = 0.9p_n + 100.$$

Based on our conjecture what should happen? Do the results of your experiments support or contradict the conjecture?

Exercise I.7.4: Investigate the linear dynamical system

$$p_{n+1} = -0.9p_n + 100.$$

Based on our conjecture what should happen? Do the results of your experiments support or contradict the conjecture?

Exercise I.7.5: Investigate the linear dynamical system

$$p_{n+1} = 0.98p_n + 100.$$

Based on our conjecture what should happen? Do the results of your experiments support or contradict the conjecture?

Exercise I.7.6: Investigate the linear dynamical system

$$p_{n+1} = -0.98p_n + 100.$$

Based on our conjecture what should happen? Do the results of your experiments support or contradict the conjecture?

*Exercise I.7.7: Investigate the linear dynamical system

$$p_{n+1} = 1.02p_n + 100.$$

Based on our conjecture what should happen? Do the results of your experiments support or contradict the conjecture?

*Exercise I.7.8: Investigate the linear dynamical system

$$p_{n+1} = -1.02p_n + 100.$$

Based on our conjecture what should happen? Do the results of your experiments support or contradict the conjecture?

§7. When Does an Equilibrium Attract?—Theory

*Exercise I.7.9: The conjecture doesn't apply to the dynamical system
$$p_{n+1} = -p_n + b.$$
What happens in this case?

We have enough evidence to justify an attempt to prove our conjecture.

Theorem 6 *If $p_{n+1} = mp_n + b$ is a linear dynamical system, $|m| < 1$, and p_1 is any initial term then*
$$\lim_{n \to \infty} p_n = \frac{b}{1-m}$$

Notice that $b/(1-m)$ is simply the equilibrium point for this dynamical system.

If $|m| > 1$ then there is no limit. In fact, the sequence $p_1, p_2, p_3, \ldots$ keeps getting farther away from the equilibrium point $b/(1-m)$. (Exception: if the initial term p_1 is actually equal to the equilibrium point, $b/(1-m)$, then every p_n is $b/(1-m)$.)

Proof:

The key idea in this proof is to look at the distance
$$\left| p_n - \frac{b}{1-m} \right|$$
from each term of the sequence to the equilibrium point.

We will begin our proof by calculating the distance $\left| p_2 - \frac{b}{1-m} \right|$ in terms of $\left| p_1 - \frac{b}{1-m} \right|$

$$\begin{aligned}
\left| p_2 - \frac{b}{1-m} \right| &= \left| mp_1 + b - \frac{b}{1-m} \right| \\
&= \left| mp_1 + \frac{b(1-m) - b}{1-m} \right| \\
&= \left| mp_1 - \frac{bm}{1-m} \right| \\
&= |m| \left| p_1 - \frac{b}{1-m} \right|
\end{aligned}$$

Entirely analogous calculations show that

$$\left| p_3 - \frac{b}{1-m} \right| = |m| \left| p_2 - \frac{b}{1-m} \right|$$
$$= |m|^2 \left| p_1 - \frac{b}{1-m} \right|$$

and, more generally, that

$$\left| p_n - \frac{b}{1-m} \right| = |m|^{n-1} \left| p_1 - \frac{b}{1-m} \right|$$

Now we can see why the theorem is true. If $|m| < 1$ then $|m|^{n-1}$ will be extremely small if n is large. Thus, for very large values of n, p_n will be very close to $b/(1-m)$. On the other hand if $|m| > 1$ then $|m|^{n-1}$ will be quite large, so that p_n will be quite far away from $b/(1-m)$ unless, of course, $p_1 = b/(1-m)$.

We have solved our problem for linear dynamical systems. We can determine whether or not the equilibrium point for a linear dynamical system, $p_{n+1} = mp_n + b$, is attracting simply by looking at m. If $|m| < 1$ then the equilibrium point is attracting. Our ultimate goal is to be able to determine whether an equilibrium point, p_*, of a more general kind of dynamical system,

$$p_{n+1} = f(p_n),$$

is attracting. The same basic criterion works. The crucial thing is the "slope" of the curve $y = f(p)$ at the point p_*. In order to work with this idea we have to explore what we mean by the "slope" of a curve that is not a straight line. We will do this in the next section. First, however, we apply the work we have already done to some interesting economic models.

The Economics of Supply and Demand

I teach mathematics and write books. I eat food and heat my house with oil. How do I exchange a book like the one you are reading for a dozen ears of corn or for a barrel of oil? The answer, of course, is

§7. WHEN DOES AN EQUILIBRIUM ATTRACT?—THEORY

money. How many copies of this book must be sold to heat my house for one winter? Who decides? and how? What determines the relative price of books and oil?

There is no one answer to these questions. For some commodities and services, prices are regulated by government action. The supply of some products is controlled by a few suppliers who are able to control the price. For example, for a number of years OPEC was able to control the price of oil. For some other commodities prices are determined by the rough-and-tumble of the marketplace. We want to develop and examine one particular model of how such a marketplace might work.

We study an idealized commodity, Byties. Byties are a particularly delicious breakfast roll. They are baked by many individual families who rise each morning before dawn to bake the morning's Byties. Around 6:00 A.M. each morning the bakers bring their Byties to the town market. Hungry potential customers come down to the market to buy Byties.

We are interested in three related things—the price of, the supply of, and the demand for Byties. The supply of Byties depends on how many people are willing to get up early to bake Byties. This will depend, in turn, on the price of Byties. If Byties are very expensive then bakers can anticipate a large profit for their early morning labor; many people will be willing to get up early to bake Byties. If the price is low then bakers cannot anticipate much profit; fewer people will be willing to get up early to bake Byties. Thus the supply of Byties is a function $S(p)$ of price that will look roughly like Figure I.63. For high prices the supply will be high; for low prices the supply will be low.

The demand for Byties also depends on the price. If Byties are very cheap many people will buy them. In fact, each customer might even buy several. But, if the price is high, then fewer people will buy Byties and those that do will buy fewer. Thus if the price is low the demand is high and if the price is high the demand is low. A graph of the demand function will look roughly like Figure I.64.

The ideal situation would be one in which the supply of Byties exactly matches the demand; then each day all the bakers would be able to sell all their Byties and each customer would be able to buy the Byties he or she wanted. The price at which this happens is called the **equilibrium price.** This point is shown in Figure I.65.

If the price is below the equilibrium price then the demand will

Chapter I. Approximation, Limits and Derivatives

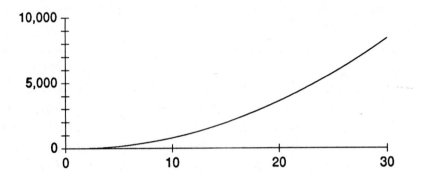

Figure I.63: Supply as a function of price

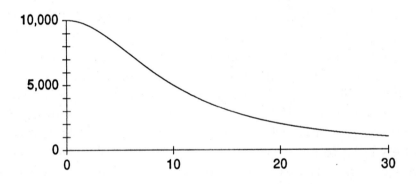

Figure I.64: Demand as a function of price

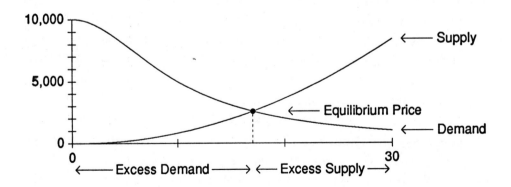

Figure I.65: Supply and demand

§7. When Does an Equilibrium Attract?—Theory

be bigger than the supply; at the end of the day there will be hungry customers milling around the marketplace and no more Byties to be had. In this case you would expect the price to rise because as the end of the morning approaches and there are only a few Byties left the bakers will be able to raise their prices.

On the other hand, if the price is above the equilibrium price then the supply will be above the demand. At the end of the day the bakers will have some leftover Byties with no customers left. In this situation you would expect the price of Byties to fall as bakers try to salvage some profit from their remaining Byties by selling them off cheap. Byties get stale very quickly and must be baked fresh each morning. Thus, bakers either have to drop their price to move their remaining Byties or literally eat their losses.

The key quantity involved is the **excess demand**, $D(p) - S(p)$. If this quantity is positive then demand is higher than supply and prices can be expected to rise. If it is negative then supply is higher than demand and prices can be expected to fall. One possible model of this kind of behavior is the dynamical system

$$p_{n+1} = p_n + k(D(p_n) - S(p_n))$$

where k is some positive constant. Notice when the excess demand is positive (i.e. when demand is bigger than supply) the price will rise and when the excess demand is negative (i.e. when demand is less than supply) the price will fall. The constant, k, depends on the psychology of the marketplace. If customers and bakers love to bargain then prices will probably change quite rapidly and k would be a relatively large number. If customers and bakers are more conservative then prices will change more slowly and k will be a relatively small number.

We work with simple examples of possible supply and demand functions,

$$S(p) = 1000p - 400$$
$$D(p) = 1000 - 500p,$$

shown in Figure I.66.

We calculate the equilibrium price as follows.

$$\begin{aligned} S(p) &= D(p) \\ 1000p - 400 &= 1000 - 500p \\ 1500p &= 1400 \end{aligned}$$

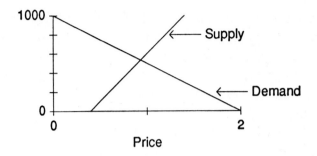

Figure I.66: $S(p) = 1000p - 400$ and $D(p) = 1000 - 500p$

$$p = 14/15.$$

The dynamics of this model are given by

$$\begin{aligned} p_{n+1} &= p_n + k(D(p_n) - S(p_n)) \\ &= p_n + k[(1000 - 500p_n) - (1000p_n - 400)] \\ &= p_n + k(1400 - 1500p_n) \\ &= p_n + 1500k\left[\left(\frac{14}{15}\right) - p_n\right] \end{aligned}$$

The following exercises ask you to explore this model.

Exercises

*Exercise I.7.10: Using the supply-demand model described above with $k = .0002$ and $p_1 = 0.50$, find $p_2, p_3, p_4, \ldots p_{10}$. Based on your calculations do you think

$$\lim_{n \to \infty} p_n = 14/15?$$

Exercise I.7.11: Using the supply-demand model described above with $k = .0002$ and $p_1 = 1.50$, find $p_2, p_3, p_4, \ldots p_{10}$. Based on your calculations do you think

$$\lim_{n \to \infty} p_n = 14/15?$$

Exercise I.7.12: Using the supply-demand model described above with $k = .0006$ and $p_1 = 0.50$, find $p_2, p_3, p_4, \ldots p_{10}$. Based on your calculations do you think

§7. When Does an Equilibrium Attract?—Theory

$$\lim_{n \to \infty} p_n = 14/15?$$

*Exercise I.7.13: Using the supply-demand model described above with $k = .0012$ and $p_1 = 0.50$, find $p_2, p_3, p_4, \ldots p_{10}$. Based on your calculations do you think

$$\lim_{n \to \infty} p_n = 14/15?$$

*Exercise I.7.14: For what values of k will

$$\lim_{n \to \infty} p_n = 14/15?$$

*Exercise I.7.15: Consider the supply-demand model described above with $k = .0001$. Suppose that $p_1 = .50$. How many days will be required before the price is within 0.05 of the equilibrium price?

Exercise I.7.16: Consider the supply-demand model described above with $k = .0001$. Suppose that $p_1 = .50$. How many days will be required before the price is within 0.01 of the equilibrium price?

Exercise I.7.17: Consider the supply-demand model described above with $k = .0004$. Suppose that $p_1 = .50$. How many days will be required before the price is within 0.05 of the equilibrium price?

Exercise I.7.18: Consider the supply-demand model described above with $k = .0004$. Suppose that $p_1 = .50$. How many days will be required before the price is within 0.01 of the equilibrium price?

Now we would like to use this basic supply-demand model to explore what might happen when the price of the ingredients used in baking Byties rises. We have been working with the supply function

$$S(p) = 1000p - 400.$$

Another possible supply function might be

$$S(p) = 1000(p - d)$$

where the constant d is the price of the ingredients used in baking Byties. This function makes sense because it says that the bakers decide how many Byties to bake based on their net profit after subtracting the

cost of ingredients. In particular, there will be no bakers baking Byties unless the price is above the basic cost of ingredients. Our original supply function,
$$S(p) = 1000p - 400 = 1000(p - 0.40),$$
is exactly this kind of function with $d = 0.40$.

For this supply function we have already calculated the equilibrium price, $p = 14/15$. Now suppose that the cost of ingredients suddenly goes up to 0.55 so that the new supply function is
$$S(p) = 1000(p - 0.55) = 1000p - 550.$$
We want to examine the impact of this change. Suppose that the price starts out at the old equilibrium (before the rise in the cost of ingredients). Thus $p_1 = 14/15$. Suppose, further, that the price changes according to the new situation
$$p_{n+1} = p_n + .0003(D(p_n) - S(p_n))$$
where $S(p)$ is the new supply function. Answer the following questions based on this model.

Exercises

Exercise I.7.19: What is the new equilibrium price?

Exercise I.7.20: How long will it be before the price is within .01 of the new equilibrium price?

Exercise I.7.21: Assume that each baker makes 20 Byties per day both before and after the price of ingredients rises. How many bakers will go out of business when the price of ingredients rises?

Exercise I.7.22: Assume that each baker makes 20 Byties per day both before and after the price of ingredients rises. Will each baker's total daily profit go up or down after the price of ingredients rises?

Exercise I.7.23: Who pays for the increased cost of ingredients— bakers or customers.

The last exercise in this problem set involved a very crucial question. The same question comes up, for example, when people talk about raising business taxes. To some extent businesses can pass on the increased cost of doing business to their customers. The extent to which this is possible depends on many factors. This last exercise illustrates one situation in which only part of the increased cost of ingredients is passed on to customers.

§8. The Tangent to a Curve   83

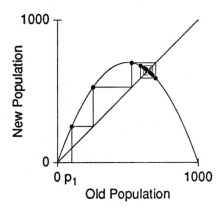

Figure I.67: $p_{n+1} = 2.8p_n(1 - .001p_n)$, $p_1 = 200$

§8. The Tangent to a Curve

In section 7 we saw that linear dynamical systems $p_{n+1} = mp_n + b$ are particularly simple. Unless $m = 1$ they have exactly one equilibrium point, which is attracting if $|m| < 1$ and not attracting if $|m| > 1$. This is a fairly common situation in mathematics. Linear functions are simple and easy to analyze. If all dynamical systems were linear, a mathematician's life would be much simpler. Most dynamical systems, however, are not linear. For example, the logistic population models

$$p_{n+1} = ap_n(1 - bp_n)$$

from section 1 are not linear and our theorem from section 7 is of no help for these models. Nonetheless, if you look at Figures I.67 and I.68, you will see that the same idea is important. If the "slope" of the curve $f(p) = ap(1 - bp)$ is too steep at the equilibrium point then the equilibrium point will not be attracting. In order to make sense out of this observation and to turn it from a general feeling into a useful theorem we need to be more precise about the "slope" of a curve.

We determine the slope of a curve, $y = f(x)$, at a particular point, $(x, f(x))$, by drawing the tangent to the curve at that point as shown in Figure I.69 and then determining the slope of the tangent. We illustrate this idea by looking at a particular point on a particular curve.

Example:

Determine the slope of the curve $y = 1 - x^2$ at the point $(0.6, 0.64)$ as shown in Figure I.70.

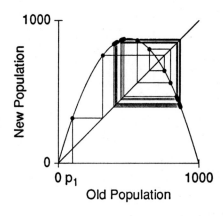

Figure I.68: $p_{n+1} = 3.5 p_n (1 - .001 p_n)$, $p_1 = 100$

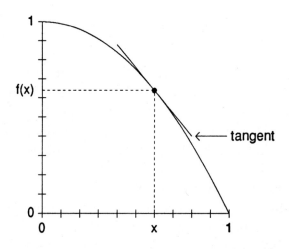

Figure I.69: The tangent to the curve $y = f(x)$ at the point $(x, f(x))$

§8. The Tangent to a Curve

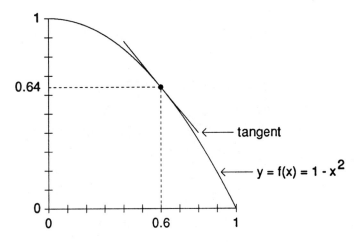

Figure I.70: The tangent to the curve $y = 1 - x^2$ at the point $(0.6, 0.64)$

Answer:

We begin by trying to *estimate* the slope of the tangent. Look at a second point $(0.7, 0.51)$ on the same curve that is close to the first point as shown in Figure I.71. Draw the **chord** or **secant** that goes through the two points $(0.6, 0.64)$ and $(0.7, 0.51)$. Notice that this chord is close to the tangent. Thus the slope of the chord is close to the slope of the tangent. It is easy to determine the slope of this chord since we know two points on the chord.

$$\begin{aligned} \text{slope} &= \frac{\text{change in y}}{\text{change in x}} \\ &= \frac{0.51 - 0.64}{0.7 - 0.6} \\ &= -1.3 \end{aligned}$$

Notice that this gives us an *estimate* for the slope of the tangent. It is apparent from Figure I.71 that the chord is quite close to the tangent, so this estimate is a good one. If we choose the second point closer to the first point we will get a chord that is closer to the tangent (so close, in fact, that it becomes difficult to draw pictures showing the tangent and the chord as being distinct lines) and the slope of the chord will be a better estimate of the slope of the tangent. The following table

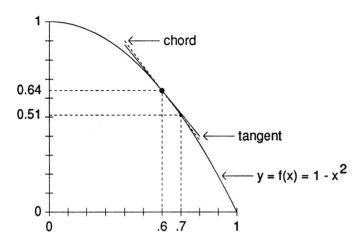

Figure I.71: A chord close to the tangent

computes more estimates for the slope of the tangent based on this idea. By choosing the second point closer and closer to the original point $(0.6, 0.64)$ we get better and better estimates.

Point	Change in y	Change in x	Slope of chord
(0.6500, 0.57750000)	−0.06250000	0.0500	−1.2500
(0.6100, 0.62790000)	−0.01210000	0.0100	−1.2100
(0.6010, 0.63879900)	−0.00120100	0.0010	−1.2010
(0.6001, 0.63987999)	−0.00012001	0.0001	−1.2001

Exercises

This series of exercises asks you to find estimates for the slope of the tangent to the circle
$$x^2 + y^2 = 1$$
or
$$y = \sqrt{1 - x^2}$$
at the point $(0.6, 0.8)$ as shown in Figure I.72.

*Exercise I.8.1: Find the slope of the chord that passes through the points, $(0.6, 0.8)$ and $(0.7, ?)$ on the curve $y = \sqrt{1 - x^2}$. Notice that

§8. The Tangent to a Curve

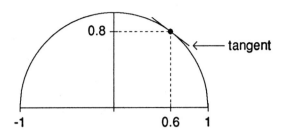

Figure I.72: Tangent to the circle $x^2 + y^2 = 1$ at the point $(0.6, 0.8)$

you will need to compute the y-coordinate of the second point by using the equation $y = \sqrt{1 - x^2}$.

Exercise I.8.2: Find the slope of the chord that passes through the points, $(0.6, 0.8)$ and $(0.65, ?)$ on the curve $y = \sqrt{1 - x^2}$.

Exercise I.8.3: Find the slope of the chord that passes through the points, $(0.6, 0.8)$ and $(0.61, ?)$ on the curve $y = \sqrt{1 - x^2}$.

*Exercise I.8.4: Find the slope of the chord that passes through the points, $(0.6, 0.8)$ and $(0.55, ?)$ on the curve $y = \sqrt{1 - x^2}$.

Exercise I.8.5: Find the slope of the chord that passes through the points, $(0.6, 0.8)$ and $(0.59, ?)$ on the curve $y = \sqrt{1 - x^2}$.

*Exercise I.8.6: Find the slope of the chord that passes through the points, $(0.4, ?)$ and $(0.41, ?)$ on the curve $y = \sqrt{1 - x^2}$.

Exercise I.8.7: Find the slope of the chord that passes through the points, $(0.4, ?)$ and $(0.39, ?)$ on the curve $y = \sqrt{1 - x^2}$.

*Exercise I.8.8: For this particular curve we can actually determine the exact slope of the tangent by using the fact that the tangent to a point on a circle and the radius from the center of the circle to the same point meet at right angles. See Figure I.73. It is easy to find the slope of the radius because we know two points on the radius. Then we can use the fact that if two lines meet at right angles the slope of one is the negative reciprocal of the slope of the other, that is, if the slope of one line is m then the slope of the other is $-1/m$. Use these facts to compare the estimates you obtained in the exercises above with the exact slope of each of the tangents.

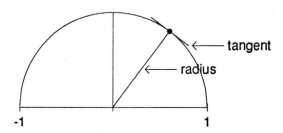

Figure I.73: Tangent to the circle $x^2 + y^2 = 1$ at the point $(0.6, 0.8)$

This technique allows us to obtain very good estimates for the slope of the tangent by choosing two points that are extremely close together on the curve. Often all that is required is an estimate for the slope of the tangent. In this case we could stop here. We have a very general method for estimating the slope of the tangent. Ideally, however, we would like to know the exact slope of the tangent. We illustrate how this can be done by continuing the example above.

Example:

Find the exact slope of the tangent to the curve $y = f(x) = 1 - x^2$ at the point $(0.6, 0.64)$.

Answer:

We have already found several good estimates. Now we want to find a general formula for the estimate obtained by looking at the chord through the points $(0.6, 0.64)$ and $(0.6 + h, f(0.6 + h))$ as shown in Figure I.74.

$$\begin{aligned}
\frac{\text{change in y}}{\text{change in x}} &= \frac{f(0.6 + h) - f(0.6)}{h} \\
&= \frac{[1 - (0.6 + h)^2] - (1 - 0.6^2)}{h} \\
&= \frac{1 - 0.36 - 1.2h - h^2 - 0.64}{h} \\
&= \frac{-1.2h - h^2}{h} \\
&= -1.2 - h
\end{aligned}$$

§8. The Tangent to a Curve

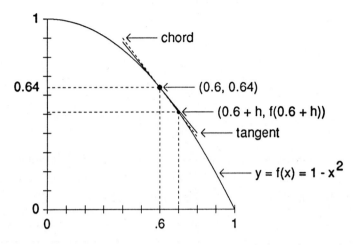

Figure I.74: The chord through the points $(0.6, 0.64)$ and $(0.6, f(0.6))$

This formula give us a general procedure for estimating the slope of the tangent. By choosing h very small, we get a chord that is very close to the tangent, so that the slope of the chord, $-1.2 - h$, is very close to the slope of the tangent. Notice that this procedure makes sense only when h is not zero. If h were zero then we would be dividing zero by zero which doesn't make sense.

In section 3 we looked at a method for obtaining a sequence of estimates for the area of a particular region. We have a similar situation here—a procedure for finding *estimates* rather than exact answers. It is, however, easy to see what the exact answer is—as h gets smaller and smaller, the chords are getting closer and closer to the tangent and our estimates are getting closer and closer to the exact answer. Now notice that as h gets smaller and smaller the estimate, $-1.2 - h$, is getting closer and closer to -1.2. Thus the slope of the tangent must be -1.2. The ideas involved here are very similar to the ideas involved in our earlier discussion of the limit of a sequence and of continuity. In fact, we use the same word *limit* to describe this situation. We can treat this idea on an intuitive level or on a more precise level. We begin at the intuitive level.

Definition:

Suppose that $F(h)$ is a function that may or may not be defined at the point $h = a$. We say the **limit of $F(h)$ as h approaches a is L** and write

$$\lim_{h \to a} F(h) = L$$

if whenever h is very close to a then $F(h)$ is very close to L. The actual value of F at a is irrelevent. In fact, we often use this idea when $F(a)$ is not defined.

Example:

Consider the curve $f(x) = x^2$. Find the slope of the tangent to this curve at the point $(2, 4)$.

Answer:

Let h be nonzero and consider the slope of the chord through the two points $(2, 4)$ and $((2 + h), (2 + h)^2)$

$$\begin{aligned} F(h) &= \frac{(2+h)^2 - 4}{h} \\ &= \frac{4 + 4h + h^2 - 4}{h} \\ &= \frac{4h + h^2}{h} \\ &= 4 + h. \end{aligned}$$

As h gets closer and closer to zero the chord is getting closer and closer to the tangent. Hence the slope of the chord is getting closer and closer to the slope of the tangent. That is,

$$\text{Slope of tangent} = \lim_{h \to 0}(4 + h).$$

But clearly

$$\lim_{h \to 0}(4 + h) = 4$$

§8. The Tangent to a Curve

since when h is close to zero then $4+h$ is close to 4. Hence the slope of the tangent is 4.

Example:

Find
$$\lim_{x \to 2} \left(\frac{x^2 - 4}{x - 2} \right).$$

Answer:

First let
$$F(x) = \frac{x^2 - 4}{x - 2}$$
and notice that $F(2) = 0/0$ doesn't make any sense. This is perfectly all right since when we evaluate the limit of $F(x)$ as x approaches 2 the value of F at 2 is irrelevent.

Now we do a little simple algebra
$$\frac{x^2 - 4}{x - 2} = \frac{(x-2)(x+2)}{x - 2}.$$
$$= x + 2$$

So,
$$\lim_{x \to 2} \left(\frac{x^2 - 4}{x - 2} \right) = \lim_{x \to 2}(x + 2) = 4$$
since when x is close to 2, $x + 2$ is close to 4.

Example:

Consider the curve $f(x) = x^2$. Find a general formula for the slope of the tangent to this curve at the point (x, x^2).

Answer:

Let h be nonzero and consider the slope of the chord through the points (x, x^2) and $((x+h), (x+h)^2)$
$$F(h) = \frac{(x+h)^2 - x^2}{h}$$

$$= \frac{x^2 + 2xh + h^2 - x^2}{h}$$
$$= \frac{2xh + h^2}{h}$$
$$= 2x + h.$$

As h gets closer and closer to zero the chord is getting closer and closer to the tangent. Hence the slope of the chord is getting closer and closer to the slope of the tangent. That is,

$$\text{Slope of tangent} = \lim_{h \to 0}(2x + h),$$

but clearly

$$\lim_{h \to 0}(2x + h) = 2x.$$

Hence the slope of the tangent is $2x$.

The following more precise definition of limit uses the same ideas that came up earlier in our discussion of continuity and in our discussion of the limit of a sequence.

Definition:

Suppose that $F(h)$ is a function that may or may not be defined at the point $h = a$. We say the **limit of $F(h)$ as h approaches a is L** and write

$$\lim_{h \to a} F(h) = L$$

if for any output tolerance $\epsilon > 0$ there is an input tolerance $\delta > 0$ such that

if $0 < |h - a| < \delta$ then $|F(h) - L| < \epsilon$.

The actual value of F at a is irrelevent. In fact, we often use this idea when $F(a)$ is not defined.

The definition of limit is designed to apply to exactly the situation we have been facing. It does not require that the function $F(h)$ be defined when $h = a$. This definition says precisely what we mean when

§8. The Tangent to a Curve

we say that our procedure, $F(h)$, can be used to obtain as good an estimate of L as might be necessary for any particular purpose.

Although the concept of limit is often used when the function $F(h)$ is not defined at the point $h = a$, it is easy to prove that if the function $F(h)$ is *continuous* at the point $h = a$ then

$$\lim_{h \to a} F(h) = F(a).$$

We are now in a position to make a precise definition of one of the most important tools in mathematics, the **derivative**.

Definition:

Suppose that f is a function and x is a point. We define the **derivative** of f at the point x, written $f'(x)$, as follows.

$$f'(x) = \lim_{h \to 0} \frac{f(x+h) - f(x)}{h}$$

Notice that $f'(x)$ is the slope of the tangent to the curve $y = f(x)$ at the point $(x, f(x))$. This slope is obtained by taking the limit of the slopes,

$$\frac{f(x+h) - f(x)}{h},$$

of chords approximating the tangent. Thus f' is a new function whose value at each point x is the slope of the tangent to the curve $y = f(x)$ at the point $(x, f(x))$.

Example:

Compute the derivative of the function $f(x) = x - x^2$.

Answer:

$$\begin{aligned} f'(x) &= \lim_{h \to 0} \frac{f(x+h) - f(x)}{h} \\ &= \lim_{h \to 0} \frac{((x+h) - (x+h)^2) - (x - x^2)}{h} \\ &= \lim_{h \to 0} \frac{x + h - x^2 - 2xh - h^2 - x + x^2}{h} \end{aligned}$$

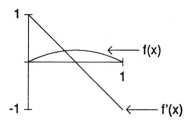

Figure I.75: Comparison of $f(x) = x - x^2$ and $f'(x) = 1 - 2x$

$$= \lim_{h \to 0} \frac{h - 2xh - h^2}{h}$$

$$= \lim_{h \to 0} (1 - 2x - h)$$

$$= 1 - 2x$$

It is worthwhile to compare the graphs of the function f and its derivative f' (see Figure I.75). Notice that when the graph of $y = f(x)$ is heading down, for example at $x = 1$, its derivative $f'(x)$ is negative because the tangent is heading down and, thus, has a negative slope. Similarly, when the graph of $y = f(x)$ is heading up, for example at $x = 0$, its derivative $f'(x)$ is positive because the tangent is heading up and, thus, has a positive slope.

Example:

Find the derivative of the function $f(x) = x^3 - 2x$. What is the slope of the tangent to the curve $y = x^3 - 2x$ at the point $(2, 4)$? Find the equation describing the tangent to the curve $y = x^3 - 2x$ at the point $(2, 4)$.

Answer:

We find the derivative as follows

$$f'(x) = \lim_{h \to 0} \frac{f(x + h) - f(x)}{h}$$

§8. The Tangent to a Curve

$$= \lim_{h \to 0} \frac{((x+h)^3 - 2(x+h)) - (x^3 - 2x)}{h}$$

$$= \lim_{h \to 0} \frac{x^3 + 3x^2h + 3xh^2 + h^3 - 2x - 2h - x^3 + 2x}{h}$$

$$= \lim_{h \to 0} \frac{3x^2h + 3xh^2 + h^3 - 2h}{h}$$

$$= \lim_{h \to 0} (3x^2 + 3xh + h^2 - 2)$$

$$= 3x^2 - 2$$

We find the slope of the tangent at the point $(2,4)$ by computing $f'(2) = 3(2^2) - 2 = 10$.

We now know the slope of the tangent at the point $(2,4)$ and also one point on the tangent—the point $(2,4)$. Whenever we know the slope, m, of a line and any point, (x_0, y_0), on the line we can determine the equation describing the line by the point-slope formula,

$$y - y_0 = m(x - x_0).$$

In this case we obtain

$$\begin{aligned} y - 4 &= 10(x - 2) \\ y - 4 &= 10x - 20 \\ y &= 10x - 16. \end{aligned}$$

Example:

Find the derivative of the function $f(x) = |x|$.

Answer:

Before beginning our calculations, it is instructive to look at a graph of this function. Recall that

$$|x| = \begin{cases} x & \text{if } x \geq 0; \\ -x & \text{if } x < 0. \end{cases}$$

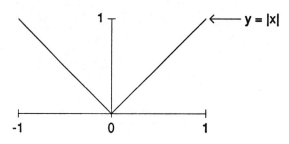

Figure I.76: $f(x) = |x|$

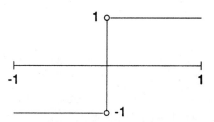

Figure I.77: The derivative of $f(x) = |x|$

So the graph of $y = |x|$ looks like Figure I.76. By looking at this figure we can see:

- If $x > 0$ then the curve $y = |x|$ at x is simply a straight line heading upward with a slope of $+1$.

- If $x < 0$ then the curve $y = |x|$ at x is simply a straight line heading downward with a slope of -1.

- At the point $x = 0$, the graph of $y = |x|$ comes to a sharp point. At a sharp point like this the idea of a tangent really doesn't make sense.

Based on this intuitive discussion, we can graph $f'(x)$ as shown in Figure I.77. Notice, particularly, that $f'(0)$ is undefined because our original function $f(x) = |x|$ does not have a well-defined tangent at the point $x = 0$.

§8. The Tangent to a Curve

Now we would like to see how our new method handles this function, $f(x) = |x|$. We need to calculate the limit

$$\lim_{h \to 0} \frac{|x+h| - |x|}{h}$$

Because the absolute value function is difficult to work with, we must perform this calculation in three steps.

1. Suppose that x is positive. In order to find the limit, we are only interested in small values of h. If x is positive and h is very small (if $|h|$ is smaller than x) then $x+h$ is also positive. Thus $|x| = x$ and $|x+h| = x+h$, so we can continue our calculations

$$\begin{aligned}\lim_{h \to 0} \frac{|x+h| - |x|}{h} &= \lim_{h \to 0} \frac{x+h-x}{h} \\ &= \lim_{h \to 0} \frac{h}{h} \\ &= +1\end{aligned}$$

Notice that this result matches our intuitive result above. When x is positive the derivative, $f'(x)$, is $+1$.

2. Suppose that x is negative. In order to find the limit, we are only interested in small values of h. If x is negative and h is very small (if $|h|$ is smaller than $|x|$) then $x+h$ is also negative. Thus $|x| = -x$ and $|x+h| = -(x+h)$, so we can continue our calculations

$$\begin{aligned}\lim_{h \to 0} \frac{|x+h| - |x|}{h} &= \lim_{h \to 0} \frac{-(x+h) - (-x)}{h} \\ &= \lim_{h \to 0} \frac{-x - h + x}{h} \\ &= \lim_{h \to 0} \frac{-h}{h} \\ &= -1\end{aligned}$$

Notice that this result matches our intuitive result above. When x is negative the derivative, $f'(x)$, is -1.

3. Suppose that x is zero. Then we want to evaluate the limit

$$\lim_{h \to 0} \frac{|0+h| - |0|}{h} = \lim_{h \to 0} \frac{|h|}{h}$$
$$= \lim_{h \to 0} \left(\begin{cases} +1 & \text{if } h > 0; \\ -1 & \text{if } h < 0. \end{cases} \right)$$

and this last limit doesn't exist!

Exercises

*Exercise I.8.9: Find the derivative of the function $f(x) = x^3$.

*Exercise I.8.10: Find the derivative of the function $f(x) = x^4$.

*Exercise I.8.11: Find the derivative of the function $f(x) = x^3 + x^4$. Do you have any observations?

*Exercise I.8.12: Find the derivative of the function $f(x) = x^5 = (x^2)(x^3)$. Do you have any observations?

*Exercise I.8.13: Find the equation of the line that is tangent to the curve $y = x^2 + 3x$ at the point $(2, 10)$.

*Exercise I.8.14:
(a) Find the nonzero equilibrium point of the dynamical system $p_{n+1} = 2.9 p_n (1 - .001 p_n)$. Call it p_*.
(b) Find the derivative of the function $f(p) = 2.9p(1 - .001p)$.
(c) Find $f'(p_*)$.
(d) Based on your work so far in this class, do you think p_* is an attracting equilibrium point.
(e) Check your answer to (d) by letting $p_1 = 50$ and computing $p_2, p_3, p_4, \ldots p_{20}$.

*Exercise I.8.15:
(a) Find the nonzero equilibrium point of the dynamical system $p_{n+1} = 3.1 p_n (1 - .001 p_n)$. Call it p_*.
(b) Find the derivative of the function $f(p) = 3.1p(1 - .001p)$.
(c) Find $f'(p_*)$.
(d) Based on your work so far in this class, do you think p_* is an attracting equilibrium point.

§8. The Tangent to a Curve

(e) Check your answer to (d) by letting $p_1 = 50$ and computing $p_2, p_3, p_4, \ldots p_{20}$.

*Exercise I.8.16: Figure I.78 shows two curves. One curve is the graph of $y = f(x)$ and the other is the graph of $y = f'(x)$. Which is which?

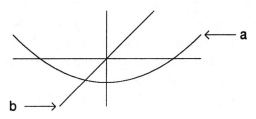

Figure I.78: $f(x)$ and $f'(x)$

*Exercise I.8.17: Figure I.79 shows two curves. One curve is the graph of $y = f(x)$ and the other is the graph of $y = f'(x)$. Which is which?

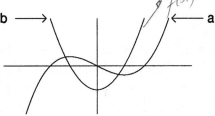

Figure I.79: $f(x)$ and $f'(x)$

*Exercise I.8.18: Figure I.80 shows two curves. One curve is the graph of $y = f(x)$ and the other is the graph of $y = f'(x)$. Which is which?

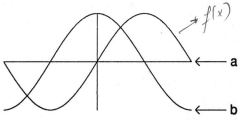

Figure I.80: $f(x)$ and $f'(x)$

§9. What is the Derivative?

Historians studying a particular period of time often notice the same name coming up over and over again in different contexts. If the same person is on the board of directors for several organizations, president of a university, and a frequent visitor to the White House then that person is likely to be a significant historical figure. The derivative is one of calculus' VIPs. It appears in an extraordinary number and variety of contexts. It is one of the most fundamental ideas in all of mathematics.

Our original motivation for developing the derivative came from our observation that linear dynamical systems

$$p_{n+1} = mp_n + b$$

are easy to understand—if $m \neq 0$ then such a system always has exactly one equilibrium point, which is attracting if $|m| < 1$ and not attracting if $|m| > 1$. We also had some evidence that for a nonlinear dynamical system

$$p_{n+1} = f(p_n)$$

the character of an equilibrium point p_* depended on the "slope" of the curve $y = f(p)$ at the point p_*. We developed the derivative as a way of getting at the idea of the slope of a curve. In this section we will see that the derivative appears in many other contexts. You will see the derivative appear in additional contexts later in this course and in the future.

The Derivative and the Microscope

Figure I.81 shows two views of the function $f(x) = x^2$. The second view is a magnified view of the gray area inside the first view. This magnified view shows what you would see if you looked through a microscope at the point under the crosshairs in the first view. The dotted straight line in this view is the tangent to the curve. Notice that the curve looks very much like the tangent. In fact if we were to use a somewhat more powerful microscope as in Figure I.82 the curve and the tangent would be almost indistinguishable.

Figure I.83 shows two views of the function $y = \sin x$. The second view is a magnification of the area around the point $(0,0)$. Notice that here again when magnified the curve looks very much like its tangent.

§9. What is the Derivative?

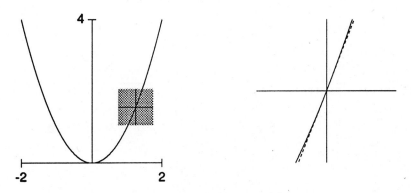

Figure I.81: Two views of $y = x^2$

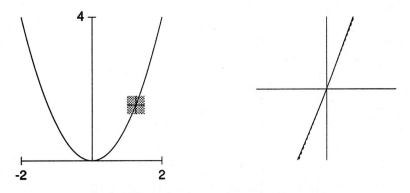

Figure I.82: Two views of $y = x^2$

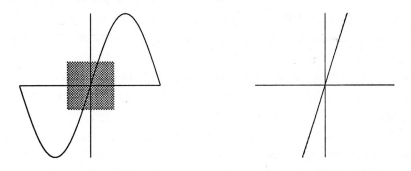

Figure I.83: Two views of $y = \sin x$

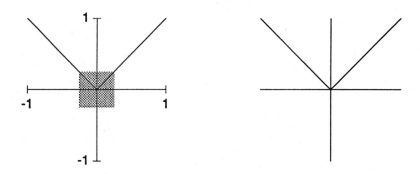

Figure I.84: Two views of $y = |x|$ at $(0,0)$

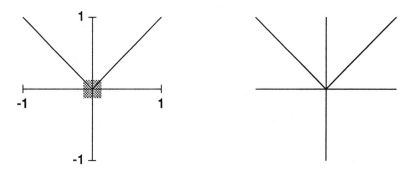

Figure I.85: Two views of $y = |x|$ at $(0,0)$

Figure I.84 shows two views of the function $y = |x|$. The second view is a magnification of the area around the point $(0,0)$. Notice that under magnification the "point" at $(0,0)$ is still very noticeable. It remains noticeable even if we use a more powerful microscope as depicted in Figure I.85.

Figures I.86 and I.87 examine the curve $y = |x|$ at the point $(.25, .25)$ under the microscope. Figure I.87 uses a more powerful microscope than Figure I.86. Notice that near the point $(.25, .25)$ the curve $y = |x|$ looks like a straight line.

We can summarize this experimental evidence by noticing that usually when one examines the area close to a point $(x, f(x))$ on a curve $y = f(x)$ under a powerful microscope the curve looks very much like its tangent. The slope of the tangent is $f'(x)$. There are exceptions to

§9. What is the Derivative?

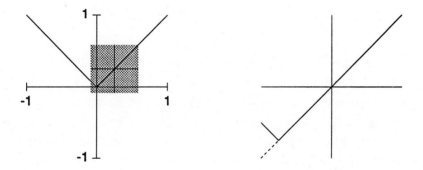

Figure I.86: Two views of $y = |x|$ at $(0.25, 0.25)$

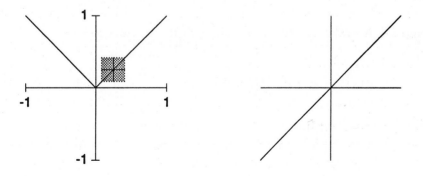

Figure I.87: Two views of $y = |x|$ at $(0.25, 0.25)$

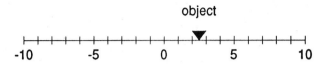

Figure I.88: Object moving along straight line

this general observation. Looking at the area close to the point $(0,0)$ on the curve $y = |x|$ we see a sharp point that does not disappear under magnification. This corresponds to the observation that the derivative of the function $y = |x|$ does not make sense at the point $x = 0$. We usually say that a function $y = f(x)$ is **smooth** at the point x if the derivative makes sense at this point. If a curve is smooth at a point then under powerful magnification the curve near that point looks very much like its tangent.

The Derivative and Velocity

Suppose that an object is moving along a straight line as shown in Figure I.88 Let $f(t)$ denote the position of the object at time t. Suppose we would like to determine the velocity of this object at time t. One possible way to estimate the object's velocity would be to note its position $f(t)$ at time t and its position $f(t+h)$ a little while later at time $t+h$ and then to divide the change in position by the elapsed time, h.

$$\frac{f(t+h) - f(t)}{h}.$$

For example, suppose that you were traveling along an interstate highway and that your speedometer was broken. Suppose at 5:00 you passed mile marker 65 and at 5:06 you passed mile marker 72. You might estimate your velocity at 5:00 by dividing

$$\frac{72 - 65}{6 \text{ minutes}} = \frac{72 - 65}{0.1 \text{ hours}} = 70 \text{ mph}.$$

You can obtain a better estimate by using a smaller value for h, that is, by looking at how far the car has moved over a shorter period

§9. WHAT IS THE DERIVATIVE?

of time. By choosing smaller and smaller values of h one can obtain better and better estimates for the exact velocity. This is exactly the kind of situation we worked with earlier. We want to estimate some quantity, in this case, the velocity of an object at time t, and we have a procedure for getting estimates

$$\text{Estimate} = \frac{f(t+h) - f(t)}{h}.$$

As h approaches 0 the estimates approach the exact velocity. In other words,

$$\text{Exact velocity} = \lim_{h \to 0} \frac{f(t+h) - f(t)}{h}.$$

But, the right hand side of this equation is just $f'(t)$, so we see that

$$\text{Velocity at time } t = f'(t).$$

Like all definitions involving limits, this definition does two important jobs. It provides a precise definition and it provides a method for estimation.

Exercises

*Exercise I.9.1: The height in feet of a ball thrown straight upward is given by $h(t) = 60t - 16t^2$ where t is the time in seconds after it was thrown.
  (a) What is its velocity at time $t = 1$?
  (b) What is its velocity at time $t = 2$?
  (c) How long after the ball is thrown will it continue to rise?
  (d) When will it hit the ground ($h = 0$)?
  (e) How fast will it be falling just before it hits the ground?

*Exercise I.9.2: Suppose that an extraterrestrial spacecraft has been observed heading in a straight line toward the earth. The following table lists the most recent observations. Each observation in this table was made at noon.

Date	Distance from Earth
September 1	200,000,000 miles
September 2	195,000,000 miles
September 3	190,100,000 miles
September 4	185,300,000 miles
September 5	180,600,000 miles
September 6	176,000,000 miles

(a) Estimate its velocity at noon on September 1.
(b) Estimate its velocity at noon on September 2.
(c) Estimate its velocity at noon on September 3.
(d) Estimate its velocity at noon on September 4.
(e) Estimate its velocity at noon on September 5.
(f) Estimate its velocity at noon on September 6.
(g) Find the best estimates you can with the given data.

Rates of Change

We live in a world of change. Many of the things we observe around us are in a constant state of change. The derivative can help us study change.

Example:

Suppose that a particular town obtains all its drinking water from a reservoir. Because of the importance of the reservoir, the town keeps careful records. Suppose the town is in the middle of a long drought and the mayor has asked the supervisor of public works how fast the amount of water in the reservoir is changing. The supervisor prepares a report based on the following portion of the town's records for one particular day.

Time	Water in Reservoir (Gallons)
8:00 A.M.	2,103,612
8:05 A.M.	2,103,511
9:00 A.M.	2,102,842
9:05 A.M.	2,102,811
1:00 P.M.	2,101,953
1:05 P.M.	2,101,911
2:00 P.M.	2,101,497
2:05 P.M.	2,101,475

Here is the supervisor's report.

In the morning the water level is dropping very rapidly. At 8:00 A.M. I estimate that the amount of water in the reservoir is changing at the rate of $-1,212$ gallons per hour. In the afternoon the water level is dropping more slowly. At

§9. What is the Derivative?

1:00 P.M. I estimate the water level is changing at the rate of -504 gallons per hour.

Notice that both figures are negative because the water level is decreasing. If it were raining and the water level was increasing then the figures would be positive. In order to see how the supervisor came up with these figures it is helpful to introduce a little notation. Let $w(t)$ denote the amount of water in the reservoir at time t. For example, $w(1{:}00) = 2,101,953$. The supervisor estimated the rate at which the water level was changing at 8:00 by computing

$$\frac{w(8{:}05) - w(8{:}00)}{5 \text{ minutes}} = \frac{2,103,511 - 2,103,612}{\frac{1}{12} \text{ hours}}$$

$$= \frac{-101}{\frac{1}{12}}$$

$$= (-101)(12)$$

$$= -1212$$

This is exactly the way we have been approximating derivatives. We could rewrite this calculation as

$$\frac{w(8{:}00 + h) - w(8{:}00)}{h} = \frac{2,103,511 - 2,103,612}{\frac{1}{12} \text{ hours}}$$

$$= \frac{-101}{\frac{1}{12}}$$

$$= (-101)(12)$$

$$= -1212$$

where $h = 1/12$ hours. This gives us an estimate of the derivative of $w'(8{:}00)$. The exact value of $w'(8{:}00)$ is

$$\lim_{h \to 0} \frac{w(8{:}00 + h) - w(8{:}00)}{h}$$

and the same limit would give us the exact rate at which the amount of water in the reservoir was changing at 8:00. Thus we can interpret the derivative $w'(t)$ as the rate at which the amount of water in the reservoir is changing at time t.

This is a very general interpretation of the derivative. If $f(t)$ is a function that describes the value of a particular quantity at time t then $f'(t)$ describes how fast the quantity is changing at time t.

Exercises

Exercise I.9.3: Using the data from the example above check the supervisor's estimate of the rate at which the amount of water in the reservoir is changing at 1:00 P.M.

*Exercise I.9.4: Using the data from the example above estimate the rate at which the amount of water in the reservoir is changing at 9:00 A.M.

Exercise I.9.5: Using the data from the example above estimate the rate at which the amount of water in the reservoir is changing at 2:00 P.M.

Sensitivity

Frequently we are interested in two quantities that are related by some function. For example, in section 7 we discussed a relationship between the demand for a particular commodity and its price. This relationship was described by a demand function, $D(p)$. When the price changes the demand will also change. In this situation we might ask how **sensitive** the demand is to changes in the price. One way to estimate this is to change the price a little bit h and compare the resulting change in the demand to the change in price. That is, we can compute

$$\frac{D(p+h) - D(p)}{h}$$

This, of course, is the same fraction we would look at if we wanted to estimate $D'(p)$. Thus the derivative $D'(p)$ tells us how fast the demand is changing with respect to the price or the **sensitivity of demand to price.**

§9. What is the Derivative?

Example:

Suppose the demand function for a particular commodity is

$$D(p) = 1000 - 200p.$$

Notice that $D'(p) = -200$. It is negative because increasing the price causes a decrease in the demand. Each dollar rise in price causes a drop of 200 in the demand.

Suppose that the supply function for a particular commodity is given by

$$S(p) = 500p - 1000.$$

Notice that $S'(p) = 500$. It is positive because increasing the price causes an increase in supply. Each dollar rise in price causes a rise of 500 in supply.

This is a very general interpretation. Whenever one quantity y is related to another quantity x by a function $y = f(x)$, the derivative $f'(x)$ describes the sensitivity of the quantity y with respect to (i.e., compared to) the quantity x.

Exercise

*Exercise I.9.6: The profit that a company makes from selling a product depends on the price in two different ways. First, a higher price means more profit on each item sold. Second, a higher price means fewer items will be sold. These two factors pull in opposite directions. The total profit at a given price is

$$T(p) = D(p)(p - d)$$

where p is the price, $D(p)$ is the demand at price p, and d is the cost of manufacturing each item. Suppose that the manufacturing cost for a particular product is \$0.20 and the demand function is $D(p) = 2000 - 1000p$.

(a) What is the sensitivity of the total profit to the price when the price is \$0.30? Based on your answer, if a company is currently selling this product for \$0.30 would you recommend that the company should increase its price?

(b) What is the sensitivity of the total profit to the price when the price is \$0.80? Based on your answer, if a company is currently selling this product for $0.80 would you recommend that the company should increase its price?

Notation and Terminology

Because the derivative is so useful in so many different ways, it has been studied by many different people. Different workers have developed different terminology and notation to express the same ideas. For example, if $y = f(x)$, people frequently write

$$\frac{dy}{dx} = f'(x)$$

to emphasize the interpretation of the derivative as the sensitivity of y with respect to x. Another common notation is y' or $y'(x)$.

Some people use the term "rate of change of y with respect to x" instead of "sensitivity of y compared to x." The two phrases mean exactly the same thing.

§10. Summary[+]

Each chapter will end with a "Summary[+]" section like this one having two purposes—to summarize the main points of the chapter and to place the work of the chapter in a larger context.

The context of calculus

Our course always begins with "applications," the context of calculus. We have discussed two major areas where mathematics can make some contribution to our understanding. The first such area is population dynamics, the study of how populations change. Naturally occurring populations can be divided into two groups—those species, like temperate zone insects, where the generations are distinct, and those species where the generations are intermingled. For species in which generations are distinct an appropriate model is a discrete dynamical system.

We studied a number of such models including exponential models, exponential models with immigration, and logistic models. Despite the simplicity of these models we observed some very complex behavior. We made a start at understanding some of this behavior. For example,

§10. Summary[+]

an exponential model with immigration is a linear dynamical system. For linear dynamical systems we can find the equilibrium point and we know when the equilibrium point will be attracting.

The linear dynamical system

$$p_{n+1} = mp_n + b$$

has exactly one equilibrium point (unless $m = 1$)

$$p = \frac{b}{1-m}.$$

This point is attracting if $|m| < 1$. Furthermore, no matter what the initial population p_1 is we always have

$$\lim_{n \to \infty} p_n = \frac{b}{1-m}.$$

If $|m| > 1$ then the equilibrium point is not attracting.

We developed a graphical method for computing the sequence produced by a given dynamical system starting with a particular p_1. Using this method we accumulated experimental evidence suggesting that the slope of the function f at an equilibrium point plays a crucial role. Thus we know theoretically that the slope determines whether or not an equilibrium point is attracting for *linear* dynamical systems and we have some experimental evidence that the same geometric concept is involved for *nonlinear* dynamical systems as well.

In the last part of this chapter we developed a mathematical tool, the derivative, that allows us to talk about the slope of a curve. In the next chapter we will make use of this new tool to study the question of whether or not an equilibrium point for a nonlinear dynamical system is attracting. We have observed many other interesting phenomena involving population models and, especially, logistic models. For example, we saw some models that exhibited cyclic behavior, where the same sequence of numbers would keep repeating over and over. Over the course of this year we will study and try to understand some of this complex behavior.

We began our study of population models with discrete population models for two reasons. First, they are extremely interesting and extraordinarily rich. The study of such models, even just simple logistic models, has been one of the most active areas both in mathematics and in biology in recent years. The other reason we began with discrete population models is that conceptually they are much easier. We can describe how a discrete model changes with a simple equation like

$$p_{n+1} = f(p_n).$$

For a continuous model such an equation doesn't make any sense. The population is changing continuously and must be described by a function $p(t)$ rather than by a sequence. We need a way to describe the rate at which such a population is changing. At the beginning of this chapter we didn't have the tools or the language to express such an idea. Now we do. An equation like

$$p'(t) = F(p(t))$$

describes how a continuously changing population changes. Our new concept, the derivative, is exactly the tool we need to study such populations. A model like this is called a **continuous dynamical system**. The study of continuous dynamical systems will be one of the main themes of this course.

The second major application we studied in this chapter was the fluctuation of prices. This is another rich area of study that we will return to over the course of the year. Here again, we began with a rather special kind of product, "Byties." Byties had a "life cycle" analogous to temperate zone insects. That is, they were always baked in the early morning, purchased in the morning and then were either eaten or went stale the same day. They have distinct "generations" just like temperate zone insects. For this reason we were able to study the price fluctuations of Byties using the same kind of mathematics, discrete dynamical systems, that we used to study population change for a species with distinct generations. Here again, the language of the derivative will allow us to investigate price fluctuations for other kinds of commodities that don't have such a rigid "life cycle."

As you review this chapter, be sure to pay special attention to the questions we have left open or unanswered. There are many such questions that might occur to you. For example, we studied population change for a single species that was not directly interacting with other species. Most real species do not exist in isolation. We will eventually develop tools for studying several interacting species.

§10. Summary†

Exercises

Exercise I.10.1: Consider the following example of an exponential model with immigration.

$$p_1 = 100, \quad p_{n+1} = 0.6p_n + 500$$

(a) Calculate $p_1, p_2, p_3, \ldots p_{10}$.
(b) What seems to be happening?
(c) What can you say based on the work in section 3 about

$$\lim_{n \to \infty} p_n?$$

(d) Does the value of p_1 make any difference in the long run behavior of such a population?
(e) What would happen if the 500 were changed to 600, or some other value?
(f) What would happen if the 0.6 were changed to 0.95, 1.2, or some other value?

Exercise I.10.2: An exponential model with *emigration* might look like

$$p_1 = 800, \quad p_{n+1} = 1.2p_n - 100$$

(a) Calculate $p_1, p_2, p_3, \ldots p_{10}$.
(b) What seems to be happening?
(c) What can you say about

$$\lim_{n \to \infty} p_n$$

(d) Does the value of p_1 make any difference in the long run behavior of such a population?
(e) What would happen if the 100 were changed to 200, or some other value?
(f) What would happen if the 1.2 were changed to 1.1, 0.9, or some other value?

Exercise I.10.3: Consider the logistic model

$$p_1 = 10, \quad p_{n+1} = 2.8p_n(1 - .001p_n)$$

(a) Calculate enough terms $p_1, p_2, p_3, \ldots$ so that you think you can see what is happening.

(b) Describe what you see.

Exercise I.10.4: Consider the logistic model

$$p_1 = 10, \qquad p_{n+1} = 3.2p_n(1 - .001p_n)$$

(a) Calculate enough terms $p_1, p_2, p_3, \ldots$ so that you think you can see what is happening.
(b) Describe what you see.

Exercise I.10.5: Consider the logistic model

$$p_1 = 10, \qquad p_{n+1} = 3.4p_n(1 - .001p_n)$$

(a) Calculate enough terms $p_1, p_2, p_3, \ldots$ so that you think you can see what is happening.
(b) Describe what you see.

Exercise I.10.6: Consider the logistic model

$$p_1 = 10, \qquad p_{n+1} = 3.6p_n(1 - .001p_n)$$

(a) Calculate enough terms $p_1, p_2, p_3, \ldots$ so that you think you can see what is happening.
(b) Describe what you see.

Exercise I.10.7: Consider the logistic model

$$p_1 = 10, \qquad p_{n+1} = 3.8p_n(1 - .001p_n)$$

(a) Calculate enough terms $p_1, p_2, p_3, \ldots$ so that you think you can see what is happening.
(b) Describe what you see.

Estimation

The most important mathematical topic for this chapter is estimation. There are many situations in which it is difficult or even impossible to find an exact value for a particular quantity but we can find very good estimates for that quantity. For example, we developed a powerful method, called the Bisection Method, that can be used to estimate a solution for equation of the form

$$f(x) = 0$$

in an interval $[a, b]$ no matter how complicated f is as long as

§10. Summary[+]

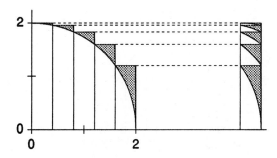

Figure I.89: Estimation error

- The function f is continuous on the interval $[a, b]$.

- Either $f(a) < 0 < f(b)$ or $f(a) > 0 > f(b)$.

- We can find the values of f.

The Bisection Method does not usually yield an exact answer but it can be used to get as good an estimate as might be necessary for any given purpose.

We also developed a method for estimating certain kinds of areas based on Figure I.89 which is a reprise of Figure I.26.

We will return to this theme throughout the semester. We will often be unable to find exact answers but we will be able to find techniques for finding very good estimates, techniques that can be used to get as good estimates as might be necessary in any given situation.

Exercises

*Exercise I.10.8: Consider the equation

$$x^3 - 10x + 1 = 0$$

(a) Let $f(x) = x^3 - 10x + 1$ and compute

$$f(-10), f(-2), f(2), \text{ and } f(10).$$

(b) Based on your computations in (a) what can you say about solutions of this equation?

(c) Estimate three different solutions of this equation. Find estimates that are within 0.05 of the exact solutions.

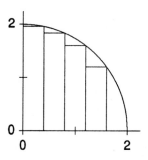

Figure I.90: Approximating strips at right edges

*Exercise I.10.9: In section 3 we estimated the area of a quarter circle of radius 2 in order to find an estimate of the number π. Our estimate was based on Figure I.89. We divided the quarter circle into strips and approximated each strip by a rectangle. The height of each rectangle was determined by using the left edge of each strip. We could equally well have used rectangles whose heights were determined by the right edge of each strip. See Figure I.90.

(a) Find an estimate of π based on Figure I.90. (Use five strips as shown in Figure I.90. (Warning: Notice that the rightmost rectangle has a height of zero and, hence, its area is also zero).

(b) What can you say about this estimate? Is it an overestimate or an underestimate? How good an estimate is it?

(c) Find the average of the estimate you obtained in (a) and the estimate we obtained in section 3 based on five strips. How does this average compare to the true value of π, $3.14159\ldots$.

*Exercise I.10.10: Another possibility would be to approximate each strip by a rectangle whose height was determined by the *middle* of each strip. See Figure I.91.

(a) Find an estimate of π based on Figure I.91. (Use five strips as shown in Figure I.91).

(b) What can you say about this estimate?

(c) Compare this estimate with the true value of $\pi = 3.14159\ldots$ and with the estimates that you have obtained earlier.

(d) This method, called the **midpoint method**, of estimating area is much better than either the "Left Edge Method" or the "Right Edge Method." Can you see intuitively why it is better?

§10. Summary†

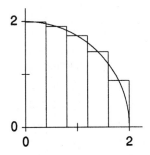

Figure I.91: Approximating strips at midpoints

Limits

The idea of estimation is closely related to the idea of limit. This idea is one of the central ideas of calculus. The notion of limit has two equally important aspects. One is theoretical. The notion of limit allows us to define a concept like the derivative rigorously. The other aspect of the notion of limit is extremely important for practical reasons. Whenever a quantity is defined in terms of a limit the definition specifies a method for estimating that quantity. In the next section we will develop a number of formulas for finding derivatives for many common functions. Often when you need a derivative you need only apply one of these formulas. Sometimes, however, one runs into an obscure or difficult function to which none of the formulas applies. In such a case one can always use the definition of the derivative as a limit to estimate the derivative. This can even be done when the function is not given by a mathematical formula. The only necessity is that we can determine values of the function.

Exercise

*Exercise I.10.11: A group of scientists drops an object into the ocean and keeps a record of its velocity as it drops through the water. They observe the following velocities.

Time	Velocity
0	00.0000
2	50.0000
4	75.0000
6	87.5000
8	93.7500
10	96.8750
12	98.4375

where the figures in the time column are seconds after the object was dropped and the figures in the velocity column are in feet per second.

(a) Find the pattern in these figures.

(b) If this pattern continues what will the velocity of the object be 14 seconds after it was dropped, 16 seconds after it was dropped.

(c) Find a general formula for the velocity of the object t seconds after it was dropped.

(d) What is the limit of the velocity?

(e) After how many seconds will the object's velocity be within 1 feet per second of this limit?

(f) After how many seconds will the object's velocity be within 0.1 feet per second of this limit?

(g) After how many seconds will the object's velocity be within 0.01 feet per second of this limit?

The Derivative

One of the most important characteristics of a linear function is its slope. For a linear function the slope is the same at every point. For a nonlinear function the slope will generally have different values at different points. In fact, a nonlinear function may have points (for example, sharp corners like the point $(0,0)$ on the graph of $f(x) = |x|$) where there is no tangent and, hence, no slope.

Figure I.92 shows the function $f(x) = x^3 - 3x + 1$ and its derivative. Notice that f' is negative where f is decreasing and that f' is positive where f is increasing.

The first three exercises below ask you to find the derivative of a function using the definition of the derivative as a limit. In the next chapter we will develop some formulas that often make it much easier to find derivatives.

§10. Summary+

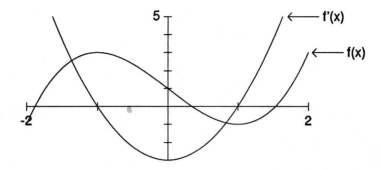

Figure I.92: $f(x) = x^3 - 3x + 1$ and $f'(x)$

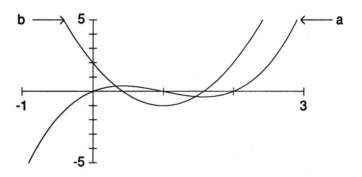

Figure I.93: A function and its derivative

Exercises

*Exercise I.10.12: Suppose $(x) = x^4$. Find $f'(x)$.

*Exercise I.10.13: Suppose $f(x) = \frac{1}{x+2}$. Find $f'(x)$.

*Exercise I.10.14: Suppose $(x) = x^4 + \frac{1}{x+2}$. Find $f'(x)$.

*Exercise I.10.15: Figure I.93 shows the graphs of a function and its derivative. Which one is the original function? Which one is the derivative?

Chapter II

Differentiation

§1. Polynomials

We begin this chapter by studying **power functions**, functions of the form $f(x) = x^n$ where n is a positive integer. Figure II.1 shows the functions $f(x) = x$, $f(x) = x^2$, $f(x) = x^3$, and $f(x) = x^4$.

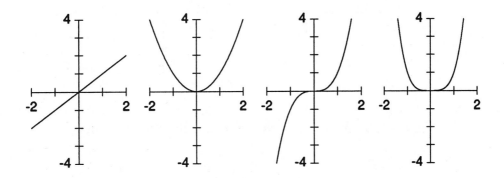

Figure II.1: $f(x) = x$, $f(x) = x^2$, $f(x) = x^3$, and $f(x) = x^4$

Notice, if n is even then for any number x, x^n and $(-x)^n$ are equal. For example, $2^4 = (-2)^4 = 16$. Figures II.2 and II.3 illustrate this observation. A mirror image of a function $f(x) = x^n$ when n is even looks exactly like the original function. This property is worthy of a name.

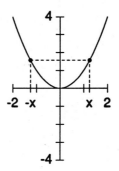

Figure II.2: An even power of x

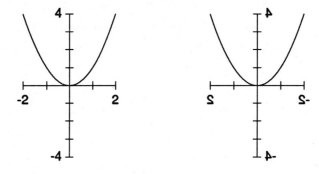

Figure II.3: An even power of x and its mirror image

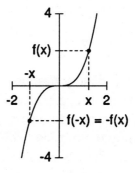

Figure II.4: An odd power of x

§1. Polynomials

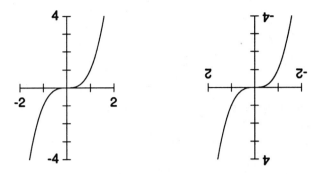

Figure II.5: An odd power of x and its rotated image

Definition:

A function $f(x)$ is said to be **symmetric about the y-axis** if for every x, $f(-x) = f(x)$.

Geometrically, this says that if the graph of $y = f(x)$ is reflected in a mirror lying along the y-axis the reflected graph and the original graph look identical.

Sometimes people use the term "**even**" rather than "symmetric about the y-axis" because the simplest examples of such functions are the functions $f(x) = x^n$ when n is even.

If n is odd then the function $f(x) = x^n$ has the property that $f(-x) = -f(x)$. Figures II.4 and II.5 illustrate this observation. If the graph of such a function is rotated $180°$ about the origin the rotated graph and the original graph look exactly alike. This property is also worthy of a name.

Definition:

A function $f(x)$ is said to be **symmetric about the origin** if for every x, $f(-x) = -f(x)$.

Sometimes people use the term "**odd**" rather than "symmetric about the origin" because the simplest examples of functions that are symmetric about the origin are the functions $f(x) = x^n$ when n is odd.

The following theorem enables us to differentiate any function of the form $f(x) = x^n$ where n is a positive integer.

Theorem 1 *(Power Rule)*
   The derivative of the function
$$f(x) = x^n$$
where n is a positive integer is
$$f'(x) = nx^{n-1}.$$

Proof:
   The proof of this theorem uses a formula that you probably learned in high school and may very well have forgotten, the binomial formula:
$$(x+y)^n = x^n + nx^{n-1}y + \frac{n(n-1)}{2}x^{n-2}y^2 + \cdots + nxy^{n-1} + y^n.$$

The proof is a straightforward calculation from the definition of the derivative.

$$\begin{aligned} f'(x) &= \lim_{h \to 0} \left(\frac{f(x+h) - f(x)}{h} \right) \\ &= \lim_{h \to 0} \left(\frac{(x+h)^n - x^n}{h} \right) \\ &= \lim_{h \to 0} \left(\frac{x^n + nx^{n-1}h + \frac{n(n-1)}{2}x^{n-2}h^2 + \cdots + nxh^{n-1} + h^n - x^n}{h} \right) \\ &= \lim_{h \to 0} \left(\frac{nx^{n-1}h + \frac{n(n-1)}{2}x^{n-2}h^2 + \cdots + nxh^{n-1} + h^n}{h} \right) \\ &= \lim_{h \to 0} \left(nx^{n-1} + \frac{n(n-1)}{2}x^{n-2}h + \cdots + nxh^{n-2} + h^{n-1} \right) \\ &= nx^{n-1} \end{aligned}$$

where the final line comes from the observation that every term in the preceding line except the first has a factor of h, so as h approaches 0 these terms approach 0.

Example:
   The derivative of the function $f(x) = x^8$ is $f'(x) = 8x^7$.

§1. Polynomials

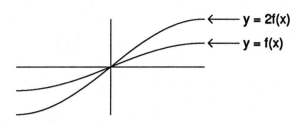

Figure II.6: Comparison of $y = f(x)$ and $y = 2f(x)$

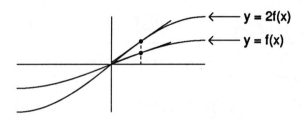

Figure II.7: Comparison of tangents

One way to modify a function is by multiplying it by a constant. For example, Figure II.6 compares a function $y = f(x)$ and the function $y = 2f(x)$. Notice that the functions $f(x)$ and $2f(x)$ are identical except that $2f(x)$ has been stretched by a factor of 2 vertically. Hence, you would expect that the slope of tangent to the graph of $y = 2f(x)$ at a point x would be twice the slope of the tangent to the graph $y = f(x)$ at the same point. See Figure II.7. The following theorem expresses this idea.

Theorem 2 *(Constant Multiple Rule)*

Suppose that $g(x) = kf(x)$ where k is a constant and that $f(x)$ is differentiable. Then
$$g'(x) = kf'(x).$$

Proof:

$$g'(x) = \lim_{h \to 0} \frac{g(x+h) - g(x)}{h}$$

$$= \lim_{h \to 0} \frac{kf(x+h) - kf(x)}{h}$$
$$= \lim_{h \to 0} k \left(\frac{f(x+h) - f(x)}{h} \right)$$
$$= kf'(x)$$

where the final line comes from noticing that as h approaches 0 the term enclosed in parentheses in the preceding line approaches $f'(x)$.

Example:

Let $g(x) = 25x^7$. Then $f'(x) = 25(7x^6) = 175x^6$.

Example:

Suppose that you are driving a car in Europe. At 2:15 your speedometer reads 90 kph (kilometers per hour). How fast are you driving in miles per hour?

Answer:

Let $f(t)$ denote your location at time t measured in kilometers from some particular reference point. Let $g(t)$ denote your location in miles from the same reference point. Since 0.62 miles equals 1 kilometer

$$g(t) = 0.62 f(t).$$

By the Constant Multiple Rule

$$g'(t) = 0.62 f'(t).$$

Now

$$g'(2\!:\!15) = 0.62 f'(2\!:\!15) = (0.62)(90) = 55.80 \text{ mph}.$$

Another way to produce new functions is by adding together old ones. For example, the function $g(x) = x^2 + x^3$ can be obtained by adding the two functions $p(x) = x^2$ and $q(x) = x^3$. The following theorem tells us how to find the derivative of a function obtained in this way.

§1. Polynomials

Theorem 3 *(Sum Rule)*

Suppose that $g(x) = p(x) + q(x)$ and that $p(x)$ and $q(x)$ are differentiable. Then
$$g'(x) = p'(x) + q'(x).$$

Proof:

$$\begin{aligned}
g'(x) &= \lim_{h \to 0} \left(\frac{g(x+h) - g(x)}{h} \right) \\
&= \lim_{h \to 0} \left(\frac{p(x+h) + q(x+h) - (p(x) + q(x))}{h} \right) \\
&= \lim_{h \to 0} \left(\frac{p(x+h) - p(x)}{h} + \frac{q(x+h) - q(x)}{h} \right) \\
&= p'(x) + q'(x)
\end{aligned}$$

where the final line follows from the observation that as h approaches 0
$$\frac{p(x+h) - p(x)}{h}$$
approaches $p'(x)$ and
$$\frac{q(x+h) - q(x)}{h}$$
approaches $q'(x)$.

Example:

If $f(x) = x^2 + x^3$ then $f'(x) = 2x + 3x^2$.

Example:

A chain store is selling a particular item in two different cities. The item is currently selling at \$4.00 in both cities. At that price the sensitivity of the demand to price is -1000 units per dollar in one city and -900 units per dollar in the other city. What is the sensitivity of the total demand in both cities at this price?

Answer:

Let $p(t)$ denote the demand in one city and $q(t)$ denote the demand in the other city. The total demand in both cities is $f(t) = p(t) + q(t)$. By the Sum Rule $f'(t) = p'(t) + q'(t)$ and, in particular,

$$f'(4) = p'(4) + q'(4) = -1000 + (-900) = -1900.$$

The previous theorem allows us to find the derivative of a sum of any number of functions by differentiating each term.

Example:

If $f(x) = x^2 + x^3 + x^4 + x^5$ then $f'(x) = 2x + 3x^2 + 4x^3 + 5x^4$.

By using these two theorems we can differentiate any polynomial.

Example:

If $f(x) = 14x^2 - 17x^3 + 21x^7$ then

$$\begin{aligned} f'(x) &= 14(2x) - 17(3x^2) + 21(7x^6) \\ &= 28x - 51x^2 + 147x^6 \end{aligned}$$

Example:

If $f(x) = (2x - 4)(x + 1)$ find $f'(x)$.

Answer:

Notice $f(x) = (2x - 4)(x + 1) = 2x^2 - 2x - 4$, so $f'(x) = 4x - 2$.

Exercises

Exercise II.1.1: Find the derivative of the function

$$f(x) = x^6 - 3x^4 + 27x^2 - 4.$$

Exercise II.1.2: Differentiate the function $g(x) = 23x^7 - 4x^5 + 6$.

Exercise II.1.3: If $y = 3x^2 + 5x - 6$ find dy/dx.

Exercise II.1.4: Let $f(t) = 1 - 3t^2 + 6t^5$. Find $f'(t)$.

§1. POLYNOMIALS

*Exercise II.1.5: Suppose that at 8:00 A.M. on a particular day the temperature is rising at the rate of ten degrees Fahrenheit per hour. How fast is it rising in degrees Celsius per hour?

Exercise II.1.6: Suppose that at 9:15 a person is on a train traveling at 60 miles per hour and that inside the train he is walking forward (relative to the train) at 3 miles per hour. How fast is he traveling relative to the ground?

*Exercise II.1.7: Find the equation of the line that is tangent to the curve $y = x(2-x)$ at the point $x = 0$.

Exercise II.1.8: Find the equation of the line that is tangent to the curve $y = x(2-x)$ at the point $x = 1$.

Exercise II.1.9: Find the equation of the line that is tangent to the curve $y = x(2-x)$ at the point $x = 2$.

*Exercise II.1.10: Prove that if $f(x)$ is symmetric about the y-axis then $f'(x)$ is symmetric about the origin.

Exercise II.1.11: Prove that if $f(x)$ is symmetric about the origin then $f'(x)$ is symmetric about the y-axis.

*Exercise II.1.12: Prove that if $p(x)$ and $q(x)$ are symmetric about the y-axis then so is $f(x) = p(x) + q(x)$.

Exercise II.1.13: Prove that if $p(x)$ and $q(x)$ are symmetric about the origin then so is $f(x) = p(x) + q(x)$.

Exercise II.1.14: Prove that if $f(x)$ is symmetric about the y-axis then so is $g(x) = kf(x)$ where k is any constant.

Exercise II.1.15: Prove that if $f(x)$ is symmetric about the origin then so is $g(x) = kf(x)$ where k is any constant.

Exercise II.1.16: Prove that if $p(x)$ is a polynomial with only even powers of x then $p(x)$ is symmetric about the y-axis.

Exercise II.1.17: Prove that if $p(x)$ is a polynomial with only odd powers of x then $p(x)$ is symmetric about the origin.

*Exercise II.1.18: Are there any functions that are both symmetric about the origin and symmetric about the y-axis?

Logistic Models Revisited

Based on our observations we have conjectured that an equilibrium point p_* of a dynamical system

$$p_{n+1} = f(p_n)$$

is attracting if $|f'(p_*)| < 1$ and not attracting if $|f'(p_*)| > 1$. Later in this chapter we will prove that this conjecture is true. In this section we would like to use our new differentiation formulas to see what this unproved conjecture predicts about logistic models. Recall that a logistic model is given by

$$p_{n+1} = f(p_n)$$

where

$$f(p) = ap(1 - bp).$$

We find the equilibrium points of such a dynamical system as follows

$$\begin{aligned} p &= ap(1 - bp) \\ p &= ap - abp^2 \\ 0 &= ap - p - abp^2 \\ 0 &= p[(a - 1) - abp] \end{aligned}$$

which gives us one equilibrium point, $p = 0$. Continuing

$$\begin{aligned} (a - 1) - abp &= 0 \\ abp &= a - 1 \\ p &= \frac{a - 1}{ab} \end{aligned}$$

we get a second equilibrium point

$$p_* = \frac{a - 1}{ab}.$$

Thus we have two equilibrium points, 0 and p_*. Our conjecture asks us to look at the derivative f' at each of these equilibrium points.

§1. Polynomials

$$f(p) = ap(1 - bp)$$
$$= ap - abp^2$$
$$f'(p) = a - 2abp$$

First we look at f' at the equilibrium point $p = 0$.

$$f'(0) = a - 2ab(0) = a.$$

Thus our conjecture tells us that the equilibrium point $p = 0$ will be attracting if $|a| < 1$ and not attracting if $|a| > 1$. For logistic models of population growth a and b are always positive, so for these models the equilibrium point $p = 0$ will be attracting if $a < 1$ and not attracting if $a > 1$.

Next we look at the equilibrium point

$$p_* = \frac{a-1}{ab}.$$

$$\begin{aligned} f'(p_*) &= a - 2abp_* \\ &= a - 2ab\left(\frac{a-1}{ab}\right) \\ &= a - 2(a-1) \\ &= a - 2a + 2 \\ &= 2 - a. \end{aligned}$$

Thus according to our conjecture the equilibrium point p_* will be attracting if $|2 - a| < 1$. That is, if

$$\begin{array}{rcl} -1 < & 2 - a & < 1 \\ -3 < & -a & < -1 \\ 3 > & a & > 1. \end{array}$$

Thus our conjecture tells us that for a logistic model of population growth

- If $a < 1$ then $p = 0$ is an attracting equilibrium point.

- If $1 < a < 3$ then $p_* = (a-1)/(ab)$ is an attracting equilibrium point.

Exercises

*Exercise II.1.19: Based on the work above what would you expect to be the longterm behavior of the model

$$p_{n+1} = 0.95 p_n(1 - .001 p_n), \quad p_1 = 50.$$

Calculate some of the terms $p_2, p_3, \ldots$ to see if your expectations are realized.

*Exercise II.1.20: Based on the work above what would you expect to be the longterm behavior of the model

$$p_{n+1} = 1.05 p_n(1 - .001 p_n), \quad p_1 = 50.$$

Calculate some of the terms $p_2, p_3, \ldots$ to see if your expectations are realized.

Exercise II.1.21: Based on the work above what would you expect to be the longterm behavior of the model

$$p_{n+1} = 2.95 p_n(1 - .001 p_n), \quad p_1 = 50.$$

Calculate some of the terms $p_2, p_3, \ldots$ to see if your expectations are realized.

Exercise II.1.22: Based on the work above what would you expect to be the longterm behavior of the model

$$p_{n+1} = 3.05 p_n(1 - .001 p_n), \quad p_1 = 50.$$

Calculate some of the terms $p_2, p_3, \ldots$ to see if your expectations are realized.

*Exercise II.1.23: Based on the work above what would you expect to be the longterm behavior of the model

$$p_{n+1} = 1.00 p_n(1 - .001 p_n), \quad p_1 = 50.$$

§2. Products and Quotients

Calculate some of the terms $p_2, p_3, \ldots$ to see if your expectations are realized.

Exercise II.1.24: Based on the work above what would you expect to be the longterm behavior of the model

$$p_{n+1} = 3.00 p_n(1 - .001 p_n), \quad p_1 = 50.$$

Calculate some of the terms $p_2, p_3, \ldots$ to see if your expectations are realized.

*Exercise II.1.25: Describe how the values of the constants a and b affect the longterm behavior of a logistic model of population growth. Try to relate your comments to the underlying biology. Hint: See the comments in chapter I section 1 about the way in which the values of the constants a and b reflect the underlying biology of the model.

*Exercise II.1.26: The following family of dynamical systems describes the population growth of species that hunt in packs.

$$R(p) = \begin{cases} .009 a p, & \text{if } p \leq 100; \\ a(1 - .001 p), & \text{if } 100 < p \leq 1000. \end{cases}$$

$$p_{n+1} = R(p_n) p_n.$$

Find all the equilibrium points for these dynamical systems and examine for what values of the constant a each equilibrium point is attracting and not attracting.

---

§2. Products and Quotients

There are many ways to build new functions from old ones. In the last section we studied two such ways, addition and multiplication by a constant. In this section we discuss two additional ways, multiplication and division.

Example:

Suppose that the function $D(p)$ describes the demand for a particular product at the price p and that the function $U(p)$ describes the profit made on each unit sold at the price p. Then the total profit a company will make if it makes exactly enough to meet the demand and sells the product at the price p is given by the product

$$T(p) = D(p)U(p).$$

Example:

Suppose that the total food production in a particular country at time t is given by the function $f(t)$ and that the country's population is given by $p(t)$. Then the per capita food production is given by quotient

$$C(t) = \frac{f(t)}{p(t)}.$$

Our next goal is to find the derivative of a product $f(t) = u(t)v(t)$ if we know the derivative of each of the factors, $u(t)$ and $v(t)$. In order to do this we need the following theorem which is of independent interest.

Theorem 4 *Suppose that a function $f(x)$ is differentiable at the point a. Then it is also continuous at the same point.*

Proof:

In order to show that $f(x)$ is continuous at the point a we must show that

$$\lim_{x \to a} f(x) = f(a)$$

or, equivalently,

$$\lim_{h \to 0} f(a + h) = f(a)$$

or

$$\lim_{h \to 0} f(a + h) - f(a) = 0.$$

But

$$f(a+h) - f(a) = \left(\frac{f(a+h) - f(a)}{h} \right) h$$

and, as h approaches 0 the first factor on the right approaches $f'(a)$ and the second factor approaches 0. So the product approaches zero, proving the theorem.

Note this is a one way theorem. The function $f(x) = |x|$, for example, is continuous but not differentiable at the point 0.

§2. Products and Quotients

Theorem 5 *(Product Rule)*
   Suppose that $f(t) = u(t)v(t)$ and that $u(t)$ and $v(t)$ are differentiable. Then
$$f'(t) = u(t)v'(t) + v(t)u'(t).$$

Proof:
   We begin our proof with the definition of the derivative
$$f'(t) = \lim_{h \to 0}\left(\frac{f(t+h) - f(t)}{h}\right).$$
The numerator is the difference $f(t+h) - f(t)$ or
$$f(t+h) - f(t) = u(t+h)v(t+h) - u(t)v(t).$$
We make use of a common algebraic trick in order to simplify this; we add and substract the same term obtaining
$$u(t+h)v(t+h) - u(t)v(t) = u(t+h)v(t+h) - u(t+h)v(t) + u(t+h)v(t) - u(t)v(t).$$
This gives us a longer, apparently more compicated, expresssion. In fact, however, it is the key step in proving this theorem because it allows us to express the change $f(x+h) - f(x)$ in terms of the changes $u(x+h) - u(x)$ and $v(x+h) - v(x)$ and eventually the derivative $f'(x)$ in terms of $u'(x)$ and $v'(x)$ as shown below.

$$\begin{aligned}
f'(t) &= \lim_{h \to 0}\left(\frac{f(t+h) - f(t)}{h}\right) \\
&= \lim_{h \to 0}\left(\frac{u(t+h)v(t+h) - u(t)v(t)}{h}\right) \\
&= \lim_{h \to 0}\left(\frac{u(t+h)v(t+h) - u(t+h)v(t) + u(t+h)v(t) - u(t)v(t)}{h}\right) \\
&= \lim_{h \to 0}\left[\frac{u(t+h)v(t+h) - u(t+h)v(t)}{h} + \frac{u(t+h)v(t) - u(t)v(t)}{h}\right] \\
&= \lim_{h \to 0}\left[u(t+h)\left(\frac{v(t+h) - v(t)}{h}\right) + v(t)\left(\frac{u(t+h) - u(t)}{h}\right)\right] \\
&= u(t)v'(t) + v(t)u'(t)
\end{aligned}$$

where the final line follows from the preceding line by noticing that

$$\lim_{h \to 0} u(t+h) = u(t),$$

$$\lim_{h \to 0} \left(\frac{v(t+h) - v(t)}{h} \right) = v'(t),$$

and

$$\lim_{h \to 0} \left(\frac{u(t+h) - u(t)}{h} \right) = u'(t).$$

We can get some insight into this theorem by thinking about it geometrically. Multiplication is often used to compute the area of a rectangle. A rectangle whose length is l and whose width is w has area $A = lw$. If the length and width of a rectangle at time t are given by the functions $l(t)$ and $w(t)$ then its area at time t is given by the function $A(t) = l(t)w(t)$. A change in t causes both $l(t)$ and $w(t)$ to change which in turn causes $A(t)$ to change. Figure II.8 illustrates these changes. Notice that the L-shaped change in A can be broken up into two pieces as shown in Figure II.9 and II.10. The change shown in Figure II.9 is the width $w(t)$ times the change in length. The change shown in Figure II.10 is the length $l(t+h)$ times the change in width. These two pieces correspond to the two terms in the formula for the derivative of a product. The mysterious algebraic trick

$$u(x+h)v(x+h) - u(x)v(x) = u(t+h)v(t+h) - u(t+h)v(t) + u(t+h)v(t) - u(t)v(t)$$

in the preceding proof mirrors the geometric splitting of the L-shaped piece into two rectangular pieces.

Example:

Let $f(t) = (t^2 - 2t + 6)(3t^3 - t^2)$ and find $f'(t)$.

Answer:

Think of $f(t)$ as a product

$$f(t) = \underbrace{(t^2 - 2t + 6)}_{u(t)} \underbrace{(3t^3 - t^2)}_{v(t)}.$$

§2. Products and Quotients

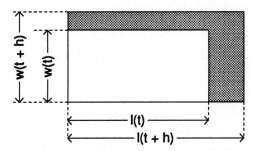

Figure II.8: The change in area caused by changes in height and width

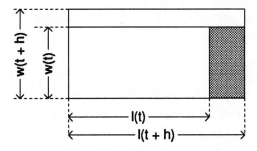

Figure II.9: One piece of the change

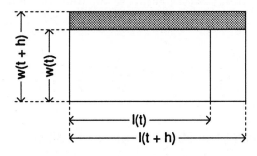

Figure II.10: The other piece of the change

Notice that $u'(t) = 2t - 2$ and $v'(t) = 9t^2 - 2t$. Thus

$$\begin{aligned}
f'(t) &= u(t)v'(t) + v(t)u'(t) \\
&= (t^2 - 2t + 6)(9t^2 - 2t) + (3t^3 - t^2)(2t - 2) \\
&= (9t^4 - 20t^3 + 58t^2 - 12t) + (6t^4 - 8t^3 + 2t^2) \\
&= 15t^4 - 28t^3 + 60t^2 - 12t.
\end{aligned}$$

Example:

Suppose the per capita daily energy consumption in a particular country is currently 800,000 BTUs and that due to energy conservation efforts it is falling at the rate of 1,000 BTUs per year. Suppose that the population is currently 200,000,000 people and is rising at the rate of 1,000,000 people per year. What is the current total daily energy consumption? Is the total daily energy consumption rising or falling? By how much?

This is an important problem. We are often faced with a complicated situation in which several factors contribute to a particular quantity. In this case total daily energy consumption is determined by two factors, population and per capita daily energy consumption. When these two factors change, the total daily energy consumption changes as well. Notice that in this problem we have two changes working in opposite directions. Population is rising, which will cause total daily energy consumption to rise, and per capita daily energy consumption is falling, which will cause total daily energy consumption to fall. The Product Rule enables us to determine the net result of these two changes.

Answer:

Let $C(t)$ denote daily per capita energy consumption and let t_* denote now. Hence,

$$C(t_*) = 800,000 = 8 \times 10^5 \text{ BTU}$$

and

$$C'(t_*) = -1,000 = -10^3 \text{ BTU per year}.$$

§2. Products and Quotients

Let $P(t)$ denote population. Hence,

$$P(t_*) = 200,000,000 = 2 \times 10^8 \text{ people}$$

and

$$P'(t_*) = 1,000,000 = 10^6 \text{ people per year.}$$

Total daily energy consumption $E(t)$ is

$$E(t) = C(t)P(t).$$

So the current total daily energy consumption is

$$E(t_*) = C(t_*)P(t_*) = (8 \times 10^5)(2 \times 10^8) = 16 \times 10^{13}$$

Using the Product Rule we get

$$E'(t) = C(t)P'(t) + P(t)C'(t)$$

and

$$\begin{aligned} E'(t_*) &= C(t_*)P'(t_*) + P(t_*)C'(t_*) \\ &= (8 \times 10^5)(10^6) + (2 \times 10^8)(-10^3) \\ &= (8 \times 10^{11}) - (2 \times 10^{11}) \\ &= 6 \times 10^{11}. \end{aligned}$$

The total daily energy consumption is currently rising at the rate of 6×10^{11} BTUs per year.

The next theorem enables us to find the derivative of a quotient.

Theorem 6 *(Quotient Rule)*
Suppose that

$$f(x) = \frac{u(x)}{v(x)}$$

and $u'(x)$ and $v'(x)$ exist. Then

$$f'(x) = \frac{v(x)u'(x) - u(x)v'(x)}{v(x)^2}.$$

Proof:

The proof is similar to the proof of the Product Rule and begins with the definition of the derivative.

$$f'(x) = \lim_{h \to 0} \left(\frac{f(x+h) - f(x)}{h} \right)$$

$$= \lim_{h \to 0} \left(\frac{\frac{u(x+h)}{v(x+h)} - \frac{u(x)}{v(x)}}{h} \right)$$

$$= \lim_{h \to 0} \left(\frac{u(x+h)v(x) - u(x)v(x+h)}{hv(x+h)v(x)} \right)$$

$$= \lim_{h \to 0} \left(\frac{u(x+h)v(x) - u(x+h)v(x+h) + u(x+h)v(x+h) - u(x)v(x+h)}{hv(x+h)v(x)} \right)$$

$$= \lim_{h \to 0} \left[\frac{1}{v(x+h)v(x)} \right] \left[-u(x+h) \left(\frac{v(x+h) - v(x)}{h} \right) + v(x+h) \left(\frac{u(x+h) - u(x)}{h} \right) \right]$$

$$= \left(\frac{1}{v(x)^2} \right) (-u(x)v'(x) + v(x)u'(x))$$

$$= \frac{v(x)u'(x) - u(x)v'(x)}{v(x)^2}$$

where the next-to-last line comes from the preceding line by noticing

$$\lim_{h \to 0} u(x+h) = u(x),$$

$$\lim_{h \to 0} v(x+h) = v(x),$$

$$\lim_{h \to 0} \left(\frac{v(x+h) - v(x)}{h} \right) = v'(x),$$

and

$$\lim_{h \to 0} \left(\frac{u(x+h) - u(x)}{h} \right) = u'(x).$$

Notice the term in the numerator of the derivative involving the derivative of the original denominator has a minus sign because as the denominator increases the quotient *decreases*. The term in the derivative involving the derivative of the original numerator has a plus sign because as the numerator increases the quotient also increases.

§2. Products and Quotients

Example:

Suppose
$$f(x) = \frac{x^2 + 4}{2x - 5}.$$

Find $f'(x)$.

Answer:

Think of $f(x)$ as a quotient

$$f(x) = \frac{x^2 + 4 \leftarrow u(x)}{2x - 5 \leftarrow v(x)}.$$

So

$$f'(x) = \frac{v(x)u'(x) - u(x)v'(x)}{v(x)^2}$$

$$= \frac{(2x - 5)(2x) - (x^2 + 4)(2)}{(2x - 5)^2}$$

$$= \frac{2x^2 - 10x - 8}{(2x - 5)^2}.$$

Example:

Suppose that the current total daily energy consumption in a particular country is 16×10^{13} BTUs and is rising at the rate of 6×10^{11} BTUs per year. Suppose that the current population is 2×10^8 people and is rising at the rate of 10^6 people per year. What is the current daily per capita energy consumption? Is it rising or falling? By how much?

Answer:

Let $E(t)$ denote total daily energy consumption and let t_* denote now. Hence,
$$E(t_*) = 16 \times 10^{13}$$

and
$$E'(t_*) = 6 \times 10^{11}.$$

Let $P(t)$ denote population. Hence,
$$P(t_*) = 2 \times 10^8$$
and
$$P'(t_*) = 10^6.$$

Per capita daily energy consumption is
$$C(t) = \frac{E(t)}{P(t)}.$$

Hence, current daily per capita energy consumption is
$$C(t_*) = \frac{E(t_*)}{P(t_*)} = \frac{16 \times 10^{13}}{2 \times 10^8} = 8 \times 10^5.$$

Using the Quotient Rule we see
$$C'(t) = \frac{P(t)E'(t) - E(t)P'(t)}{P(t)^2}$$
and
$$\begin{aligned}
C'(t_*) &= \frac{P(t_*)E'(t_*) - E(t_*)P'(t_*)}{P(t_*)^2} \\
&= \frac{(2 \times 10^8)(6 \times 10^{11}) - (16 \times 10^{13})(10^6)}{(2 \times 10^8)^2} \\
&= \frac{(12 \times 10^{19}) - (16 \times 10^{19})}{4 \times 10^{16}} \\
&= \frac{-4 \times 10^{19}}{4 \times 10^{16}} \\
&= -1,000
\end{aligned}$$

So, per capita daily energy consumption is dropping at the rate of 1,000 BTUs per year.

§2. Products and Quotients

The Power Rule Revisited

In section 1 we showed that for any positive integer n

$$\frac{d}{dx}x^n = nx^{n-1}.$$

The following theorem shows that the same formula applies if n is a negative integer.

Theorem 7 *(The Power Rule (Extended))*
Suppose that n is any integer then

$$\frac{d}{dx}x^n = nx^{n-1}.$$

Proof:
If n is positive then the result follows from the version of the Power Rule in section 1. If $n = 0$ then $x^n = x^0 = 1$ and

$$\frac{d}{dx}1 = 0.$$

So our only concern is for n negative.
For n negative

$$x^n = \frac{1}{x^{-n}} \begin{array}{l} \leftarrow u(x) \\ \leftarrow v(x) \end{array}$$

and we can apply the Quotient Rule noting that

$$u'(x) = 0$$

and

$$v'(x) = (-n)x^{-n-1} = -nx^{-n-1}$$

since $-n$ is positive.
Hence,

$$\begin{aligned}\frac{d}{dx}x^n &= \frac{x^{-n}(0) - 1(-nx^{-n-1})}{(x^{-n})^2} \\ &= \frac{nx^{-n-1}}{x^{-2n}} \\ &= nx^{n-1}\end{aligned}$$

Example:

Differentiate the function $f(x) = 1/x^3$.

Answer:

$$f(x) = x^{-3}.$$

So

$$f'(x) = -3x^{-4} = \frac{-3}{x^4}.$$

Exercises

Differentiate each of the following functions.

Exercise II.2.1: $f(x) = (x^2 + 4)(x^2 - 4) = x^4 - 16$

Exercise II.2.2: $p(t) = \frac{4}{t^2} + 3t^2 - t^3$.

Exercise II.2.3: $f(x) = (26 + x^5)(x^2 - 6x^3)$

Exercise II.2.4: $f(x) = x^4(x^3 + x^2 + 3)$

Exercise II.2.5: $f(t) = (t^3 - 12t + 1)(t^2 + t + 1)$

Exercise II.2.6: $f(x) = (x^3 - 3x)(x^3 - 3x^2 + 1)(x^3 - 4x^2 + 6)$

Exercise II.2.7: $g(x) = (x^2 - 3)(x^2 + 1)(x^2 - 2)$

Exercise II.2.8: $f(x) = (x^3 - 2x + 14)(x - 6x^4)$

Exercise II.2.9: $f(x) = (-2x^3 - x^2 + 162)(x^5 + x^4 + x)$

Exercise II.2.10: $f(t) = (t^4 - 12t^3 + t - 16)(t^2 + t + 1)$

Exercise II.2.11: $f(x) = (x^3 + x)(x^4 - 3x + 1)(x^3 + 3x^2 + 2)$

Exercise II.2.12: $g(x) = x^2(x^2 + 1)(x^2 - 2)$

Exercise II.2.13:

$$\frac{3x^3 - 2x^2 + 4}{x^2 - 12}$$

Exercise II.2.14:

$$\frac{x^3 + 1}{3x^2 + 2x + 4}$$

§2. Products and Quotients

Exercise II.2.15:
$$x^3 + \frac{x^2 - 4}{x - 3}$$

Exercise II.2.16:
$$\frac{(x^2 + 1)(x^3 - 1)}{x^2 - 7}$$

Exercise II.2.17:
$$\frac{x^3 + x^2 - 14}{x - 12x^3}$$

Exercise II.2.18:
$$\frac{x^2 + x - 1}{x^2 - 3x + 1}$$

Exercise II.2.19:
$$x^2 + \frac{x - 1}{x + 1}$$

Exercise II.2.20:
$$\frac{(x + 1)(x - 1)}{x^3 - x^2 + 5}$$

Exercise II.2.21:
$$f(x) = 2x^{-3} + x^{-5}$$

Exercise II.2.22:
$$f(x) = \frac{1}{x} + \frac{2}{x^2}$$

Exercise II.2.23:
$$p(t) = \frac{4}{t} - \frac{6}{t^2} + \frac{1}{2t^4}$$

Exercise II.2.24:
$$f(x) = 3x^{-2} + 4x^{-4}$$

Exercise II.2.25:
$$f(x) = \frac{1}{x} + 52x^3$$

Exercise II.2.26:

$$p(t) = \frac{1}{t} + \frac{1}{t^2} + \frac{2}{t^4}$$

*Exercise II.2.27: The population of a particular country is 15,000,000 people and is growing at the rate of 10,000 people per year. In the same country the per capita yearly expenditure for energy is $1,000 per person and is growing at the rate of $8 per year. What is the country's current total yearly energy expenditure? How fast is the country's total yearly energy expenditure growing?

*Exercise II.2.28: An explorer is marooned on an iceberg. The iceberg is shaped like a square with sides of length 100 feet. What is the area of the iceberg? The length of the sides of the iceberg are shrinking at the rate of two feet per day. How fast is the area of the iceberg shrinking? In five days the iceberg will have sides of length 90 feet. Assuming the sides are still shrinking at the rate of two feet per day how fast will the area then be shrinking then?

*Exercise II.2.29: The iceberg of the preceding exercise is actually a cube not a square. How fast is the volume of the cube shrinking now? How fast will it be shrinking in five days?

*Exercise II.2.30: The population of a particular country is 30,000,000 and rising at the rate of 4,000 people per year. The total yearly personal income in the country is $20,000,000,000 and rising at the rate of $500,000,000 per year. What is the current per capita personal income? How fast is it rising or falling? Is it rising? or falling?

The last three exercises in this set of exercises are more challenging. We have been looking at the ordinary derivative. If $y = f(t)$ describes a quantity y in terms of time t then $f'(t)$ tells us how fast y is changing in absolute terms. For example, if $f(t)$ is the amount of a pollutant in a particular lake at time t where t is measured in years and $f(t)$ is measured in gallons then $f'(t)$ is measured in gallons per year. For many applications this absolute measure of rate of change is not the most interesting quantity. For example, suppose someone tells you that the price of a new building is going to be $12,000 higher than expected. This fact alone doesn't give you a real idea of how "big" a jump has occurred. If the building was originally expected to cost $120,000 then the cost override is very substantial, 1/10 or 10%, of the expected cost. But if the building was originally expected to cost $12,000,000 then the cost overrun is much less substantial, only 1/1000 or 0.1%, of the original cost. For this reason rates of change are often reported not in absolute terms but in relative terms. For example, one reads that "The price of oil has reason by 1/10 this year" or "The number of people without jobs has dropped by 1/5 this year."

We can define a new kind of "derivative" called the **relative derivative** and denoted $f^*(x)$ by

$$f^*(x) = \frac{f'(x)}{f(x}$$

$f^*(x)$ measures the relative rate of change of the function $f(x)$.

*Exercise II.2.31: Suppose that

$$f(x) = u(x)v(x)$$

Find a formula for $f^*(x)$ in terms of $u^*(x)$ and $v^*(x)$.

*Exercise II.2.32: Suppose that

$$f(x) = \frac{u(x)}{v(x)}$$

Find a formula for $f^*(x)$ in terms of $u^*(x)$ and $v^*(x)$.

*Exercise II.2.33: Suppose that

$$f(x) = u(x) + v(x)$$

Find a formula for $f^*(x)$ in terms of $u^*(x)$ and $v^*(x)$.

---

§3. The Trigonometric Functions

We begin this section with a brief trigonometry review.

Figure II.11: Measuring an angle in radians

Trigonometry Review

For everyday use we usually measure angles in **degrees**. Although this system is fine for everyday use, for mathematical and scientific purposes another system, based on **radians**, is far superior. *In this class we will always measure angles in radians.*

To measure an angle in radians place the vertex of the angle at the center of a circle of radius one as shown in Figure II.11 and then measure the length of the portion of the circumference of the circle cut off by the two lines forming the angle. The length of this portion of the circumference is the size of the angle in radians. Because the length of the entire circumference is 2π, one can convert from degrees to radians by the formula

$$\text{radians} = 2\pi \times \text{degrees}/360$$

and from radians to degrees by the formula

$$\text{degrees} = 360 \times \text{radians}/2\pi.$$

For example, a right angle (90 degrees) is $\pi/2$ radians.

We measure angles starting at the positive side of the x-axis with positive angles measured in the counterclockwise direction and negative angles measured in the clockwise direction as shown in Figure II.12.

The same physical angle has many different labels. For example, the angle $\pi/4$ is identical to the angle $-7\pi/4$ as shown in Figure II.13. Imagine yourself walking around the circle labeling angles as you walk. Because you are walking in circles each angle will get many labels. As you walk counterclockwise the right angle $\pi/2$ radians will be labeled $\pi/2, 5\pi/2, 9\pi/2, 13\pi/2$ and so forth. As you walk clockwise the same angle will receive the labels $-3\pi/2, -7\pi/2, -11\pi/2$ and so forth.

§3. The Trigonometric Functions

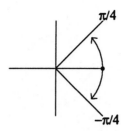

Figure II.12: Measuring positive and negative angles

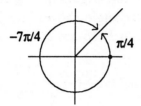

Figure II.13: Two names for the same angle

We define the trigonometric functions—sine, cosine, tangent, cosecant, secant and cotangent—by looking at a circle of radius 1 and at the coordinates of a point (x, y) that corresponds to a given angle θ as shown in Figure III.1.4.

Definition:

$$\begin{aligned} \sin \theta &= y \\ \cos \theta &= x \\ \tan \theta &= y/x = (\sin \theta)/(\cos \theta) \\ \csc \theta &= 1/y = 1/\sin \theta \\ \sec \theta &= 1/x = 1/\cos \theta \\ \cot \theta &= x/y = (\cos \theta)/(\sin \theta) \end{aligned}$$

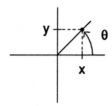

Figure II.14: A point on the circle of radius 1

One often thinks about the trigonometric functions in terms of the lengths of the sides of the triangle shown in Figure II.15. Because the triangles in Figures II.14 and II.15 are similar we can write

$$\sin\theta = \frac{\text{opposite side}}{\text{hypotenuse}}$$

$$\cos\theta = \frac{\text{adjacent side}}{\text{hypotenuse}}$$

$$\tan\theta = \frac{\text{opposite side}}{\text{adjacent side}}$$

$$\csc\theta = \frac{\text{hypotenuse}}{\text{opposite side}}$$

$$\sec\theta = \frac{\text{hypotenuse}}{\text{adjacent side}}$$

$$\cot\theta = \frac{\text{adjacent side}}{\text{opposite side}}$$

Example:

Consider the isosceles right triangle shown in Figure II.16. Since this triangle is half of a square the angle at the base is one-half of a right angle. A right angle is $\pi/2$ radians, so the angle at the base of this isosceles triangle is $\pi/4$ radians and the values of the trig functions for the angle $\pi/4$ radians are:

§3. The Trigonometric Functions

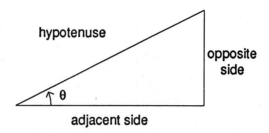

Figure II.15: The sides of a triangle

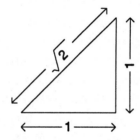

Figure II.16: An isosceles right triangle

$$
\begin{aligned}
\sin(\pi/4) &= \text{opposite side/hypotenuse} = 1/\sqrt{2} \\
\cos(\pi/4) &= \text{adjacent side/hypotenuse} = 1/\sqrt{2} \\
\tan(\pi/4) &= \text{opposite side/adjacent side} = 1/1 = 1 \\
\csc(\pi/4) &= \text{hypotenuse/opposite side} = \sqrt{2}/1 = \sqrt{2} \\
\sec(\pi/4) &= \text{hypotenuse/adjacent side} = \sqrt{2}/1 = \sqrt{2} \\
\cot(\pi/4) &= \text{adjacent side/opposite side} = 1/1 = 1
\end{aligned}
$$

Example:

Consider the triangle that is formed by taking half of an equilateral triangle as shown in Figure II.17. Each of the sides of the original triangle has length 2. Each of its angles measures $\pi/3$ radians. The shaded triangle is obtained by cutting the equilateral triangle in half. Its hypotenuse is the same as one of the sides of the original triangle

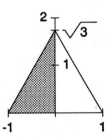

Figure II.17: A 30–60–90 triangle

and has length 2. The base is half of one of the original sides and has length 1. By the Pythagorean Theorem the third side has length $\sqrt{2^2 - 1^2} = \sqrt{3}$. The angle at the base measures $\pi/3$ radians since it is one of the original angles from the equilateral triangle. The smaller angle at the top of the shaded triangle is one-half of one of the original angles and, hence, measures $\pi/6$ radians. This sometimes called a "30-60-90" degree triangle because those are the measurements of its angles in degrees. Based on the shaded triangle we can see the following

$$\begin{aligned}
\sin(\pi/3) &= \text{opposite side/hypotenuse} = \sqrt{3}/2 \\
\cos(\pi/3) &= \text{adjacent side/hypotenuse} = 1/2 \\
\tan(\pi/3) &= \text{opposite side/adjacent side} = \sqrt{3}/1 = \sqrt{3} \\
\csc(\pi/3) &= \text{hypotenuse/opposite side} = 2/\sqrt{3} \\
\sec(\pi/3) &= \text{hypotenuse/adjacent side} = 2/1 = 2 \\
\cot(\pi/3) &= \text{adjacent side/opposite side} = 1/\sqrt{3}
\end{aligned}$$

and

$$\begin{aligned}
\sin(\pi/6) &= \text{opposite side/hypotenuse} = 1/2 \\
\cos(\pi/6) &= \text{adjacent side/hypotenuse} = \sqrt{3}/2 \\
\tan(\pi/6) &= \text{opposite side/adjacent side} = 1/\sqrt{3} \\
\csc(\pi/6) &= \text{hypotenuse/opposite side} = 2/1 \\
\sec(\pi/6) &= \text{hypotenuse/adjacent side} = 2/\sqrt{3} \\
\cot(\pi/6) &= \text{adjacent side/opposite side} = \sqrt{3}/1 = \sqrt{3}.
\end{aligned}$$

§3. The Trigonometric Functions

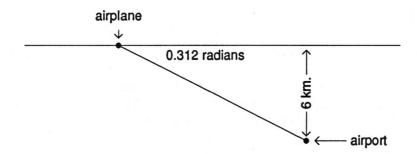

Figure II.18: Airplane and airport

The trigonometric functions are frequently used to determine the length of the sides of a right triangle given the length of one side and one angle other than the right angle.

Example:

An airplane is flying at an altitude of 6 kilometers. A passenger looking out the window notices that the line of sight to the airport is 0.312 radians below the horizontal as shown in Figure II.18. How far is the plane from the airport?

Answer:

We want to find the length of the hypotenuse of the triangle in Figure II.18. Since

$$\frac{\text{hypotenuse}}{\text{opposite side}} = \csc(0.312),$$

hypotenuse = (opposite side) $\csc(0.312) = (6)(3.2577) = 19.5463$ km.

Exercises

*Exercise II.3.1: A surveyor stands 50 feet away from a tree on level ground and measures the angle between his line of sight to the top of the tree and the ground. This angle is 0.623 radians. The surveyor's eye is five feet above the ground. How tall is the tree?

Second Quadrant In this quadrant x is negative and y is positive	First Quadrant In this quadrant both x and y are positive
Third Quadrant In this quadrant both x and y are negative	Fourth Quadrant In this quadrant x is positive and y is negative

Figure II.19: Four quadrants

Exercise II.3.2: A rope 20 feet long is hanging from a pole that is 12 feet high. A child holds the free end of the rope against the ground so that the rope is taut. What angle does the rope make with the ground? How far is the end of the rope from the base of the pole?

Exercise II.3.3: Which of the trigonometric functions is symmetric about the y-axis?

Exercise II.3.4: Which of the trigonometric functions is symmetric about the origin?

The two axes divide the plane into four quadrants as shown in Figure II.19. The signs of the trig functions depend on the quadrant in which the angle θ is located. For example, in the fourth quadrant (when $-\pi/2 < \theta < 0$) $\cos\theta$ is positive since x is positive, and $\sin\theta$ is negative since y is negative.

Figures II.20–II.25 show graphs of the six trig functions. Notice that there are some points at which some of these functions are not defined. For example, $\tan(\pi/2)$ is undefined since when $\theta = \pi/2$, $y/x = 1/0$ which doesn't make sense. Notice also that the fact that each angle has many different names causes the trig functions to keep repeating. For example, each trig function looks exactly the same on the interval

§3. The Trigonometric Functions 155

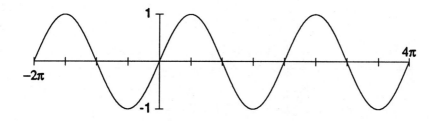

Figure II.20: $y = \sin x$

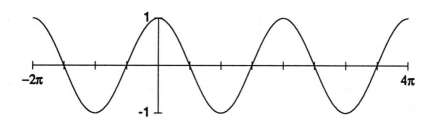

Figure II.21: $y = \cos x$

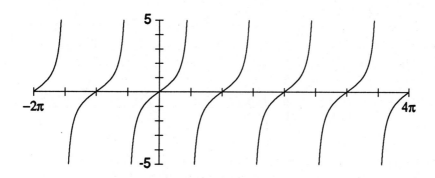

Figure II.22: $y = \tan x$

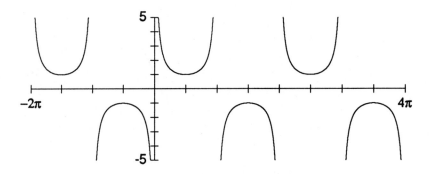

Figure II.23: $y = \csc x$

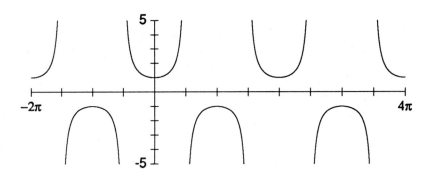

Figure II.24: $y = \sec x$

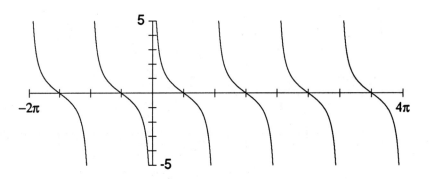

Figure II.25: $y = \cot x$

§3. The Trigonometric Functions

$[2\pi, 4\pi]$ as on the interval $[0, 2\pi]$. Symbolically we can express this by

$$\sin(x + 2\pi) = \sin x.$$

A function that repeats in this way is said to be **periodic**. The trig functions all have **period** 2π since they repeat every 2π. The functions $\tan \theta$ and $\cot \theta$ also have period π.

There are a number of identities involving the trigonometric functions. The following is a short list of some of the most useful ones.

Trigonometric Identities

$$\sin(\pi/2 - \theta) = \cos \theta$$
$$\tan(\pi/2 - \theta) = \cot \theta$$
$$\sin(-\theta) = -\sin \theta$$
$$\cos(-\theta) = \cos \theta$$
$$\tan(-\theta) = -\tan \theta$$
$$\sin^2 \theta + \cos^2 \theta = 1$$
$$\cos \theta = \sqrt{1 - \sin^2 \theta}$$
$$\tan^2 \theta + 1 = \sec^2 \theta$$
$$\sin(\alpha + \beta) = \sin \alpha \cos \beta + \sin \beta \cos \alpha$$
$$\sin 2\alpha = 2 \sin \alpha \cos \alpha$$
$$\cos(\alpha + \beta) = \cos \alpha \cos \beta - \sin \alpha \sin \beta$$
$$\cos 2\alpha = \cos^2 \alpha - \sin^2 \alpha = 2\cos^2 \alpha - 1 = 1 - 2\sin^2 \alpha$$
$$\tan(\alpha + \beta) = \frac{\tan \alpha + \tan \beta}{1 - \tan \alpha \tan \beta}$$
$$\tan 2\alpha = \frac{2 \tan \alpha}{1 - \tan^2 \alpha}$$
$$\sin^2 \theta = \frac{1 - \cos 2\theta}{2}$$
$$\cos^2 \theta = \frac{1 + \cos 2\theta}{2}$$

Exercises

*Exercise II.3.5: Find a formula for $\sin(\alpha - \beta)$.

Exercise II.3.6: Find a formula for $\cos(\alpha - \beta)$.

Derivatives of the trigonometric functions

Theorem 8 *If $f(x) = \sin x$ then $f'(x) = \cos x$.*

Proof:

$$\begin{aligned}
f'(x) &= \lim_{h \to 0} \frac{f(x+h) - f(x)}{h} \\
&= \lim_{h \to 0} \frac{\sin(x+h) - \sin x}{h} \\
&= \lim_{h \to 0} \frac{\sin x \cos h + \sin h \cos x - \sin x}{h} \\
&= \lim_{h \to 0} \left(\frac{\sin x \cos h - \sin x}{h} + \frac{\sin h \cos x}{h} \right) \\
&= \lim_{h \to 0} \left(\frac{\sin x \cos h - \sin x}{h} \right) + \lim_{h \to 0} \left(\frac{\sin h \cos x}{h} \right) \\
&= \lim_{h \to 0} \left[(\sin x) \left(\frac{\cos h - 1}{h} \right) \right] + \lim_{h \to 0} \left[(\cos x) \left(\frac{\sin h}{h} \right) \right] \\
&= (\sin x) \left[\lim_{h \to 0} \left(\frac{\cos h - 1}{h} \right) \right] + (\cos x) \left[\lim_{h \to 0} \left(\frac{\sin h}{h} \right) \right]
\end{aligned}$$

In order to complete this calculation we need to know two limits.

$$\lim_{h \to 0} \left(\frac{\sin h}{h} \right) \quad \text{and} \quad \lim_{h \to 0} \left(\frac{\cos h - 1}{h} \right)$$

Before giving formal proofs evaluating these limits we look at some experimental evidence. The following table gives values of

$$\frac{\sin h}{h} \quad \text{and} \quad \frac{\cos h - 1}{h}$$

for some small values of h.

h	$(\sin h)/h$	$(\cos h - 1)/h$
.10000000	.99833417	−.04995835
−.10000000	.99833417	.04995835
.01000000	.99998333	−.00499996
−.01000000	.99998333	.00499996
.00100000	.99999983	−.00050000
−.00100000	.99999983	.00050000

§3. The Trigonometric Functions

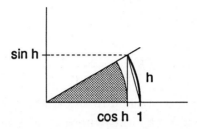

Figure II.26: Examining $(\sin h)/h$

Based on this evidence we conjecture that

$$\lim_{h \to 0} \left(\frac{\sin h}{h} \right) = 1 \quad \text{and} \quad \lim_{h \to 0} \left(\frac{\cos h - 1}{h} \right) = 0$$

and finish the computation for the derivative of the sine function as follows.

$$f'(x) = (\sin x) \left[\lim_{h \to 0} \left(\frac{\cos h - 1}{h} \right) \right] + (\cos x) \left[\lim_{h \to 0} \left(\frac{\sin h}{h} \right) \right] = \cos x$$

To complete this proof we need to prove that

$$\lim_{h \to 0} \left(\frac{\sin h}{h} \right)$$

and

$$\lim_{h \to 0} \left(\frac{\cos h - 1}{h} \right) = 0.$$

We give geometric proofs based on Figure II.26. The angle at the center is h radians. Because we are measuring this angle in radians the length of the heavy arc, the portion of the circle of radius 1 cutoff by the angle h, is h. The length of the vertical line is $\sin h$. Since this line is shorter than the heavy arc we see that

$$\sin h \leq h$$

$$\frac{\sin h}{h} \leq 1$$

The area of an entire circle of radius r is πr^2. Hence, the radius of an entire circle of radius $\cos h$ is $\pi(\cos h)^2$. The shaded portion of this circle is the fraction $h/2\pi$ of the whole circle and, hence, its area is

$$\frac{h}{2\pi}\pi(\cos h)^2 = \frac{h(\cos h)^2}{2}.$$

This portion of the circle is contained in the triangle whose vertexes are at $(0,0)$, $(1,0)$, and $(\cos h, \sin h)$ and whose area is $(\sin h)/2$. Hence,

$$\frac{h(\cos h)^2}{2} \leq \frac{\sin h}{2}$$

$$(\cos h)^2 \leq \frac{\sin h}{h}$$

Putting this all together we see that

$$(\cos h)^2 \leq \frac{\sin h}{h} \leq 1.$$

Since

$$\lim_{h \to 0} \cos h = 1$$

we see that

$$\lim_{h \to 0} \left(\frac{\sin h}{h}\right) = 1$$

proving the first of our two limits.

Notice that $1 - \cos h$ is the length in Figure II.26 of the short line segment along the x-axis between 1 and $\cos h$. It is clear from this picture that as h gets very small the ratio

$$\frac{1 - \cos h}{h}$$

gets close to zero. Thus

$$\lim_{h \to 0} \left(\frac{\cos h - 1}{h}\right) = 0,$$

the second of our two limits.

The formulas for differentiating trigonometric functions are collected below. The remaining proofs are left as exercises.

§3. The Trigonometric Functions

Differentiation Formulas for Trigonometric Functions

$$(\sin x)' = \cos x$$
$$(\cos x)' = -\sin x$$
$$(\tan x)' = \sec^2 x$$
$$(\sec x)' = \sec x \tan x$$
$$(\csc x)' = -\csc x \cot x$$
$$(\cot x)' = -\csc^2 x$$

Exercises

Exercise II.3.7: Prove $(\cos x)' = -\sin x$.

*Exercise II.3.8: Prove $(\tan x)' = \sec^2 x$.

Exercise II.3.9: Prove $(\sec x)' = \sec x \tan x$.

Exercise II.3.10: Prove $(\csc x)' = -\csc x \cot x$.

Exercise II.3.11: Prove $(\cot x)' = -\csc^2 x$.

*Exercise II.3.12: Find the derivative of the function

$$f(x) = 2x \cos x.$$

Exercise II.3.13: Find the derivative of the function $g(t) = t^3 \tan t$.

Exercise II.3.14: Find the derivative of the function

$$f(\theta) = \frac{\sin \theta + \cos \theta}{1 - \theta^4}.$$

Exercise II.3.15: Find the derivative of the function

$$f(t) = (\sin t + \cos t)(t^3 - 4).$$

*Exercise II.3.16: Find the derivative of the function $y = 2(\sin \theta)(\cos \theta)$.

*Exercise II.3.17: Find the derivative of the function $y = \sin 2\theta$. Compare your answer to this exercise with the previous exercise. Explain.

Exercise II.3.18: Find the derivative of the function
$$y = \sin^2 \theta + \cos^2 \theta.$$
There is a short way to do this problem and an obvious longer way. Do the problem both ways and compare your answers.

Exercise II.3.19: Find the derivative of the function $f(t) = (\sin t)^2$.

Exercise II.3.20: Find the derivative of the function $g(t) = -(\cos t)^2$. Compare your answer to this problem with the preceding problem. Explain.

Exercise II.3.21: Find the derivative of the function
$$f(t) = \frac{\sin t}{t}.$$

---

§4. Optimization, I

In section I.7 we studied a product that was produced by many producers and purchased by many consumers with no collusion among either producers or consumers. Now we want to look at a product made by one manufacturer who can choose whatever price she wants without fear of competition. The demand for this product is determined by its price.

As an example, we work with the linear demand function
$$D(p) = 1000 - 500p$$
shown in Figure II.27. Notice that the maximum possible demand is 1000 when the the product is free and that when the price is 2 that the demand is 0. Thus we are interested in a range of prices $0 \leq p \leq 2$.

Now suppose that the manufacturing cost for each unit of the product is 0.20. Thus the unit profit at price p is $(p - 0.20)$ and the total profit $T(p)$ is the product of the number of units sold at the price p (that is, the demand $D(p)$) and the unit profit
$$T(p) = D(p)(p - 0.20) = (1000 - 500p)(p - 0.20).$$
Figure II.28 shows a graph of this function.

§4. Optimization, I

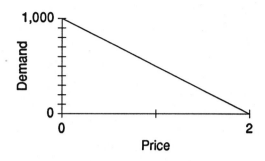

Figure II.27: $D(p) = 1000 - 500p$

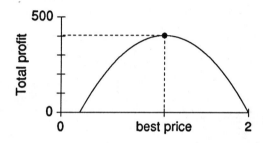

Figure II.28: The best price

When the price is low, near the manufacturing cost, 0.20, then the total profit is low because the unit profit is low. When the price is high, near 2.00, then the total profit is low because the demand is low. Someplace in between the total profit is at its highest as shown in Figure II.28. This problem is an example of an important group of problems called **Optimization Problems**, problems in which one wants to find the best of something.

We can find the best price with the help of the derivative. Look at Figure II.29. Notice that at a point where a function is increasing the tangent is sloping upward and, hence, the derivative is positive. At a point where the function is decreasing the tangent is sloping downward and, hence, the derivative is negative. At a place where the curve is at a maximum (the top of a hill) or at a minimum (the bottom of a valley) the tangent will be horizontal and, hence, the derivative will be zero. We state these observations as a theorem without giving a proof.

Theorem 9 *Suppose that $f(x)$ is a smooth function. Then*

1. *If $f'(a) > 0$ the curve $y = f(x)$ is increasing near the point a.*

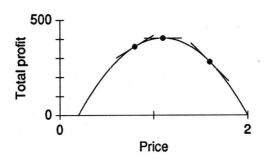

Figure II.29: Some tangents

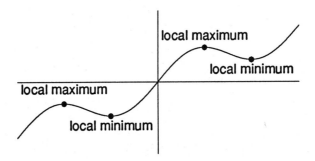

Figure II.30: Several local maximums and minimums

2. If $f'(a) < 0$ the curve $y = f(x)$ is decreasing near the point a.

3. If the curve $y = f(x)$ has a local minimum or local maximum at the point a then $f'(a) = 0$.

The last clause requires some explanation. A **local maximum** is the top of a hill on the graph. A **local minimum** is the bottom of a valley on the graph. There may be several local maximums and several local minimums as shown in Figure II.30. Usually we are interested in the **global maximum** (i.e., the highest point on the curve) or the **global minimum** (i.e., the lowest point on the curve). The global maximum can be either the highest local maximum or, if we are interested in a function on an interval of the form $[a, b]$, it might be at one of the endpoints, a or b. Similarly, the global minimum can be either the lowest local minimum or it might be at one of the endpoints.

We can use these observations to find the maximum of the function

§4. Optimization, I

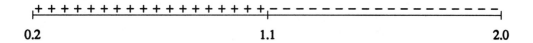

Figure II.31: The sign of the derivative

$T(p) = (1000 - 500p)(p - 0.20)$ on the interval $[0, 2]$. The first step is to differentiate the function $T(p)$.

$$\begin{aligned} T'(p) &= (1000 - 500p)(1) + (p - 0.20)(-500) \\ &= 1000 - 500p - 500p + 100 \\ &= 1100 - 1000p \end{aligned}$$

Since we are looking for the top of a hill we begin by asking where $T'(p) = 0$. That is, we solve the equation

$$\begin{aligned} T'(p) &= 0 \\ 1100 - 1000p &= 0 \\ 1000p &= 1100 \\ p &= 1100/1000 \\ p &= 1.10 \end{aligned}$$

Thus $T'(p)$ is zero at $p = 1.10$. Notice, in addition, that $T'(p)$ is positive if $p < 1.10$ and $T'(p)$ is negative if $p > 1.10$. See Figure II.31. Thus we can see that $T(p)$ is increasing from $p = 0$ to $p = 1.10$ since $T'(p)$ is positive on the interval $(0, 1.10)$. At $p = 1.10$ the slope of the derivative is zero. Then from $p = 1.10$ to $p = 2.00$, $T(p)$ is decreasing. We summarize this information about $T(p)$ in Figure II.32 in which upward pointing arrows indicate the function is increasing and downward pointing arrows indicate the function is decreasing. The function $T(p)$ is increasing from $p = 0$ to $p = 1.10$ where it reaches its maximum value $T(1.10) = 405$. Then from $p = 1.10$ to $p = 2.00$ the function $T(p)$ is decreasing.

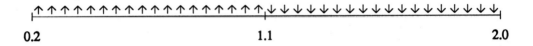

Figure II.32: The behavior of the function

Exercises

*Exercise II.4.1: Suppose that the unit manufacturing cost for the product we have been discussing rises to 0.30. What price should the manufacturer charge to maximize her profit now? How much of the rise in the manufacturing cost was passed on to the consumer?

*Exercise II.4.2: Suppose the demand function for a particular product is $D(p) = 2000 - 500p$ and that the unit manufacturing cost is 0.30. What price should the manufacturer charge to maximize profit? Suppose that the unit manufacturing cost rises to 0.50. What price should the manufacturer charge to maximize her profit now? How much of the rise in the manufacturing cost was passed on to the consumer?

Exercise II.4.3: Suppose the demand function for a particular product is $D(p) = 1500 - 100p$ and that the unit manufacturing cost is 0.30. What price should the manufacturer charge to maximize profit? Suppose that the unit manufacturing cost rises to 0.50. What price should the manufacturer charge to maximize her profit now? How much of the rise in the manufacturing cost was passed on to the consumer?

You may have noticed an interesting phenomenon in the three exercises above. In each case exactly half of the rise in the unit manufacturing cost was passed on to the consumer. When I worked out these examples I was very surprised. It seemed unlikely that I got exactly the same result for all three exercises by pure chance. So I asked whether it could possibly be true that a manufacturer always passes half the cost of a rise in unit manufacturing costs on to her customers. For my first approach to this question I tried a general *linear* demand function. That is, I used the demand function

$$D(p) = a - bp$$

§4. Optimization, I

where a and b are arbitrary positive constants. I let the constant c represent the unit manufacturing cost for the product. Thus the profit per unit is $(p - c)$ and the total profit is

$$T(p) = (p - c)D(p) = (p - c)(a - bp)$$

where a, b and c are all positive constants.

To find the best price we calculate

$$\begin{aligned} T'(p) &= (p - c)(-b) + (a - bp)(1) \\ &= -bp + bc + a - bp \\ &= -2bp + bc + a \end{aligned}$$

To find the maximum we set $T'(p) = 0$ and solve

$$\begin{aligned} -2bp + bc + a &= 0 \\ 2bp &= bc + a \\ p &= \frac{bc + a}{2b} \\ &= \frac{a}{2b} + \frac{c}{2} \end{aligned}$$

Since the graph of $T'(p)$ is a straight line with negative slope, $T'(p)$ is positive when p is less than

$$\frac{a}{2b} + \frac{c}{2}$$

and $T'(p)$ is negative when p is greater than

$$\frac{a}{2b} + \frac{c}{2}.$$

Thus $T(p)$ is increasing until p reaches

$$\frac{a}{2b} + \frac{c}{2}$$

at which price the total profit is maximized. Above this price $T(p)$ decreases.

This answers our question with a resounding and, to me, very surprising "yes!" *for any linear demand function.* Notice that in the formula for the best price
$$\frac{a}{2b} + \frac{c}{2}$$
exactly half of the unit manufacturing cost c is passed on to the consumer. The obvious next question is whether this is true for every demand function. The following exercise asks you to investigate this question.

Exercise

*Exercise II.4.4:
Suppose that the demand for a particular commodity is given by the function
$$D(p) = \frac{1000}{1 + p^2}$$
and that the unit manufacturing cost is 0.50. What is the best price for the producer to charge?

If the unit manufacturing cost rises to 1.00, what is the best price for the producer to charge? How much of the increase is passed on to the consumer?

If the unit manufacturing cost rises to 1.50, what is the best price for the producer to charge? How much of this additional increase is passed on to the consumer?

In view of this exercise what can you say about our earlier question?

An Optimization Recipe:

In this section we develop a recipe for optimizing a function f that is continuous on an interval $[a, b]$. We begin with a theorem that tells us this kind of optimization problem always has a solution.

Theorem 10 *(The Extreme Value Theorem)*

If $f(x)$ is a continuous function on an interval $[a, b]$ then there are points $u, v \in [a, b]$ such that for every $x \in [a, b]$,

$$f(u) \leq f(x) \leq f(v).$$

§4. Optimization, I

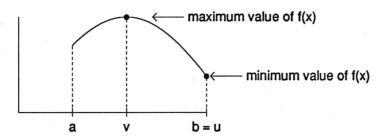

Figure II.33: The global maximum and minimum of $f(x)$ on $[a, b]$

Notice that $f(u)$ is the minimum value of f on the interval and $f(v)$ is the maximum value of f on the interval. See Figure II.33.

There may be several points at which $f(x)$ attains its maximum value and several points at which $f(x)$ attains its minimum value. For example, consider the function $f(x) = \cos x$ on the interval $[0, 3\pi]$—its maximum value, 1, is attained at $x = 0$ at $x = 2\pi$ and its minimum value, -1, is attained at $x = \pi$ and $x = 3\pi$.

The rest of this section is devoted to a recipe for finding the global minimum and the global maximum of a continuous function $f(x)$ on a closed interval $[a, b]$. This recipe is based on the observation that at a global maximum there are three possibilities.

1. It is the top of a smooth hill.

2. It is the top of a pointy hill (i.e., a sharp corner).

3. It is at an endpoint.

These three possibilities are shown in Figure II.34
Similarly, at a global minimum there are three possibilities.

1. It is the bottom of a smooth valley.

2. It is the bottom of a pointy valley (i.e., a sharp corner).

3. It is at an endpoint.

At the top of a smooth hill or at the bottom of a smooth valley the derivative is zero. At the top of a pointy hill or at the bottom of a pointy valley there is no derivative.

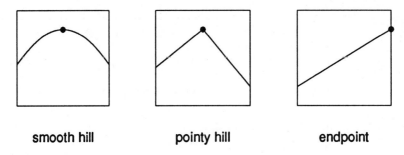

Figure II.34: Three possibilities at a maximum

Definition:

> A point c is said to be a **critical point** for the function f if either $f'(c) = 0$ or if f has no derivative at the point c.

In view of our observations above, there is a simple recipe for finding the global minimum and global maximum of a function $f(x)$ on an interval $[a, b]$.

An Optimization Recipe

1. Find all the critical points of f on the interval $[a, b]$.

2. Make a table showing the value of the function f at each critical point and at each of the two endpoints.

3. Read off the global minimum and the global maximum from this table.

   We recommend two additional steps.

4. Make a chart showing where the derivative $f'(x)$ is positive and, hence, the original function $f(x)$ is increasing; and where the derivative $f'(x)$ is negative and, hence, the original function $f(x)$ is decreasing.

5. Use the information obtained above to make a rough sketch of the graph $y = f(x)$.

§4. Optimization, I

Example:

Find the global minimum and global maximum of the function
$$f(x) = x^3 - 6x^2 + 9x + 1$$
on the interval $[0, 5]$.

Answer:

1. Find the critical points by looking at the derivative
$$f'(x) = 3x^2 - 12x + 9.$$

   Notice first that there are no points at which the derivative is not defined. Hence, the only critical points will be at points at which the derivative is zero.
$$\begin{aligned} 3x^2 - 12x + 9 &= 0 \\ x^2 - 4x + 3 &= 0 \\ (x-1)(x-3) &= 0 \end{aligned}$$

   The critical points are at $x = 1$ and $x = 3$.

2. Make a table showing the value of the function f at each of the endpoints and each of the critical points.

Type of point	$x =$	$f(x) =$
endpoint	0	1
critical point	1	5
critical point	3	1
endpoint	5	21

3. Read off the global minimum—$f(x) = 1$, attained at $x = 0$ and at $x = 3$—and the global maximum—$f(x) = 21$, attained at $x = 5$.

4. Make a chart showing the sign of the derivative and, hence, where the function is increasing and decreasing.
$$f'(x) = 3x^2 - 12x + 9 = 3(x^2 - 4x + 3) = 3(x-1)(x-3).$$

Figure II.35: The sign of the derivative $f'(x) = 3x^2 - 12x + 9$

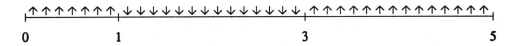

Figure II.36: The behavior of $f(x) = x^3 - 6x^2 + 9x + 1$

To the left of 1 (i.e., when $x < 1$) the last two factors are negative. Since the product of two negative numbers is positive, the derivative is positive. Between 1 and 3 only the second factor is negative; so the derivative is negative. To the right of 3 all factors are positive and, hence, the derivative is positive. See Figures II.35 and II.36

5. Use the information above to make a rough sketch of the graph $y = f(x)$. There are two steps involved—first, plot the values of the function computed in Step 2 and, then, draw lines connecting those values, respecting the results of Step 4. See Figure II.37. Notice that some of the information is redundant. For example,

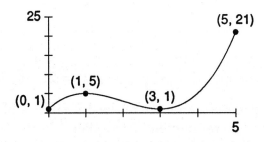

Figure II.37: A sketch of $y = x^3 - 6x^2 + 9x + 1$

§4. Optimization, I

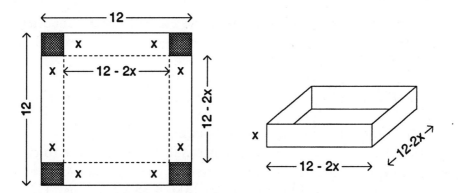

Figure II.38: Making an open box

one must connect the point at $(0,1)$ to the point at $(1,5)$ with a curve that is increasing. This is redundant. The only way to go from $(0,1)$ to $(1,5)$ is by increasing. This redundancy is extremely valuable since it helps check the work. For example, if we had miscomputed $f(0)$ as 10, we would have caught that mistake when we tried to draw an *increasing* line from $(0,10)$ to $(1,5)$.

Example:

Suppose that we have a square piece of thin metal that is twelve inches by twelve inches and that we would like to make a box with a square base by cutting off the four corners and bending up the sides along the dotted lines as shown in Figure II.38. This will produce a box with metal sides and a metal bottom but no top. We would like to build the largest possible box where by "largest" we mean largest in volume.

Before we can use our recipe we must first translate the problem into mathematics. The first step is to find two variables—one over which we have control and another that we want to maximize. Then we want to find a function that describes the variable that we want to maximize in terms of the variable over which we have control. The variable that we want to maximize is the volume of the box. The variable over which we have control is the length (and width) of the piece we will cut off each corner. We will call this variable x.

Chapter II. Differentiation

Looking at Figure II.38 we see that the box will have a square base whose sides are $(12 - 2x) \times (12 - 2x)$. The height of the box will be x. Thus its volume will be

$$V(x) = x(12-2x)(12-2x) = x(144 - 48x + 4x^2) = 144x - 48x^2 + 4x^3.$$

This is the function that we want to maximize.

Notice that x cannot be less than zero. It makes no sense to cut off a piece of negative length. Also x cannot be greater than 6. It makes no sense to cut a piece of length greater than 6 off both ends. Thus we consider only values of x in the interval $[0,6]$. We now have a purely mathematical problem to solve:

Find the maximum of the function $V(x) = 144x - 48x^2 + 4x^3$ on the interval $[0, 6]$.

1. Find the critical points by looking at the derivative

$$V'(x) = 144 - 96x + 12x^2.$$

Notice first that there are no points where the derivative is not defined. Hence, the only critical points will be at points where the derivative is zero. So we solve the equation

$$12x^2 - 96x + 144 = 0$$
$$x^2 - 8x + 12 = 0$$
$$(x-2)(x-6) = 0$$

to see that the only critical points are at $x = 2$ and $x = 6$. Notice that one of these critical points ($x = 6$) is also an endpoint.

2. Make a table showing the value of the function V at each of the endpoints and each of the critical points.

Type of point	$x =$	$V(x) =$
endpoint	0	0
critical point	2	128
critical point and endpoint	6	0

§4. OPTIMIZATION, I   175

Figure II.39: The sign of the derivative $V' = 144 - 96x + 12x^2$

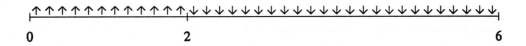

Figure II.40: The behavior of the function $V = 144x - 48x^2 + 4x^3$

3. Read off the global minimum, $V(x) = 0$, at $x = 0$ and at $x = 6$, and the global maximum, $V(x) = 128$, at $x = 2$. Thus, we should cut a piece 2 inches square off each of the four corners. The resulting box will have a volume of 128 cubic inches.

4. Make a chart showing the sign of the derivative and, hence, where the function is increasing and decreasing. See Figures II.39 and II.40.

5. Use the information above to make a rough sketch of the graph $y = V(x)$. There are two steps involved—first, plot the values of the function computed in Step 2 and then draw lines connecting those values respecting the results of Step 4. See Figure II.41.

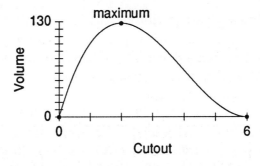

Figure II.41: A sketch of the function $V(x) = 144x - 48x^2 + 4x^3$

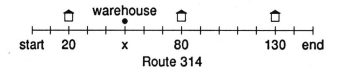

Figure II.42: Three stores and a warehouse on Route 314

Example:

This is an example of a very important class of problems called **location problems**. Suppose that a particular company is deciding where to locate a warehouse. The company plans to locate the warehouse someplace along Route 314 because they already have three retail stores located on the same road. A map of Route 314 is shown in Figure II.42. The road is 150 miles long and the three existing retail stores are located at mile markers 20, 80 and 130. The company can choose to locate the warehouse anyplace along Route 314. Let x denote the mile marker of a possible location. Thus we are interested in x in the interval $[0, 150]$.

The company wants to minimize the total distance from the warehouse to its three retail stores. This distance is

$$D(x) = |x - 20| + |x - 80| + |x - 130|.$$

We want to minimize this function on the interval $[0, 150]$.

We begin by looking for critical points. This time there are several places where the derivative doesn't exist. (Whenever you see absolute values you should worry about this possibility.) We must analyze the function $D(x)$ very carefully. Notice

$$D(x) = \begin{cases} (20 - x) + (80 - x) + (130 - x), & \text{if } 0 \leq x \leq 20; \\ (x - 20) + (80 - x) + (130 - x), & \text{if } 20 \leq x \leq 80; \\ (x - 20) + (x - 80) + (130 - x), & \text{if } 80 \leq x \leq 130; \\ (x - 20) + (x - 80) + (x - 130), & \text{if } 130 \leq x \leq 150. \end{cases}$$

§4. Optimization, I

or
$$D(x) = \begin{cases} (230 - 3x), & \text{if } 0 \leq x \leq 20; \\ (190 - x), & \text{if } 20 \leq x \leq 80; \\ (30 + x), & \text{if } 80 \leq x \leq 130; \\ (3x - 230), & \text{if } 130 \leq x \leq 150. \end{cases}$$

The derivative $D'(x)$ is

$$D'(x) = \begin{cases} -3, & \text{if } 0 < x < 20; \\ -1, & \text{if } 20 < x < 80; \\ 1, & \text{if } 80 < x < 130; \\ 3, & \text{if } 130 < x < 150. \end{cases}$$

Notice the three critical points at $x = 20$, $x = 80$, and $x = 130$ where this function has no derivative. Now our recipe proceeds as follows.

1. There are three critical points $x = 20$, $x = 80$, and $x = 130$.

2. Make a table showing the value of the function D at each of the endpoints and each of the critical points.

Type of point	$x =$	$D(x) =$
endpoint	0	230
critical point	20	170
critical point	80	110
critical point	130	160
endpoint	150	220

3. Read off the global minimum, $D(x) = 110$, at $x = 80$. The company should locate its warehouse at the same location as the middle retail store.

4. Make a chart showing the sign of the derivative and, hence, where the function is increasing and decreasing. See Figures II.43 and II.44.

5. Use the information above to make a rough sketch of the graph. See Figure II.45.

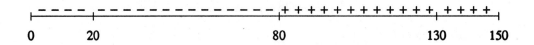

Figure II.43: The sign of the derivative $D'(x)$

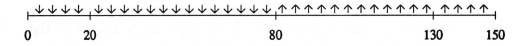

Figure II.44: The behavior of the function $D(x)$

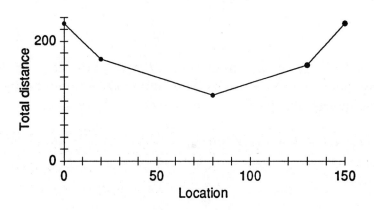

Figure II.45: The function $D(x)$

§4. Optimization, I

Further Remarks

- The most difficult part of an optimization problem is often extracting the mathematical information needed to solve the problem. As you read the problem look for two variables—one which you would like to maximize or minimize and another over which you have control. Then look for the information that describes how the first or **dependent** variable depends on the second or **independent** variable. This information yields the function which must be maximized or minimized. You should also look for the range of possible values for the independent variable.

- The fourth step in the optimization recipe for a function $f(x)$ on an interval $[a, b]$ required the determination of the sign of the derivative $f'(x)$. One way to do this makes use of the Intermediate Value Thoerem. The only way that a function (in this case $f'(x)$) can change sign is either by passing through zero or by passing a point at which it is discontinuous or undefined. Thus the only way the derivative can change sign is by crossing a critical point. This means that the critical points break the interval $[a, b]$ up into pieces on each of which $f'(x)$ is either always positive or always negative. For example, the metal box example required us to find the maximum of the function

$$V(x) = 144x - 48x^2 + 4x^3$$

on the interval $[0, 6]$. The derivative

$$V'(x) = 144 - 96x + 12x^2$$

is zero at $x = 2$ and $x = 6$. This divides the interval $[0, 6]$ into two pieces, $[0, 2]$ and $[2, 6]$ as shown in Figure II.46

On each of these two pieces $V'(x)$ is either always positive or always negative. To determine which, we need only test the sign of $V'(x)$ at one point in each interval. For example, if we compute $V'(x)$ at the point $x = 1$ in the piece $[0, 2]$ we get

$$V'(1) = 144 - 96(1) + 12(1)^2 = 60,$$

which is positive. So $V'(x)$ is positve on the entire piece $[0, 2]$. Similarly, computing $V'(x)$ at the point $x = 3$ in the piece $[2, 6]$

Figure II.46: Breaking the interval $[0,6]$ into two pieces

Figure II.47: The sign of the derivative $V' = 144 - 96x + 12x^2$

we obtain

$$V'(3) = 144 - 96(3) + 12(3)^2 = -36,$$

which is negative. So $V'(x)$ is negative on the entire interval $[2,6]$. These calculations give us Figure II.47.

Exercises

This set of exercises is divided into three parts. The first part consists of simple differentiation problems like the ones in the previous section. These routine problems are included again here because the best way to memorize the differentiation formulas is by using them. The formulas should become second nature to you. The second part consists of mathematical optimization problems. These problems are intended to help you practice the recipe that was developed in this section. The final part is the most interesting, optimization problems that require two steps. The first step is to express the problem mathematically. The second step is to solve the resulting mathematical optimization problem.

Part I: Differentiate each of the following.

*Exercise II.4.5: $y = x^2 - 32x^2 + 27x$

§4. Optimization, I

Exercise II.4.6: $w = 3s^5 - 2s^4 + 42s^2 + 12s + 1$

Exercise II.4.7: $v = 2t^{-2} - 3t^{-1} + 4t^4 - 12t^6$

*Exercise II.4.8: $w = (s^4 - s^2 - 2s)(s^2 + s - 6)$

Exercise II.4.9: $y = (x^3 - 2x^2 + 17)(x^4 + 3x^3 + 12)$

*Exercise II.4.10:
$$y = \frac{t^2 - 3t + 12}{1 - t^3}$$

Exercise II.4.11:
$$w = \frac{x^3 - 2x^2}{3x^2 + x + 2}$$

Exercise II.4.12:
$$y = \frac{\left(\frac{x^2-1}{x^2+1}\right)(x^4 - 3x^2 + 4)}{x^4 + x^3 + x + 1}$$

Exercise II.4.13: $z = 4t^6 + 6t + \cos t + \sec t$.

Exercise II.4.14: $T(x) = x^2 \sin x$.

Exercise II.4.15:
$$f(x) = \frac{x^2 + \cos x}{x^3 + \tan x}.$$

Part II: Find the global maximum and global minimum of each of the following functions on the indicated interval.

*Exercise II.4.16: $y = x^3 - 3x^2 + 3x + 2$, $\quad [-2, 2]$

*Exercise II.4.17: $y = x^3 - 6x + 4$, $\quad [-3, 3]$

Exercise II.4.18: $y = x^3 + 7x + 4$, $\quad [-10, 10]$

Exercise II.4.19: $y = x^3 + 7x + 4$, $\quad [0, 10]$

*Exercise II.4.20:
$$y = \frac{x}{x^2 + 1}, \quad [-5, 5]$$

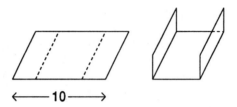

Figure II.48: Making a gutter

Part III:

*Exercise II.4.21: A homeowner has a long piece of thin metal from which she wants to make a gutter. The piece of metal is ten inches wide and she plans to make a gutter whose cross section is rectangular by folding up the sides as shown in Figure II.48. The gutter will have an open top. She would like to make a gutter with the largest possible cross sectional area. What should she do?

*Exercise II.4.22: What is the largest rectangle that can be fit into the region enclosed by the curve $y = 4 - x^2$ and the x-axis.

*Exercise II.4.23: A farmer has 100 feet of fencing with which she wants to enclose a rectangular field. What dimensions should she make the field in order to have the largest possible area?

*Exercise II.4.24: A farmer has 100 feet of fencing with which she wants to enclose a rectangular field. Her land has a straight river along one edge. She plans to put the field next to the river. The river will be one side of the field so that she will need to put fencing only along its other three sides. What dimensions should she make the field in order to have the largest possible area?

*Exercise II.4.25: A farmer has 100 feet of fencing with which he wants to enclose a rectangular field. The field will have a fence running down the middle to divide it into two equal parts. What dimensions should the field have to maximize its total area?

*Exercise II.4.26: Find a general rule that tells a company where to locate its warehouse along a road if it has three retail stores already located on the road at a_1, a_2, and a_3.

§5. Curve-Sketching and Optimization

*Exercise II.4.27: Find a general rule that tells a company where to locate its warehouse along a road if it has four retail stores already located on the road at $a_1, a_2, a_3,$ and a_4.

Exercise II.4.28: Find a general rule that tells a company where to locate its warehouse along a road if it has five retail stores already located on the road at a_1, a_2, a_3, a_4 and a_5.

*Exercise II.4.29: Find a general rule that tells a company where to locate its warehouse along a road if it has n retail stores already located on the road at $a_1, a_2, \ldots a_n$.

---

§5. Curve-Sketching and Optimization

Computer programs like *Mathematica*, Maple, and DERIVE have two features, graphics and routines for finding solutions of equations, that are particularly useful for solving optimization problems. The main purpose of this section is to illustrate the ways in which these features can be used. First, however, we wish to look at a cautionary example. Many people believe that with the widespread availability of computer graphics and graphing calculators, it is no longer necessary to learn techniques for curve-sketching like those developed in the previous section. This belief is incorrect for two reasons—first, unless they are used intelligently, computer graphics can produce misleading results and, second, in order to understand graphs it is important to understand the concepts involved in curve-sketching. The following example illustrates the first point.

Example:

Consider the function $f(x) = \sin x + \sin(0.80x)$. The following *Mathematica* statement produces the graph shown in Figure II.49.

```
f[x_] := Sin[x] + Sin[0.80 x]
Plot[f[x], {x, -Pi, Pi}]
```

This graph is perfectly correct. It is, however, very misleading because we only asked *Mathematica* to produce a graph on the interval $[-\pi, \pi]$. If we look at the larger interval $[-15\pi, 15\pi]$ using the *Mathematica* statement

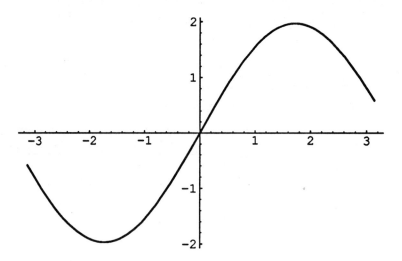

Figure II.49: First *Mathematica* graph of $f(x) = \sin x + \sin(0.80x)$

`Plot[f[x], {x, -15 Pi, 15 Pi}]`

we see Figure II.50 which shows some interesting behavior that is not apparent from Figure II.49. As you can see, unthinking use of computer graphics can produce correct but misleading results. Now we look at some ways in which computers and calculators can help solve optimization problems.

Example:

Suppose that a man has a twenty foot ladder and leans it against a brick wall ten feet high and one foot thick as shown in Figure II.51. He would like to place the ladder in such a way that the top of the ladder extends as far as possible on the other side of the wall. That is, he wants to choose an angle θ so that the length y shown in Figure II.51 is as large as possible.

First, notice that $y + x + 1 = 20\cos\theta$ so that

$$y = 20\cos\theta - x - 1,$$

and, since $x = 10\cot\theta$,

$$y = 20\cos\theta - 10\cot\theta - 1$$

§5. Curve-Sketching and Optimization

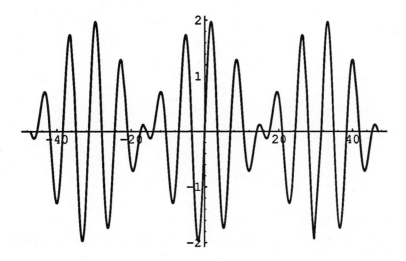

Figure II.50: Second *Mathematica* graph of $f(x) = \sin x + \sin(0.80x)$

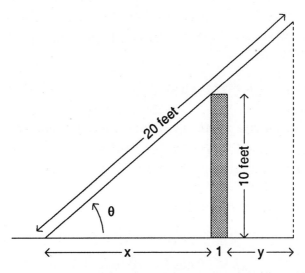

Figure II.51: Ladder and wall

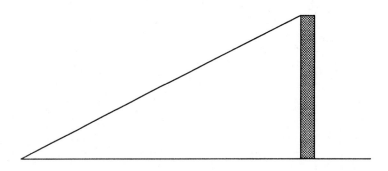

Figure II.52: An endpoint for θ

This is the function we want to minimize.

Next we must consider the range of possible values for θ. The largest reasonable value for θ is $\pi/2$, which corresponds to placing the ladder so that it is pointing straight up. The smallest reasonable value for θ can be seen from Figure II.52, in which the tip of the ladder is just resting on top of the wall and the ladder doesn't extend at all into the other side. Since the ladder is 20 feet long and the wall is 10 feet high, the sine of this angle is $10/20 = 1/2$. We can find this angle by asking a computer algebra system like *Mathematica* to solve[1] the equation

$$\sin\theta = 1/2$$

or by using a numerical method like the Bisection Method to estimate a solution.

```
Solve[Sin[theta] == 1/2,theta]
Solve::ifun:
    Warning: Inverse functions are being used by Solve, so
      some solutions may not be found.
            Pi
{{theta -> --}}
            6
```

Thus, we want to optimize the function

$$y = 20\cos\theta - 10\cot\theta - 1$$

---

[1] When we study the inverse trig functions used by *Mathematica* to solve this equation we will see why *Mathematica* issued a warning.

§5. Curve-Sketching and Optimization

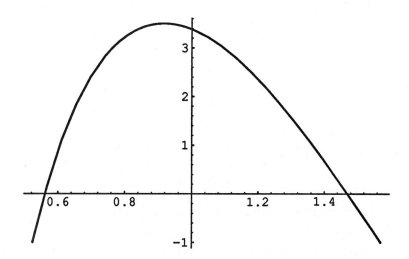

Figure II.53: $20\cos\theta - 10\cot\theta - 1$

on the interval $[\pi/6, \pi/2]$. The following *Mathematica* statement produces Figure II.53 from which we can see that the maximum value of y is roughly 3.5 and is attained when θ is roughly 0.9.

```
Plot[20 Cos[theta] - 10 Cot[theta] - 1, {theta, Pi/6, Pi/2}]
```

To find the exact maximum, the next step is to find the derivative.

$$\frac{dy}{d\theta} = -20\sin\theta + 10\csc^2\theta.$$

To find the critical points, we solve

$$-20\sin\theta + 10\csc^2\theta = 0.$$

A numerical procedure like the Bisection Method or the *Mathematica* procedure `FindRoot` can be used to estimate a solution to this equation.

```
FindRoot[-20 Sin[theta] + 10 Csc[theta]^2 == 0, {theta, 0.9}]
{theta -> 0.916868}
```

Thus, the maximum possible value of y occurs at $\theta \approx 0.916868$ and at this point the ladder extends

$$20\cos(0.916868) - 10\cot(0.916868) - 1 = 3.50196 \text{ feet}$$

on the other side of the wall.

Example:

Sketch a graph of the function
$$f(\theta) = \frac{\theta}{2} + \sin\theta$$
on the interval $[-2\pi, 2\pi]$.

Answer:

For this problem a computer or graphing calculator can do essentially all the work. In fact, I used such a program to produce Figure II.55. Nonetheless it is worthwhile to work out this example using the methods that we developed in the last section.

The first step is to analyze the derivative $f'(\theta)$.

$$f(\theta) = \frac{\theta}{2} + \sin\theta$$

$$f'(\theta) = \frac{1}{2} + \cos\theta.$$

The derivative is always defined, so the only critical points are when the derivative is zero. We find these as follows.

$$\frac{1}{2} + \cos\theta = 0$$

$$\cos\theta = -1/2$$

Notice that $\cos\theta = -1/2$ when $\theta = 2\pi/3$, when $\theta = 4\pi/3$, when $\theta = -2\pi/3$, and when $\theta = -4\pi/3$ giving us four critical points in the interval $[-2\pi, 2\pi]$. The table of critical points and endpoints is

Type of point	θ	$f(\theta)$
Endpoint	-2π	$-\pi$
Critical Point	$-4\pi/3$	-1.2284
Critical Point	$-2\pi/3$	-2.0944
Critical Point	$2\pi/3$	2.0944
Critical Point	$4\pi/3$	1.2284
Endpoint	2π	π

§5. Curve-Sketching and Optimization

Figure II.54: Where $\theta/2 + \sin\theta$ is increasing and decreasing

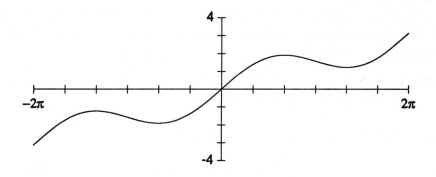

Figure II.55: $y = \theta/2 + \sin\theta$

We determine where f is increasing and decreasing in the usual way (Figure II.54) and sketch the graph as shown in Figure II.55.

We might have used a computer algbera system like *Mathematica* to find the critical points by solving the equation

$$\cos\theta = -1/2$$

as follows

```
Solve[Cos[theta] == -1/2, theta]
Solve::ifun:
   Warning: Inverse functions are being used by Solve, so
     some solutions may not be found.
            2 Pi
{{theta -> ----}}
             3
```

Notice that *Mathematica* found only one critical point, although it did issue a warning message.

Exercises

*Exercise II.5.1:
 Sketch a graph of the function $y = \tan x - x$ on the interval $[0, 2\pi]$. Compare your graph with one produced by either a graphing calculator or a computer.

*Exercise II.5.2:
 Sketch a graph of the function $y = \sin x + \cos x$ on the interval $[-\pi, \pi]$. Compare your graph with one produced by either a graphing calculator or a computer.

Exercise II.5.3:
 Sketch a graph of the function

$$f(t) = t \sin t$$

on the interval $[-4\pi, 4\pi]$. Compare your graph with one produced by either a graphing calculator or a computer.

*Exercise II.5.4:
 Sketch a graph of the function

$$y = \sin t + \cos 2t$$

on the interval $[-2\pi, 2\pi]$. Compare your graph with one produced by either a graphing calculator or a computer.

*Exercise II.5.5:
 Sketch a graph of the function $y = \tan x - x$ on the interval $[0, 2\pi]$. Compare your graph with one produced by either a graphing calculator or a computer.

*Exercise II.5.6:
 Sketch a graph of the function $y = \sin x + \cos x$ on the interval $[-\pi, \pi]$. Compare your graph with one produced by either a graphing calculator or a computer.

Exercise II.5.7:
 Sketch a graph of the function

$$f(t) = t \sin t$$

§6. Optimization, II

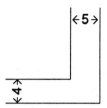

Figure II.56: Hallways in a house

on the interval $[-4\pi, 4\pi]$. Compare your graph with one produced by either a graphing calculator or a computer.

*Exercise II.5.8:
Sketch a graph of the function

$$y = \sin t + \cos 2t$$

on the interval $[-2\pi, 2\pi]$. Compare your graph with one produced by either a graphing calculator or a computer.

*Exercise II.5.9:
Figure II.56 is a diagram showing the hallways in a particular house. The homeowner wants to slide a piece of furniture around the corner. As she slides it, the piece of furniture will be flat against the floor. Assume that the furniture has no thickness. How long a piece of furniture can she slide around the corner?

---

§6. Optimization, II

For many problems involving optimizing a function $f(x)$ there are some restrictions on the possible values of the variable x. For example, an employer is required by law to pay at least minimum wage, so if she were trying to maximize a function $f(w)$ where w represented the hourly wage she paid her workers she would have to maximize $f(w)$ subject to the restriction

$$\text{minimum wage} \leq w.$$

We begin this section by introducing some terminology and notation for working with this kind of situation.

Definition:

> When we are interested in a function $f(x)$ for a particular set D of possible values of x we call the set D the **domain** of $f(x)$.

We often use notation like the following to describe a particular set.

$$\{x \text{ s.t. } a \leq x \leq b\}$$

This notation is shorthand for the set of all real numbers x such that $a \leq x \leq b$. The same set is also denoted $[a, b]$. If we wish to exclude either or both of the endpoints from the set we use parentheses instead of brackets. That is,

$$\begin{aligned} [a, b) &= \{x \text{ s.t. } a \leq x < b\} \\ (a, b] &= \{x \text{ s.t. } a < x \leq b\} \\ (a, b) &= \{x \text{ s.t. } a < x < b\}. \end{aligned}$$

We are sometimes interested in sets with a restriction on only one end. For example,

$$\{x \text{ s.t. } a \leq x\},$$

the set of all real numbers x such that $a \leq x$. We use the following notation for such sets.

$$(a, +\infty) = \{x \text{ s.t. } a < x\}$$

$$[a, +\infty) = \{x \text{ s.t. } a \leq x\}$$

$$(-\infty, b] = \{x \text{ s.t. } x \leq b\}$$

$$(-\infty, b) = \{x \text{ s.t. } x < b\}$$

Finally, we use the notation $(-\infty, +\infty)$ for the set of all real numbers.

These sets are called **intervals**. Intervals of the form $[a, b]$, $[a, b)$, (a, b) and $(a, b]$ are called **bounded intervals**. The other intervals are called **unbounded intervals**.

In the last section we developed techniques for finding the maximum or minimum of a function $f(x)$ when the domain was of the form $[a, b]$.

§6. Optimization, II

In this section we are interested in situations where the domain is of the form $(-\infty, +\infty)$, $[a, +\infty)$, or $(-\infty, b]$. On these domains we do not know if there is a global maximum (or a global minimum). If a global maximum (or a global minimum) does exist, however, it must be at a critical point or an endpoint. This is a general principle: maximums and minimums always occur at critical points or at endpoints.

Example:

Find the global maximum and global minimum *if they exist* of the function
$$f(x) = \frac{1}{1+x^2}$$
on the interval $(-\infty, +\infty)$.

Answer:

The first step is to look at the derivative
$$\begin{aligned} f'(x) &= \frac{(1+x^2)(0) - (1)(2x)}{(1+x^2)^2} \\ &= \frac{-2x}{(1+x^2)^2}. \end{aligned}$$

Notice that $f'(x)$ is always defined, so the only critical points will be when $f'(x) = 0$. The only place where $f'(x) = 0$ is at $x = 0$, so we have a single critical point. There are no endpoints, so our table of critical points and endpoints is particularly simple.

Type of point	$x =$	$f(x) =$
critical point	0	1

If there is a global maximum it must be at the point $(0, 1)$. *If there is a global minimum it must be at the point $(0, 1)$.* However, because the domain is unbounded there is no guarantee that there is a global maximum or a global minimum.

Looking at the derivative we see that the denominator is always positive; so the sign of the derivative is determined by the sign of the numerator. The numerator is positive when x is negative and negative

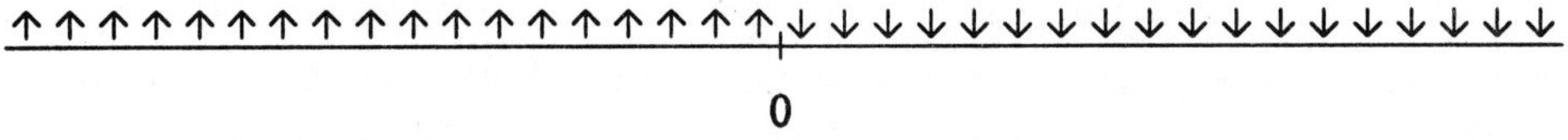

↑↑↑↑↑↑↑↑↑↑↑↑↑↑↑↑↓↓↓↓↓↓↓↓↓↓↓↓↓↓↓↓
　　　　　　　　　0

Figure II.57: The behavior of $f(x) = 1/(1+x^2)$

when x is positive. Thus $f'(x)$ is positive and $f(x)$ is increasing when x is negative, and $f'(x)$ is negative and $f(x)$ is decreasing when x is positive. See Figure II.57.

From this information we can see that the point $(0,1)$ is the global maximum because $f(x)$ is increasing until it reaches $(0,1)$ after which it is decreasing. We can also see that there is no global minimum.

Now notice that the original function

$$f(x) = \frac{1}{1+x^2}$$

is always positive. If x is a very large number then x^2 is also very large, so $1+x^2$ is even larger and $f(x) = 1/(1+x^2)$ will be very small. Thus, for very large values of x, $f(x)$ will be very close to zero. For example,

$$f(10) = 1/101$$

and

$$f(-100) = 1/10,001.$$

Figure II.58 shows a graph of this function. Notice the global maximum at the point $(0,1)$ and the way that $f(x)$ is very close to zero when x is very large. There is no global minimum.

Terminology

We need to clarify the use of the English words "large" and "small." These adjectives can be ambiguous when they are attached to *negative* numbers. In this book these words will always refer to the *absolute value* of a number. Thus, 10,000 and $-10,000$ are both *large* numbers and 0.0001 and -0.0001 are both *small* numbers. The reader should be aware, however, that this usage is far from universal.

§6. Optimization, II

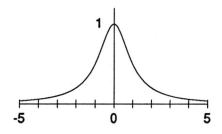

Figure II.58: $f(x) = 1/(1+x^2)$

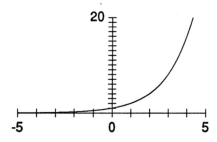

Figure II.59: $f(x) = 2^x$

One of the key properties of a function $f(x)$ on an unbounded interval is its behavior for very large values of x. In the preceding example $f(x)$ was very close to zero for very large values of x. In this particular example it didn't make any difference whether x was large and positive or x was large and negative. The following examples show that this is not always true.

Example:

Investigate the behavior of the function $f(x) = 2^x$ when x is very large.

We must make a distinction between very large *positive* values of x and very large *negative* values of x. If x is very large and positive than 2^x is also very large. For example, $f(10) = 2^{10} = 1024$. On the other hand if x is very large and negative then $f(x)$ will be very small. For example, $f(-10) = 2^{-10} = 1/2^{10} = 1/1024$. Figure II.59 shows a graph of this function. Notice the difference between the behavior of $f(x)$ for very large positive values of x and for very large negative values of x.

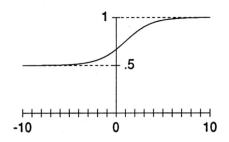

Figure II.60: $y = (1 + 2^x)/(2 + 2^x)$

Example:

Find the behavior of the function
$$f(x) = \frac{1 + 2^x}{2 + 2^x}$$
for very large values of x.

First, by dividing both the numerator and the denominator by 2^x we see that
$$f(x) = \frac{1 + 2^x}{2 + 2^x} = \frac{\frac{1}{2^x} + 1}{\frac{2}{2^x} + 1}.$$

If x is very large and positive so is 2^x. Thus $1/2^x$ and $2/2^x$ are very small. Thus the numerator is very close to 1 and the denominator is very close to 1. So the quotient is very close to 1. Thus for very large positive values of x, $f(x)$ is very close to 1.

On the other hand if x is very large and negative then 2^x is very small, so the numerator of the original formula for $f(x)$ is close to 1 and the denominator is close to 2. Thus, for very large negative values of x, $f(x)$ is very close to $1/2$. Figure II.60 shows this behavior.

We use the following vocabulary to describe the behavior of a function $f(x)$ for very large positive and for very large negative values of x.

Definition:

- We say that the **limit as x approaches $+\infty$ of $f(x)$ is equal to L** and write
$$\lim_{x \to +\infty} f(x) = L$$

§6. Optimization, II

if for very large positive values of x, $f(x)$ is very close to L.

- We say that the **limit as x approaches $-\infty$ of $f(x)$ is equal to L** and write

$$\lim_{x \to -\infty} f(x) = L$$

if for very large negative values of x, $f(x)$ is very close to L.

Graphically,

$$\lim_{x \to +\infty} f(x) = L$$

says that the part of the graph $y = f(x)$ far out to the right is close to the horizontal line $y = L$. Similarly,

$$\lim_{x \to -\infty} f(x) = L$$

says that the part of the graph $y = f(x)$ far out to the left is close to the horizontal line $y = L$. See Figure II.60.

To emphasize the graphical meaning of these limits we often say $f(x)$ **has a horizontal asymptote L as x goes to $+\infty$** if

$$\lim_{x \to +\infty} f(x) = L,$$

or $f(x)$ **has a horizontal asymptote L as x goes to $-\infty$** if

$$\lim_{x \to -\infty} f(x) = L.$$

The idea involved here is very similar to the idea of the limit of a sequence that we discussed in sections I.2 and I.3. The definition above is fine if we are interested in the general behavior of $f(x)$ for large values of x. Many practical problems, however, require more precise information—how large must x be to ensure that $f(x)$ is sufficiently close to L? The following definition captures this idea.

Definition:

- We say that the **limit as x approaches $+\infty$ of $f(x)$ is equal to L** and write
$$\lim_{x \to +\infty} f(x) = L$$
if for every positive tolerance ϵ there is a B such that for every $x > B$, $|f(x) - L| < \epsilon$.

- We say that the **limit as x approaches $-\infty$ of $f(x)$ is equal to L** and write
$$\lim_{x \to -\infty} f(x) = L$$
if for every positive tolerance ϵ there is a B such that for every $x < B$, $|f(x) - L| < \epsilon$.

Example:

Suppose that federal law requires that bottled water contains less than 5 mg of Agent YK per gallon. A bottler obtains water from a spring that has no Agent YK at all. Unfortunately, however, he has bought a tank containing 1,000,000 gallons of water from another supplier. This tank also contains 10,000 gm (10,000,000 mg) of Agent YK or 10 mg per gallon. Rather than discard the contaminated water the bottler decides to dilute it with his own water. If he adds x gallons of his own water to the contaminated water the concentration of Agent YK will be
$$C(x) = \frac{10,000,000}{1,000,000 + x} \text{ mg per gallon}$$
since the original 10,000,000 mg of Agent YK is now contaminating $(1,000,000 + x)$ gallons of water.

Notice that for very large values of x the denominator is very large so $C(x)$ is very small. Thus
$$\lim_{x \to +\infty} C(x) = 0$$
and the supplier can meet the federal standard or any standard, no matter how strict, by diluting the contaminated water with enough water.

§6. OPTIMIZATION, II

How much water will the bottler need to add to the contaminated water in order to meet the federal standard?

Answer:

We want
$$\frac{10,000,000}{1,000,000 + x} < 5$$

$$10,000,000 < 5,000,000 + 5x$$
$$5,000,000 < 5x$$
$$1,000,000 < x$$

So the bottler must add more than 1,000,000 gallons of uncontaminated water.

Exercises

Exercise II.6.1: How much uncontaminated water would the bottler need to add if federal law required the concentration of Agent YK to be below 3 mg per gallon?

Exercise II.6.2: How much uncontaminated water would the bottler need to add if federal law required the concentration of Agent YK to be below 1 mg per gallon?

Exercise II.6.3: Suppose that the bottler's normal supply of water contains 0.5 mg Agent YK per gallon. Find a formula $G(x)$ for the concentration of Agent YK after the bottler adds x gallons of this water to the contaminated water. What is the limit

$$\lim_{x \to +\infty} G(x)?$$

In this situation how much water would the bottler need to add to the contaminated water to meet each of the three standards above? Suppose federal law required the concentration of Agent YK to be below 0.3 mg per gallon. Could the bottler meet this standard?

There are many different possibilities for the behavior of a function $f(x)$ as x goes to $+\infty$ or to $-\infty$. We have introduced terminology

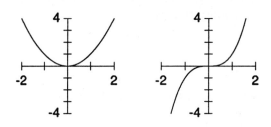

Figure II.61: $f(x) = x^2$ and $f(x) = x^3$

for one possibility—horizontal asymptotes. Another possibility is that a function grows larger and larger without any bound. Consider, for example, the functions $y = x^2$ and $y = x^3$ shown in Figure II.61. As x goes off to the right the function $y = x^2$ is growing larger and larger without any bound. As x goes off to the right the function $y = x^3$ is also growing larger and larger without any bound. As x goes off to the left the function $y = x^2$ is growing larger and larger without any bound. As x goes off to the left the function $y = x^3$ is growing larger and larger *in the negative direction* without any bound. We introduce the following terminology to describe this kind of behavior.

Definition:

- We say the **limit of $f(x)$ as x approaches $+\infty$ is $+\infty$** and write
$$\lim_{x \to +\infty} f(x) = +\infty$$
if for very large positive values of x, $f(x)$ is large and positive.

- We say the **limit of $f(x)$ as x approaches $+\infty$ is $-\infty$** and write
$$\lim_{x \to +\infty} f(x) = -\infty$$
if for very large positive values of x, $f(x)$ is large and negative.

- We say the **limit of $f(x)$ as x approaches $-\infty$ is $+\infty$** and write
$$\lim_{x \to -\infty} f(x) = +\infty$$

§6. Optimization, II

if for very large negative values of x, $f(x)$ is large and positive.

- We say the **limit of $f(x)$ as x approaches $-\infty$ is $-\infty$** and write
$$\lim_{x \to -\infty} f(x) = -\infty$$
if for very large negative values of x, $f(x)$ is large and negative.

As usual for practical problems we need more precise definitions.

Definition:

- We say the **limit of $f(x)$ as x approaches $+\infty$ is $+\infty$** and write
$$\lim_{x \to +\infty} f(x) = +\infty$$
if for every B there is an M such that for all $x > M$, $f(x) > B$.

- We say the **limit of $f(x)$ as x approaches $+\infty$ is $-\infty$** and write
$$\lim_{x \to +\infty} f(x) = -\infty$$
if for every B there is an M such that for all $x > M$, $f(x) < B$.

- We say the **limit of $f(x)$ as x approaches $-\infty$ is $+\infty$** and write
$$\lim_{x \to -\infty} f(x) = +\infty$$
if for every B there is an M such that for all $x < M$, $f(x) > B$.

- We say the **limit of $f(x)$ as x approaches $-\infty$ is $-\infty$** and write
$$\lim_{x \to -\infty} f(x) = -\infty$$
if for every B there is an M such that for all $x < M$, $f(x) < B$.

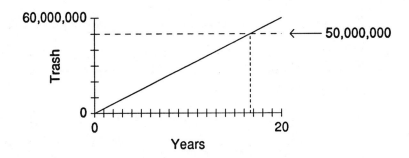

Figure II.62: Smithville's landfill

Example:

The town of Smithville has just opened up a new landfill. The landfill has a capacity of 50,000,000 tons of trash. The residents of Smithville are generating trash at the rate of 3,000,000 tons per year. In how many years will the capacity of the landfill be exceeded?

Answer:

After t years the landfill will have $3,000,000t$ tons of trash. The problem here is that

$$\lim_{t \to +\infty} 3,000,000t = +\infty$$

so that town will eventually surpass the storage capacity of any landfill. In particular this landfill, whose capacity is 50,000,000 tons, will be full when

$$3,000,000t > 50,000,000$$

$$t > \frac{50,000,000}{3,000,000}$$

$$t > 16.67$$

as shown in Figure II.62.

§6. Optimization, II

Exercises

*Exercise II.6.4: Suppose that the town had built a landfill with a capacity of 70,000,000 tons of trash. In how many years would the capacity of the landfill have been exceeded?

Exercise II.6.5: Suppose that the town had built a landfill with a capacity of 90,000,000 tons of trash. In how many years would the capacity of the landfill have been exceeded?

Now we are ready to solve some graph sketching and optimization problems for functions whose domain is unbounded.

Example:

Find the best price (i.e., the price which yields the highest total profit) for the product described in Exercise II.4.4, that is, find the maximum of the function

$$T(p) = (p - 0.50)\left(\frac{1000}{1 + p^2}\right)$$

on the interval $[0.50, +\infty)$.

The first step is to find the derivative of the function $T(p)$.

$$\begin{aligned}
T(p) &= (p - 0.50)\left(\frac{1000}{1 + p^2}\right) \\
T(p) &= \frac{1000p - 500}{1 + p^2} \\
T'(p) &= \frac{(1 + p^2)1000 - (1000p - 500)(2p)}{(1 + p^2)^2} \\
&= \frac{1000 + 1000p^2 - 2000p^2 + 1000p}{(1 + p^2)^2} \\
&= \frac{-1000p^2 + 1000p + 1000}{(1 + p^2)^2} \\
&= (-p^2 + p + 1)\left(\frac{1000}{(1 + p^2)^2}\right)
\end{aligned}$$

Notice that the derivative is always defined so the only critical points will be points at which $T'(p) = 0$. We find these by solving the equation

$$(-p^2 + p + 1)\left(\frac{1000}{(1+p^2)^2}\right) = 0$$

$$-p^2 + p + 1 = 0$$

using the Quadratic Formula to obtain

$$p = \frac{-1 \pm \sqrt{1+4}}{-2} = 0.5 \pm \frac{\sqrt{5}}{2}.$$

So there are two critical points, one at $p = 0.5 + \sqrt{5}/2 = 1.6180$ and the other at $p = 0.5 - \sqrt{5}/2 = -0.6180$. The second of these is irrelevant for our problem since we are only interested in the interval $[0.50 + \infty)$.

Next, we make a table showing the value of the function $T(p)$ at each of the endpoints and critical points. In this example we have one endpoint, $p = 0.50$, and one critical point, $p = 0.5 + \sqrt{5}/2 = 1.6180$.

Type of point	Price	Total Profit
endpoint	0.50	0
critical point	1.6180	309.0170

From this table we can see that *if there is a global maximum* it must be at the price $p = 1.6180$ and *if there is a global minimum* it must be at the price $p = 0.50$. Unfortunately, however, we don't know if there is a global maximum or if there is a global minimum. We need to do a little more analysis.

The next step is to analyze the sign of the derivative

$$T'(p) = (-p^2 + p + 1)\left(\frac{1000}{(1+p^2)^2}\right).$$

The derivative $T'(p)$ is continuous on the interval $[0.50, +\infty)$ and it is zero only at $p = 0.5 + \sqrt{5}/2 = 1.6180$. Hence, $T'(p)$ must be either all positive or all negative on the whole interval $[0.50, 1.6180)$ and either all positive or all negative on the whole interval $(1.6180, +\infty)$. Thus to

§6. Optimization, II

```
↑↑↑↑↑↑↑↑↑↑↑  ↓↓↓↓↓↓↓↓↓↓↓↓↓↓↓↓↓↓↓↓
0.50              1.6180
```

Figure II.63: Diagram showing where $T(p)$ is increasing and decreasing

find the sign of $T'(p)$ on each of these two pieces we need only evaluate $T'(p)$ at one sample point in each piece. For example, in the piece $[0.50, 1.6180)$ we can pick the sample point $p = 1$ and calculate

$$T'(1) = (-1 + 1 + 1)\left(\frac{1000}{(1 + 1^2)^2}\right) = 250$$

and in the interval $(1.6180, +\infty)$ we can pick the point $p = 3$ and calculate

$$T'(3) = (-9 + 3 + 1)\left(\frac{1000}{(1 + 3^2)^2}\right) = -50.$$

Hence, on the interval $[0.50, 1.6180)$, $T'(p)$ is positive and $T(p)$ is increasing and on the interval $(1.6180, +\infty)$, $T'(p)$ is negative and $T(p)$ is decreasing. This information is indicated in Figure II.63.

Now we can answer our original question. There is a global maximum and it is located at the point $(1.6180, 309.0170)$. We know this is the global maximum because our function is increasing from $p = 0.50$ to $p = 1.6180$ and then, to the right of $p = 1.6180$, it is decreasing.

Before graphing the function $T(p)$ we would like to find out its behavior for very large positive values of p. We are working in the domain $[0.50, +\infty)$, so we are interested only in

$$\lim_{p \to +\infty} \left((p - 0.50)\left(\frac{1000}{1 + p^2}\right)\right).$$

Notice that

$$\left((p - 0.50)\left(\frac{1000}{1 + p^2}\right)\right) = \frac{1000p - 500}{1 + p^2} = \frac{\frac{1000}{p} - \frac{500}{p^2}}{\frac{1}{p^2} + 1}$$

and for very large values of p, $1000/p$, $500/p^2$ and $1/p^2$ are all very small, so that the numerator is very close to zero and the denominator

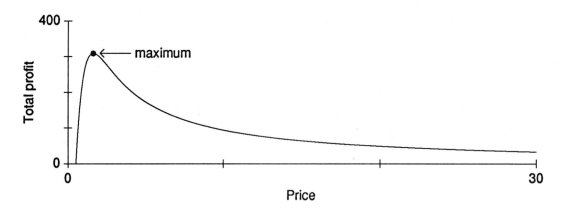

Figure II.64: $y = T(p)$

is very close to 1. Thus the quotient is very close to zero. Hence

$$\lim_{p \to +\infty} T(p) = 0.$$

Based on all this information we sketch the graph of $T(p)$ in Figure II.64.

We can summarize the steps involved in sketching the graph of a function on an unbounded domain as follows.

Sketching a Graph on an Unbounded Interval

1. Analyze the derivative.

   Find out where the derivative is zero or undefined. These points are the critical points. They are possible local maximums and local minimums. Find out where the derivative is positive (and, hence, the function is increasing) and where the derivative is negative (and, hence, the function is decreasing). The result of this analysis of the derivative should be a chart like Figure II.63.

   Make a table showing the value of the function at each of the critical points and at any endpoints. From this table you can identify all the local maximum(s) and all the local minimum(s). In addition, you can see where the global maximum will be *if there is a global maximum* and where the global minimum will be *if there is a global minimum*.

§6. Optimization, II

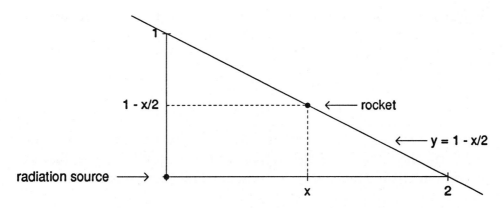

Figure II.65: Rocket passing source of radiation

2. Look for horizontal asymptotes.

   If the domain is an interval of the form $[a, +\infty)$ look for
   $$\lim_{x \to +\infty} f(x).$$
   If the domain is an interval of the form $(-\infty, a]$ look for
   $$\lim_{x \to -\infty} f(x).$$
   If the domain is the whole real line look for both
   $$\lim_{x \to +\infty} f(x) \quad \text{and} \quad \lim_{x \to -\infty} f(x).$$

3. Sketch the graph using the information above.

Example:

Suppose that a rocket ship is traveling on a straight line past a source of radiation as shown in Figure II.65. Notice that the radiation source is located at the point $(0,0)$ and a typical point on the path of the rocket has coordinates $(x, 1 - x/2)$. Thus the distance from the radiation source to a typical point on the path of the rocket is

$$\sqrt{x^2 + \left(1 - \frac{x}{2}\right)^2}$$

In general, the intensity of radiation at a point is inversely proportional to the square of the distance from the point to the source of radiation. Thus the intensity of the radiation is given by the function

$$I(x) = \frac{c}{x^2 + \left(1 - \frac{x}{2}\right)^2}$$

where c is some positive constant. At what point on the rocket ship's path will the intensity of the radiation be strongest? That is, find the maximum of the function $I(x)$ on the interval $(-\infty, +\infty)$.

1. Analyze the derivative.

    Notice

$$I(x) = \frac{c}{x^2 + \left(1 - \frac{x}{2}\right)^2}$$

$$I(x) = \frac{c}{x^2 + 1 - x + \frac{x^2}{4}}$$

$$I(x) = \frac{c}{\frac{5x^2}{4} - x + 1}$$

$$I'(x) = \frac{c - \frac{5cx}{2}}{\left(\frac{5x^2}{4} - x + 1\right)^2}$$

The derivative $I'(x)$ is always defined, so the only critical points are points at which $I'(x) = 0$. We find them as follows.

$$\frac{c - \frac{5cx}{2}}{\left(\frac{5x^2}{4} - x + 1\right)^2} = 0$$

$$c - \frac{5cx}{2} = 0$$

$$\frac{5cx}{2} = c$$

$$x = 2/5$$

We have only one critical point and no endpoints, so our chart of critical points and endpoints is particularly simple.

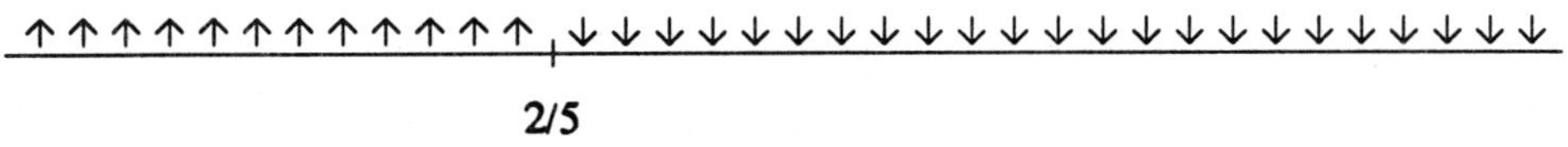

§6. OPTIMIZATION, II

↑↑↑↑↑↑↑↑↑↑↑ ↓↓↓↓↓↓↓↓↓↓↓↓↓↓↓↓↓↓↓↓↓↓
              2/5

Figure II.66: Where $I'(x)$ is increasing and decreasing

Type of point	$x =$	$I(x) =$
critical point	2/5	$5c/4$

We determine the sign of $I'(x)$ by evaluating $I'(x)$ at one sample point to the left of 2/5 and one sample point to the right of 2/5.

$$I'(0) = c \quad \text{and} \quad I'(1) = -0.96c.$$

This leads to Figure II.66 showing where $I(x)$ is increasing and decreasing and we see that the point $(2/5, 5c/4)$ is the global maximum.

2. We continue by looking for horizontal asymptotes.

   Notice that if x is very large then x^2 is very large and positive. Also, $(1 - x/2)^2$ is very large and positive. Thus the denominator of $I(x)$ is the sum of two very large positive numbers and, hence, is itself very large and positive. Thus $I(x)$ is very small and positive. Thus,

   $$\lim_{x \to +\infty} I(x) = 0 \quad \text{and} \quad \lim_{x \to -\infty} I(x) = 0.$$

3. Now we sketch the graph in Figure II.67.

The physical interpretation of the fact that

$$\lim_{x \to +\infty} I(x) = 0 \quad \text{and} \quad \lim_{x \to -\infty} I(x) = 0$$

is that when the rocket is very far away from the radiation source then the intensity of the radiation that it receives is very low. This agrees with our physical intuition.

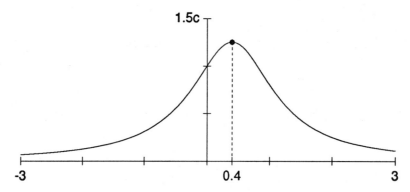

Figure II.67: The function $I(x)$

Curve sketching techniques can help determine the solutions of an equation of the form
$$f(x) = 0.$$

Example:

Find all the solutions of the equation
$$x^3 - 15x + 1 = 0.$$

Answer:

We begin by sketching a graph of the function
$$f(x) = x^3 - 15x + 1.$$
We are looking for places at which the graph crosses the x-axis.

1. First, we analyze the derivative
$$f'(x) = 3x^2 - 15.$$

The derivative is always defined so the only critical points are when $f'(x) = 0$. Setting $f'(x) = 0$ we see
$$\begin{aligned} 3x^2 - 15 &= 0 \\ 3x^2 &= 15 \\ x^2 &= 5 \end{aligned}$$

§6. Optimization, II

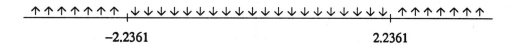

Figure II.68: Where $f(x)$ is increasing and decreasing

$$x = \pm\sqrt{5}$$

Thus there are two critical points, at $x = \sqrt{5}$ and $x = -\sqrt{5}$. There are no endpoints, so our critical and endpoint table is

Type of point	$x =$	$f(x) =$
critical point	$-\sqrt{5}$	23.3607
critical point	$\sqrt{5}$	-21.3607

These two critical points break the real line up into three pieces. We determine the sign of $f'(x)$ by evaluating $f'(x)$ at one point in each of the pieces: $f'(-3) = 12$, $f'(0) = -15$ and $f'(3) = 12$. This leads to Figure II.68 showing us where $f(x)$ is increasing and decreasing. Thus, $(-\sqrt{5}, 23.3607)$ is a local maximum and $(\sqrt{5}, -21.3607)$ is a local minimum.

2. Next we look for the behavior of $f(x)$ for very large and very small values of x.

$$f(x) = x^3 - 15x + 1 = x^3\left(1 - \frac{15}{x^2} + \frac{1}{x^3}\right).$$

If x is very large and positive then the first factor is large and positive and the second factor is close to 1. Hence, $f(x)$ will be large and positive. Thus

$$\lim_{x \to +\infty} f(x) = +\infty.$$

If x is very large and negative then the first factor is large and negative and the second factor is close to 1. Hence, $f(x)$ will be

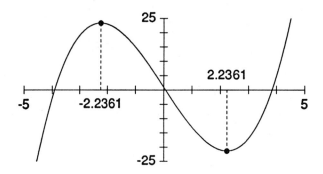

Figure II.69: $f(x) = x^3 - 15x + 1$

large and negative. Thus

$$\lim_{x \to -\infty} f(x) = -\infty.$$

Thus $f(x)$ heads off to $+\infty$ as x heads off to $+\infty$, and $f(x)$ heads off to $-\infty$ as x heads off to $-\infty$.

3. Now we sketch the graph in Figure II.69.

This graph tells us what we need to know. There are three places where $f(x)$ crosses the x-axis—once someplace to the left of $-\sqrt{5}$, once between $-\sqrt{5}$ and $+\sqrt{5}$ and once someplace to the right of $\sqrt{5}$. Hence, our original equation has three solutions.

Now that we know generally where the solutions are we can use a numerical method like the Bisection Method to get good estimates for each of the three solutions.

$$x \approx -3.9059 \quad \text{and} \quad x \approx 0.0667 \quad \text{and} \quad x \approx 3.8392.$$

Exercises

Part I: Sketch a graph for each of the following functions. Indicate clearly all the important features; critical points, local maximum(s) and local minimum(s), the global maximum and the global minimum, and behavior near $+\infty$ and $-\infty$.

§6. Optimization, II

*Exercise II.6.6:
$$y = \frac{1}{x^2 + (2-x)^2}$$

*Exercise II.6.7:
$$y = x^4 - x^2$$

*Exercise II.6.8:
$$y = \frac{x}{x^2 + 4}$$

*Exercise II.6.9:
$$y = \frac{x^2}{x^2 + 4}$$

*Exercise II.6.10:
$$y = \frac{x^3}{x^2 + 4}$$

Part II: Find estimates for all of the solutions of each of the following equations.

*Exercise II.6.11:
$$x^3 - x + 1 = 0$$

*Exercise II.6.12:
$$x^3 + x + 1 = 0$$

*Exercise II.6.13:
$$3x^5 - 10x^3 + 15x + 1 = 0$$

*Exercise II.6.14:
$$x^5 - 5x^3 + 10x = 1$$

*Exercise II.6.15:
$$\sin x = \frac{x}{2}$$

*Exercise II.6.16:
$$\sin x = \frac{x}{4}$$

*Exercise II.6.17:
$$\sin x = \frac{x}{8}$$

Part III:

We often want to find the value of a particular quantity. For example, we might want to find out the percentage of registered voters planning to vote for a particular candidate or we might want to find the total mass of the moon. In order to measure the quantity in question we might do an experiment or a series of experiments. If we do several experiments we will usually obtain several values for the quantity in question. Usually these values will not all be the same. Thus we often have several experimental estimates $a_1, a_2, \ldots a_n$ for the same quantity. The problem is to find the best estimate possible for the real value of this quantity based on our experimental data. Before reading on answer the following questions intuitively.

Question 1: Suppose that you have done four experiments to determine the value of a particular quantity. Your results are: 102, 102.5, 100.7, 102.2. What is your best guess for the true value of the quantity in question?

Question 2: Suppose that you have done four experiments to determine the value of a particular quantity. Your results are: 102, 102.5, 100.7, 120.2. What is your best guess for the true value of the quantity in question?

The purpose of this series of exercises is to explore means of obtaining the "best" estimate possible for the true value of a given quantity based on several experimentally determined values. You probably already have several ideas. For example, the following ideas are often used.

Idea 1: Given experimental values $a_1, a_2, \ldots a_n$ for a particular quantity the "best" estimate of the true value is the "average" or "mean" $(a_1 + a_2 + \ldots + a_n)/n$.

Idea 2: Given experimental values $a_1, a_2, \ldots a_n$ for a particular quantity the "best" estimate of the true value is the median. The median is obtained by writing the experimental values in order from smallest to largest and then taking the middle experimental value. For example the median of 2, 1, 4, 3, 5 is 3. If there are an odd number of

§6. Optimization, II

experimental values this makes sense because there is a middle value. With an even number there are two middle numbers. Use the average of these two. For example the median of 1, 3, 5, 4, 6, 8 is 4.5.

In order to analyze these ideas we need to discuss what we mean by the "best" estimate of the true value of the quantity. The basic idea is that the estimate will be "good" if it is close to our experimental results. One possible measure of how good or bad an estimate x is is the sum

$$|x - a_1| + |x - a_2| + |x - a_3| + \ldots + |x - a_n|.$$

If this sum is small then the estimate is quite good. If this sum is large then the estimate is quite poor. The "best estimate" is the one that minimizes this sum. This leads to the following series of exercises.

*Exercise II.6.18:   Given experimental values 12.1, 12.3 and 12.4 what value of x minimizes the sum

$$|x - 12.1| + |x - 12.3| + |x - 12.4|?$$

*Exercise II.6.19:   Given experimental values a_1, a_2, a_3 what value of x minimizes the sum

$$|x - a_1| + |x - a_2| + |x - a_3|?$$

*Exercise II.6.20:   Given experimental values a_1, a_2, a_3, a_4 what value of x minimizes the sum

$$|x - a_1| + |x - a_2| + |x - a_3| + |x - a_4|?$$

*Exercise II.6.21:   Given experimental values $a_1, a_2, a_3, \ldots a_n$ what value of x minimizes the sum

$$|x - a_1| + |x - a_2| + |x - a_3| + \ldots + |x - a_n|?$$

The idea of minimizing the sum

$$|x - a_1| + |x - a_2| + |x - a_3| + \ldots + |x - a_n|$$

is a good one. There are, however, other possibilities. The most common criterion for how close an estimate x is to the experimental values $a_1, a_2, \ldots a_n$ is the sum

$$(x - a_1)^2 + (x - a_2)^2 + \ldots + (x - a_n)^2.$$

This criterion has several nice properties. One of the most intuitive is that it penalizes large discrepancies very strongly. Intuitively, if you were asked to estimate a quantity whose true value was $1.00 an estimate of $1.20 is much worse than an estimate of $1.10. A mistake of $0.20 is at least twice as bad as a mistake of $0.10. In fact, some people might argue that one mistake of $0.20 is worse than two mistakes of $0.10. By squaring the differences between the experimental values $a_1, a_2, \ldots a_n$ and x, the formula

$$(x - a_1)^2 + (x - a_2)^2 + \ldots (x - a_n)^2$$

is essentially saying that large discrepancies are much worse than small ones. This leads us to the following series of exercises.

Exercise II.6.22: Given experimental values 12.1, 12.3 and 12.4 what value of x minimizes the sum

$$(x - 12.1)^2 + (x - 12.3)^2 + (x - 12.4)^2?$$

Exercise II.6.23: Given experimental values a_1, a_2, a_3 what value of x minimizes the sum

$$(x - a_1)^2 + (x - a_2)^2 + (x - a_3)^2?$$

Exercise II.6.24: Given experimental values $a_1, a_2, a_3, \ldots a_n$ what value of x minimizes the sum

$$(x - a_1)^2 + (x - a_2)^2 + (x - a_3)^2 + \ldots + (x - a_n)^2?$$

§7. Inverse Functions

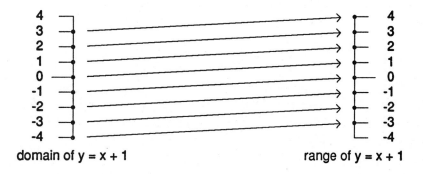

Figure II.70: Visualizing $y = x + 1$

§7. Inverse Functions

The formula for the area A of a circle of radius r, $A = \pi r^2$, defines a function that determines the area of a circle from its radius. Often, however, we have the reverse problem. We know the area of a circle and we would like to determine its radius. We can find a function that goes backward, computing the radius of a circle given its area, by solving the equation $A = \pi r^2$ for r as follows.

$$A = \pi r^2$$

$$\pi r^2 = A$$
$$r^2 = A/\pi$$
$$r = \sqrt{A/\pi}$$

This new function is called the **inverse** of the original function. The process of finding the inverse of a function is called **inverting** the function.

We will discuss two nice ways of visualizing a function and its inverse. The first way begins as follows. Use one vertical line to represent the domain of f and another vertical line to represent the **range of f**, the set of possible values of $f(x)$. Then draw an arrow going from each x in the domain to its value $f(x)$ in the range, for example, the function $y = x + 1$ is pictured in Figure II.70.

Now one can visualize the inverse f^{-1} of the function f by following the same arrows backward. For example, Figure II.71 shows the inverse

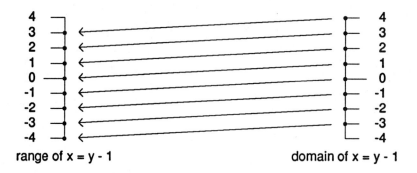

Figure II.71: Visualizing $x = y - 1$ as the inverse of $y = x + 1$

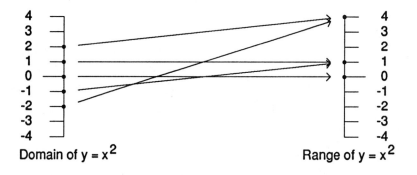

Figure II.72: Visualizing $y = x^2$

$x = y - 1$ of the function $y = x + 1$ shown in Figure II.70.

Consider the function $f(x) = x^2$ shown in Figure II.72. If we try to compute $f^{-1}(4)$ by following an arrow backward from 4, we have a problem—there are two arrows going backward from 4. In this case $f^{-1}(4)$ is ambiguous. It could equally well be $+2$ or -2. A function is said to be **one–to–one** if there is no ambiguity, that is, if no two distinct points in the domain are mapped to the same point in the range. A function does not have a well-defined inverse if it is not one-to-one. The lack of one-to-one-ness of the function $f(x) = x^2$ can cause some difficulty. For practical problems like finding the radius of a circle of given area, however, the practical facts often remove any ambiguity.

Because the nonnegative square root of a number is most commonly used in practical problems we use the notation $\sqrt{t}$ for the *nonnegative*

§7. Inverse Functions

square root of t. If we desire the negative square root we write $-\sqrt{t}$ and if we want to indicate both possibilities we write $\pm\sqrt{t}$.

Another problem is evident in Figure II.72—one cannot find the square root of a negative number because none of the arrows in Figure II.72 end at a negative number. The domain of the function $f(t) = \sqrt{t}$ is the interval $[0, +\infty)$.

Definition:

We have been using the word "range" somewhat informally to mean "the set of all possible values" of a function f. Mathematicians often use two words "range" and "image" more precisely. The **range** of a function describes the set of all possible values of the function in a general sense. Most of the functions that we have been discussing have the same "range" in this general sense, namely the set of all real numbers.

We use the word **image** in the context of one particular function to indicate the set of all actual values of the function. For example, the range of the function $f(t) = t^2$ is the set of all real numbers, but its image is the interval $[0, +\infty)$.

In general, the domain of the inverse f^{-1} of a one-to-one function f is the image of the function f. If the image of f is the whole real line then we say that f is **onto**. If f is onto then the domain of f^{-1} is the entire real line.

The terms "independent" and "dependent" are sometimes used for the two variables involved in a function. The adjective "dependent" is meant to suggest that the dependent variable is determined by or depends on the independent variable.

Definition:

Consider the function $y = f(x)$. The variable x is called the **independent variable** and the variable y is called the **dependent variable**.

Now we want to look at another way of visualizing the inverse of a function f. Start with a graph of the original function $w = f(t)$ as

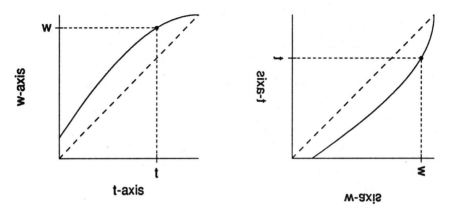

Figure II.73: $w = f(t)$ and its inverse $t = f^{-1}(w)$

shown, for example, on the left side of Figure II.73. To find the graph of the inverse f^{-1} we want to exchange the roles of the variables t and w.

In the function $w = f(t)$, w is the dependent variable and t is the independent variable. One usually uses the vertical axis for the dependent variable and the horizontal axis for the independent variable. In the function $t = f^{-1}(w)$ the two roles are reversed. Now t is the dependent variable whose value is determined by the independent variable w. Thus for the graph of f^{-1} we put the variable w on the horizontal axis and the variable t on the vertical axis. We are looking at the same basic relationship between t and w from two different viewpoints. To get the graph of f^{-1} we simply pick the graph of f up and flip it over around the line $w = t$, exchanging the horizontal and vertical axes. I made the right side of Figure II.73 by using a "flip" command in my computer graphics program. I left the labels on the graph exactly the way they would be after this flipping operation to emphasize that the original graph was flipped over to get the graph of the inverse.

Exercises

*__Exercise II.7.1:__ Sketch a graph of the function $f(x) = x^3 + x + 1$. Is it one-to-one? If so, sketch its inverse. What is the domain of the inverse?

*__Exercise II.7.2:__ Sketch a graph of the function $f(x) = x^3 - x + 1$.

§7. INVERSE FUNCTIONS

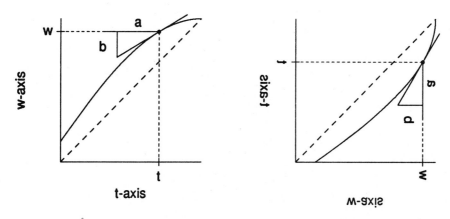

Figure II.74: Comparing $f'(t)$ and $(f^{-1})'(w)$

Is it one-to-one? If so, sketch its inverse. What is the domain of the inverse?

*Exercise II.7.3: Use computer graphics or a graphing calculator to graph the function $f(x) = 2^x$. Is it one-to-one? If so, sketch its inverse. What is the domain of the inverse?

Whenever we develop a new way of obtaining functions we need a corresponding differentiation technique. Now we want a formula that will enable us to obtain the derivative of the inverse f^{-1} of a function. Look at Figure II.74.

We want to compute the derivative of the function f^{-1} at the point (w, t). That is, we want to find the slope of the tangent in the righthand graph (the graph of f^{-1}) in Figure II.74. We assume that we already know the derivative of the function f. Thus we know the slope, $f'(t)$, of the tangent at the point (t, w) in the lefthand graph of Figure II.74. If we draw a triangle like the one shown in Figure II.74 (lefthand side) and label its sides as shown, then the slope of the tangent is

$$f'(t) = b/a.$$

When we flip the lefthand side of Figure II.74 over to get the righthand side of Figure II.74 the old tangent to the curve f at the point (t, w) becomes the tangent to the curve f^{-1} at the point (w, t). The triangle is also flipped over. The slope of the tangent in the righthand graph is

$$(f^{-1})'(w) = \frac{a}{b} = \frac{1}{b/a} = \frac{1}{f'(t)}.$$

This is the formula we seek

$$(f^{-1})'(w) = \frac{1}{f'(t)} = \frac{1}{f'(f^{-1}(w))}.$$

where the last equality follows from the fact that $t = f^{-1}(w)$.

Theorem 11 *(Inverse Function Rule)*
Suppose that $y = f(x)$ is a one-to-one function. Then

$$(f^{-1})'(y) = \frac{1}{f'(f^{-1}(y))}.$$

This formula works at every point (x, y) on the curve $y = f(x)$ at which $f'(x)$ exists and is nonzero.

Using different notation, we sometimes write

$$\frac{dx}{dy} = \frac{1}{dy/dx}.$$

Example:

Find the derivative of the function $t = \sqrt{w}$

Answer:

Let $w = f(t) = t^2$, so $t = f^{-1}(w) = \sqrt{w}$. Thus, by the Inverse Function Rule,

$$\frac{dt}{dw} = \frac{1}{dw/dt} = \frac{1}{2t} = \frac{1}{2\sqrt{w}} = \frac{1}{2}w^{-\frac{1}{2}}.$$

Recall the Power Rule:

"Suppose that n is a positive integer and $f(x) = x^n$ then $f'(x) = nx^{(n-1)}$."

§7. INVERSE FUNCTIONS

The preceding example shows the same formula applies when $n = 1/2$. The exercises below ask you to show that the same formula works for $n = 1/q$ where q is any positive integer.

Exercises

If q is a positive integer then $a^{1/q}$ is the qth root[2] of a, that is, the function $b = a^{1/q}$ is the inverse of the function $a = b^q$.

*Exercise II.7.4: Show that if $y = x^{\frac{1}{3}}$ then $y' = \frac{1}{3}x^{-\frac{2}{3}}$.

Exercise II.7.5: Show that if $y = x^{\frac{1}{4}}$ then $y' = \frac{1}{4}x^{-\frac{3}{4}}$.

*Exercise II.7.6: Show that if $y = x^{\frac{1}{q}}$ then $y' = \left(\frac{1}{q}\right)x^{\left(\frac{1}{q}-1\right)}$.

*Exercise II.7.7: Suppose that $f(t)$ is the function that converts a temperature in degrees Fahrenheit into degrees Celsius. That is,

$$f(t) = \frac{5}{9}(t - 32).$$

Find f', f^{-1}, and $(f^{-1})'$.

Another Interpretation of the Derivative

One of the reasons that the derivative is so important is that it has many different interpretations. In the following paragraphs we discuss an interpretation that can shed additional light on our work in this and the following section.

Suppose we label each point on a spring or piece of elastic with a real number as shown in Figure II.75. We use a spring or piece of elastic because we want to use the metaphor of stretching or compressing the real line. Suppose, as shown in Figure II.76, that directly under the piece of elastic we have a piece of graph paper whose points have been labeled with real numbers in the same way as the elastic.

Now suppose that we pick up the piece of elastic and place it on the piece of graph paper as shown in Figure II.77. In the process of doing this we stretch and compress the piece of elastic. A typical point originally marked x on the piece of elastic will be placed at a point marked $f(x)$ on the piece of graph paper as shown in Figure II.78.

---

[2] If q is even choose the nonnegative qth root.

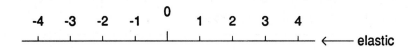

Figure II.75: A piece of elastic

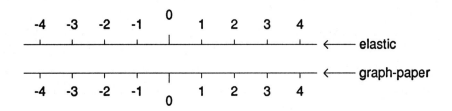

Figure II.76: Elastic and graph paper

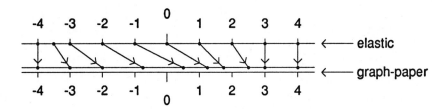

Figure II.77: Elastic on graph paper

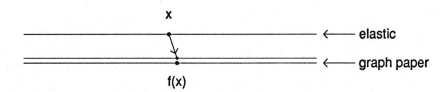

Figure II.78: Elastic on graph paper

§7. INVERSE FUNCTIONS

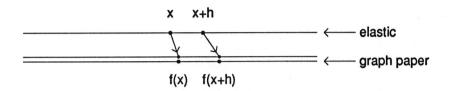

Figure II.79: Estimating the amount of stretching or compression

We would like to measure how much the elastic at a particular point labeled x has been stretched or compressed. One way to estimate this is to compare the new position $f(x)$ of the point x with the new position $f(x+h)$ of a nearby point $x+h$ as shown in Figure II.79. The fraction

$$\frac{f(x+h) - f(x)}{h}$$

provides an estimate of how much the elastic at the point x has been stretched or compressed. If h is very small then this fraction will give us a very good estimate. To find the exact amount of stretching or compression we take the limit

$$\lim_{h \to 0} \frac{f(x+h) - f(x)}{h}.$$

This is just the derivative $f'(x)$ and we see that the derivative measures the amount by which the elastic has been stretched or compressed at each point. If $f'(x) > 1$ then at the point x the elastic has been stretched. If $0 < f'(x) < 1$ then at the point x it has been compressed. If $f'(x) < 0$ then, as shown in Figure II.80, the elastic was reversed as well as stretched or compressed.

To visualize the inverse of the function f think of this process reversed with the elastic on the bottom and placed on the graph paper on top by following the original arrows backwards. Where the original picture compressed the old elastic the new picture will stretch the new elastic and vice versa. For example, if f compressed the elastic at the point x by a factor of $1/2$ then f^{-1} will stretch the elastic at the point $y = f(x)$ by a factor of 2. Figures II.81 and II.82 illustrate this idea.

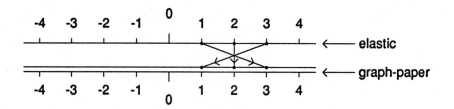

Figure II.80: Reversal

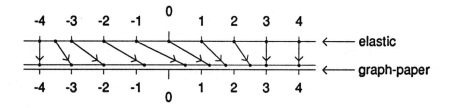

Figure II.81: The function f

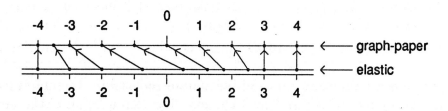

Figure II.82: The function f^{-1}

§7. Inverse Functions

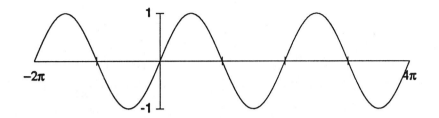

Figure II.83: $y = \sin \theta$

The Inverse Trigonometric Functions

We often must solve equations like

$$\sin \theta = c$$

where c is a real number between -1 and $+1$ or

$$\tan x = c$$

where c is any real number. This leads us to six new functions, the inverses of the six trigonometric functions. Most scientific calculators have these functions built in so that one can solve equations like

$$\sin \theta = -0.64$$

or

$$\tan \theta = 1.23$$

at the touch of a button. We begin by looking at the inverse of the sine function. Figure II.83 shows a graph of the sine function.

We have the same two problems with this function as with the function $y = x^2$. First, the sine function is not one-to-one. The problem here is even worse than it was for the function $y = x^2$. There were only two choices for the square root of a positive number and we solved the problem by choosing the positive square root. For the sine function we have an infinite number of choices. For example, all of the angles $\pi/2$, $5\pi/2$, $9\pi/2, \ldots$ and $-3\pi/2$, $-7\pi/2$, $-11\pi/2 \ldots$ have sine $+1$, so there are an infinite number of possibilities for the inverse sine of $+1$. We will choose the angle that is between $-\pi/2$ and $+\pi/2$. The second problem is that the values of the sine function are always between -1

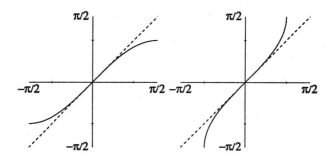

Figure II.84: The function $y = \sin\theta$ and its inverse $y = \arcsin x$

and $+1$. For this reason the domain of the inverse of the sine function is the interval $[-1, 1]$. This is analogous to the situation for the square root function whose domain is the interval $[0, \infty)$ because the square of a number is never negative. Figure II.84 shows the function $y = \sin x$ and its inverse. We call the inverse of the sine function the **arcsine**, written "arcsin." Some calculators use "asin" to denote the arcsine.

When one uses the arcsine function to solve an equation like, for example,
$$\sin\theta = 1/2,$$
one must remember that the solution
$$\theta = \arcsin 0.5 = \pi/6$$
is only one of many possible solutions. All of the angles

$$\pi/6, 13\pi/6, 25\pi/6, \ldots$$
$$-11\pi/6, -23\pi/6, -35\pi/6, \ldots$$
$$5\pi/6, 17\pi/6, 29\pi/6, \ldots$$
$$-7\pi/6, -19\pi/6, -31\pi/6, \ldots$$

have the same sine, $1/2$.

The first of these possibilities, $\pi/6$, is obtained from the arcsine function. The solution $5\pi/6$ is obtained from $\pi/6$ by the identity
$$\sin(\pi - \theta) = \sin\theta,$$

§7. INVERSE FUNCTIONS

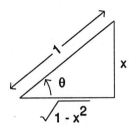

Figure II.85: Finding cos(arcsin x)

and the remainder from the periodicity of the sine function

$$\sin\theta = \sin(\theta + 2\pi).$$

When a computer algebra program like *Mathematica* solves an equation like $\sin x = 1/2$, it uses the inverse trig functions. For this reason it does not find all the solutions of this equation. See, for example, the following *Mathematica* session.

```
Solve[Sin[x] == 1/2, x]

Solve::ifun:
   Warning: Inverse functions are being used by Solve, so
    some solutions may not be found.

          Pi
{{x -> --}}
           6
```

The derivative of the arcsine function is found using the Inverse Function Rule.

$$\left(\frac{d}{dx}\right)\arcsin x = \frac{1}{\cos(\arcsin x)}$$

We can write this in a form that is more easy to work with by considering the triangle shown in Figure II.85. The hypotenuse of this triangle has length 1 and its height has length x. If we label the angle θ as shown then $x = \sin\theta$, so $\theta = \arcsin x$. By the Pythagorean Theorem the length of the base of the triangle is $\sqrt{1-x^2}$ and, hence

$$\cos\theta = \sqrt{1-x^2}.$$

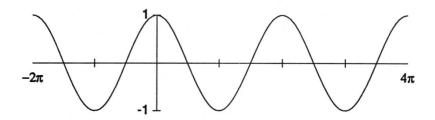

Figure II.86: $y = \cos\theta$

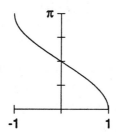

Figure II.87: $y = \arccos x$

This relation holds for any θ in the image, $[-\pi/2, \pi/2]$, of the arcsine function. So
$$\cos(\arcsin x) = \sqrt{1-x^2}$$
and
$$\left(\frac{d}{dx}\right)\arcsin x = \frac{1}{\cos(\arcsin x)} = \frac{1}{\sqrt{1-x^2}}.$$

Next we consider the cosine function and its inverse the **arccosine**, written "arccos" or, on some calculators, "acos." Figure II.86 is a graph of the cosine function. To overcome the fact that the cosine is not one-to-one we choose angles in the interval $[0, \pi]$. Since the cosine is always between -1 and $+1$ the domain of the arccosine is the interval $[-1, 1]$. Figure II.87 shows a graph of the arccosine.

We differentiate the arccosine using the Inverse Function Rule.
$$\left(\frac{d}{dx}\right)\arccos x = \frac{1}{-\sin(\arccos x)}$$

Using Figure II.88 we see that
$$\sin\theta = \sin(\arccos x) = \sqrt{1-x^2}.$$

§7. Inverse Functions

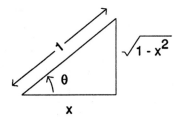

Figure II.88: Finding arccos(sin x)

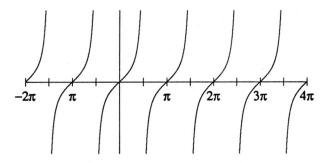

Figure II.89: $y = \tan \theta$

This relation holds for any θ in the image, $[0, \pi]$, of the arccosine. So

$$\left(\frac{d}{dx}\right) \arccos x = \frac{1}{-\sin(\arccos x)} = -\frac{1}{\sqrt{1-x^2}}$$

Next we consider the tangent function and its inverse, the **arctangent**, written "arctan" or, sometimes, "atan." Figure II.89 shows a graph of the tangent function. To deal with the fact that the tangent function is not one-to-one we choose angles in the interval $[-\pi/2, \pi/2]$. Figure II.90 shows a graph of the function $y = \arctan x$.

We differentiate arctan using the Inverse Function Rule.

$$\left(\frac{d}{dx}\right) \arctan x = \frac{1}{(\sec(\arctan x))^2}$$

and looking at Figure II.91 we see that

$$\sec \theta = \sec(\arctan x) = \sqrt{1+x^2}.$$

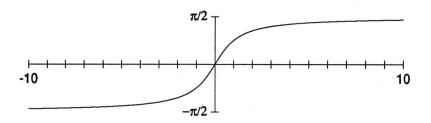

Figure II.90: $y = \arctan x$

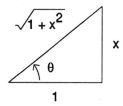

Figure II.91: Finding $\sec(\arctan x)$

This relation holds for any θ in the image, $(-\pi/2, \pi/2)$, of the arctangent. So
$$(\sec(\arctan x))^2 = 1 + x^2$$
and
$$\left(\frac{d}{dx}\right) \arctan x = \frac{1}{(\sec(\arctan x))^2} = \frac{1}{1+x^2}$$

Next we consider the inverse of the secant function, the **arcsecant**, written "arcsec." Figure II.92 shows a graph of the secant. This time we choose angles between 0 and π. Notice that the domain of the arcsecant is the set of all real numbers whose absolute value is at least 1. Figure II.93 shows a graph of the arcsecant function.

We differentiate arcsec using the Inverse Function Rule.
$$\left(\frac{d}{dx}\right) \operatorname{arcsec} x = \frac{1}{(\sec(\operatorname{arcsec} x))(\tan(\operatorname{arcsec} x))}$$

Of course, $\sec(\operatorname{arcsec} x) = x$. Looking at Figure II.94 we see that
$$\tan(\operatorname{arcsec} x) = \sqrt{x^2 - 1}$$

§7. Inverse Functions

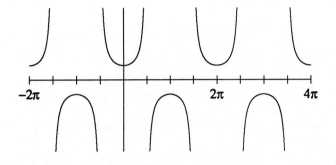

Figure II.92: $y = \sec\theta$

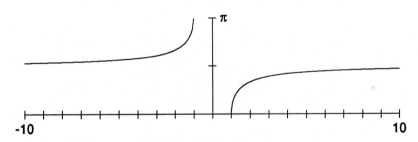

Figure II.93: $y = \operatorname{arcsec} x$

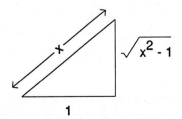

Figure II.94: Finding $\tan(\operatorname{arcsec} x)$

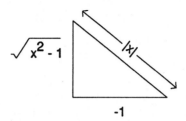

Figure II.95: Finding tan(arcsec x)

but we need to be careful about the sign. If x is negative this picture doesn't make any sense. We need to consider Figure II.95 instead. In this case we see that

$$\tan(\operatorname{arcsec} x) = -\sqrt{x^2 - 1}$$

Thus

$$\left(\frac{d}{dx}\right) \operatorname{arcsec} x = \frac{1}{(\sec(\operatorname{arcsec} x))(\tan(\operatorname{arcsec} x))}$$

$$= \begin{cases} \frac{1}{x\sqrt{x^2-1}}, & \text{if } x \geq 1; \\ -\frac{1}{x\sqrt{x^2-1}}, & \text{if } x \leq -1. \end{cases}$$

$$= \frac{1}{|x|\sqrt{x^2 - 1}}$$

There are two more inverse trigonometric functions, the inverse cotangent (arccotangent) and the inverse cosecant (arccosecant). They are left for exercises.

Finally, we should note that sometimes the notation "$\sin^{-1}$," "$\cos^{-1}$," etc. is used instead of "arcsin," "arccos," etc. to denote the inverse trig functions.

Exercises

Exercise II.7.8:
Discuss the inverse of the cotangent function. This function is called the **arccotangent** and written "arccot." Find

$$\left(\frac{d}{dx}\right) \operatorname{arccot} x.$$

§7. Inverse Functions

*Exercise II.7.9:
Discuss the inverse of the cosecant function. This function is called the **arccosecant** and written "arccsc." Find
$$\left(\frac{d}{dx}\right) \operatorname{arccsc} x.$$

*Exercise II.7.10: Find all the solutions of the equation
$$\sin x = 0.75.$$

Exercise II.7.11: Find all the solutions of the equation
$$\sin x = 1.25.$$

*Exercise II.7.12: Find all the solutions of the equation
$$\cot x = 0.80.$$

*Exercise II.7.13: Find all the solutions of the equation
$$\sin x = \cos x.$$

Exercise II.7.14: Find all the solutions of the equation
$$\sin x = 3 \cos x.$$

*Exercise II.7.15: Find the derivative of the function
$$y = \arcsin x + \arccos x$$

*Exercise II.7.16: Find the derivative of the function
$$y = \arcsin x - \arccos x$$

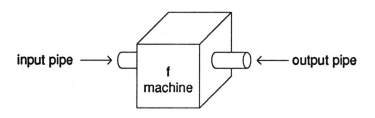

Figure II.96: The f-machine

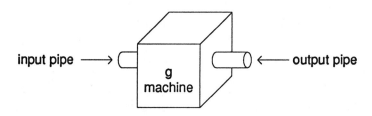

Figure II.97: The g-machine

§8. The Chain Rule

Composing Functions

Consider two functions, $f(x) = x^2 - 1$ and $g(y) = \sqrt{y}$. Think of these two functions, as shown in Figure II.96 and II.97, as machines, each with an input pipe and an output pipe. If one puts the input x into the input pipe of the f-machine, the machine will grind and beep, and after a short time the output $x^2 - 1$ will emerge from the output pipe. Similarly, if one puts the input y into the input pipe of the g machine, the machine will grind and beep, and after a short time the output $\sqrt{y}$ will emerge from the output pipe.

Now connect the output pipe of the f-machine directly to the input pipe of the g machine as shown in Figure II.98. If we put the input x into the input pipe of this combined machine the f-machine will produce the output $x^2 - 1$ which will go directly into the g-machine. The g-machine will then compute $\sqrt{x^2 - 1}$ and send it out the output pipe of the combined machine. The combined machine computes the

§8. The Chain Rule

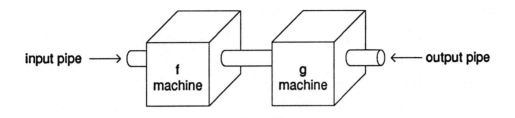

Figure II.98: The combined machine

function $h(x) = \sqrt{x^2 - 1}$. This new function is called the **composition** of f and g and is written

$$(g \circ f)(x) = g(f(x)) = \sqrt{x^2 - 1}.$$

Exercises

*Exercise II.8.1: Suppose $f(x) = x^2$ and $g(y) = y + 1$. Compute the following.

$$(g \circ f)(4)$$

$$(g \circ f)(9)$$

$$(f \circ g)(4).$$

*Exercise II.8.2: Suppose that $f(t)$ is the temperature measured in degrees Fahrenheit of a particular cup of coffee at time t. Suppose that c is a function that converts a temperature in degrees Fahrenheit into degrees Celsius. Express the temperature of the cup of coffee at time t in degrees Celsius using these two functions.

*Exercise II.8.3: Suppose that $p(t)$ is the pressure exerted by a particular gas in particular container where t is the temperature of the gas in degrees Fahrenheit. Suppose that c is a function that converts a temperature expressed in degrees Fahrenheit into degrees Celsius. Find a function that produces the pressure exerted by the same gas in the same container given its temperature in degrees Celsius.

*Exercise II.8.4: Break the function $z = \sqrt{x^2 + 2x + 4}$ up into the composition of two simpler functions.

Exercise II.8.5: Break the function $z = (x^4 - 3x^3 + 17x^2 + 14)^3$ up into the composition of two simpler functions.

*Exercise II.8.6: Suppose $f(x) = x^2 - 4$ and $g(y) = y^3$. What is $g(f(x))$?

*Exercise II.8.7: Suppose that f is any one-to-one function and that $g = f^{-1}$ is its inverse. What is $g(f(x))$?

The Chain Rule

Our next job is to find a technique for finding the derivative of the composition of two functions if we know the derivative of each of the original functions. For example, we can think of the function $z = \sqrt{1-x^2}$ as the composition of the two functions $y = 1-x^2$ and $z = \sqrt{y}$. We know that

$$\frac{dz}{dy} = \left(\frac{1}{2}\right) y^{-\frac{1}{2}} \quad \text{and} \quad \frac{dy}{dx} = -2x$$

We would like to find the derivative dz/dx. If dz/dy, dy/dx, and dz/dx were actually fractions we could say

(II.1) $$\frac{dz}{dx} = \left(\frac{dz}{dy}\right)\left(\frac{dy}{dx}\right).$$

Because they are not fractions it is not immediately obvious that the equation above is correct. In some ways the notation dy/dx is misleading because it looks like a fraction but isn't truly a fraction. Even though derivatives are not fractions, however, they are the limits of fractions. They often behave as if they were fractions. We have already seen one example of this, the Inverse Function Rule

$$\frac{dx}{dy} = \frac{1}{dy/dx}.$$

Formula II.1 is correct and is another example in which derivatives behave as if they were fractions. This rule is called the Chain Rule.

Theorem 12 *(Chain Rule)*
  Suppose that
$$y = f(x) \quad \text{and} \quad z = g(y)$$

§8. The Chain Rule

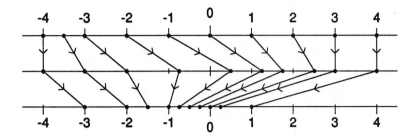

Figure II.99: Composing two functions with elastic

are two smooth functions. Then so is the composition

$$z = (g \circ f)(x) = g(f(x))$$

and

$$\frac{dz}{dx} = \left(\frac{dz}{dy}\right)\left(\frac{dy}{dx}\right) = g'(f(x))f'(x).$$

Proof:

We give an intuitive proof of the Chain Rule based on the interpretation of the derivative developed in the last section. Think of x as a point on a piece of elastic which is picked up, stretched and compressed and placed on top of a second piece of elastic so that the point x is placed at the point $y = f(x)$. The derivative $f'(x)$ tells us how much the the original piece of elastic at the point x has been stretched or compressed. Now glue the first piece of elastic in its current position to the second piece of elastic. Next think of the second function g as picking up the second piece of elastic, stretching and compressing it and laying it down on top of a piece of graph paper, so that the point y on the second piece of elastic is placed on top of the point $z = g(y)$ on the piece of graph paper. See Figures II.99 and II.100.

As the second piece of elastic is placed on top of the graph paper, it carries with it the first piece of elastic. The derivative dz/dx measures how much the *first piece* of elastic has been stretched or compressed as a result of these two combined operations. The derivative $dy/dx = f'(x)$ tells us how much the first piece of elastic at the point x was stretched or compressed by the first operation. The derivative $g'(y)$

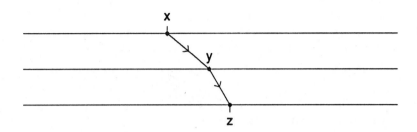

Figure II.100: Composing two functions with elastic

tells us how much the *second piece* of elastic at the point y was stretched or compressed by the second operation. The point x on the first piece of elastic is stretched or compressed twice, once by each operation. The net result produced by the two operations is the product $g'(y)f'(x) = g'(f(x))f'(x)$. This is what we wanted to show.

Example:

Find the derivative of the function

$$h(x) = \sqrt{1 - x^2}.$$

Answer:

The first step is to think of this function as the composite

$$h(x) = g(f(x))$$

where

$$y = f(x) = 1 - x^2 \quad \text{and} \quad z = g(y) = \sqrt{y}.$$

Now, by the Chain Rule

$$\frac{dz}{dx} = \left(\frac{dz}{dy}\right)\left(\frac{dy}{dx}\right)$$

$$= \left(\frac{1}{2}y^{-\frac{1}{2}}\right)(-2x)$$

§8. The Chain Rule

$$= \left(\frac{1}{2}(1-x^2)^{-\frac{1}{2}}\right)(-2x)$$

$$= \frac{-x}{\sqrt{1-x^2}}.$$

where the second-to-last line follows from the fact that $y = 1 - x^2$.

Example:

Find the derivative of the function

$$h(x) = (1 + \sqrt{x})^4.$$

Answer:

The first step is to think of this function as the composite

$$h(x) = g(f(x))$$

where

$$y = f(x) = 1 + \sqrt{x} \quad \text{and} \quad z = g(y) = y^4.$$

Now, by the Chain Rule

$$\begin{aligned}
\frac{dz}{dx} &= \left(\frac{dz}{dy}\right)\left(\frac{dy}{dx}\right) \\
&= (4y^3)\left(\frac{1}{2}x^{-\frac{1}{2}}\right) \\
&= 4(1+\sqrt{x})^3\left(\frac{1}{2}x^{-\frac{1}{2}}\right) \\
&= \frac{2(1+\sqrt{x})^3}{\sqrt{x}}
\end{aligned}$$

where the second-to-last line follows from the fact that $y = 1 + \sqrt{x}$.

Exercises

Find the derivative of each of the following functions.

*Exercise II.8.8: $z = (12 - 3x^7)^8$

*Exercise II.8.9: $y = (12 - 3x^7)^8$

Exercise II.8.10: $y = \sqrt{1 - 3x + x^2}$

*Exercise II.8.11: $y = \sqrt{1 - x^2} + \sqrt{1 + x^2}$

*Exercise II.8.12:
$$y = \sqrt{\frac{1 + x^2}{1 - x}}$$

Exercise II.8.13:
$$y = \left(\frac{1 + x}{1 - x}\right)^5$$

*Exercise II.8.14: $\sin(3x + 1)$.

Exercise II.8.15: $\cos(x^2 - 7)$.

*Exercise II.8.16: $x^2 \tan(1 - 2x^3)$.

Exercise II.8.17: $\arctan(4x)$.

*Exercise II.8.18: $\sin(\cos x)$.

Exercise II.8.19: $(\sin(2x))(\cos(3x))$.

*Exercise II.8.20: $\sin(\cos(x^3))$.

We have already shown that if c is either an integer or a fraction of the form $c = 1/q$ where q is a positive integer, then the derivative of the function $y = x^c$ is $y' = cx^{(c-1)}$. The next theorem shows that this same formula works for any rational number c.

Theorem 13 *(Power Rule, extended)*
  Suppose that $y = x^c$ where c is a rational number. Then $y' = cx^{(c-1)}$

Proof:
  Since c is a rational number we can write $c = \frac{p}{q}$ where q is a positive integer and p is an integer. Thus,
$$y = x^{\frac{p}{q}} = \left(x^{\frac{1}{q}}\right)^p$$

We can differentiate this function by thinking of it as the composition of the two functions
$$u = x^{\frac{1}{q}} \quad \text{and} \quad y = u^p.$$

§8. The Chain Rule

Thus

$$\frac{dy}{dx} = \left(\frac{dy}{du}\right)\left(\frac{du}{dx}\right)$$

$$= pu^{(p-1)}\left(\frac{1}{q}\right)x^{\left(\frac{1}{q}-1\right)}$$

$$= p\left(x^{\frac{1}{q}}\right)^{(p-1)}\left(\frac{1}{q}\right)x^{\left(\frac{1}{q}-1\right)}$$

$$= p\left(x^{\frac{p-1}{q}}\right)\left(\frac{1}{q}\right)x^{\left(\frac{1}{q}-1\right)}$$

$$= \left(\frac{p}{q}\right)x^{\left(\frac{p-1}{q}+\frac{1}{q}-1\right)}$$

$$= \left(\frac{p}{q}\right)x^{\left(\frac{p}{q}-1\right)}$$

$$= cx^{(c-1)}$$

Exercises

Find the derivative of each of the following functions.

*Exercise II.8.21:
$$y = x^{\frac{2}{3}}$$

*Exercise II.8.22:
$$y = (1 - 2x)^{\frac{3}{4}}$$

Exercise II.8.23:
$$y = x^{-\frac{3}{5}}$$

Exercise II.8.24:
$$v = (u^2 - u)^{\frac{7}{4}}$$

*Exercise II.8.25: Two cities, Elmville and Oakville, decide to cooperate in building a water treatment plant to purify their drinking water supply. They can build the plant anywhere along the Green River. They would like to locate the plant at the place that will minimize the total amount of pipe they need to connect the water treatment plant to the two cities. Where should they locate it? Figure II.101 is map

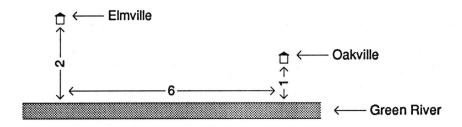

Figure II.101: Map of the Green River

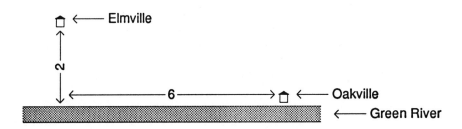

Figure II.102: Map of the Green River

showing the Green River and the two cities. Notice that Elmville is two miles from the Green River, Oakville is one mile from the Green River, and the distance between the closest points on the Green River to each of the two cities is six miles.

Exercise II.8.26: Two cities, Elmville and Oakville, decide to cooperate in building a water treatment plant to purify their drinking water supply. They can build the plant anywhere along the Green River. They would like to locate the plant at the place that will minimize the total amount of pipe they need to connect the water treatment plant to the two cities. Where should they locate it? Figure II.102 is map showing the Green River and the two cities. Notice that Elmville is two miles from the Green River, Oakville is on the Green River, and the distance between the closest point on the Green River to Elmville and Oakville is 6 miles.

Exercise II.8.27: Three cities, Mapleville, Elmville and Oakville, decide to cooperate in building a water treatment plant to purify their drinking water supply. They can build the plant anywhere along the

§8. The Chain Rule

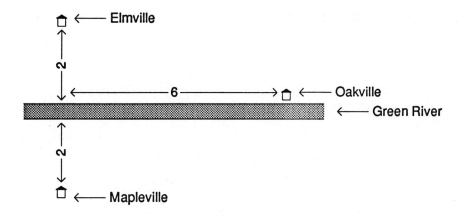

Figure II.103: Map of the Green River

Green River. They would like to locate the plant at the place that will minimize the total amount of pipe they need to connect the water treatment plant to the three cities. Where should they locate it? Figure II.103 is map showing the Green River and the two cities. Notice that Elmville and mapleville are each two miles from the Green River, Oakville is on the Green River, and the distance between the closest point on the Green River to Elmville and Mapleville and Oakville is 6 miles. Ignore the width of the Green River.

*Exercise II.8.28: Three cities, Mapleville, Elmville and Oakville, decide to cooperate in building a water treatment plant to purify their drinking water supply. They can build the plant anywhere along the Green River. They would like to locate the plant at the place that will minimize the total amount of pipe they need to connect the water treatment plant to the three cities. Where should they locate it? Figure II.104 is map showing the Green River and the two cities. Notice that Elmville and mapleville are each two miles from the Green River, Oakville is on the Green River, and the distance between the closest point on the Green River to Elmville and Mapleville and Oakville is A miles. Ignore the width of the Green River. Does the distance A make any difference? Be careful how you answer this question.

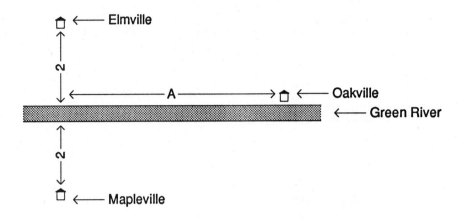

Figure II.104: Map of the Green River

2-Cycles

In chapter I section 6 we looked at the dynamical system

$$p_{n+1} = 2.8p_n(1 - .001p_n)$$

and saw that this model had two equilibrium points 0, and $p_* = 2.8/.0018$ and that, for any nonzero initial population p_1,

$$\lim_{n \to \infty} p_n = p_*.$$

Now consider a slightly different dynamical system:

$$p_{n+1} = 3.2p_n(1 - .001p_n).$$

We can compute the equilibrium points for this dynamical system as follows.

$$\begin{aligned} p &= 3.2p(1 - .001p) \\ p &= 3.2p - .0032p^2 \\ 0 &= 2.2p - .0032p^2 \\ 0 &= p(2.2 - .0032p). \end{aligned}$$

So

$$p = 0$$

§8. The Chain Rule

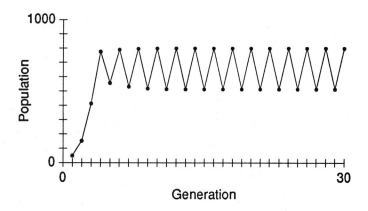

Figure II.105: $p_{n+1} = 3.2p_n(1 - .001p_n)$ $p_1 = 50$

or
$$2.2 - .0032p = 0$$

and there are two equilibrium points, 0 and $2.2/.0032$. If we do some calculations for this dynamical system with $p_1 = 50$ we obtain the results shown in Figure II.105. Notice that in this example the sequence $p_1, p_2, p_3, \ldots$ appears to be settling down into a pattern, called a **2-cycle**, in which the population bounces back-and-forth between the same two numbers.

Recall that we find equilibrium points for a dynamical system
$$p_{n+1} = f(p_n)$$
by solving the equation
$$p = f(p)$$
because an equilibrium point is a point p such that if $p_n = p$ then p_{n+1} will also be p.

Similarly, a point p is part of a 2-cycle if whenever $p_n = p$ then p_{n+2} is also equal to p. We compute p_{n+2} from p_n by applying the function f twice. That is,
$$p_{n+2} = f(f(p_n))$$
or
$$p_{n+2} = (f \circ f)(p_n).$$

Hence, in order to find 2-cycles we need to solve the equation
$$p = (f \circ f)(p).$$

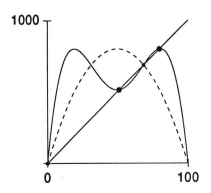

Figure II.106: Looking for 2-cycles

Geometrically, this means finding the points at which the graph $y = (f \circ f)(p)$ intersects the line $y = p$. Figure II.106 shows the functions $y = f(p)$ (the dashed curve) and $y = (f \circ f)(p)$ (the solid curve) and the line $y = p$ for the function $f(p) = 3.2p(1-.001p)$. Notice that the curve $y = f(p)$ crosses the line $y = p$ at the two equilibrium points. The curve $y = (f \circ f)(p)$ crosses the line $y = p$ at these same two points (since an equilibrium point never changes) and at two new points. These two new points are the two points of the 2-cycle shown in Figure II.105.

Exercises

For each of the following models use the computer to examine graphs of the functions f and $(f \circ f)$. What can you say about 2-cycles as a results of your computer investigations? Compute $p_1, p_2, \ldots p_{20}$ for each model to see whether the 2-cycle actually appears. Note that just as an equilibrium point might or might not be attracting, a 2-cycle might or might not be attracting.

*Exercise II.8.29:

$$p_{n+1} = 2.5p_n(1 - .001p_n), \quad p_1 = 50.$$

*Exercise II.8.30:

$$p_{n+1} = 3.3p_n(1 - .001p_n), \quad p_1 = 50.$$

§9. Optics

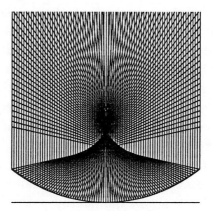

Figure II.107: Light rays bouncing off a circular mirror

*Exercise II.8.31:

$$p_{n+1} = 3.4p_n(1 - .001p_n), \quad p_1 = 50.$$

*Exercise II.8.32:

$$p_{n+1} = 3.5p_n(1 - .001p_n), \quad p_1 = 50.$$

*Exercise II.8.33:

$$p_{n+1} = 3.6p_n(1 - .001p_n), \quad p_1 = 50.$$

---

§9. Optics

This section is about everyday optical phenomena like the patterns called **light caustics** that are formed in coffee cups (see Figure II.107 and the way in which the appearance of an underwater object is "distorted." Our work in this section is based on **Fermat's Principle**. The simplest, but unfortunately not entirely correct, formulation of this principle says that light traveling from one point to another will follow the fastest path. This formulation sounds natural. If a light ray were intelligent then it might very well choose the shortest path from one point to another. Unfortunately, however, light rays do not always choose the fastest path. Sometimes they choose the slowest path. Besides seeming unnatural this choice at first seems totally opposite to the choosing the fastest path. Are there two kinds of light rays—fast track light rays in a hurry that always choose the fastest path and laid back light rays that always choose the slowest path? Calculus enables us to see how the same mechanism can make choices that seem at first glance to be totally contradictory. The correct formulation of Fermat's Principle is that given a choice of possible paths a light ray will choose a path that is a critical point. Since a critical point can be a local minimum, or a local maximum a chosen path can be faster than all the nearby paths or slower than all the nearby paths. We begin by using the simplest formulation of Fermat's Principle—light follows the fastest possible path.

Philosophically, this principle is rather curious. It does not say why light follows the fastest path or by what mechanism it chooses the fastest path. It simply says that light does follow the fastest path. It is an example of a principle that is useful for predicting what will happen in a given situation but doesn't say anything at all about the *how* or the *why* of the way that light travels. Such principles are nonetheless extremely important. Scientists trying to explain the how and the why of the way that light travels need to explain this one rather general principle rather than many individual experimental results.

Reflections in a flat mirror

Example:

Suppose that we have the experimental setup pictured in Figure II.108, a mirror located along the x-axis with an object at the point $(200, 50)$ and an eye at the point $(0, 100)$. We are interested in paths that light might follow traveling from the object to the eye by way of the mirror. A typical such path is shown in Figure II.109. It bounces off the mirror at the point $(x, 0)$. The length of such a path is given by the function

$$L(x) = \sqrt{x^2 + 100^2} + \sqrt{(200 - x)^2 + 50^2}.$$

The first term is the length of the hypotenuse of the right triangle whose vertices are at the points $(0, 100)$, $(0, 0)$ and $(x, 0)$. The second term is

§9. Optics

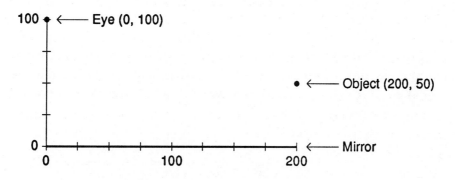

Figure II.108: Light bouncing off a mirror

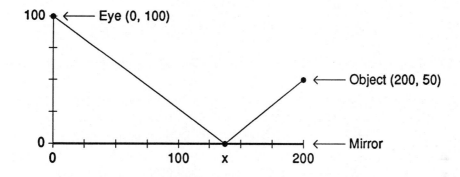

Figure II.109: A typical bouncing path

the length of the hypotenuse of the right triangle whose vertices are at the points $(x, 0)$, $(200, 50)$ and $(200, 0)$.

According to Fermat's Principle, light will choose the path for which $L(x)$ is a minimum. We begin by computing the derivative.

$$L(x) = \sqrt{x^2 + 100^2} + \sqrt{(200-x)^2 + 50^2}$$

$$L(x) = [x^2 + 100^2]^{\frac{1}{2}} + [(200-x)^2 + 50^2]^{\frac{1}{2}}$$

$$L(x) = (x^2 + 10{,}000)^{\frac{1}{2}} + (x^2 - 400x + 42{,}500)^{\frac{1}{2}}$$

$$L'(x) = \left(\frac{1}{2}\right)(x^2 + 10{,}000)^{-\frac{1}{2}}(2x) + \left(\frac{1}{2}\right)(x^2 - 400x + 42{,}500)^{-\frac{1}{2}}(2x - 400)$$

$$= \frac{x}{\sqrt{x^2 + 10{,}000}} + \frac{x - 200}{\sqrt{x^2 - 400x + 42{,}500}}.$$

Since neither denominator is ever zero, the derivative is always defined and the only critical points will be points at which the derivative is equal to zero. We set the derivative equal to zero and solve the resulting equation as follows.

$$0 = \frac{x}{\sqrt{x^2 + 10{,}000}} + \frac{x - 200}{\sqrt{x^2 - 400x + 42{,}500}}$$

$$-\frac{x}{\sqrt{x^2 + 10{,}000}} = \frac{x - 200}{\sqrt{x^2 - 400x + 42{,}500}}$$

$$\frac{x^2}{x^2 + 10{,}000} = \frac{x^2 - 400x + 40{,}000}{x^2 - 400x + 42{,}500}$$

$$(x^2)(x^2 - 400x + 42{,}500) = (x^2 - 400x + 40{,}000)(x^2 + 10{,}000)$$

$$x^4 - 400x^3 + 42{,}500x^2 = x^4 - 400x^3 + 50{,}000x^2 - 4{,}000{,}000x + 400{,}000{,}000$$

$$42{,}500x^2 = 50{,}000x^2 - 4{,}000{,}000x + 400{,}000{,}000$$

$$0 = 7{,}500x^2 - 4{,}000{,}000x + 400{,}000{,}000$$

$$0 = 3x^2 - 1{,}600x + 160{,}000$$

Solving this equation by the Quadratic Formula, we get two solutions, $x = 400/3$ and $x = 400$, but we have to be a little careful. To obtain the third line of the preceding calculation we squared the second line. Squaring an equation can introduce extraneous roots. So we need to try each of these two roots back in the original equation

$$\frac{x}{\sqrt{x^2 + 10{,}000}} + \frac{x - 200}{\sqrt{x^2 - 400x + 42{,}500}} = 0.$$

If you do this substitution you will see that $x = 400/3$ is a solution of the original equation but that $x = 400$ is not. Thus our original function has one critical point $x = 400/3$. In order to find out where the derivative is positive and where it is negative we need to compute the derivative at one sample point to the left of $x = 400/3$ and one

§9. OPTICS   253

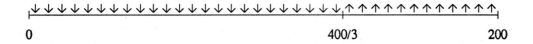

Figure II.110: Behavior of $L(x)$

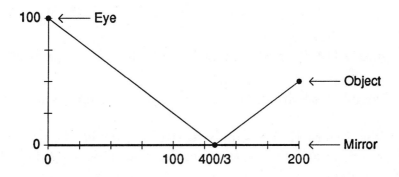

Figure II.111: Path followed by light

sample point to the right of $x = 400/3$. We choose $x = 0$ for our sample point to the left of $x = 400/3$

$$L'(0) = \frac{-200}{\sqrt{42,500}}$$

and $x = 200$ for our sample point to the right of $x = 400/3$.

$$L'(200) = \frac{200}{\sqrt{50,000}}$$

Thus we see that $L'(x)$ is negative to the left of $x = 400/3$ and positive to the right of $x = 400/3$, and $L(x)$ behaves as shown in Figure II.110. Thus $x = 400/3$ is the global minimum of the function $L(x)$ and a light ray between the two points will follow the path shown in Figure II.111.

Problems involving flat mirrors like the example above can be solved using elementary geometry instead of calculus. We illustrate this simpler method below. The simpler method, however, cannot be used

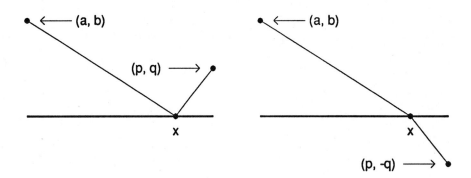

Figure II.112: Two representations of a flat mirror problem

in more complicated situations like those in the computer lab for this section.

Consider Figure II.112. Suppose that we would like to find the path followed by a light ray traveling from an object located at the point (p, q), bouncing off a mirror located along the x-axis, and then traveling to an eye located at the point (a, b). A typical such path is shown on the left side of Figure II.112. The right side of the same figure represents the same situation in a different way. We place a dot representing the object at the point $(p, -q)$ rather than at the point (p, q) and the part of the light ray's path from the mirror to the object below the mirror rather than above it. Because the mirror is located along the x-axis the picture on the right is an accurate representation of the same situation. In particular, the line representing the path on the left and the line representing the path on the right have the same length. In the right hand picture it is clear that the fastest path is a straight line from the point (a, b) to the point $(p, -q)$ as shown in Figure II.113

The two triangles in either side of Figure II.113 are similar and, hence,

$$\frac{x - a}{b} = \frac{p - x}{q}$$
$$qx - aq = bp - bx$$
$$qx + bx = aq + bp$$
$$x = \frac{aq + bp}{q + b}$$

§9. Optics

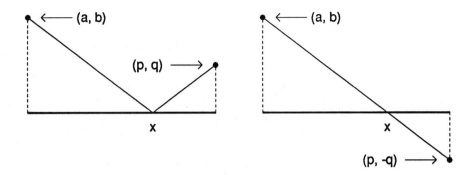

Figure II.113: Two representations of the fastest path

In our first example, $a = 0, b = 100, p = 200$, and $q = 50$, so that $x = 20,000/150 = 400/3$, not surprisingly the same answer we obtained earlier. It is clear from this picture that when a light ray bounces off a flat mirror the angle of incidence (the angle between the light ray from the object to the mirror and the mirror) and the angle of coincidence (the angle between the light ray from the eye to the mirror and the mirror) are equal.

Light Caustics

In this section we use the observation above

> The angle of incidence equals the angle of coincidence.

together with some trigonometry to examine a rather pretty optical phenomenon. We begin with a simple problem. Consider Figure II.114 in which we have a flat mirror whose slope is m. An incoming light ray travels straight downward and bounces off the mirror. We want to determine the slope of the reflected or outgoing light ray. The angle that the mirror makes with the horizontal is denoted β. Notice that $\tan \beta = m$. The angle of incidence, i.e., the angle that the incoming light ray makes with the mirror is denoted α and the angle that the outgoing light ray makes with the horizontal is labeled θ. We need to determine $\tan \theta$.

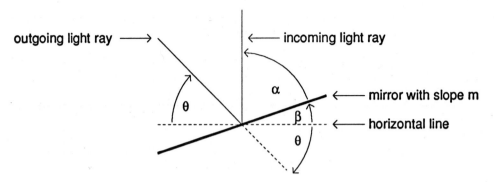

Figure II.114: Light bouncing off a flat mirror

Since the angle of incidence and the angle of coincidence are equal

$$\alpha = \beta - \theta$$
$$\theta = \beta - \alpha.$$

Notice that

$$\alpha = \frac{\pi}{2} - \beta,$$

so that

$$\tan \alpha = \cot \beta = 1/m.$$

Thus,

$$\tan \theta = \tan[\beta + (-\alpha)]$$
$$= \frac{\tan \beta + \tan(-\alpha)}{1 - (\tan \beta)(\tan(-\alpha))}$$
$$= \frac{\tan \beta - \tan \alpha}{1 - (\tan \beta)(-\tan \alpha)}$$
$$= \frac{m - \frac{1}{m}}{2}$$
$$= \frac{m^2 - 1}{2m}$$

§9. Optics

The following exercises ask you to answer some simple questions with and about this formula.

Exercises

*__Exercise II.9.1:__ Find the slope of the outgoing line if the mirror has a slope of 1/3. Draw a diagram of this situation.

__Exercise II.9.2:__ Find the slope of the outgoing line if the mirror has a slope of 1. Draw a diagram for this situation.

__Exercise II.9.3:__ Find the slope of the outgoing line if the mirror has a slope of 2. Draw a diagram of this situation.

__Exercise II.9.4:__ Explain when the slope of the outgoing line is positive and when it is negative.

__Exercise II.9.5:__ Find the slope of the outgoing line if the slope of the mirror is $-1/2$. Draw a diagram of this situation.

__Exercise II.9.6:__ Find the slope of the outgoing line if the slope of the mirror is -1. Draw a diagram of this situation.

__Exercise II.9.7:__ Find the slope of the outgoing line if the slope of the mirror is 2. Draw a diagram of this situation.

*__Exercise II.9.8:__ If necessary refine your answer to Exercise II.9.4.

We can use this work to find out what happens when a light ray bounces off a curved mirror. We are interested in a situation like that shown in Figure II.115 which shows eleven light rays coming straight down and bouncing off the circular mirror described by the equation

$$y = 8 - \sqrt{64 - x^2}.$$

These light rays bounce off the curved mirror exactly the same way as they would off a flat mirror.

Example:

Suppose a light ray comes straight down along the line $x = 2$, strikes the curved mirror described by

$$y = f(x) = 8 - \sqrt{64 - x^2}$$

and bounces off. Find the equation of the line that the outgoing light ray follows after bouncing off the mirror.

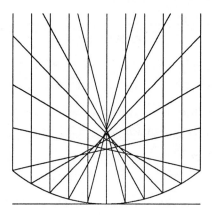

Figure II.115: Light rays bouncing off a concave mirror

Answer:

The first step is to find the slope of the outgoing line. By our work above this slope is given by the formula

$$\frac{f'(2)^2 - 1}{2f'(2)}$$

Notice that the slope of the flat mirror in our original formula has been replaced by the value of the derivative of f (i.e., by the slope of the tangent to the curve $y = f(x)$) at the point $x = 2$ where the light ray strikes the mirror.

Now

$$f(x) = 8 - \sqrt{64 - x^2}$$

$$f'(x) = \frac{x}{\sqrt{64 - x^2}}$$

$$f'(2) = \frac{2}{\sqrt{64 - 4}}$$

$$f'(2) = 0.2582,$$

§9. Optics

so that the slope of the outgoing line is

$$\frac{(0.2582)^2 - 1}{2(0.2582)} = -1.8074.$$

Thus the outgoing line is described by an equation of the form

$$y = -1.8074x + b.$$

We also know the point at which the outgoing line hits the mirror,

$$(2, f(2)) = (2, 0.2540).$$

Using these two pieces of information we can find the y-intercept b by solving the equation

$$0.2540 = -1.8074(2) + b.$$

So

$$b = 0.2540 + 1.8074(2) = 3.8688$$

and the equation of the outgoing line is

$$y = -1.8074x + 3.8688.$$

Exercises

*Exercise II.9.9:
Suppose a light ray comes straight down along the line $x = -2$, strikes the curved mirror described by

$$y = 8 - \sqrt{64 - x^2}$$

and bounces off. Find the equation of the line that the outgoing light ray follows after bouncing off the mirror.

Exercise II.9.10:
Suppose a light ray comes straight down along the line $x = 4$, strikes the same curved mirror and bounces off. Find the equation of the line that the outgoing light ray follows after bouncing off the mirror.

Exercise II.9.11:
Suppose a light ray comes straight down along the line $x = 0$, strikes the same curved mirror and bounces off. Find the equation of the line that the outgoing light ray follows after bouncing off the mirror.

Exercise II.9.12:

Sketch a graph similar to Figure II.115 showing the results of your work above.

Figure II.107 shows the results of similar computations for 81 equally spaced light rays traveling straight down and striking the usual mirror. Notice the very dark "cusp" shape formed in the middle of the picture. You can observe this phenomenon in a coffee cup made out of shiny material. Fill the cup part way with your favorite noncarbonated beverage and hold it so that light from a distant light source comes over the lip of the cup, bounces off the inside of the cup and strikes the top of the liquid. You will see a bright image similar to Figure II.107 on the surface of the liquid.

The next set of exercises asks you to explore what happens when a different shaped mirror is used.

Exercises

Consider a mirror whose shape is given by the function $y = x^2$

*Exercise II.9.13: Suppose a light ray travels straight down along the line $x = 1$ and bounces off this mirror. Find the equation of the outgoing line. Draw a diagram showing this situation.

Exercise II.9.14: Suppose a light ray travels straight down along the line $x = -1$ and bounces off this mirror. Find the equation of the outgoing line. Draw a diagram showing this situation.

Exercise II.9.15: Suppose a light ray travels straight down along the line $x = 2$ and bounces off this mirror. Find the equation of the outgoing line. Draw a diagram showing this situation.

Exercise II.9.16: Suppose a light ray travels straight down along the line $x = 3$ and bounces off this mirror. Find the equation of the outgoing line. Draw a diagram showing this situation.

Exercise II.9.17: Suppose a light ray travels straight down along the line $x = 0$ and bounces off this mirror. Find the equation of the outgoing line. Draw a diagram showing this situation.

Exercise II.9.18: Do you notice anything interesting about these five outgoing lines? If so, what? Does this experimental evidence suggest a conjecture?

§9. OPTICS   261

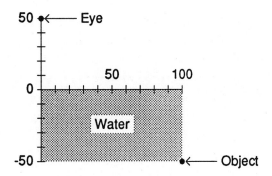

Figure II.116: Pool

Exercise II.9.19: Test your conjecture from the preceding exercise by looking at several more incoming light rays. Does this new evidence support your conjecture.

*Exercise II.9.20: If the answer to the preceding question was yes, try to prove the conjecture.

Refraction

Now we want to investigate what happens to light rays as they pass from one medium to another, for example, from water to air. This phenomenon, called **refraction**, is responsible for the way that underwater objects appear to be distorted and for eyeglasses, binoculars, telescopes, microscopes and other optical tools.

Light travels at different speeds in different mediums. It travels so quickly that it is convenient for us to measure time in nanoseconds rather than seconds. One **nanosecond** is one-billionth of a second. The speed of light in air is 30 cm./ns. and the speed of light in water is 22.5 cm./ns.

Consider the situation shown in Figure II.116 with an object in a large pool of water and an observer looking at the object from a position above the surface of the water. A light ray traveling from the object to the observer's eye will travel part way through water and then the rest

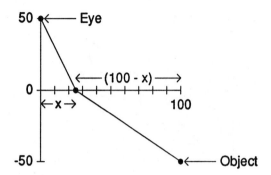

Figure II.117: A typical path

of the way through air. Figure II.117 shows a typical path that a light ray might follow from the object to the eye.

In order to save some work we solve a more general problem than the one shown in Figures II.116 and II.117. Suppose that the eye is located at the point (a, b) and the underwater object is located at the point (p, q). In Figure II.116, $(a, b) = (0, 50)$ and $(p, q) = (100, -50)$. Figure II.118 is similar to Figure II.117 except that we have used the general coordinates (a, b) and (p, q).

We can break the path up into two pieces. The first piece is in the water from the object to the surface of the water. Its length is

$$\sqrt{(p-x)^2 + q^2}.$$

Since the speed of light in water is 22.5 cm./ns. the time (in nanoseconds) required to travel from the object to the surface of the water is

$$\frac{\sqrt{(p-x)^2 + q^2}}{22.5}.$$

The second piece is in air from the surface of the water to the eye. Its length is

$$\sqrt{(x-a)^2 + b^2}.$$

Since the speed of light in air is 30 cm./ns. the time necessary to travel

§9. Optics

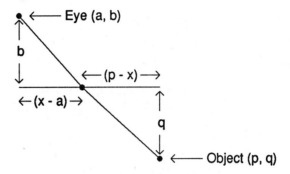

Figure II.118: A typical path

from the surface of the water to the eye is

$$\frac{\sqrt{(x-a)^2 + b^2}}{30}.$$

Thus the total time required to travel from the target to the eye is

$$T(x) = \frac{\sqrt{(p-x)^2 + q^2}}{22.5} + \frac{\sqrt{(x-a)^2 + b^2}}{30}.$$

In order to find the actual path followed by a light ray traveling from the target to the eye, we need to find the minimum of this function. As usual we begin by differentiating the function $T(x)$. Real problems, as opposed to the kinds of simplified problems often found in textbooks, are often very messy. They can lead to long complicated calculations. This problem is real and the calculations are complicated.

$$T(x) = \frac{\sqrt{(p-x)^2 + q^2}}{22.5} + \frac{\sqrt{(x-a)^2 + b^2}}{30}$$

$$T(x) = \frac{1}{22.5}[(p-x)^2 + q^2]^{\frac{1}{2}} + \frac{1}{30}[(x-a)^2 + b^2]^{\frac{1}{2}}$$

$$T'(x) = \frac{1}{22.5}\frac{1}{2}[(p-x)^2 + q^2]^{-\frac{1}{2}}2(p-x) + \frac{1}{30}\frac{1}{2}[(x-a)^2 + b^2]^{-\frac{1}{2}}2(x-a)$$

$$= \frac{1}{22.5}[(x-p)^2 + q^2]^{-\frac{1}{2}}(x-p) + \frac{1}{30}[(x-a)^2 + b^2]^{-\frac{1}{2}}(x-a)$$

$$= \frac{x-p}{22.5\sqrt{(x-p)^2+q^2}} + \frac{x-a}{30\sqrt{(x-a)^2+b^2}}$$

Since both denominators are never zero the derivative is always defined and the only critical points will be when $T'(x) = 0$. We solve the equation $T'(x) = 0$ as follows.

$$\frac{x-p}{22.5\sqrt{(x-p)^2+q^2}} + \frac{x-a}{30\sqrt{(x-a)^2+b^2}} = 0$$

So

$$\frac{x-p}{22.5\sqrt{(x-p)^2+q^2}} = -\frac{x-a}{30\sqrt{(x-a)^2+b^2}}$$

and

$$30(x-p)\sqrt{(x-a)^2+b^2} = -22.5(x-a)\sqrt{(x-p)^2+q^2}.$$

Dividing by 7.5 we obtain

$$4(x-p)\sqrt{(x-a)^2+b^2} = -3(x-a)\sqrt{(x-p)^2+q^2}.$$

Now, squaring both sides,

$$16(x-p)^2[(x-a)^2+b^2] = 9(x-a)^2[(x-p)^2+q^2]$$
$$16(x^2-2px+p^2)(x^2-2ax+a^2+b^2) = 9(x^2-2ax+a^2)(x^2-2px+p^2+q^2)$$

and, after a lot of algebra,

$$7x^4 + (-14p - 14a)x^3 +$$
$$(7a^2 + 7p^2 + 16b^2 - 9q^2 + 28ap)x^2 +$$
$$(-14ap^2 - 14a^2p - 32pb^2 + 18aq^2)x +$$
$$(7a^2p^2 + 16b^2p^2 - 9a^2q^2) = 0.$$

This is a horrible equation. Fortunately, we do not have solve it algebraically. For a particular observer located at the point (a, b) and

§9. Optics

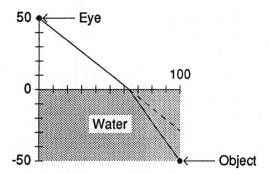

Figure II.119: Apparent path

a particular target located at the point (p, q) we substitute the values of a, b, p and q into the equation above. This gives us an equation with numeric coefficients. For example, if the observer is located at $(0, 50)$ and the target is located at $(100, -50)$ as shown in Figure II.116, then $a = 0$, $b = 50$, $p = 100$, and $q = -50$ and we can calculate the coefficients of the polynomial on the left side of our equation to obtain

$$7x^4 - 1{,}400x^3 + 87{,}500x^2 - 8{,}000{,}000x + 400{,}000{,}000 = 0.$$

As a fourth degree equation, this equation can have up to four roots. It is obvious looking at our diagrams that the root we seek is between $x = 0$ and $x = 100$. In general, it will between $x = a$ and $x = p$, the x-coordinates of the observer and the target, respectively. Now, it is a simple matter to solve this equation numerically. For this example, the solution is $x = 63.5253$.

Figure II.119 shows the path followed by the light ray from the target to the observer. The dashed line shows the **apparent path** followed by the light ray. This is the path that the eye assumes was followed by the light ray.

This particular example is typical of what happens when an observer above the surface of the water looks at an object underneath the surface of the water. It can give us some insight into what we see when we look at an object under the surface of water. If, for example, you stand at the edge of a pool looking at an object on the bottom of the pool you

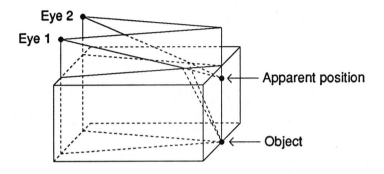

Figure II.120: A normal observer

usually hold your head upright. Your nose is pointed at the underwater object and both eyes are level, one on each side of your nose as shown in Figure II.120. If we analyze what happens for each individual eye we get a picture very much like Figure II.119 with the apparent path of the light ray above its the true path. When we put the two pictures together as shown in Figure II.120 we see that the two apparent lines intersect at a point directly above the true location of target. This point of intersection is the point at which the observor believes the object to be located. Thus when you look at an object under the surface of the water from above the water it will appear to be much higher than it actually is. The process by which an observor determines the apparent position of an object based on the apparent paths seen by each of his or her two eyes is called **binocular vision.**

The techniques developed above can answer many questions about the appearance of an underwater object as viewed from above the water. Use the following procedure to find the path followed by a light ray traveling from an underwater object or target to an observer above the water. Because we are dealing with real questions the computations can be lengthy and rather difficult.

Ray-Tracing for Underwater Targets

1. Draw a sketch of the situation.

§9. Optics

2. Calculate the coefficients for each term of the polynomial on the left side of the following equation.

$$\begin{aligned}
7x^4 + (-14p - 14a)x^3 &+ \\
(7a^2 + 7p^2 + 16b^2 - 9q^2 + 28ap)x^2 &+ \\
(-14ap^2 - 14a^2p - 32pb^2 + 18aq^2)x &+ \\
(7a^2p^2 + 16b^2p^2 - 9a^2q^2) &= 0
\end{aligned}$$

This is the equation we found by setting $T'(x) = 0$.

3. Use a numerical method to estimate the solution of the equation obtained in Step 2. The solution will be between a and p.

4. Find the equation of the apparent path.

5. Find out where the apparent path intersects the vertical line through the target (i.e., the line $x = p$).

6. Add your new information to the sketch. Indicate the real and apparent path of the light ray.

Example:

Suppose that you are standing at the deep end of a fifty foot pool looking at the shallow end. The shallow end is three feet deep. Your eyes are six feet above the surface of the water. How deep does the shallow end appear?

Answer:

1. Figure II.121 shows a sketch of this situation with the origin at your feet. Thus the location of the observer is $(0, 6)$ and the location of the underwater object is $(50, -3)$.

2. After some calculation we see that our infamous equation becomes

$$7x^4 - 700x^3 + 17{,}995x^2 - 57{,}600x + 1{,}440{,}000 = 0$$

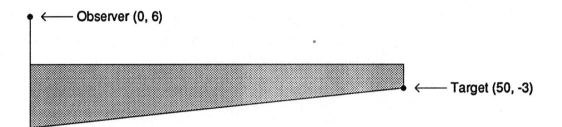

Figure II.121: Looking at the shallow end of a pool

3. The Bisection Method or some other numerical method yields $x = 46.6608$.

4. We now know two points on the apparent path followed by the light ray from the target to your eye: $(0, 6)$ and $(0, 46.6608)$. Thus the slope of the apparent path is

$$\text{Slope of apparent path} = \frac{6 - 0}{0 - 46.6608} = -0.1286.$$

Since the y-intercept of the apparent path is $(0, 6)$ the equation of the apparent path is

$$y = -0.1286x + 6.$$

5. The vertical line through the target is the line $x = 50$. So the apparent line intersects this vertical at

$$y = -0.1286(50) + 6 = -0.4294$$

This answers the question. It appears as if the pool is only 0.4296 feet deep at the shallow end.

6. Finally, we add our new information to our sketch. See Figure II.122.

§9. Optics

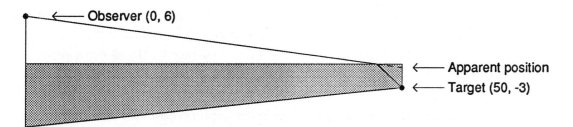

Figure II.122: Looking at the shallow end of a pool

Exercises

The first two exercises are meant to explore whether the apparent depth of the shallow end of the pool depends on the length of the pool. Compare your answers to these two exercises to each other and to the last example.

*Exercise II.9.21:

Suppose that you are standing at the deep end of a thirty foot pool looking at the shallow end. The shallow end is three feet deep. Your eyes are six feet above the surface of the water. How deep does the shallow end appear?

Exercise II.9.22:

Suppose that you are standing at the deep end of a seventy foot pool looking at the shallow end. The shallow end is three feet deep. Your eyes are six feet above the surface of the water. How deep does the shallow end appear?

The next two exercises are meant to explore whether the apparent depth of the shallow end of the pool depends on the height of the observer. Compare your answers to these two exercises to each other and to the last example.

*Exercise II.9.23:

Suppose that you are standing at the deep end of a fifty foot pool looking at the shallow end. The shallow end is three feet deep. Your eyes are four feet above the surface of the water. How deep does the shallow end appear?

Exercise II.9.24:

Suppose that you are standing on a diving platform at the deep end of a fifty foot pool looking at the shallow end. The shallow end is three feet deep. Your eyes are ten feet above the surface of the water. How deep does the shallow end appear?

The next two exercises are meant to explore how the amount by which the shallow end appears to rise depends on the depth of the shallow end. Compare your answers to these exercises to each other and to the last example.

*Exercise II.9.25:

Suppose that you are standing at the deep end of a fifty foot pool looking at the shallow end. The shallow end is five feet deep. Your eyes are six feet above the surface of the water. How deep does the shallow end appear?

Exercise II.9.26:

Suppose that you are standing at the deep end of a fifty foot pool looking at the shallow end. The shallow end is two feet deep. Your eyes are six feet above the surface of the water. How deep does the shallow end appear?

The next two exercises are more complicated. To solve these exercises you will need to put together several different techniques that we have developed in this section. These two exercises involve an observer with two eyes. To make the drawings easier we assume that the two eyes are farther apart (ten feet!) than normal human eyes.

*Exercise II.9.27:

For this exercise we have a contortionist observer. Instead of standing in the normal way while he is looking at an underwater object, he holds his head on its side so that his two eyes are arranged as shown in Figure II.123. Find the apparent position of the target.

*Exercise II.9.28:

For this exercise we have a contortionist observer. Instead of standing in the normal way while he is looking at an underwater object, he holds his head facing down and looks sideways at the target so that his two eyes are arranged as shown in Figure II.124. Find the apparent position of the target.

§9. Optics

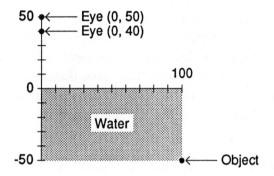

Figure II.123: Sore neck and vertical eyes

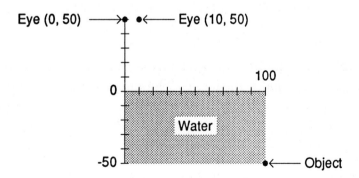

Figure II.124: Sore neck and horizontal eyes

Compare the apparent position of an underwater object as viewed by different kinds of contortionist observers and by a normal observer. Notice that the apparent position seems to change depending on the way the observer holds her head. The first time the author noticed this phenomenon he thought he must have made some mathematical mistake. However, this rather odd result is actually true. You can verify it experimentally at a swimming pool or with an aquarium. These exercises looked at somewhat special cases—either the two eyes and the object were all located in the same plane, perpendicular to the surface of the water, or the two eyes were the same distance above the surface of the water and from the object. What do you think would happen if the two eyes were located in a more haphazard way?

*Exercise II.9.29: For this exercise we return to the law:

"The angle of incidence equals the angle of coincidence."

Use Fermat's Principle to derive this law by the following strategy.

Consider the experimental setup shown in Figure II.125. Find an equation describing the point x at which the light ray traveling from the target at (L, b) will bounce off the mirror (located on the x-axis) on its way to the eye at $(0, a)$. Show that the angle of incidence is equal to the angle of coincidence by showing that

$$\frac{x}{\sqrt{a^2 + x^2}} = \frac{L - x}{\sqrt{(L - x)^2 + b^2}}$$

and, thus, that the two obvious triangles are similar.

*Exercise II.9.30: At the beginning of this section we discussed the proper formulation of Fermat's Principle. The purpose of this exercise is to illustrate this proper formulation. Suppose that you are standing in a circular room with a radius of 10 feet at a point two feet away from the center. Suppose that the walls are mirrored. Where would your reflection appear? See Figure II.126.

*Exercise II.9.31: If you know Snell's Law, prove it using the techniques developed in this section.

We have just begun to scratch the surface of Geometric Optics. You can use these techniques to answer many extremely interesting

§9. Optics

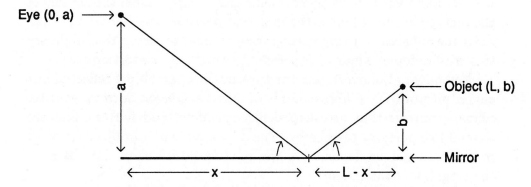

Figure II.125: Angle of incidence = angle of coincidence

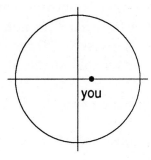

Figure II.126: A mirrored circular room

questions. You may very well encounter some extremely difficult calculations. If so, use the computer lab.

One interesting set of questions to consider is the appearance of an underwater object, for example, a rectangle. You might draw a pool with an underwater rectangle on some graph paper, label the corners of the rectangle, and compute the apparent position of each corner. Then draw the apparent rectangle using the apparent corners. You might try this with different kinds of observors, normal and contortionist.

You might also investigate the appearance of an object reflected in a curved mirror. Try different kinds of mirrors, convex mirrors, concave mirrors, mirrors that are shaped like cylinders, and mirrors that are shaped like portions of spheres. The appearance of an object reflected in a curved mirror depends on the relative positions of the observer, the target, and the mirror.

§10. The Hare: Newton's Method

Real problems frequently lead to complicated equations. For example, in the last section we needed to solve the equation

$$7x^4 - 1,400x^3 + 87,500x^2 - 8,000,000x + 400,000,000 = 0.$$

We already know one numerical method, the Bisection Method, for estimating solutions to an equation of this type. In this section we develop another method, called **Newton's Method**, for solving such equations. This method is the "hare" of numerical methods and the Bisection Method is the "tortoise." Bisection is slow, steady, and reliable. The hare is not as reliable as the tortoise but when it works it is extremely fast.

Newton's Method is based on two observations.

- It is easy to solve a linear equation $mx + b = 0$. The solution is $x = -b/m$.

- Under high magnification near a point, a smooth curve looks very much like the tangent to the curve at the point.

Graphically, we are trying to find the point at which the curve $y = f(x)$ crosses the x-axis as shown in Figure II.127. Like the Bisection

§10. The Hare: Newton's Method

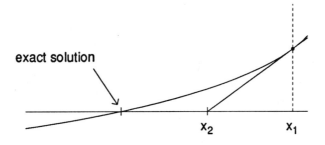

Figure II.127: The tangent, the curve, and a better estimate

Method, Newton's Method produces a sequence of estimates rather than an exact answer. The method begins with a rough estimate x_1 for a solution as shown in Figure II.127. Consider the tangent to the curve $y = f(x)$ at the point x_1. The idea behind Newton's Method is that this tangent is close to the actual curve. Hence, the point at which the tangent crosses the x-axis will be close to the point at which the curve crosses the x-axis. Because the tangent is a straight line it is easy to find the point x_2 at which the tangent crosses the x-axis. The point x_2 is usually a better estimate than the original estimate x_1.

Next we look at the tangent to the curve $y = f(x)$ at the point x_2 and determine the point x_3 at which it crosses the x-axis as shown in Figure II.128. This new point is usually an even better estimate for the solution than x_2. This procedure can be repeated again and again to produce a sequence of estimates for the solution.

Figure II.129 illustrates the basic step involved in Newton's Method. Each step begins with an estimate x_n and produces a new estimate x_{n+1}. The slope of the tangent to the curve $y = f(x)$ at the point $(x_n, f(x_n))$ is $f'(x_n)$. Since the tangent goes through the point $(x_n, f(x_n))$ its equation is

$$y = f(x_n) + f'(x_n)(x - x_n).$$

To find the point at which the tangent crosses the x-axis we solve the equation

$$f(x_n) + f'(x_n)(x - x_n) = 0$$
$$f'(x_n)(x - x_n) = -f(x_n)$$

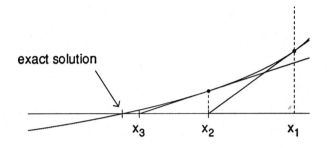

Figure II.128: An even better estimate

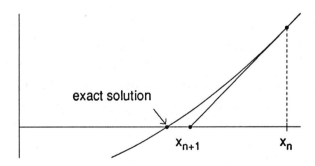

Figure II.129: The basic step in Newton's Method

§10. The Hare: Newton's Method

$$x - x_n = -f(x_n)/f'(x_n)$$
$$x = x_n - f(x_n)/f'(x_n)$$

giving us a new estimate x_{n+1}. That is,

$$x_{n+1} = x_n - \frac{f(x_n)}{f'(x_n)}.$$

This equation is the heart of Newton's Method. Starting with an initial estimate x_1 one computes a second estimate

$$x_2 = x_1 - \frac{f(x_1)}{f'(x_1)}.$$

and then a third estimate

$$x_3 = x_2 - \frac{f(x_2)}{f'(x_2)}.$$

and as many additional estimates as are needed to obtain a sufficiently precise estimate for the true solution.

Example:

Solve the equation $x^2 = 10$ using Newton's Method with an initial estimate $x_1 = 3$.

Answer:

The first step is to rewrite the equation as $x^2 - 10 = 0$. We want to find out where the function $f(x) = x^2 - 10$ crosses the x-axis.

The basic formula is

$$x_{n+1} = x_n - \frac{f(x_n)}{f'(x_n}.$$

Notice

$$f(x) = x^2 - 10, \quad \text{so} \quad f(x_n) = x_n^2 - 10,$$

and

$$f'(x) = 2x, \quad \text{so} \quad f'(x_n) = 2x_n.$$

So

$$x_{n+1} = x_n - \frac{x_n^2 - 10}{2x_n}.$$

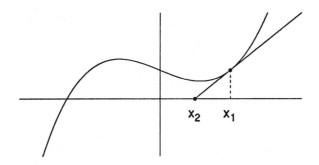

Figure II.130: The result of a poor initial estimate

Thus

$$x_2 = x_1 - \frac{x_1^2 - 10}{2x_1} = 3 - \frac{9 - 10}{6} = \frac{19}{6} = 3.16666667,$$

and

$$x_3 = x_2 - \frac{x_2^2 - 10}{2x_2} = 3.16666667 - \frac{3.16666667^2 - 10}{(2)(3.16666667)} = 3.16228070,$$

and

$$x_4 = x_3 - \frac{x_3^2 - 10}{2x_3} = 3.16228070 - \frac{3,16228070^2 - 10}{(2)(3.16228070)} = 3.16227766.$$

This last estimate *after only three iterations of Newton's Method* is already within 0.00000001 of the exact solution to the equation $x^2 = 10$.

As you can see, Newton's Method can be extremely fast. The catch is that the initial estimate must be reasonably good. For example, suppose we were to try to apply Newton's Method to the curve shown in Figure II.130 with the indicated initial estimate x_1. As you can see, the second estimate x_2 is nowhere near the place where the curve actually crosses the x-axis. Even worse, the third estimate x_3 would be way off in left field as shown in Figure II.131.

Newton's Method and the Bisection Method are a good team. Newton's Method is fast *when it works* and the Bisection Method is sure and reliable but slow. Sometimes one uses the two in combination, using the Bisection Method to get an estimate that is in the general

§10. The Hare: Newton's Method

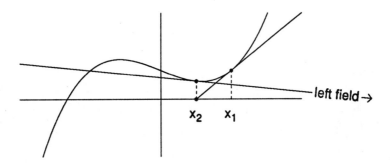

Figure II.131: The result of a poor initial estimate

neighorhood of the actual answer and then using Newton's Method to get an extremely good estimate very quickly.

Newton's Method also works extremely well in partnership with computer graphics. One can use computer graphics to find a good initial estimate for a solution of $f(x) = 0$ from which Newton's Method can produce an extremely good estimate in just a few steps.

Exercises

Exercise II.10.1: Use Newton's Method to estimate a solution to the equation $x^5 - x^4 + 3x^3 + 2x - 1 = 0$.

Exercise II.10.2: Use Newton's Method to estimate a solution to the equation $x^3 - x = 1$.

Exercise II.10.3: Use Newton's Method to estimate a solution to the equation

$$\frac{1}{x^2 + 1} = x$$

Exercise II.10.4: Before reading this section or even taking this course one might come up with the following idea for finding the square root of a number a. If x_n is an estimate for $\sqrt{a}$ then a/x_n would be another estimate which would be too small if x_n were too big, and too big if x_n were too small. Thus another estimate for $\sqrt{a}$ would be the average

$$\frac{x_n + \frac{a}{x_n}}{2}.$$

This gives us a formula for finding a new estimate

$$x_{n+1} = \frac{x_n + \frac{a}{x_n}}{2}.$$

This formula can be used in the same way as Newton's Method to find a sequence of estimates starting with some initial estimate.

Show that Newton's Method leads to exactly the same formula. Recall that $\sqrt{a}$ is the solution of the equation $x^2 - a = 0$.

Exercise II.10.5: Use Newton's Method to find an estimate of the positive solution of the equation

$$\sin x = x/2.$$

Exercise II.10.6: Use Newton's Method to find an estimate of any positive solution of the equation

$$\tan x = x.$$

*Exercise II.10.7: We can think of Newton's Method

$$x_{n+1} = F(x_n) = x_n - \frac{f(x_n)}{f'(x_n)}$$

as a dynamical system.

- What are the equilibrium point(s) of this dynamical system?

- We conjecture that an equilibrium point, x_*, of a dynamical system $x_{n+1} = F(x_n)$ is attracting if $|F'(x_*)| < 1$. Based on this conjecture, what can you say about Newton's Method?

---

§11. When Does an Equilibrium Attract?

One of the first things we noticed in our study of population models in chapter I was that some models approached a limit. For a dynamical system described by a continuous function

$$p_{n+1} = f(p_n)$$

§11. When Does an Equilibrium Attract?

if a sequence approaches a limit then that limit is an equilibrium point. We have, however, seen many examples in which there is an equilibrium point that does not "attract" the sequence. The question that we want to answer in this section is "When does an equilibrium point attract?" Before continuing with this section, you may want to review the experimental evidence that we accumulated in chapter I about this question. In section I.7 we answered this question for linear dynamical systems with the following theorem.

Theorem 14 *A linear dynamical system $p_{n+1} = mp_n + b$ with $m \neq 1$ has exactly one equilibrium point*

$$p_* = \frac{b}{1-m}.$$

- *If $|m| < 1$ and p_1 is any initial term then*

$$\lim_{n \to \infty} p_n = p_*.$$

- *If $|m| > 1$ then there is no limit unless $p_1 = p_*$.*

Nonlinear dynamical systems are more complicated than linear dynamical systems. Nonlinear dynamical systems may have many equilibrium points and the longterm behavior of a model may depend on the initial p_1.

The theorem we prove in this section is based on two observations.

- We know all about linear dynamical systems.

- Under high magnification near a point, a smooth curve looks very much like the tangent to the curve at the point. The slope of this tangent is given by the derivative.

These are the same two ideas that were the basis for Newton's Method—the linear situation is easy, and a smooth function near a point looks linear. We will eventually prove that if p_* is an equilibrium point of the dynamical system $p_{n+1} = f(p_n)$, $|f'(p_*)| < 1$, and if p_1 is sufficiently close to p_*, then

$$\lim_{n \to \infty} p_n = p_*.$$

This is exactly the kind of theorem we should expect as a result of our experience. The key factor that determines whether or not an equilibrium point p_* is attracting is the size of the derivative $f'(p_*)$. In contrast to linear dynamical systems, however, we need another clause. The initial term p_1 makes a difference. This theorem says that *if p_1 is sufficiently close to p_** then the sequence will approach p_*. This agrees with our experimental results and our knowledge about smooth functions—*close to a point* a smooth function behaves very much like its tangent at that point. If we start out *close* to an equilibrium point at which the tangent has slope less than 1 in absolute value the equilibrium point will attract the sequence.

In order to prove this theorem we need a theorem, the Mean Value Theorem.

The Mean Value Theorem

Definition:

Suppose that $y = f(x)$ is a function on an interval $[a, b]$. The **average rate of change** of $f(x)$ on this interval is

$$\frac{f(b) - f(a)}{b - a}.$$

Geometrically, the average rate of change of the function f on the interval $[a, b]$ is the slope of the chord through the points $(a, f(a))$ and $(b, f(b))$ as shown in Figure II.132.

If the function $f(t)$ describes the location of an object at time t then we usually say **average velocity** rather than average rate of change.

Example:

Suppose that a car is traveling on a thruway. At eight o'clock in the morning it is at mile marker 24. One-quarter hour later, at 8:15, it is at mile marker 41. Its average velocity is

$$\frac{41 - 24}{8\frac{1}{4} - 8} = 68 \text{ miles per hour.}$$

§11. When Does an Equilibrium Attract?

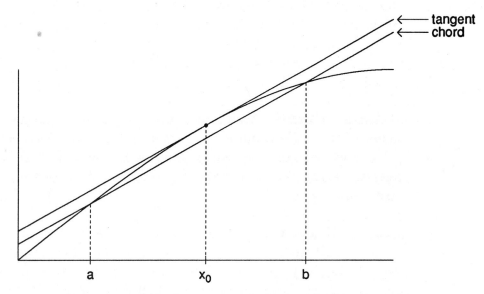

Figure II.132: The tangent and the chord

Theorem 15 *(The Mean Value Theorem)*
Suppose that $y = f(x)$ is a smooth function on an interval $[a, b]$. Then there is a point x_0 in the interval $[a, b]$ such that

$$f'(x_0) = \frac{f(b) - f(a)}{b - a}.$$

That is, there is some point x_0 at which the derivative $f'(x_0)$ is equal to the average rate of change on the interval.

Geometrically, the Mean Value Theorem says that there is a point x_0 in the interval $[a, b]$ at which the tangent is parallel to the chord through the points $(a, f(a))$ and $(b, f(b))$ as shown in Figure II.132.

Before proving this theorem we look at an example.

Example:

> **Prosecutor:** Your honor, Mr, Smith is charged with driving 80 miles per hour in a 55 mile per hour zone on the turnpike. Our evidence is incontrovertible. He entered the turnpike at mile marker 100. When he entered the turnpike he received a card stamped with the time, 3:00 P.M. He left

the turnpike at mile marker 260 at 5:00 P.M. As you can see his velocity was

$$\frac{260-100}{5-3} = \frac{160}{2} = 80 \text{ miles per hour}$$

Defense Attorney: Your honor, my learned colleague has made an error. He is quite correct that my client's *average* velocity was 80 miles per hour. The law, however, does not refer to average velocity. The law refers to velocity at a given time.

Prosecuting Attorney: Unfortunately, your honor, my learned and hair-splitting colleague apparently never studied calculus. By the Mean Value Theorem, if his client's average velocity during the period 3:00 P.M. to 5:00 P.M. was 80 miles per hour then at some time during that period his velocity must have been 80 miles per hour.

Now we are ready to prove the Mean Value Theorem

Proof:

Let m denote the average rate of change of the function f on the interval $[a, b]$.

$$m = \frac{f(b) - f(a)}{b - a}.$$

We are looking for a point x_0 at which the derivative is equal to m. Our proof is based on two ideas. First, it is easy to find places at which a derivative is equal to zero. Second, by a bit of sleight of hand we can convert our original problem into another problem whose solution requires finding a point where a derivative is equal to zero.

The sleight of hand involves defining at a new function

$$g(x) = f(x) - mx.$$

Notice that

$$g'(x) = f'(x) - m.$$

Hence,

$$f'(x) = g'(x) + m.$$

§11. When Does an Equilibrium Attract?

So if we can find a point x_0 at which $g'(x_0) = 0$ then $f'(x_0) = m$ and we will have solved our original problem.

Calculating $g(a)$ and $g(b)$ we see that

$$g(a) = f(a) + ma$$

and

$$g(b) = f(b) + mb.$$

So that

$$\begin{aligned} g(b) - g(a) &= f(b) + mb - (f(a) - ma) \\ &= f(b) - f(a) - m(b-a) \\ &= f(b) - f(a) - \left(\frac{f(b) - f(a)}{b - a}\right)(b - a) \\ &= f(b) - f(a) - (f(b) - f(a)) \\ &= 0. \end{aligned}$$

Thus

$$g(a) = g(b).$$

Now by the Extreme Value Theorem the function g has a maximum and a minimum on the interval $[a, b]$. That is, there are points u and v in the interval $[a, b]$ such that for every $x \in [a, b]$

$$g(u) \leq g(x) \leq g(v).$$

There two possibilities.

1. At least one of the two points u and v is not an endpoint. Suppose u is not an endpoint. Then since u is a minimum of g and u is not an endpoint, u must be a critical point of g. But g is smooth, so $g'(u) = 0$ completing the proof. Exactly the same argument works if v is not an endpoint.

2. Both u and v are endpoints. But then $g(u) = g(v)$ since $g(a) = g(b)$ and thus the function g is constant. Hence, at every point x, $g'(x) = 0$ completing the proof.

We now have the mathematical tools needed to prove two theorems about when an equilibrium point is attracting.

Theorem 16 *Suppose that p_* is an equilibrium point of the dynamical system*
$$p_{n+1} = f(p_n).$$
Suppose that $L < 1$.
Suppose that $\delta > 0$ and that on the interval $(p_* - \delta, p_* + \delta)$
$$|f'(x)| \leq L.$$
Then, if $p_1 \in (p_* - \delta, p_* + \delta)$,
$$\lim_{n \to \infty} p_n = p_*.$$

This theorem is exactly the kind of theorem we should expect as a result of our experiments. The key factor that determines whether or not an equilibrium point p_* is attracting is the size of the derivative $f'(p_*)$. In contrast to linear dynamical systems, however, we need another clause. The initial term p_1 makes a difference. This theorem says that if p_1 is sufficiently close to p_* then the sequence will approach p_*. This matches our experimental results.

Proof:

To prove this theorem we want to compare $|p_{n+1} - p_*|$ with $|p_n - p_*|$. The key idea is to notice that $p_{n+1} = f(p_n)$ and that $p_* = f(p_*)$ so the Mean Value Theorem implies that there is some point x_0 between p_n and p_* such that
$$\frac{p_{n+1} - p_*}{p_n - p_*} = \frac{f(p_n) - f(p_*)}{p_n - p_*} = f'(x_0).$$
Thus
$$p_{n+1} - p_* = f'(x_0)(p_n - p_*)$$
and
$$|p_{n+1} - p_*| = |f'(x_0)||p_n - p_*|.$$

Now, if p_1 is in the interval $(p_* - \delta, p_* + \delta)$ then, since the point x_0 is between p_1 and p_*, it is in the same interval and therefore $|f'(x_0)| \leq L$. Thus
$$|p_2 - p_*)| = |f'(x_0)||p_1 - p_*| \leq L|p_1 - p_*|.$$

§11. When Does an Equilibrium Attract?

Since $L < 1$, p_2 is closer to p_* then p_1 is to p_*.

The same argument shows that

$$|p_3 - p_*| \leq L|p_2 - p_*| \leq LL|P_1 - p_*| = L^2|p_1 - p_*|$$

and, in general,

$$|p_n - p_*| \leq L^{n-1}|p_1 - p_*|.$$

Since $L < 1$, L^{n-1} approaches zero as n goes to ∞. Thus

$$\lim_{n \to \infty} p_n = p_*$$

completing the proof of the theorem.

Theorem 17 *Suppose that p_* is an equilibrium point of the dynamical system*

$$p_{n+1} = f(p_n),$$

f' is continuous at the point p_*, and that $|f'(p_*)| < 1$.

Then, if p_1 is sufficiently close to p_*,

$$\lim_{n \to \infty} p_n = p_*.$$

Proof:

The idea behind this proof is that the assumptions of this theorem imply the assumptions of the previous theorem.

Let $\epsilon = (1 - |f'(p_*)|)/2$ and notice that $\epsilon > 0$. Since f' is continuous at the point p_* there is a $\delta > 0$ such that for every $x \in (p_* - \delta, p_* + \delta)$ we have $|f'(x) - f'(p_*)| < \epsilon$. This implies that for every $x \in (p_* - \delta, p_* + \delta)$ we have $|f'(x))| < 1 - \epsilon/2$. If we let $L = 1 - \epsilon/2$, the conclusion follows by the previous theorem.

If $|f'(p_*)| > 1$ then the same kinds of calculations show that whenever a sequence gets close to p_* it is pushed away. Thus if $f'(p_*)| > 1$ the equilibrium point is not attracting. We need to add some words of exception. If a sequence ever lands exactly on p_* it will remain there.

Definition:

Suppose that p_* is an equilibrium point of the dynamical system $p_{n+1} = f(p_n)$.

If $|f'(p_*)| < 1$ we say that p_* is a **stable** equilibrium.

If $|f'(p_*)| > 1$ we say that p_* is an **unstable** equilibrium.

Stable equilibriums will attract if the initial term p_1 is sufficiently close to the equilibrium. They are also correcting in the sense that after a small deviation from the equilibrium due to "outside" factors the sequence will be pulled back toward the stable equilibrium. Unstable equilibriums usually do not attract (unless the equilibrium is hit exactly by some term of the sequence). They are not correcting. A small deviation due to "outside" factors can be magnified.

We can use our new results to learn more about logistic models. We want to look at the family of models

$$p_{n+1} = f(p_n) = ap_n(1 - bp_n).$$

In chapter I we asked many questions about this family of models. For example,

1. When does the population die out?

2. When does the population tend to a limit?

3. Does the initial population p_1 make any difference?

4. Does the constant a make any difference?

5. Does the constant b make any difference?

We can now make a start at answering some of these questions? Question 3 is the hardest. We will not discuss it here, although you may be able to answer it yourself looking at graphical techniques like those we discussed in section I.6.

We can shed some light on Question 1 by noticing that $p = 0$ is always an equilibrium. The question is whether it is stable.

$$f(p) = ap(1 - bp) = ap - abp^2.$$

So

$$f'(p) = a - 2abp$$

and

$$f'(0) = a.$$

Thus zero will be a stable equilibrium point if $a < 1$ and unstable if $a > 1$. This implies that a sufficiently small initial population will die

§11. When Does an Equilibrium Attract?

out if $a < 1$ and will not die out if $a > 1$. Notice that the constant b is irrelevant.

Exercises

*Exercise II.11.1: Experiment with the model

$$p_{n+1} = 0.9p_n(1 - .005p_n); \qquad p_1 = 10.$$

Do your results agree with our conclusions above? Try other initial populations. Does the initial population seem to make any difference?

*Exercise II.11.2: Experiment with the model

$$p_{n+1} = 1.1p_n(1 - .005p_n); \qquad p_1 = 10.$$

Do your results agree with our conclusions above? Try other initial populations. Does the initial population seem to make any difference?

*Exercise II.11.3: Consider a population model of the form

$$p_{n+1} = R(p_n)p_n.$$

Show that if $R(0) < 1$ then 0 is a stable equilibrium. Explain why this is to be expected.

To respond to the other questions we begin by finding the equilibrium points in the usual way.

$$\begin{aligned} p &= ap(1 - bp) \\ p &= ap - abp^2 \\ 0 &= (a-1)p - abp^2 \\ 0 &= p[(a-1) - abp]. \end{aligned}$$

So that the nonzero equilibrium point is

$$p_* = \frac{a-1}{ab}.$$

Notice p_* is above zero and meaningful for population models if $a > 1$.
Now

$$f'(p_*) = a - 2ab\left(\frac{a-1}{ab}\right) = a - 2(a-1) = 2 - a.$$

So
$$|f'(p_*)| < 1$$
if $1 < a < 3$. Thus p_* will be a stable equilibrium point if $1 < a < 3$. This implies that for initial populations sufficiently close to p_* the limit will be p_*.

We can summarize our results so far by

- If $a < 1$ then 0 is a stable equilibrium point.

- If $1 < a < 3$ then p_* is a stable equilibium point and 0 is not.

Exercises

*Exercise II.11.4: Experiment with the model
$$p_{n+1} = 2.9 p_n (1 - .005 p_n); \qquad p_1 = 10.$$
Do your results agree with our conclusions above? Try other initial populations. Does the initial population seem to make any difference?

*Exercise II.11.5: Experiment with the model
$$p_{n+1} = 3.1 p_n (1 - .005 p_n); \qquad p_1 = 10.$$
Do your results agree with our conclusions above? Try other initial populations. Does the initial population seem to make any difference?

*Exercise II.11.6: Consider the dynamical systems
$$p_{n+1} = a p_n (1 - .000001 p_n^2).$$
Discuss the behavior of models based on this family of dynamical systems based on our work in this section.

*Exercise II.11.7: Let $R(p)$ be the function
$$R(p) = \begin{cases} .0243 p, & \text{if } p \leq 100; \\ 2.7(1 - .001 p), & \text{if } 100 \leq p \leq 1000. \end{cases}$$
shown in Figure II.133. Consider the dynamical systems
$$p_{n+1} = a p_n R(p_n).$$
This kind of dynamical system might describe a species that hunts in packs. Discuss the behavior of models based on this family of dynamical systems based on our work in this section.

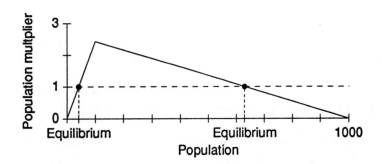

Figure II.133: Population multiplier for pack-hunters

§12. Summary⁺

The Differentiation Formulas

The differentiation formulas are the musical scales of calculus. By themselves they aren't terribly interesting. They are fundamental, however, to much of our work. By now they should be second nature to you. If not, it is important to do enough exercises, boring as they may be, so that they do become second nature. The formulas we have obtained so far are summarized below. The summary is followed by some exercises.

Differentiation Formulas

- The derivative of a linear function $f(x) = mx + b$ is $f'(x) = m$.

- **The Power Rule:** If c is a rational number and $f(x) = x^c$ then $f'(x) = cx^{c-1}$.

- **Derivatives of Trigonometric Functions**

$$\begin{aligned}(\sin x)' &= \cos x \\ (\cos x)' &= -\sin x \\ (\tan x)' &= \sec^2 x \\ (\cot x)' &= -\csc^2 x \\ (\sec x)' &= \sec x \tan x\end{aligned}$$

$$(\csc x)' = -\csc x \cot x$$

- **The Sum Rule:** If $u(x)$ and $v(x)$ are differentiable then so is $f(x) = u(x) + v(x)$ and $f'(x) = u'(x) + v'(x)$.

- **Constant Multiple Rule:** If $u(x)$ is differentiable and k is a constant then $f(x) = ku(x)$ is differentiable and $f'(x) = ku'(x)$.

- **Product Rule:** If $u(x)$ and $v(x)$ are differentiable then so is $f(x) = u(x)v(x)$ and $f'(x) = u(x)v'(x) + v(x)u'(x)$.

- **Quotient Rule:** If $u(x)$ and $v(x)$ are differentiable then so is $f(x) = u(x)/v(x)$ (at every point where $v(x) \neq 0$) and

$$f'(x) = \frac{v(x)u'(x) - u(x)v'(x)}{v(x)^2}.$$

- **Inverse Function Rule:** Suppose that $y = f(x)$ is a one-to-one function. Then

$$(f^{-1})'(y) = \frac{1}{f'(f^{-1}(y))}.$$

This formula works at every point (x, y) on the curve $y = f(x)$ at which $f'(x)$ exists and is nonzero.

Using different notation we sometimes write

$$\frac{dx}{dy} = \frac{1}{dy/dx}.$$

- **Derivatives of the Inverse Trigonometric Functions**

$$(\arcsin x)' = \frac{1}{\sqrt{1-x^2}}$$

$$(\arccos x)' = -\frac{1}{\sqrt{1-x^2}}$$

$$(\arctan x)' = \frac{1}{1+x^2}$$

$$(\text{arccot } x)' = -\frac{1}{1+x^2}$$

§12. Summary†

$$(\text{arcsec } x)' = \frac{1}{|x|\sqrt{x^2-1}}$$

$$(\text{arccsc } x)' = -\frac{1}{|x|\sqrt{x^2-1}}$$

- **Chain Rule:** Suppose that $y = f(x)$ and $z = g(y)$. Let $h = g \circ f$. That is, $z = h(x) = g(f(x))$. If $f'(x)$ and $g'(y)$ exist then so does $h'(x)$ and

$$h'(x) = g'(f(x))f'(x)$$

or, using different notation,

$$\frac{dz}{dx} = \left(\frac{dz}{dy}\right)\left(\frac{dy}{dx}\right).$$

Exercises

Find the derivative of each of the following.

*Exercise II.12.1:
$$y = 5x^3 + \sqrt{x}$$

*Exercise II.12.2:
$$A = \pi r^2$$

*Exercise II.12.3:
$$f(t) = 4t^{\frac{2}{3}} - 2t + 1/t^2$$

*Exercise II.12.4:
$$D(p) = \frac{1}{1+p^2}$$

*Exercise II.12.5:
$$y = (x - \sqrt{x})^4$$

*Exercise II.12.6:
$$z = \sqrt{\frac{t+1}{t}}$$

*Exercise II.12.7:
$$y = \sqrt{\frac{1-x^2}{2x^3 - 12x + 3}}$$

*Exercise II.12.8:
$$z = \left(\frac{1 - 2x^5}{x + x^2}\right)^{\frac{2}{5}}$$

*Exercise II.12.9:
$$F(t) = \sqrt{(1-x)^2 + (4-x)^2} + \sqrt{\frac{1-x}{1+x}}$$

*Exercise II.12.10:
$$y = \frac{x + \sqrt{x}}{2x - \sqrt{x}}$$

*Exercise II.12.11:
$$f(\theta) = \cos(\theta^2 + 1)$$

*Exercise II.12.12:
$$y = \sin(\cos x)$$

*Exercise II.12.13:
$$y = (4 + \tan x)^3$$

*Exercise II.12.14:
$$y = \arctan(\sin x)$$

*Exercise II.12.15:
$$w = \arcsin(4x + 1)$$

*Exercise II.12.16:
$$w = \arcsin(5x - 1) + \arccos(5x - 1)$$

§12. Summary[†]

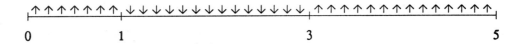

Figure II.134: The behavior of the function $f(x) = x^3 - 6x^2 + 9x + 1$

Curve Sketching

One of the most important tools that a mathematician has is curve sketching. Looking at the graph of a function one can immediately grasp its most important features. The main steps involved in sketching a graph are the following.

- Analyze the derivative: Look for points at which the derivative is either zero or undefined. These points are called **critical points** and are potential local maximums and local minimums. Find out where the derivative is positive (and, hence, the original function is increasing) and negative (and, hence, the original function is decreasing). Make a chart similar to Figure II.134 summarizing this information. From this chart you can immediately see where the local maximums and local minimums are.

- Evaluate the function at any endpoints and at any critical points: Make a table showing these values. If the domain is an interval of the form $[a, b]$ then the function will have a global minimum and a global maximum. Otherwise the function may or may not have a global minimum and may or may not have a global maximum. In either case if there is a global minimum it must be at an endpoint or at a critical point. Similarly, if there is a global maximum it must be at an endpoint or at a critical point.

- Look for the behavior of the function $f(x)$ for very large positive values of x and for very large negative values of x: That is, examine the limits
$$\lim_{x \to +\infty} f(x)$$
and
$$\lim_{x \to -\infty} f(x).$$

- Sketch the graph using all the information obtained above: As you sketch the graph you may discover earlier mistakes. There is a great deal of redundancy in the information one obtains by following the steps above. This redundancy is useful because it provides some check on your work. If, as you sketch the graph, you find yourself trying to follow contradictory information (e.g., "Draw a curve from the point $(0, 4)$ to the point $(3, 2)$ increasing along the way.") then go back and check your work.

Exercises

Sketch a graph of each of the following functions on the indicated interval.

*Exercise II.12.17:
$$y = x^3 + 2x - 4, \qquad [-3, 3]$$

*Exercise II.12.18:
$$y = x^3 + 2x - 4, \qquad (-\infty, +\infty)$$

*Exercise II.12.19:
$$y = x^3 - 2x - 4, \qquad [-3, 3]$$

*Exercise II.12.20:
$$y = x^3 - 2x - 4, \qquad (-\infty, +\infty)$$

*Exercise II.12.21:
$$y = \sqrt{(x-1)^2 + 1} - \sqrt{(x+1)^2 + 1}, \qquad [-3, 3]$$

*Exercise II.12.22:
$$y = \sqrt{(x-1)^2 + 1} - \sqrt{(x+1)^2 + 1}, \qquad (-\infty, +\infty)$$

*Exercise II.12.23:
$$z = \frac{2x}{x^2 + 1}, \qquad (-\infty, +\infty)$$

*Exercise II.12.24:
$$z = \frac{2x^2}{x^2+1}, \quad (-\infty, +\infty)$$

*Exercise II.12.25:
$$z = \frac{2x^3}{x^2+1}, \quad (-\infty, +\infty)$$

*Exercise II.12.26:
$$y = \sin(x^2 - 1)$$

*Exercise II.12.27:
$$y = \arctan(\sin x)$$

Optimization

One of the most useful applications of calculus is finding the "best" of something. The basic mathematical steps involved are the same as those for curve sketching. Before we can apply our mathematical tools to a practical problem, however, we must first express the problem mathematically. This step is often the most difficult because it depends as much on the practical situation as on the mathematics. You should look over the various examples and exercises in this chapter. The following exercises can give you some additional practice.

Exercises

*Exercise II.12.28: A lifeguard can run on the sand twice as fast as she can swim. She sees a swimmer drowning. Figure II.135 shows the locations of the drowning swimmer and of the lifeguard. What path should she take to reach the swimmer as quickly as possible?

*Exercise II.12.29: Find the dimensions of the largest rectangle that can be inscribed in a circle of radius R.

*Exercise II.12.30: Find the point on the curve $y = 2x - 1$ that is closest to the origin. Call this point (x_*, y_*) where $y_* = 2x_* - 1$. Notice that the line from the origin to this point is perpendicular to the curve $y = 2x - 1$. Can you explain this observation geometrically?

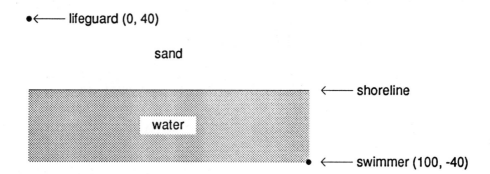

Figure II.135: Lifeguard and drowning swimmer

Exercise II.12.31: Find the point on the curve $y = mx + b$ that is closest to the origin. Call this point (x_*, y_*) where $y_* = mx_* + b$. Notice that the line from the origin to this point is perpendicular to the curve $y = mx + b$.

Exercise II.12.32: Prove that if $y = f(x)$ is any smooth function and (x_*, y_*) is the point on the curve $y = f(x)$ that is closest to the origin then the line from the origin to the point (x_*, y_*) is perpendicular the curve (i.e., to the tangent to the curve at the point (x_*, y_*)).

Exercise II.12.33: A company wants to manufacture an open box with a square base that will hold 1 cubic foot. What dimensions should it have to minimize the cost of materials (i.e., to minimize the total area of the four sides and the bottom)?

Exercise II.12.34: A company wants to manufacture a closed box with a square base that will hold one cubic foot. The bottom must be extra strong. So, material for the bottom is three times as expensive as material for the sides and the top. What dimensions should the box have to minimize the total cost of materials?

Stable vs. Unstable Equilibrium Points

We made some progress understanding some of the behavior we observed in population models. We looked at the question of when an equilibrium point is attracting and when it is not attracting. We al-

§12. Summary[+]

ready knew that for a linear dynamical system

$$p_{n+1} = mp_n + b$$

the situation is particularly simple. If $m \neq 1$ then there is exactly one equilibrium point

$$p_* = \frac{1}{1-m}.$$

- If $|m| < 1$ the equilibrium point p_* is attracting and for any initial p_1 we always have

$$\lim_{n \to \infty} p_n = p_*.$$

- if $|m| > 1$ the equilibrium point is not attracting.

A nonlinear dynamical system may have several equilibrium points and the behavior of the sequence $p_1, p_2, p_3, \ldots p_n, \ldots$ may depend on the initial p_1. An equilibrium point p_* is said to be *stable* if $|f'(p_*)| < 1$ and *unstable* if $|f'(p_*)| > 1$. If p_* is stable and p_1 is sufficiently close to p_* then

$$\lim_{n \to \infty} p_n = p_*.$$

We still have many questions about population models and dynamical systems. For example, we have observed situations in which the population after many generations settled into a pattern in which it bounced back and forth or "cycled" among several different numbers. We would like to understand this behavior.

All the dynamical systems we have discussed so far are discrete dynamical systems, for example, population models for temperate zone insects in which there are distinct generations. We now have some of the tools needed to study "continuous" dynamical systems in which, for example, population is described by a continuously changing function of time $p(t)$ rather than a sequence of numbers $p_1, p_2, p_3, \ldots p_n, \ldots$ and the way that population changes is described by an equation like

$$p'(t) = g(p).$$

We will return to this topic later in the semester.

Newton's Method

Newton's Method is a fast but not always reliable method for solving equations of the form $f(x) = 0$. There is an important common subtext for Newton's Method and the analysis of when an equilibrium point is attracting. In general, a smooth function looks very much like a linear function very near a point. That is why Newton's method is so effective *if we have a good initial estimate for a solution*. It is also why a stable equilibrium point p_* is attracting *if p_1 is sufficiently close to p_**.

Exercise

*Exercise II.12.35: Use Newton's Method to find a good estimate for a solution of the equation $x^5 + 2x^3 - 3x^2 + 4 = 0$. You will need to do some preliminary work to find a reasonable starting estimate.

*Exercise II.12.36: Use Newton's Method to find a good estimate for a solution of the equation $\sin x = 3 \cos x$. You will need to do some preliminary work to find a reasonable starting estimate.

Geometric Optics

This will be one of the contexts for calculus to which we will return throughout this course. We study Geometric Optics by using our new knowledge about optimization together with prior knowledge about linear functions.

Exercise

*Exercise II.12.37: A fish whose eyes are five feet below the surface of the water is looking at a diving board that is fifty feet away. The true height of the diving board is six feet. How high does it appear to the fish?

Advanced Topics

One way to study 2-cycles for a dynamical system like

$$p_{n+1} = f(p_n)$$

is to study the function $f \circ f$ since

$$p_{n+2} = (f \circ f)(p_n)$$

§12. Summary†

and, thus, two points q_1 and q_2 in a 2-cycle satisfy the conditions

$$q_1 = (f \circ f)(q_1)$$

and

$$q_2 = (f \circ f)(q_2).$$

We used this observation to find 2-cycles in section 8. Once we have determined that a particular dynamical system has a 2-cycle the next question is whether that 2-cycle is attracting. One can show that a 2-cycle is attracting if

$$|(f \circ f)'(q_1)| < 1.$$

This situation is analogous to the situation for an equilibrium point.

Exercise

*Exercise II.12.38: For what values of a does the logistic model

$$p_{n+1} = ap_n(1 - bp_n)$$

have an attracting 2-cycle?

Chapter III

Differential Equations

When we study a particular *quantity* or number we are interested in information about that quantity. For example, we might know that x is a number whose square is 2. This information can be expressed by the equation
$$x^2 = 2.$$

When we study a particular *function* we are interested in information about that function. For example, we might know that $f(t)$ is a function whose derivative is $\sin t$. This information can be expressed by the equation
$$\frac{df}{dt} = \sin t.$$

This is an example of a **differential equation**, an equation involving a function and one or more of its derivatives. Differential equations occur naturally in many different contexts. For example, in economics the fact that prices are rising at the rate of 2% per year can be expressed by the differential equation
$$\frac{dp}{dt} = .02p$$
and in physics the fact that a falling body is accelerating at the rate of -32 feet per second^2 can be expressed by the differential equation
$$y'' = -32.$$

Notice this differential equation involves the second derivative. We begin this chapter by discussing the second derivative, its interpretation as acceleration and the information it gives us about the graph of a function.

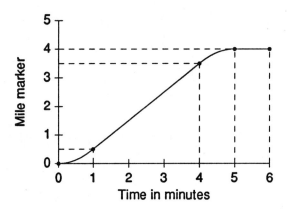

Figure III.1: A car trip

§1. Acceleration—the Second Derivative

Figure III.1 shows the graph of a function $y = p(t)$ that describes the location of a car on a short trip. The car is initially parked at mile marker zero on a highway. At time $t = 0$ the driver presses down on the accelerator. She accelerates steadily until time $t = 1$ minute at which time she has reached mile marker 0.5. Then she drives at a steady velocity for three minutes until time $t = 4$ minutes. At time $t = 4$ minutes she takes her foot off the accelerator and coasts to a stop at time $t = 5$ minutes. By this time she has reached mile marker 4.0. Then she turns off the engine and leaves the car. We can read most of this information on Figure III.1.

The most obvious information on the graph is the location of the car at each time t. Its location at time t is the coordinate $p(t)$ of the point $(t, p(t))$ on the curve. For example, at time $t = 1$ we can see that the car is at mile marker 0.5. We can also see from the graph that the car is not moving from time $t = 5$ to time $t = 6$ since on the interval $[5, 6]$ the curve is a straight horizontal line.

The derivative $p'(t)$ can be interpreted in two ways. Physically, $p'(t)$ is the velocity of the car at time t. Geometrically, $p'(t)$ is the slope of the tangent to the curve $y = p(t)$ at the point t. Thus the slope of the curve is the velocity of the car. The section of the curve from $t = 1$ to $t = 4$ is a straight line. Thus during the interval $[1, 4]$ the velocity $p'(t)$ is constant; the car is neither accelerating or decelerating.

In Figure III.2 we've added two tangents to the graph, at the points

§1. Acceleration—the Second Derivative

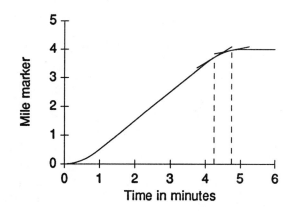

Figure III.2: Two tangents

$t = 4.25$, and $t = 4.75$. The tangent at the point $p = 4.25$ is much steeper than the one at the point $p = 4.75$. This tells us that the car is traveling faster at time $t = 4.25$ than it is at time $t = 4.75$. If you run a ruler along the curve $y = p(t)$ from $t = 4$ to $t = 5$ keeping the ruler tangent to the curve you will see that the slope of the tangent is steadily decreasing from time $t = 4$ to time $t = 5$. Thus we can see from this graph that the car is decelerating between time $t = 4$ and time $t = 5$. A similar experiment will show that the car is accelerating between time $t = 0$ and time $t = 1$.

If we are interested in the rate at which the object is accelerating or decelerating we want to know how fast the velocity is changing. To find this we examine the derivative of the velocity—the derivative $p''(t)$ of the derivative $p'(t)$. This is called the **second derivative** of the original function $p(t)$.

Example:

Suppose that the position of a moving object is described by the function $y = 90t$ where y is its position in kilometers and t is the time in hours. How fast is the object moving? How fast is it accelerating?

Answer:

To find the velocity we differentiate the function $y = p(t)$ that describes the object's position.

$$y = 90t$$
$$y' = 90$$

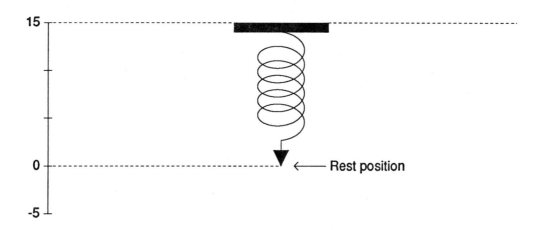

Figure III.3: Weight hanging from spring

The velocity of the object is a steady 90 kilometers per hour.
To find the acceleration we differentiate the derivative.

$$y' = 90$$
$$y'' = 0$$

and we see that the acceleration is zero, exactly what we would expect since the velocity is constant.

Example:

Figure III.3 shows a weight hanging at the end of a piece of elastic. The weight is at rest. It is stretched just enough so the the force exerted by the elastic exactly balances the force exerted by gravity. Next to the weight we have placed a scale that can be used to describe the weight's position. When the weight is at rest it is opposite 0 on this scale. This is called the **rest position** of the weight. We are interested in the motion of the weight if we pull it down and then release it. We represent the position of the elastic at time t by the function $y(t)$ and we let $t = 0$ be the time at which the weight is released. Figure III.4 shows the function $y(t)$. Figure III.5 shows the function $y'(t)$ and Figure III.6 shows $y''(t)$.

§1. Acceleration—the Second Derivative

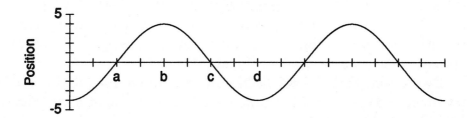

Figure III.4: $y(t)$ = position of weight at time t

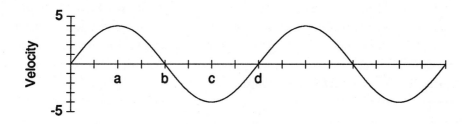

Figure III.5: $y'(t)$ = velocity of weight at time t

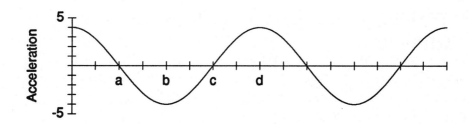

Figure III.6: $y''(t)$ = acceleration of weight at time t

At time $t = 0$ we see that $y(0) = -4$ (Figure III.4). Time $t = 0$ is the time at which we release the weight. Its velocity $y'(0)$ (Figure III.5) is zero. The elastic exerts a force on the weight that is stronger than the force exerted by gravity which causes the weight to accelerate upward. Thus $y''(0)$ is positive (Figure III.6).

By the time that the weight has returned to its rest position $y = 0$ (at time $t = a$) it is traveling quite rapidly. If you look at $y'(a)$ (Figure III.3.5) you will see that it is positive. In fact, this is the time at which the weight is traveling at its fastest. At this point the force exerted by the elastic is exactly balanced by the force exerted by gravity. As a result the acceleration $y''(a)$ is zero (Figure III.3.6).

Between time $t = a$ and time $t = b$ the weight is still traveling upward. Hence, its velocity is positive. Because it is above its rest position, the elastic is not stretched enough to overcome the force exerted by gravity. Thus the force of gravity is stronger than the force exerted by the spring, producing a negative (downward) acceleration. By time $t = b$ the weight has reached the top of its travels. Its velocity is now zero and its acceleration is a large negative number; it is accelerating rapidly in the downward direction.

Exercises

*Exercise III.1.1:

Looking at Figures III.4-III.6 consider the point $t = c$. Describe what is happening. Where is the weight? How fast and in which direction is it moving? Is it accelerating? Is the acceleration upward or downward? Explain why in terms of the forces exerted by gravity and by the elastic.

*Exercise III.1.2:

Looking at Figures III.4-III.6 consider the point $t = d$. Describe what is happening. Where is the weight? How fast and in which direction is it moving? Is it accelerating? Is the acceleration upward or downward? Explain why in terms of the forces exerted by gravity and by the elastic.

The second derivative also describes the shape of a curve. Figure III.7 shows a curve. One side of this curve is called the concave side and the other is called the convex side. Now look at Figure III.8. This

§1. Acceleration—the Second Derivative

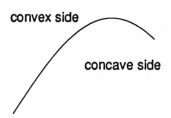

Figure III.7: Two sides of a curve

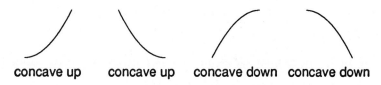

Figure III.8: Several curves

figure shows several curves. We can describe these curves by using he terms **concave up** and **concave down**.

Looking back at Figure III.1 we see that the graph is concave up between $t = 0$ and $t = 1$; then it is flat between $t = 1$ and $t = 4$; and then it is concave down between $t = 4$ and $t = 5$. If you slide a ruler along a curve moving from left to right keeping the ruler tangent to the curve, you will see that where the curve is concave down the derivative is decreasing, hence, the second derivative is negative; where the curve is concave up the derivative is increasing and, hence, the second derivative is positive. The sign of the second derivative tells us where a curve is concave up and where it is concave down.

- A curve is **concave up** on an interval if the second derivative is positive on that interval.

- A curve is **concave down** on an interval if the second derivative is negative on that interval.

Example:

Consider the function

$$y = 2x^3 - 3x^2 + 1.$$

↑↑↑↑↑↑↑↑↑↑↑↑↑↑↑↑↑↓↓↓↓↓↓↓↓↓↑↑↑↑↑↑↑↑↑
   0   1

Figure III.9: Where $y = 2x^3 - 3x^2 + 1$ is increasing and decreasing

∩∩∩∩∩∩∩∩∩∩∩∩∩∩∩∩∩∪∪∪∪∪∪∪∪∪∪
              1/2

Figure III.10: Where $y = 2x^3 - 3x^2 + 1$ is concave up and concave down

Sketch a graph of this function paying particular attention to where it is concave up and where it is concave down.

Answer:

The first step in sketching the graph of any function is examining its derivative.

$$y' = 6x^2 - 6x = 6x(x-1)$$

There are two critical points, at $x = 0$ and $x = 1$. Figure III.9 shows where the function is increasing and where it is decreasing based on looking at the first derivative.

To determine where the curve is concave up and where it is concave down we look at the second derivative

$$y'' = 12x - 6.$$

There is one point at which the second derivative is zero, namely $x = 1/2$. Figure III.10 shows where the second derivative is positive (and, hence, the curve is concave up) and where it is negative (and, hence, the curve is concave down).

There are three important points on the curve, two critical points at $x = 0$ and $x = 1$, and the point $x = 1/2$ at which the curve changes

§1. Acceleration—The Second Derivative

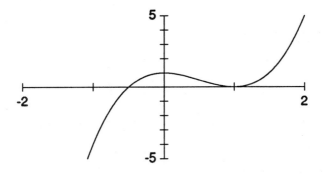

Figure III.11: $y = 2x^3 - 3x^2 + 1$

from being concave down to concave up. The first step in graphing the curve is to plot these three points.

x	y
0	1
1	0
0.5	0.5

Putting all this information together we are able to sketch the graph shown in Figure III.11.

One of the most important features a curve can have is a point (like the point $(0.5, 0.5)$ in the preceding example) at which it is S-shaped, that is, its shape changes from concave up to concave down or vice versa. At such a point the second derivative will either be zero or discontinuous.

Definition:

> A **point of inflection** of a function $f(x)$ is a point at which the function changes from being concave upward to concave downward or vice versa.

Exercises

*Exercise III.1.3:
Sketch a graph of the function
$$y = x^3 - 3x^2 + 4$$
paying particular attention to where it is concave up and concave down.

Exercise III.1.4:
Sketch a graph of the function
$$y = x^4 - 2x^2 - 1$$
paying particular attention to where it is concave up and concave down.

Exercise III.1.5:
Sketch a graph of the function
$$y = \frac{x}{2} + \sin x$$
paying particular attention to where it is concave up and concave down.

Exercise III.1.6:
Sketch a graph of the function
$$y = \frac{1}{1 + x^2}$$
paying particular attention to where it is concave up and concave down.

*Exercise III.1.7:
Sketch a graph of the function
$$y = \sqrt[3]{x^2}$$
paying particular attention to where it is concave up and concave down.

Exercise III.1.8:
Sketch a graph of the function
$$y = \sin x + \cos x$$
paying particular attention to where it is concave up and concave down.

§1. ACCELERATION—THE SECOND DERIVATIVE 313

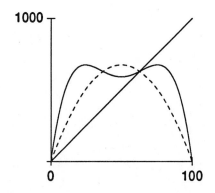

Figure III.12: Looking for 2-cycles, $p_{n+1} = 2.7p_n(1 - .001p_n)$

Period Doubling

Analyzing the shape of a curve can help us understand a phenomenon called **period-doubling** that occurs frequently in discrete dynamical systems like the logistic models that we studied at the beginning of the course. At the end of chapter II section 8 we looked at several dynamical systems of the form

$$p_{n+1} = f(p) = ap_n(1 - .001p_n)$$

for different values of a. We look for equilibrium points for such dynamical system by looking at the curve $y = f(p)$ to see where it intersects the diagonal line $y = p$ and we look for 2-cycles by looking at the curve $y = f(f(p))$ to see where it intersects the diagonal line $y = p$. Figures III.12–III.14 show the curves $y = f(p)$, $y = f(f(p))$, and the diagonal line $y = p$ for the three models

$$\begin{aligned} p_{n+1} &= 2.7p_n(1 - .001p_n) \\ p_{n+1} &= 3.0p_n(1 - .001p_n) \\ p_{n+1} &= 3.3p_n(1 - .001p_n) \end{aligned}$$

Recall that an equilibrium point p_* is attracting if $|f'(p_*)| < 1$ and repelling if $|f'(p_*)| > 1$. For these models the nonzero equilibrium point is

$$p_* = \frac{a-1}{.001a}$$

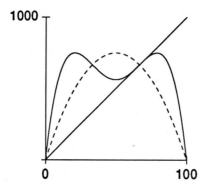

Figure III.13: Looking for 2-cycles, $p_{n+1} = 3.0p_n(1 - .001p_n)$

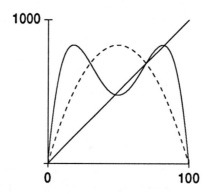

Figure III.14: Looking for 2-cycles, $p_{n+1} = 3.3p_n(1 - .001p_n)$

§1. Acceleration—the Second Derivative

and the derivative $f'(p)$ is

$$f'(p) = a - .002ap.$$

So

$$f'(p_*) = 2 - a$$

and as a goes from 2.7 through 3.0 to 3.3, $f'(p_*)$ goes from -0.7 through -1.0 to -1.3. Thus, this equilibrium point goes from attracting to repelling as a goes through 3.0.

We can see this same phenomenon in the three graphs. For $a = 2.7$, $|f'(p_*)| < 1$; for $a = 3.0$, $|f'(p_*)| = 1$ (in fact, the line $y = p$ is tangent to the curve $y = f(p)$ at p_*); and for $a = 3.3$, $|f'(p_*)| > 1$. The most interesting part of these three graphs is the S-shaped part of the curve $y = f(f(p))$ which seems to be pivoting about the equilibrium point p_*. At the precise moment ($a = 3.0$) at which the equilibrium point p_* becomes unstable the S-shaped curve is tangent at the point p_* to the curve $y = f(p)$. If we analyze the second derivative of the function $y = f(f(p))$ we see

$$(f \circ f)'(p) = f'(f(p))f'(p).$$

So

$$(f \circ f)''(p) = f''(f(p))f'(p)^2 + f'(f(p))f''(p)$$

and

$$\begin{aligned}(f \circ f)''(p_*) &= f''(f(p_*))f'(p_*)^2 + f'(f(p_*))f''(p_*) \\ &= f''(p_*)f'(p_*)^2 + f'(p_*)f''(p_*) \\ &= f''(p_*)(f'(p_*)^2 + f'(p_*))\end{aligned}$$

where the second-to-last line follows from the fact that $f(p_*) = p_*$. But now when $a = 3.0$, $f'(p_*) = -1$ so $(f \circ f)''(p_*) = 0$. In fact, the point p_* is a point of inflection for the curve $y = f(f(p))$ confirming our visual impression from Figure III.13. As a increases past 3.0 the S-shaped curve rotates to develop two intersections with the line $y = p$. These are the two points corresponding to the 2-cycle in the dynamical system

$$p_{n+1} = ap_n(1 - .001p_n)$$

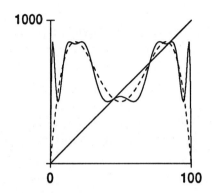

Figure III.15: Looking for 4-cycles, $p_{n+1} = 3.4p_n(1 - .001p_n)$

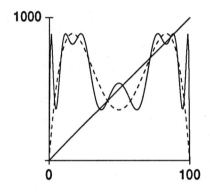

Figure III.16: Looking for 4-cycles, $p_{n+1} = 3.55p_n(1 - .001p_n)$

for a slightly above 3.0. The transition from an attracting equilibrium point at p_* to a repelling equilibrium point that occurs at the same time that an attracting 2-cycle appears is caused geometrically by the rotating S-shaped curve centered at $a = 3.0$ at the point p_*. This phenomenon is very general. It occurs in many different families of discrete dyanmical systems and it occurs for different periods. For example at the same time that the period 2-cycle becomes unstable a new period 4-cycle appears. This transition is illustrated by Figures III.15–III.16.

These figures show the curves

$$y = f(f(p)) \text{ (dashed) and}$$
$$y = f(f(f(f(p)))) \text{ (solid)}$$

along with the diagonal line $y = p$ because we are interested in 2-cycles and 4-cycles. As a increases from 3.4 to 3.55 we see a pair of S-shaped portions of the curve $y = f(f(f(f(p))))$ pivoting around the points in the 2-cycle. At the precise moment when the 2-cycle becomes unstable, a new 4-cycle appears. In the same way the 4-cycle eventually doubles introducing an 8-cycle, and then a 16-cycle, followed by a 32-cycle, and so forth. Later on odd cycles appear that, in turn, will undergo period doubling. For example, eventually a 7-cycle appears that becomes unstable at the precise moment that an attracting 14-cycle appears, followed by a 28-cycle, and so forth.

§2. Gravity

Falling Bodies

In this section we study the motion of an object under the influence of gravity. Despite the fact that this is a very common phenomenon it took an enormously long time for people to understand even roughly what happens when an object falls. Aristotle (384–322 B.C.) thought that the velocity of a falling body was proportional to its weight. Almost 2,000 years later, Galileo (1564–1642) obtained results contradicting the teachings of Aristotle. According to legend, he experimented by dropping objects from the Leaning Tower of Pisa. Although this particular experiment was actually due to Simon Steven, Galileo did perform many crucial experiments involving gravity and motion, often using very simple equipment. At the age of 19, for example, he timed the oscillations of a pendulum using his own pulse as a timing device. We begin our study of motion and gravity with a simple example.

Example:

Suppose that an object is dropped from a height five feet above the ground. Let t denote the time in seconds after the object is released and let $y(t)$ denote the height of the object in feet at time t. Since the object is dropped from a height of five feet, $y(0) = 5$. Since the object is released with no initial velocity, $y'(0) = 0$. We are interested in a relatively dense object dropped from a moderate height. So the only important force is gravity and we can ignore the effects of air resistance.

The force of gravity produces a constant acceleration of −32 feet per second per second. Notice the minus sign which reflects the fact that the acceleration is in the downward direction. Symbolically,

$$y''(t) = -32.$$

This problem is the opposite problem from the problem at which we looked in the last section. In the last section we knew the position $p(t)$ of an object and we wanted to find its velocity $p'(t)$ and its acceleration $p''(t)$. Now we know the acceleration $y''(t)$ of an object and we want to find its velocity $y'(t)$ and its position (or height) $y(t)$. Since acceleration is the derivative of velocity we know the velocity is a function whose derivative is −32. There are many possibilities, for example

$$f(t) = -32t \quad \text{or} \quad g(t) = -32t + 1.$$

In fact, any function of the form

$$f(t) = -32t + b$$

where b is a constant will do. This leaves us with two questions.

1. Is there any other function $f(t)$ whose derivative is -32t?

2. Which of these functions is the one that actually describes the velocity.

The answer to the first question is "no." We will prove this later in this section. To answer the second question we use the fact that the initial velocity of the object was 0.

$$\begin{aligned} y' &= -32t + b \\ y'(0) &= -32(0) + b \\ 0 &= b \end{aligned}$$

Hence, the function we're looking for is

$$y'(t) = -32t.$$

To summarize, we used two facts

$$y''(t) = -32$$

§2. GRAVITY

and
$$y'(0) = 0$$
to determine the function $y'(t)$. The first fact told us that $y'(t)$ must be a linear function whose slope is -32. That is, that
$$y'(t) = -32t + b.$$
The second fact told us that the constant b had to be 0.

We use the same ideas to find the function $y(t)$. We already know that
$$y'(t) = -32t.$$
So that $y(t)$ must be a function whose derivative is $-32t$. Once again, there are several possibilities. Any function of the form
$$y(t) = -16t^2 + c$$
where c is a constant will do. This raises the same two questions raised above. First, are there any other possibilities (the answer again is "no" and will be proved later in this section) and which of these possibilities actually describes the height of the object. We answer the second question using the fact that $y(0) = 5$.

$$\begin{aligned} y &= -16t^2 + c \\ y(0) &= -16(0)^2 + c \\ 5 &= c \end{aligned}$$

So the function we seek is
$$y(t) = -16t^2 + 5.$$

We have successfully solved our problem. We know the height $y(t)$ and velocity $y'(t)$ of the object at each time t. These two functions apply as long as the object is still in the air. When it hits the ground it will stop.

This is an example of an **Initial Value Problem**.

Definition:

An **initial value problem** involves two elements:

- A differential equation. In the example above the differential equation was
$$y''(t) = -32.$$

- Values of the function and its derivatives at a particular point. These data are called the **initial conditions**. In the example above the initial conditions were
$$y(0) = 5 \quad \text{and} \quad y'(0) = 0.$$

Example:

Suppose that you throw a baseball straight upward. At the moment that you release the ball it is six feet above the ground and is traveling with a velocity of 60 feet per second. Describe the subsequent motion of the ball. When will it hit the ground? How fast will it be traveling just before it hits the ground? What will be the highest point in its travels?

Answer:

Since a baseball is fairly dense and traveling at moderate speeds we ignore air resistance. The only important force acting on the ball will be gravity. As in the previous example we denote the height of the ball t seconds after it is released by the function $y(t)$. The initial conditions are
$$y(0) = 6$$
since the ball is released at a height of six feet and
$$y'(0) = 60$$
since its initial velocity is 60 feet per second.

As in the first example
$$y''(t) = -32.$$

Thus $y'(t)$ must be a linear function whose slope is -32.
$$y'(t) = -32t + b$$

§2. Gravity

We determine the value of the constant b using the initial condition $y'(0) = 60$.

$$\begin{aligned} y'(t) &= -32t + b \\ y'(0) &= -32(0) + b \\ 60 &= b \end{aligned}$$

Hence the velocity is given by

$$y'(t) = -32t + 60.$$

Next we must find a function whose derivative is $y'(t) = -32t + 60$. Any function of the form

$$y(t) = -16t^2 + 60t + c$$

where c is a constant will do.

We determine the value of the constant c using the initial condition $y(0) = 6$.

$$\begin{aligned} y(t) &= -16t^2 + 60t + c \\ y(0) &= 16(0)^2 + 60(0) + c \\ 6 &= c \end{aligned}$$

Thus the function we seek is

$$y(t) = -16t^2 + 60t + 6.$$

We have answered the first question. We have a function that describes the height of the ball t seconds after its release. The next question asks when the ball will hit the ground. To answer this question we must solve the equation

$$\begin{aligned} y(t) &= 0 \\ -16t^2 + 60t + 6 &= 0 \\ 8t^2 - 30t - 3 &= 0 \end{aligned}$$

Using the Quadratic Formula we find two solutions of this equation.

$$t = \frac{30 \pm \sqrt{900 + 96}}{16}$$

or

$$t = 3.8475 \quad \text{and} \quad t = -0.0975.$$

The second solution is negative and meaningless for this problem since we are only studying times $t \geq 0$. The ball will strike the ground 3.8475 seconds after it has been released.

To answer the next question (How fast is the ball traveling the instant before it hits the ground?) we evaluate

$$y'(3.8475) = -32(3.8475) + 60 = -63.1189 \text{ feet per second.}$$

The last question asks hows high the ball will be at the highest point in its travels. That is, we want to find the maximum of the function $y(t)$ on the interval $[0, 3.8475]$. We need to examine the critical point(s) of this function and the two endpoints. At the first endpoint $t = 0$ we have just released the ball and it is traveling upward. So this will not be the maximum height. At the other endpoint $t = 3.8475$ the ball has just hit the ground. So the only possibility is at a critical point. We must solve the equation

$$y'(t) = 0.$$

In other words, we are looking for the point at which the velocity is zero. Physically, this is the point at which the ball has stopped rising and is about to begin falling.

$$\begin{aligned} y'(t) &= 0 \\ -32t + 60 &= 0 \\ -32t &= -60 \\ t &= 60/32 \\ &= 15/8 = 1.875 \end{aligned}$$

So the ball will reach its highest point at time $t = 1.875$. At that time its height will be

$$y(1.875) = -16(1.875)^2 + 60(1.875) + 6 = 62.25 \text{ feet.}$$

§2. Gravity

Exercises

*Exercise III.2.1:

Suppose another person throws a baseball straight up. She throws the baseball twice as fast as in the preceding example, but everything else is the same. Thus

$$y'(0) = 120 \quad \text{and} \quad y(0) = 6$$

Describe the motion of the ball. When will it hit the ground? How fast will it be traveling just before it hits the ground? What will be the highest point in its travels? Compare your answers to this question to the example above.

*Exercise III.2.2:

A man drops a baseball from the top of a 100 foot tower. How long will it take before the ball hits the ground? How fast will it be traveling just before it hits the ground?

Exercise III.2.3:

A man drops a baseball from the top of a 200 foot tower. How long will it take before the ball hits the ground? How fast will it be traveling just before it hits the ground?

*Exercise III.2.4:

Energy is the capacity to do work. The energy it takes to carry an object from the ground to the top of a tower depends on the mass of the object and the height of the tower. The formula for the energy required to carry an object from the ground to the top of a tower whose height is x feet is $E = kx$ where k is a constant depending on the mass of the object and the units used for the quantities involved. This formula is not very surprising. It says that the energy required to carry a weight to the top of a tower whose height is x feet is x times the energy required to carry the same weight to the top of a tower whose height is one foot.

A moving object has energy by virtue of its motion. This energy is called **kinetic energy**. When a baseball is thrown straight upward its initial kinetic energy is used to lift it. At first, as the ball rises, it slows down as its kinetic energy is used lifting the ball against the pull of gravity. Later, as the ball falls, the work done by the pull of gravity is converted back into kinetic energy and the ball speeds up. The amount of kinetic energy a moving object has depends on its velocity and its

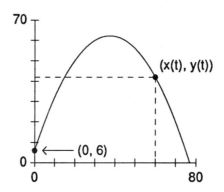

Figure III.17: Path of a baseball

mass. Try to use the work we've done so far to find a formula for the kinetic energy of a moving object with velocity v. Your formula will involve the constant k. The exact value of this constant is irrelevant for this exercise.

Motion in Two Dimensions

In the first part of this section we discussed one-dimensional motion. We looked at an object that was traveling straight up and down. In this section we want to discuss motion in two dimensions. We are interested in studying what happens when an object, for example a baseball, is thrown not straight up but at an angle so that it is traveling both horizontally and vertically. To describe the motion of such an object we use two functions. The first function $x(t)$ describes how far the object is from its starting point in a horizontal direction. The second function $y(t)$ describes the height of the object. We continue to measure distance in feet and time in seconds. We choose coordinates so that the x-axis is the ground and the object's initial position is on the y-axis.

Example:

Figure III.17 shows the trajectory or path followed by an object that is thrown from an initial height of six feet with a velocity of 60 feet per second in the vertical direction and 20 feet per second in the horizontal direction.

Consider the horizontal motion. There is no acceleration in the horizontal direction. The only horizontal force might come from air

§2. Gravity

resistance or wind, but a baseball is sufficiently dense that we may ignore this force. Since the initial velocity is 20 feet per second in the horizontal direction and there is no horizontal force, the horizontal velocity will remain constant.

$$x'(t) = 20$$

Thus $x(t)$ must be of the form $x(t) = 20t + b$. The constant b is zero since at time zero $x(0) = 0$. Thus the function $x(t)$ is

$$x(t) = 20t.$$

Now consider the vertical motion. We have exactly the same situation as in the second example in the first part of this section. Thus

$$y(t) = -16t^2 + 60t + 6.$$

Using these two functions we see, for example, that at time $t = 3$ the position of the baseball will be at

$$(x(3), y(3)) = (60, 42).$$

This point is marked by a dot in Figure III.17.

One way to sketch a graph like Figure III.17 showing the trajectory or path followed by a object is as follows. The position of the object is described by the two functions

$$\begin{aligned} x(t) &= 20t \\ y(t) &= -16t^2 + 60t + 6. \end{aligned}$$

We solve the equation

$$x = 20t$$

for t to obtain

$$t = x/20.$$

Then we can express y as a function of x by replacing t by $x/20$ in the equation

$$y(t) = -16t^2 + 60t + 6$$

to obtain

$$y = -16t^2 + 60t + 6$$

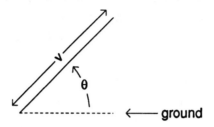

Figure III.18: Initial speed and angle of elevation

$$\begin{aligned} &= -16\left(\frac{x}{20}\right)^2 + 60\left(\frac{x}{20}\right) + 6 \\ &= -\frac{x^2}{25} + 3x + 6. \end{aligned}$$

This is the function whose graph is shown in Figure III.17. Notice that one needs to go back to the original pair of functions to find out *when* the baseball can be found at each point on its path.

Exercises

*Exercise III.2.5:
Draw a graph showing the trajectory followed by a baseball that is thrown from a height of five feet with a velocity of 30 feet per second in the horizontal direction and 40 feet per second in the vertical direction. Label the highest point on the trajectory. Label the position of the baseball one second after it is thrown. Where does it hit the ground?

Exercise III.2.6:
Draw a graph showing the trajectory followed by a baseball that is thrown from a height of five feet with a velocity of 40 feet per second in the horizontal direction and 30 feet per second in the vertical direction. Label the highest point on the trajectory. Label the position of the baseball one second after it is thrown. Where does it hit the ground?

Frequently a person describes the way in which a baseball is thrown by giving its speed and the angle of elevation. For example, one might say that a baseball is thrown at an angle of elevation of θ radians with

§2. GRAVITY

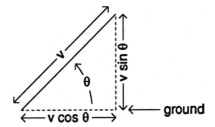

Figure III.19: Initial horizontal and vertical velocity

a speed of v feet per second. Figure III.18 shows what is meant by this description. We can express this information in terms of horizontal and vertical velocity as shown in Figure III.19.

$$\begin{aligned} \text{horizontal velocity} &= v \cos \theta \\ \text{vertical velocity} &= v \sin \theta \end{aligned}$$

Exercises

*Exercise III.2.7:
Suppose that a baseball is thrown with a speed of 45 feet per second at an angle of elevation of $\pi/4$ radians. Find the initial horizontal and vertical velocities.

*Exercise III.2.8:
Suppose that a cannon always fires a projectile with an initial speed of 100 feet per second and that the projectile is fired from ground level. Find the trajectory the projectile will follow if it is fired at an elevation of $\pi/6$ radians. Sketch a graph of this trajectory. How long will it remain in the air before hitting the ground? How far will it have traveled horizontally when it hits the ground?

Exercise III.2.9:
Suppose that a cannon always fires a projectile with an initial speed of 100 feet per second and that the projectile is fired from ground level. Find the trajectory the projectile will follow if it is fired at an elevation of $\pi/4$ radians. Sketch a graph of this trajectory. How long will it remain in the air before hitting the ground? How far will it have traveled horizontally when it hits the ground?

Exercise III.2.10:

Suppose that a cannon always fires a projectile with an initial speed of 100 feet per second and that the projectile is fired from ground level. Find the trajectory the projectile will follow if it is fired at an elevation of $\pi/3$ radians. Sketch a graph of this trajectory. How long will it remain in the air before hitting the ground? How far will it have traveled horizontally when it hits the ground?

*Exercise III.2.11:

What angle of elevation should be used so that the projectile will travel as far as possible horizontally before it hits the ground? What is the farthest possible horizontal distance for the projectile to travel? This is called the *range* of the cannon.

*Exercise III.2.12:

How does the range of the cannon depend on the initial speed of the projectile?

*Exercise III.2.13:

What angle of elevation should be used so that the projectile will remain in the air as long as possible?

A Theorem

In the first example in this section we noticed that any function of the form $v(t) = -32t + b$ was a solution of the differential equation $v' = -32$. We raised the question of whether all the solutions of this differential equation were of this form. We answered the question "yes" with no justification. In this section we want to justify this answer. We begin with a lemma.

Lemma 1 *Consider the initial value problem*

$$y'(t) = 0 \qquad y(0) = 0$$

The only solution to this problem is the function

$$y(t) = 0$$

Proof:

§2. Gravity

Suppose there is another solution
$$y(t) = g(t)$$
which is different from 0. Then there is some point a such that $g(a) \neq 0$. By the Mean Value Theorem there is some point x_0 between 0 and a such that
$$g'(x_0) = \frac{g(a) - g(0)}{a - 0}$$
But $g(0)$ is 0 since $y = g(t)$ is a solution of the initial value problem above. Since $g(a) \neq 0$ the right hand side of the equation above is not zero. Hence
$$g'(x_0) \neq 0$$
which contradicts the assumption that $y = g(t)$ is a solution of the initial value problem.

Theorem 1 *Suppose that*
$$y' = w(t)$$
is any differential equation and that
$$y = f(t)$$
is any solution of this differential equation.

Suppose that
$$y = g(t)$$
is another solution of the same differential equation.

Then there is a constant c, namely $g(0) - f(0)$, such that
$$g(t) = f(t) + c.$$

Proof:
Consider the function
$$h(t) = g(t) - (f(t) + c) = g(t) - f(t) - c = g(t) - f(t) - (g(0) - f(0)).$$
Notice that
$$h(0) = 0$$
and
$$h'(t) = g'(t) - f'(t) = w(t) - w(t) = 0$$
Thus by the previous lemma, $h(t) = 0$ and $g(t) = f(t) + c$ as we wanted to show.

§3. Visualizing Differential Equations

Some Terminology

We began this course by looking at population models and, in particular, at models of the growth of a species like temperate zone insects for which the population is broken up into distinct generations. Many species, however, do not have distinct generations. For these species it is appropriate to describe the population by a function $p(t)$ that gives the population at each time t and to describe the way in which it changes using words like

> "The population is increasing at the rate of 100 individuals per year"

or

> "The population is increasing at the rate of two percent per year"

We can express the first sentence by the differential equation

$$\frac{dp}{dt} = 100$$

and the second sentence by the differential equation

$$\frac{dp}{dt} = .02p.$$

Notice that in this second differential equation the rate at which the population p is changing depends on the current population.

In some situations the rate at which a variable p changes depends on both p and t. For example, in the temperate zones many species find winter a rough season; there is little food available and the weather is often harsh; death rates rise and birth rates fall. For such a species the population growth might be described verbally as follows.

> "The population is falling at the rate of one percent per year *during the winter* and rising at the rate of two percent per year during the other three seasons."

§3. VISUALIZING DIFFERENTIAL EQUATIONS

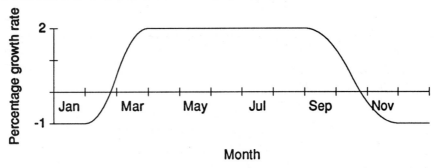

Figure III.20: Percentage rate of population growth

This sentence is an oversimplification. In fact, the percentage rate of growth would probably be a function like the one shown in Figure III.20—negative during the winter, high from late spring through early autumn, with intermediate values during the early spring and late fall. This function would be periodic with period one year; it depends on the season and not on the year. If we denote this function by $R(t)$ then this situation would be described by the differential equation

$$\frac{dp}{dt} = 0.01 R(t) p.$$

Notice that the right hand side of this equation involves both the variable p and the variable t. The following terms are useful in describing differential equations.

Definition:

- The differential equation

$$\frac{dp}{dt} = f(t, p)$$

  is said to be **autonomous** if the right hand side does not involve the variable t.

- The differential equation

$$\frac{dp}{dt} = f(t, p)$$

  is said to be **nonautonomous** if the right hand side does involve the variable t. Sometimes the adjective **time-dependent** is used instead of nonautonomous.

The definition above was phrased in terms of the variables p and t. The particular letters used to denote the variables is unimportant. For example, the differential equation

$$\frac{dy}{dx} = ky$$

is *autonomous* because the *independent* variable x does not appear on the right hand side.

Examples:

1. The differential equation

$$\frac{dp}{dt} = p(1000 - p)$$

is autonomous.

2. The differential equation

$$\frac{dz}{dx} = 2x + 3z$$

is nonautonomous.

Definition:

A function $y = g(x)$ is called a **solution to the initial value problem**

$$\frac{dy}{dx} = f(x, y), \quad y(x_0) = y_0$$

if

- $$\frac{dg}{dx} = f(x, y) \quad \text{and}$$

- $$g(x_0) = y_0.$$

§3. Visualizing Differential Equations

A function $y = g(x)$ is called a **solution of the differential equation**

$$\frac{dy}{dx} = f(x, y)$$

if

$$\frac{dg}{dx} = f(x, y).$$

Example:

The function

$$y = \frac{1}{1-x}$$

is a solution to the initial value problem

$$\frac{dy}{dx} = y^2, \quad y(0) = 1.$$

Proof:

$$\begin{aligned} y &= (1-x)^{-1} \\ y' &= -(1-x)^{-2}(-1) \\ &= (1-x)^{-2} \\ &= y^2 \end{aligned}$$

and

$$\begin{aligned} y(0) &= \frac{1}{1-0} \\ &= 1 \end{aligned}$$

Definition:

- A **first order** differential equation involves only a function and its first derivative.
- A **second order** differential equation involves only a function and its first and second derivatives.

Visualizing a Differential Equation

The main purpose of this section is to develop techniques for visualizing a first order differential equation of the form

$$\frac{dy}{dx} = f(x,y).$$

At each point (x, y) this differential equation tells us what the slope of a solution would be *if it were to pass through that point.*

For example, consider the differential equation

$$\frac{dy}{dx} = -\frac{x}{y}.$$

At the point $(3, 4)$ we see that

$$\frac{dy}{dx} = -\frac{3}{4}$$

so if a particular solution to this differential equation passes through this point then at this point its slope will be $-3/4$.

The following table shows similar calculations for several points for this differential equation.

x	y	slope $= f(x,y) = -x/y$
3	4	$-3/4$
4	3	$-4/3$
-4	3	$4/3$

Figure III.21 shows a solution of this differential equation that passes through the three points in the table. Notice at each point listed in the table the tangent has the slope given in the table.

We can visualize a differential equation by drawing a line at each point (x, y) whose slope is $f(x, y)$. Figure III.22 illustrates this for the differential equation

$$\frac{dy}{dx} = -\frac{x}{y}.$$

Notice that since the right hand side of this differential equation is undefined when $y = 0$ there are no slopes indicated for points on the

§3. VISUALIZING DIFFERENTIAL EQUATIONS 335

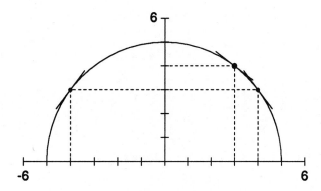

Figure III.21: Three tangents on the solution

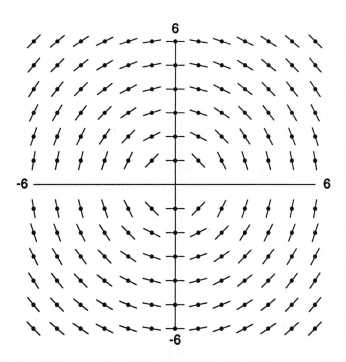

Figure III.22: The slope field for $dy/dx = -x/y$

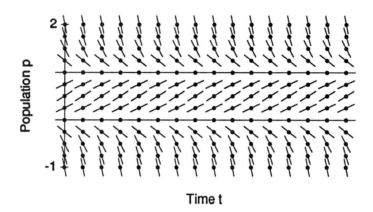

Figure III.23: The slope field for $p' = p(1-p)$

x-axis. A graph like the one shown in Figure III.22 is called a **slope field** or a **tangent field**.

One way to visualize the solution of an initial value problem is to think of yourself as walking about on a slope field starting at the point indicated by the initial condition. At each point there is a slope line like the slope lines in Figure III.22. As you walk along you must always walk in the direction indicated by the slope line. You walk forward to determine future[1] values of y and backward to determine past values of y. Thus the path you follow will look like the curve shown in Figure III.21. As this path passes through each point its tangent will line up with the slope line at that point.

We can learn a lot about a differential equation and about a situation that is described by a differential equation by looking at the slope field.

Example:

Consider the differential equation

$$p' = p(1-p)$$

Figure III.23 shows the slope field for this differential equation. This figure uses the vertical axis for the variable p and the horizontal axis for the variable t. It was drawn by a computer. You will use a computer

---

[1]The words "future" and "past" are most appropriate when the variable x represents time. We will use them even when x represents something else.

§3. Visualizing Differential Equations

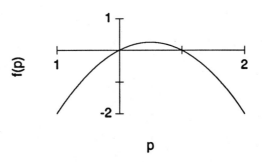

Figure III.24: The function $f(p) = p(1-p)$

program to sketch similar slope fields in the computer laboratory. One can, however, make a rough sketch like Figure III.23 without using a computer by examining the function on the right side of the differential equation

$$\frac{dp}{dt} = p(1-p).$$

If we sketch a graph of this function we obtain Figure III.24

Warning: We are looking at several very different graphs here. Figure III.24 is a graph of the function $f(p) = p(1-p)$ on the right hand side of the differential equation

$$\frac{dp}{dt} = p(1-p).$$

The horizontal axis of this graph represents the variable p and the vertical axis represents the value $f(p)$.

Figure III.23 is very different. It is a slope field. The horizontal axis represents the variable t and the vertical axis represents the variable p.

In Figure III.24 we see that the right hand side of the differential equation

$$\frac{dp}{dt} = p(1-p)$$

is positive when $0 < p < 1$ and negative when $p < 0$ or $1 < p$. Thus in Figure III.23 the slope lines are pointing upward when $0 < p < 1$ and downward when $p < 0$ or $1 < p$.

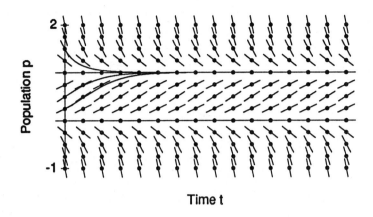

Figure III.25: Slope field and three solutions

The sign of f in Figure III.24 tells us the direction of the slope lines in Figure III.23.

In Figure III.24 we see that the right hand side of the differential equation is zero when $p = 0$ or $p = 1$. Correspondingly, in Figure III.23 the slope lines are horizontal when $p = 0$ or $p = 1$. This information gives a general idea of the appearance of Figure III.23. With this information you can make a rough sketch of the slope field that captures its most important qualitative features. This method works very well for autonomous differential equations. It is much harder to capture the essential elements of slope field of a nonautonomous differential equation without computer graphics.

Figure III.25 is identical to Figure III.23 except that we have added curves for the solutions of three initial value problems—with the initial conditions: $y(0) = 0.25, y(0) = 0.5$, and $y(0) = 1.50$. Notice that the tangents to these curves match the slope field.

By looking at slope fields like Figure III.23 one can tell a lot about the solutions of initial value problems. For example, in Figure III.23 we see that if the initial value is between $p = 0$ and $p = 1$ then as it goes to the right the solution will be directed upward away from $p = 0$ toward $p = 1$. Similarly, if the initial value of p is greater than 1 then as it goes to the right the solution will be directed downward toward $p = 1$. This behavior is reminscent of behavior we saw in chapter 1. We use similar terminology in this situation.

§3. Visualizing Differential Equations

Definition:

Consider an autonomous differential equation

$$\frac{dp}{dt} = f(p).$$

- A point p_* is said to be an **equilibrium point** if $f(p_*) = 0$. Notice that if p_* is an equilibrium point then an initial value problem that starts at p_* will have a solution that remains at p_*.

    In the preceding example there are two equilibrium points, 0 and 1.

- An equilibrium point p_* is said to be **attracting** if any initial value problem that starts sufficiently close to p_* will have

$$\lim_{t \to +\infty} p(t) = p_*.$$

    The equilibrium point $p_* = 1$ in the preceding example is attracting.

- An equilibrium point p_* is said to be **repelling** if any initial value problem that starts out sufficiently close to p_* is pushed away from p_*.

    The equilibrium point $p_* = 0$ in the preceding example is repelling.

Our next example is a little more complicated.

Example:

We want to look at the population of a species that hunts in packs. We looked at such species in chapter I in the context of models for species with distinct generations. Now we would like to look at a species without distinct generations that hunts in packs. The simplest population models look like

$$p' = Gp$$

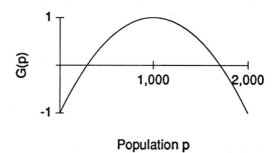

Figure III.26: $G(p)$ for a species that hunts in packs

where the constant G represents the "percentage" rate at which the population is changing. For example, a population that is increasing at the constant rate of two percent per year would be described by

$$p' = .02p$$

and a population that is declining at the rate of three percent per year would be described by

$$p' = -.03p.$$

However, models in which the "percentage" rate of of change is constant are unrealistic. These are very similar to the exponential models that we studied in section I.1. Usually, G will be a function of p rather than a constant. For a species that hunts in packs G would be a function similar to the one shown in Figure III.26.

Looking at this graph we see a number of things about the species. Notice that when the population p is close to 1000 then $G(p)$ is very large, close to 1. This says that when the population is close to 1000 it is increasing at a rate of about 100 percent per year. When p is very small notice that $G(p)$ is negative. This corresponds to the fact that when the population is small there are not enough individuals to form efficient hunting packs. As a result the animals have difficulty catching enough food to eat and the population declines. When the population is very large $G(p)$ is negative. Now the population is declining because there are too many individuals for the available food, water, and shelter. The function $G(p)$ shown in Figure III.26 is given by the equation

$$G(p) = 1 - \frac{(p - 1000)^2}{500,000}.$$

§3. Visualizing Differential Equations

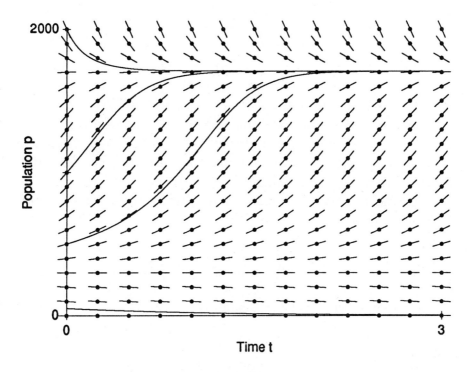

Figure III.27: Slope field and solutions

The population model corresponding to this example is described by the differential equation

$$p' = \left(1 - \frac{(p-1000)^2}{500,000}\right)p.$$

Notice that the right hand side of this differential equation is zero when $p = 0$ or when $G(p) = 0$. Except for the point $p = 0$ the sign of the right hand side is the same as the sign of $G(p)$ since p cannot be negative in a population model. Figure III.27 shows the slope field corresponding to this differential equation with solutions for the initial conditions: $p(0) = 50, p(0) = 500, p(0) = 1000$, and $p(0) = 2000$.

Looking at these figures we see exactly what we would expect. When the population is very low, too low for individuals to form efficient hunting packs, the slopes are pointing downward, indicating that the population will decline. When the population is moderate, there are enough individuals to form efficient hunting packs and there is enough food, water, and shelter to go around. As a result the slopes are pointing up, indicating that the population will rise. When the population is extremely large, there is no longer enough food, water, and shelter. As a result the slopes are pointing down, indicating that the population will decline. The four solutions of initial value problems show examples of what will happen with various initial populations. If $p(0) = 50$ there are so few individuals that the animals are unable to form efficient hunting packs. As a result the population dies out. For $p(0) = 500$ or $p(0) = 1000$ the population rises rapidly at first and then appears to be leveling off. For $p(0) = 2000$ the population drops rather dramatically and then appears to be leveling off.

To obtain a general idea about the behavior of solutions to the differential equation

$$p' = \left(1 - \frac{(p-1000)^2}{500,000}\right)p$$

we need to know when the right hand side

$$\left(1 - \frac{(p-1000)^2}{500,000}\right)p$$

is positive and when it is negative. We begin by solving the equation

$$\left(1 - \frac{(p-1000)^2}{500,000}\right)p = 0$$

$$1 - \frac{(p-1000)^2}{500,000} = 0 \quad \text{or} \quad p = 0$$

$$\frac{(p-1000)^2}{500,000} = 1$$

$$(p-1000)^2 = 500,000$$

$$p - 1000 = \pm\sqrt{500,000} = \pm 707.107$$

$$p = 1000 \pm 707.107$$

giving us three solutions 0, 292.893 and 1,707.107. In the usual way we can make a diagram (Figure III.28) showing where the right hand

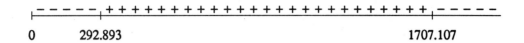

Figure III.28: The sign of $p(1 - (p - 1000)^2/500{,}000)$

side is positive and where it is negative. Notice how this information corresponds to the direction of the slopes in Figure III.27. This differential equation has three equilibrium points: 0, 292.893, and 1,707.107. The equilibrium point 292.893 is repelling. The population 292.893 is sometimes called a **threshold** because if the initial population is above this threshold then the species will grow and survive, but if the initial population is below this threshold then the species will die out. The equilibrium point 1,707.107 is attracting. This is sometimes called the **carrying capacity** for this model since this is the level the population seems to seek. At this level there is sufficient food and other resources to sustain the population.

Exercises

*Exercise III.3.1:
Consider the differential equation

$$p' = 1$$

Draw a slope field for this differential equation. Find all its equilibrium points and determine the character (attracting, repelling, or neither) of each. What can you say about the functions $p(t)$ that satisfy this differential equation?

Exercise III.3.2:
Consider the differential equation

$$p' = -1$$

Draw a slope field for this differential equation. Find all its equilibrium points and determine the character (attracting, repelling, or neither) of each. What can you say about the functions $p(t)$ that satisfy this differential equation?

*Exercise III.3.3:
Consider the differential equation
$$p' = p(p-1)(p-2).$$
Draw a slope field for this differential equation. Find all its equilibrium points and determine the character (attracting, repelling, or neither) of each. What can you say about the functions $p(t)$ that satisfy this differential equation?

*Exercise III.3.4:
Consider the differential equation
$$p' = (p-1)^2.$$
Draw a slope field for this differential equation. Find all its equilibrium points and determine the character (attracting, repelling, or neither) of each. What can you say about the functions $p(t)$ that satisfy this differential equation?

Exercise III.3.5:
Consider the differential equation
$$p' = (p-1)^2(p-2).$$
Draw a slope field for this differential equation. Find all its equilibrium points and determine the character (attracting, repelling, or neither) of each. What can you say about the functions $p(t)$ that satisfy this differential equation?

*Exercise III.3.6:
Consider the differential equation
$$p' = \sin p$$
Draw a slope field for this differential equation. Find all its equilibrium points and determine the character (attracting, repelling, or neither) of each. What can you say about the functions $p(t)$ that satisfy this differential equation?

*Exercise III.3.7:
Consider the differential equation
$$p' = t(t-1)$$

Draw a slope field for this differential equation.

Exercise III.3.8:
   Consider the differential equation
$$p' = t^2 - 1$$
Draw a slope field for this differential equation.

*Exercise III.3.9:
   Consider the differential equation
$$p' = \frac{p}{t}$$
Draw a slope field for this differential equation.

---

§4. Euler's Method

Euler's Method is a numerical method for *approximating* the solution of an initial value problem of the form

$$\frac{dy}{dt} = f(t, y), \quad y(t_0) = y_0.$$

It is based on the fact that on small intervals smooth functions look very much like linear functions.

The single most important picture in this book is Figure III.29 which shows two views of a function $y(t)$. The first view, on the left, provides an overall view of the function. The second view, on the right, shows a magnified view of the shaded area in the overall view. Both views show the function itself and the tangent to the function at a particular point t_0. The tangent has slope $y'(t_0)$ and goes through the point $(t_0, y(t_0))$. Hence, it can be described by the linear function

$$L(t) = y'(t_0)(t - t_0) + y(t_0).$$

Notice that in the magnified view the tangent is very close to the curve itself, but in the overall view at points some distance from the point t_0 the tangent and the curve are quite far apart. This figure illustrates the key idea.

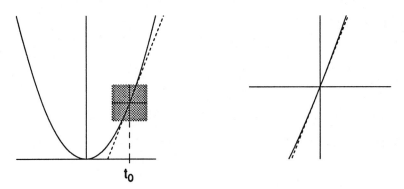

Figure III.29: A function and a tangent

Close to a point a smooth function looks very much like a linear function.

This picture motivates Euler's Method. If we know the value of a function $y(t)$ at a point t_0 and we know the derivative of the function at t_0 then we can *estimate* the value of y at a *nearby* point t by using the linear function

$$L(t) = y'(t_0)(t - t_0) + y(t_0).$$

If t is very close to t_0 then $L(t)$ will be very close to $y(t)$. Symbolically, we write

$$y(t) \approx y'(t_0)(t - t_0) + y(t_0).$$

The symbol "$\approx$" is read "approximately equal to." This approximation procedure is very natural. Consider the following example.

Example:

Suppose that at 12:00 noon a leaky bucket contains 2 liters of water. The rate at which water is leaking from the bucket is varying. In particular, as the water level drops, water leaks out more slowly. At 12:00 water is leaking out at the rate of 0.1 liters per minute. How much water would the bucket contain at 12:02? How much water would it contain at 12:10?

Answer:

Based on the information in this problem we cannot give an exact answer. We can, however, estimate that after two minutes 0.2 liters

§4. Euler's Method

will have leaked out leaving 1.8 liters at 12:02. We can write down the calculation as follows.

Let $y(t)$ denote the amount of water in the bucket at time t. We know that
$$y(12:00) = 2 \text{ liters}$$
and that
$$y'(12:00) = -0.1 \text{ liters per minute.}$$
Thus we can estimate that if t is relatively close to noon
$$y(t) \approx -0.1(t - 12:00) + 2.$$
Since the first question asked us to predict the amount of water in the bucket after a relatively short time, 2 minutes, we were able to estimate that
$$y(12:02) \approx -0.1(2) + 2.0 = 1.8 \text{ liters.}$$

The second question is more difficult. It asks us to predict how much water will remain in the bucket after 10 minutes. If we use the same procedure as above we obtain
$$y(12:10) \approx -0.1(10) + 2.0 = 1.0 \text{ liters.}$$

This estimate is not very good. As the water level in the bucket drops, the rate at which water leaks from the bucket will also drop. After ten minutes our estimate shows that the bucket will have half the water it did at noon. If this were true then water would be leaking out of the bucket at a much slower rate than it was at noon. Thus an estimate based on the assumption that water continued to leak out of the bucket at a rate of 0.1 liters per minute for the entire ten minutes would be way off. The approximation
$$y(t) \approx -0.1(t - 12:00) + 2$$
is only good for values of t close to 12:00.

Euler's Method: An Example

In this subsection we look at one particular initial value problem and show how the ideas discussed above can be used to estimate a solution. Consider the initial value problem
$$\frac{dy}{dt} = y + t, \quad y(0) = 1$$

and suppose that we are interested in approximating the solution of this initial value problem on the interval $[0,2]$. We can get a very rough approximation based on the ideas above. The slope of the tangent to the curve $y(t)$ at the point $t = 0$ is $y'(0)$. Using the initial condition $y(0) = 1$ and the differential equation

$$\frac{dy}{dt} = y + t$$

we see that

$$y'(0) = y(0) + 0 = 1 + 0 = 1.$$

The tangent goes through the point $(0, 1)$, so it is described by the linear function

$$L(t) = t + 1.$$

This gives us an approximation

$$y(t) \approx t + 1$$

for the function $y(t)$. For small values of t this will be a relatively good approximation but for larger values of t it will not be very good.

Figure III.30 shows this approximation and the actual solution. The curved line is the exact solution and the straight line is the approximation. The dot marks the initial condition. As you can see, for values of t some distance from 0 this approximation is really terrible. By looking at the graph you can see why. The slope of the exact solution is changing but the approximation has the same slope for the entire interval $[0, 2]$.

One way to improve this approximation is to make a "midcourse correction." Instead of approximating the function in one giant step we will do it in two smaller steps. The first step is to use the approximation above, that is,

$$y(t) \approx t + 1$$

on the interval $[0, 1]$. At the end of this interval we see that

$$y(1) \approx 2$$

and we use this information to estimate $y'(1)$.

$$y'(1) \approx 2 + 1 = 3.$$

§4. Euler's Method

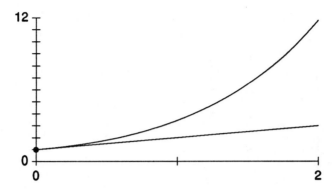

Figure III.30: A very rough approximation

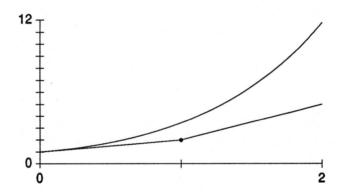

Figure III.31: A slightly better approximation

This leads us to an estimate for the tangent to the curve $y(t)$ at the point $t = 1$

$$L(t) = 3(t-1) + 2$$

which, in turn, gives us an approximation to the function $y(t)$ near the point $t = 1$

$$y(t) \approx 3(t-1) + 2.$$

We use this new approximation on the interval $[1, 2]$ together with our original approximation on the interval $[0, 1]$. Figure III.31 puts these two approximations together. Notice the dot at the point $t = 1$ where we make the "midcourse correction."

This second approximation is only slightly better than the first one. It points the way, however, to a series of better and better approxima-

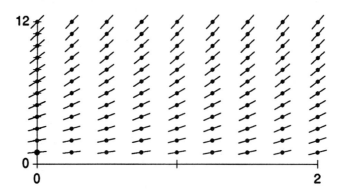

Figure III.32: Slope field for $y' = y + t$

tions. By making more "midcourse corrections" we can obtain better approximations. The basic idea is that tangents are close to smooth curves only over small intervals. By using each tangent as an approximation on only a very small interval we will stay reasonably close to the curve. This is **Euler's Method**.

One way to conceptualize Euler's Method is to imagine that you are walking on a piece of graph paper with the slope field for the differential equation. Figure III.32 shows the slope field for the differential equation

$$\frac{dy}{dt} = y + t.$$

You start walking at the point indicated by the initial condition. In Figure III.32 this point is marked by a large dot. Using Euler's Method you start at the dot and begin walking in the direction indicated by the slope field. After a short time you stop, look at the slope indicated by the slope field at your current point, and set off in this new direction. After walking for a short time in the new direction you stop, check the slope indicated by the slope field at your current point, and set off in this new direction. By stopping and changing directions very frequently you can obtain a very good approximation to the actual solution of the given initial value problem.

Notice that Euler's Method does not give us the exact solution to an initial value problem. It gives us a sequence of approximations. By making enough "midcourse corrections" one can obtain an approximation that is good enough for any particular purpose. The number of "midcourse corrections" that is necessary depends on how good the ap-

§4. Euler's Method

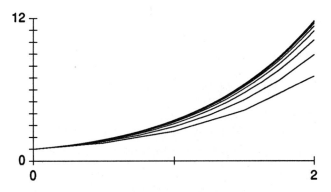

Figure III.33: More and better approximations

proximation must be. If the exact solution is denoted by $y(t)$ and the approximation obtained by using n "midcourse corrections" is denoted by $p_n(t)$ then we can say

$$\lim_{n \to \infty} p_n(t) = y(t).$$

Figure III.33 shows the actual solution $y(t)$ and the approximations $p_4(t), p_8(t), \ldots p_{512}(t)$. Notice how the approximations are getting closer and closer to the actual solution. The calculations involved in the estimate $p_4(t)$ are shown in the table below.

	beginning of interval			end of interval	
t	y		$f(t,y)$	t	y
0.00	1.00 (exact)	1.00 (exact)		0.50	1.500 (estimate)
0.50	1.50 (estimate)	2.00 (estimate)		1.00	2.500 (estimate)
1.00	2.50 (estimate)	3.50 (estimate)		1.50	4.250 (estimate)
1.50	4.25 (estimate)	5.75 (estimate)		2.00	7.125 (estimate)

Euler's Method

In this section we introduce some notation and write down carefully the formulas involved in Euler's method.

Suppose that we want to approximate a solution to the initial value problem

$$\frac{dy}{dt} = f(t,y), \quad y(a) = y_0$$

Figure III.34: $[a, b]$ broken into n subintervals

on the interval $[a, b]$ based on n "midcourse corrections." We divide the interval $[a, b]$ into n equal subintervals each of length

$$h = (b - a)/n.$$

We use the notation

$$\begin{aligned} t_0 &= a \\ t_1 &= a + h \\ t_2 &= a + 2h \\ &\vdots \\ t_i &= a + ih \\ &\vdots \\ t_n &= a + nh = b \end{aligned}$$

as shown in Figure III.34.

We use $y(t_i)$, as usual, to denote the value of the actual solution of the initial value problem at the point t_i. We use w_i to denote the estimate given by Euler's Method for $y(t_i)$. This is an important distinction. Usually we do not know $y(t_i)$, so we need to work with the estimates w_i. On each interval $[t_{i-1}, t_i]$ we approximate $y(t)$ by the linear function

$$L_i(t) = f(t_{i-1}, w_{i-1})(t - t_{i-1}) + w_{i-1}.$$

This is our best estimate for the actual tangent

$$A_i(t) = \left(\frac{dy}{dt}\right)(t_{i-1}) = f(t_{i-1}, y(t_{i-1}))(t - t_{i-1}) + y(t_{i-1})$$

§4. Euler's Method

since w_{i-1} is our best estimate for $y(t_{i-1})$. Thus, in particular, we obtain each w_i based on w_{i-1} by

$$\begin{aligned} w_i &= L_i(t_i) \\ &= f(t_{i-1}, w_{i-1})(t_i - t_{i-1}) + w_{i-1} \\ &= f(t_{i-1}, w_{i-1})h + w_{i-1} \end{aligned}$$

We summarize Euler's Method as follows

Euler's Method

Given an initial value problem

$$\frac{dy}{dt} = f(t, y), \quad y(a) = y_0,$$

an interval $[a, b]$, and a positive integer n, compute the following

- $h = (b - a)/n$.
- $w_0 = y_0$.
- $t_i = a + ih$, for $i = 0, 1, 2, \ldots, n$.
- $w_i = f(t_{i-1}, w_{i-1})h + w_{i-1}$, for $i = 1, 2, \ldots, n$.
- $p_n(t) = f(t_{i-1}, w_{i-1})(t - t_{i-1}) + w_{i-1}$ on the interval $[t_{i-1}, t_i]$.

Example:

Use Euler's Method to estimate the solution of the initial value problem

$$\frac{dy}{dt} = 2t + 3y, \quad y(1) = 0.$$

on the interval $[1, 4]$ using $n = 6$.

Answer:

First, notice that $h = (4 - 1)/6 = 0.5$. The following table shows the other calculations. Figure III.35 shows the function $p_6(t)$.

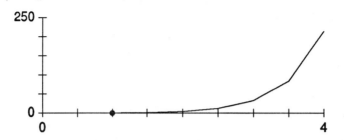

Figure III.35: The function p_6

i	t_{i-1}	w_{i-1}	$f(t_{i-1}, w_{i-1})$	t_i	w_i
1	1.0	.0000	2.0000	1.5	1.0000
2	1.5	1.0000	6.0000	2.0	4.0000
3	2.0	4.0000	16.0000	2.5	12.0000
4	2.5	12.0000	41.0000	3.0	32.5000
5	3.0	32.5000	103.5000	3.5	84.2500
6	3.5	84.2500	259.7500	4.0	214.1250

Exercises

*Exercise III.4.1:
Use Euler's Method to estimate the solution of the initial value problem
$$p' = p, \quad p(0) = 1$$
on the interval $[0, 1]$ using $n = 4$.

*Exercise III.4.2:
Use Euler's Method to estimate the solution of the initial value problem
$$p' = p + t/2, \quad p(1) = 1$$
on the interval $[1, 2]$ using $n = 5$.

Variations on a Theme by Euler

Suppose that we are interested in the initial value problem
$$p' = p(1 - p), \quad p(5) = 0.5$$

§4. Euler's Method

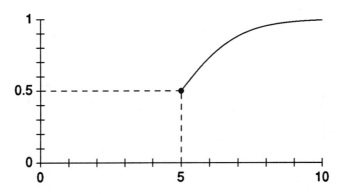

Figure III.36: Estimate for $p' = p(1-p)$, $p(5) = 0.5$ on $[5, 10]$

Figure III.36 shows an estimate for the solution of this equation on the interval $[5, 10]$ that was obtained using Euler's Method. The initial point $(5, 0.5)$ is indicated by a dot.

Euler's Method in its original incarnation allows us to predict what will happen in the future from current data. For example, using the initial data $p(5) = 0.5$, Figure III.36 shows us what will happen for $5 \leq t \leq 10$. We are frequently interested, however, in deducing what happened in the past from current data. A slight variation on Euler's Method allows us to work backward rather than forward. If we start knowing $p(t_n)$ we can estimate $p(t_{n-1})$ by

$$p(t_{n-1}) \approx p(t_n) - f(t_n, p(t_n))h$$

and, more generally, we can rewrite Euler's Method to go backward:

Euler's Method (backward)

Given an initial value problem

$$\frac{dy}{dt} = f(t, y), \quad y(b) = y_0,$$

an interval $[a, b]$, and a positive integer n, compute the following

- $h = (b - a)/n$.

- $w_n = y_0$.

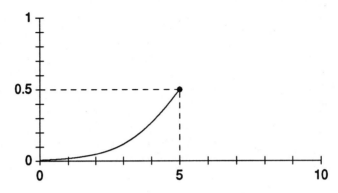

Figure III.37: Estimate for $p' = p(1-p)$, $p(5) = 0.5$ on $[0, 5]$

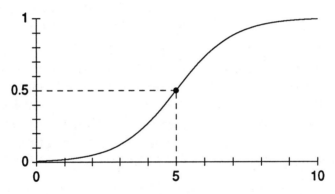

Figure III.38: Estimate for $p' = p(1-p)$, $p(5) = 0.5$ on $[0, 10]$

- $t_i = a + ih$, for $i = 0, 1, 2, \ldots, n$.
- $w_{i-1} = w_i - f(t_i, w_i)h$, for $i = 1, 2, \ldots n$.
- $p_n(t) = f(t_i, w_i)(t - t_i) + w_i$ on the interval $[t_{i-1}, t_i]$.

If we apply this variation of Euler's Method to the initial value problem above for the interval $[0, 5]$ we obtain Figure III.37. The initial point is indicated by a dot. We can combine these two figures to obtain Figure III.38. This figure both predicts the future and postdicts the past. The initial point $(0, 0.5)$ is indicated by a dot.

Our experience so far with Euler's Method might very well lead us to the following conjecture.

§4. Euler's Method

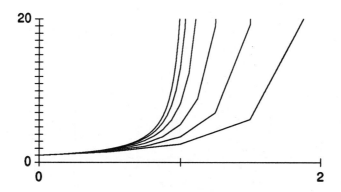

Figure III.39: Estimates for the solution of $p' = p^2$

Conjecture:

Given an initial value problem

$$p' = f(t,p), \quad p(a) = p_0$$

There is always a unique solution and that solution can be found by taking the limit of the estimates produced by Euler's Method.

Unfortunately this conjecture is not quite true. There are two problems. The first problem can be illustrated by looking at the initial value problem

$$p' = p^2, \quad p(0) = 1.$$

Figure III.39 shows several estimates for the solution of this problem using Euler's Method with $n = 4, 8, 16, 32, 64$ and 128. Notice how rapidly the solution seems to be growing. This phenomenon is more and more evident for better and better approximations to the exact solution. Figure III.40 is exactly the same as Figure III.39 except we have changed the scale so that we can see values of p up to 1000. We have also added additional estimates with $n = 256, 512$ and 1024.

Notice how dramatically the function $p(t)$ (or at least our estimates for the function $p(t)$) is increasing as t gets close to 1. From this figure you might guess that the function $p(t)$ is doing something a little odd as

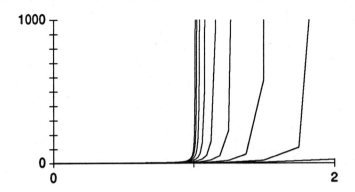

Figure III.40: Estimates for the solution of $p' = p^2$

t gets close to 1. In fact, the exact solution of this initial value problem is the function

$$p(t) = \frac{1}{1-t}.$$

If t is very close to 1 then $p(t)$ will be huge. If t is exactly equal to 1 then $p(t)$ is undefined. As we can see from this example, we cannot always expect to obtain a solution to an initial value problem that will be defined for all values of t. We must refine our expectations. The following conjecture makes this refinement.

Conjecture:

Given an initial value problem

$$p' = f(t,p), \quad p(a) = p_0,$$

there is always a unique solution and that solution can be found by taking the limit of the estimates produced by Euler's Method. The solution may not be defined for all values of t. However, there will be an interval (c, d) containing the point $t = a$ on which $p(t)$ will be defined.

This conjecture is almost right. To make it 100% correct requires some conditions on the function $f(t,p)$. The problem is that we want to use the tangent to the curve $p(t)$ at each point t_i but in order to find the slope of that tangent we need to compute

$$f(t_i, p(t_i)).$$

§4. Euler's Method

We do not know $p(t_i)$ exactly, so we estimate the slope of the tangent by using our estimate w_i for $p(t_i)$

$$f(t_i, w_i).$$

We need to know that

$$f(t_i, w_i) \approx f(t_i, p(t_i)).$$

Since $w_i \approx p(t_i)$ we need to know that the function $f(t, p)$ is not too sensitive to changes in the variable p. This leads us to the following idea.

Definition:

> The **partial derivative** of the function $f(t, p)$ with respect to the variable p is
>
> $$\frac{\partial f}{\partial p} = \lim_{h \to 0} \frac{f(t, p+h) - f(t, p)}{h}.$$

This is analogous to the ordinary derivative. You should think of this partial derivative as measuring the sensitivity of the function $f(t, p)$ to changes in the variable p in the same way that the ordinary derivative dy/dx measures the sensitivity of the variable y to changes in the variable x. Partial derivative are easy to compute. One computes $\partial f/\partial p$ in exactly the same way that one computes df/dp *thinking of the variable t as being a constant.*

Example:

Suppose $f(t, p) = 3pt + pt^2 + 2p^2$. Find $\partial f/\partial p$.

Answer:

$$\partial f/\partial p = 3t + t^2 + 4p.$$

Using this concept we can finally state a true theorem.

Theorem 2 *(Euler's Theorem for Differential Equations)*
Given an initial value problem

$$p' = f(t, p), \quad p(a) = p_0$$

where $f(t, p)$ is continuous and the partial derivative $\partial f / \partial p$ exists and is continuous, there is always a unique solution and that solution can be found by taking the limit of the estimates produced by Euler's Method. The solution may not be defined for all values of t. However, there will be an interval (c, d) containing the point $t = a$ on which $p(t)$ will be defined.

In any practical problem, it would be very surprising if the two conditions on the function $f(t, p)$ were not satisfied.

§5. The Exponential Function

The most common differential equations are those of the form

$$p' = kp$$

where k is some constant. These differential equations model situations in which the quantity p is changing with a constant *relative* rate. Such situations are found everywhere.

- The simplest model for population growth for a continuously changing population is $p' = kp$ where k is the rate of change. For example, a population growing at the rate of 2% per year would be modeled by $p' = 0.02p$.

- Radioactive substances decay at a constant relative rate. Thus an appropriate model for the amount A of a radioactive element is $A' = kA$ where k is a negative constant. People frequently write $A' = -cA$ where c is a positive constant to emphasize the fact that A is decreasing.

- The amount of money a bank account earns is proportional to the amount deposited. Some bank accounts pay interest continuously. A bank account paying interest at the rate of 6% per year continuously, computes the value of a deposit using the differential equation

$$v' = 0.06v.$$

§5. The Exponential Function

- The air pressure P of the atmosphere at a height h satisfies the differential equation
$$\frac{dP}{dh} = -kP$$
where k is a positive constant.

An Interesting Example

Example:

We begin this section by looking at the way that banks compute interest on both bank accounts and loans. We choose this example because it illustrates some of the dollars and cents implications of what a naive customer might dismiss as uninteresting accounting fine points. One of these accounting fine points caused the author several difficult days during which he doubted the laws of arithmetic.

Bank advertisements often look something like this.

Tinseltown Bank and Trust

- Annual rate of interest 6%.

- Compounded monthly.

- Effective rate of interest 6.17%.

The first item seems very straightforward. The bank pays 6% interest per year. If you deposit $100.00 for one year then at the end of the year you will have $106.00. Right? Wrong! Mathematically this is the simplest way to compute interest; each year add 6% to the account. The biggest problem with this is that people often make deposits for odd fractions of a year. If interest were only paid once each year then a depositor who withdrew her money after 11 months would receive nothing. To avoid this problem banks usually compute and pay interest more frequently. The advertisement above advertises interest *compounded* monthly. This means that the bank computes interest each month and credits it to the account.

Since this particular account pays interest at the rate of 6% per year and there are 12 months in a year the interest rate is 6%/12 = 0.5% per month. The following table shows the interest computations for one year for a bank account earning 6% annual interest compounded monthly.

Month	Start	Interest	End
1	$100.0000	.5000	$100.5000
2	$100.5000	.5025	$101.0025
3	$101.0025	.5050	$101.5075
4	$101.5075	.5075	$102.0151
5	$102.0151	.5101	$102.5251
6	$102.5251	.5126	$103.0378
7	$103.0378	.5152	$103.5529
8	$103.5529	.5178	$104.0707
9	$104.0707	.5204	$104.5911
10	$104.5911	.5230	$105.1140
11	$105.1140	.5256	$105.6396
12	$105.6396	.5282	$106.1678

Notice that at the end of the year the account contains $106.17. It has effectively earned 6.17% interest. This is the meaning of the term "*effective* rate of interest." The reason that the effective rate of interest is higher than the original rate of interest is that the interest earned each month itself earns interest in each succeeding month. The difference between the original rate of interest and the effective rate can be very signficant. Banks do the same computations for loans.

The effective rate of interest can be computed much more quickly than we did in the previous table. Let R denote the annual interest rate as a decimal. For example, if the interest rate is 6% then $R = 0.06$. If interest is compounded n times per year then each time it is compounded the interest rate is R/n. Thus each time you compound the interest you compute

$$v + \left(\frac{R}{n}\right)v = \left(1 + \frac{R}{n}\right)v$$

where v is the current deposit. This computation is done n times during the course of a year. So, if the original deposit is v, after n years it will

§5. The Exponential Function

be worth
$$\left(1 + \frac{R}{n}\right)^n v.$$
For our example above this works out to
$$\left(1 + \frac{0.06}{12}\right)^{12} v = 1.061678v$$
and the effective interest rate is 6.1678%.

Many banks now compound interest daily. Some even compound interest *continuously*. The value of a deposit in an account with interest compounded continuously at the rate of 6% per year, for example, grows according to the differential equation

$$v' = 0.06v.$$

Exercises

Exercise III.5.1: Many credit cards charge interest at an anuual rate of 18%. If this rate were compounded monthly what would the effective rate be?

Exercise III.5.2: In fact many credit cards compound interest daily. What is the effective rate of interest for 18% interest compounded daily? Assume that there are 365 days in a year.

Exercise III.5.3: You may have laughed when you read the last exercise. "Assume that there are 365 days in a year." These mathematicians sure are weird. Of course there are 365 days in a year except for leap years. Well, bankers are even more weird. Every banker knows that each year has 12 months and every month is 30 days long, so a year obviously has 360 days. When bankers compute interest they find the daily rate of interest by dividing the annual rate of interest by 360. For example, if the annual rate of interest is 18% then the daily rate of interest is 0.05%. Find the effective rate of interest for 18% compounded 360 times per year.

Exercise III.5.4: BUT bankers do business 365 days a year, so after they've computed the daily rate of interest by dividing the annual interest rate by 360, they actually compute interest using this daily rate every day. They compound the interest 365 times. Find the effective rate of interest if the annual rate of interest is 18% and the computations are done by bankers. First, compute the daily rate by dividing the annual rate by 360 and then compute interest using this daily rate 365 times.

Exercise III.5.5:

The last exercise may seem like crazy mathematics. It is! But, it's clever banking. Banks earn a lot of money using this trickery. This trickery cost me several very confused days when I bought my first calculator. I played around computing various effective rates of interest corresponding to advertisements like:

Tinseltown Bank and Trust

- Annual rate of interest 6%.

- Compounded daily.

- Effective rate of interest 6.2716%.

I computed interest the way any normal person would, dividing the annual rate by 365 to obtain the daily rate and then compounding the interest 365 times. My answers did not agree with the banks' advertisements. I began to doubt arithmetic. Fortunately, one of my students worked for an insurance company and was able to explain to me how bankers computed interest.

Find the effective rate of interest for an annual rate of 6% compounded daily the way a normal person would do the computations. Find the effective rate of interest for an annual rate of 6% compounded daily the way a banker would do the computations. Compare your two answers.

§5. THE EXPONENTIAL FUNCTION

The Exponential Function

In this section we examine the solution to the simplest initial value problem of the form
$$y' = ky, \quad y(0) = a,$$
the initial value problem
$$y' = y, \quad y(0) = 1.$$

Ideally, we would like to find a simple formula for the solution to this initial value problem. We can't. If you try all the functions that you know none of them will work. For example, suppose that you guessed the solution was a polynomial, a function of the form
$$p(t) = a_n t^n + a_{n-1} t^{n-1} + \cdots + a_1 t + a_0.$$

Differentiating this function we see that
$$p'(t) = a_n t^{n-1} + (n-1) a_{n-1} t^{n-2} + \cdots + a_1.$$

This function is also a polynomial but its highest power is t^{n-1} and the highest power of the original polynomial was t^n so $p'(t)$ and $p(t)$ are different. No polynomial can satisfy the differential equation $y' = y$.

We can get some intuition about the solution by thinking about a discrete model. The equation $y' = y$ says that y is growing at the rate of 100% per year (if t is time in years). If we were to compute $y(1), y(2), y(3), \ldots$ in the most naive way starting with $y(0) = 1$ we would get

$$\begin{aligned} y(0) &= 1 \\ y(1) &= y(0) + y(0) = 2 \\ y(2) &= y(1) + y(1) = 4 = 2^2 \\ y(3) &= y(2) + y(2) = 8 = 2^3 \\ &\vdots \\ y(n) &= 2^n \\ &\vdots \end{aligned}$$

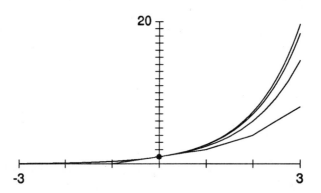

Figure III.41: Euler's Method and $y' = y$, $y(0) = 1$

This naive computation might lead us to conjecture that $y = 2^t$ was the solution to our initial value problem. This conjecture is wrong.

Exercise

*Exercise III.5.6:
Consider the function $y(t) = 2^t$. Estimate $y'(0)$. Estimate $y'(1)$. Notice that $y'(0) \neq y(0)$ and that $y'(1) \neq y(1)$.

We do know that our initial value problem does have a solution and we know how to approximate that solution. Use Euler's Method. Figure III.41 shows a sequence of approximations to the solution of our initial value problem obtained by Euler's Method. Figure III.42 shows the limit of this sequence of approximations.

Euler's method gives us a way of computing the solution to our initial value problem. Thus, in effect, we know the solution. We can compute it. We can even find its derivative. Since it is a solution of the differential equation $y' = y$, it is its own derivative. The only thing this function lacks is a name. We hereby give it a name.

Definition:

> The **exponential function**, written $\exp t$, is the solution of the initial value problem
>
> $$y' = y, \quad y(0) = 1.$$

Since the exponential function is the solution of our initial value problem we immediately have the following theorem.

§5. The Exponential Function

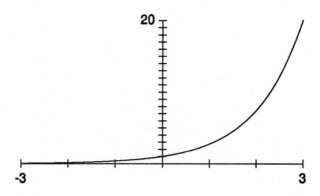

Figure III.42: The solution of $y' = y$, $y(0) = 1$

Theorem 3

- $\exp(0) = 1$.

- $(\exp t)' = \exp t$.

The exponential function is so important that most calculators have it built in along with more familiar functions like the sine and cosine. As we have seen above, the exponential function can be computed by Euler's Method, but this method is rather slow even for a calculator. In chapter VI we will discuss another, quicker way of computing it.

Properties of the Exponential Function

We want to find out more about the exponential function. Our first goal is to compute $\exp 1$. We do this by using Euler's Method to approximate $\exp t$ on the interval $[0, 1]$. The following table shows the computations for p_2. It uses the same notation as the preceding section. The end result of this computation is $p_2(1) = 2.25$.

i	t_{i-1}	t_i	w_{i-1}	w_i
1	.0000	.5000	1.0000	1.5000
2	.5000	1.0000	1.5000	2.2500

The following table shows the computations for p_{10}.

i	t_{i-1}	t_i	w_{i-1}	w_i
1	.0000	.1000	1.0000	1.1000
2	.1000	.2000	1.1000	1.2100
3	.2000	.3000	1.2100	1.3310
4	.3000	.4000	1.3310	1.4641
5	.4000	.5000	1.4641	1.6105
6	.5000	.6000	1.6105	1.7716
7	.6000	.7000	1.7716	1.9487
8	.7000	.8000	1.9487	2.1436
9	.8000	.9000	2.1436	2.3579
10	.9000	1.0000	2.3579	2.5937

The end result of this computation is $p_{10}(1) = 2.5937$.

These computations are exactly the same as the computations for one year for a bank deposit of \$1.00 in an account earning interest at the rate of 100% per year. The approximation $p_2(1)$ is the result of compounding twice a year. The approximation $p_{10}(1)$ is the result of compounding 10 times per year. Notice that

$$p_2(1) = \left(1 + \frac{1}{2}\right)^2$$

and that

$$p_{10}(1) = \left(1 + \frac{1}{10}\right)^{10}.$$

More generally, if we compute $p_n(1)$ (that is, compound n times per year), we obtain

$$p_n(1) = \left(1 + \frac{1}{n}\right)^n.$$

This gives us a sequence of estimates $p_n(1)$ for $\exp 1$. The following table shows some of these estimates.

n	$p_n(1)$
2	2.2500
4	2.4414
10	2.5937
50	2.6916
100	2.7048
200	2.7115

§5. The Exponential Function

The number $\exp 1$ is as important as the number π. It has its own name, e, and, like π, its value has been computed to many thousands of digits. The first few digits are

$$e = 2.718281828459045\ldots$$

Definition:

$$e = \exp 1 = \lim_{n \to \infty} \left(1 + \frac{1}{n}\right)^n = 2.718281828459045\ldots$$

We started out with the modest goal of solving just one initial value problem

$$y' = y, \quad y(0) = 1$$

but the following theorem enables us to solve any initial value problem of the form

$$y' = ky, \quad y(t_0) = C.$$

Theorem 4 *The solution to the initial value problem*

$$y' = ky, \quad y(t_0) = C$$

is

$$y = C \exp(k(t - t_0))$$

Proof:
First, notice $y(t_0) = C \exp(k(t_0 - t_0)) = C \exp 0 = C$, so the initial condition is satisfied.

Second, differentiating y by the Chain Rule we see

$$y' = C(\exp(k(t - t_0)))(k) = ky.$$

This completes the proof.

The following theorem states a very important property of the exponential function.

Theorem 5 *If a and b are any two real numbers then $\exp(a + b) = (\exp a)(\exp b)$*

Proof:
Consider the two functions
$$y_1(t) = \exp(a+t)$$
and
$$y_2 = (\exp a)(\exp t).$$
Since a is a constant, so is $\exp a$. Notice that
$$y_1(0) = \exp a \quad \text{and} \quad y_2(0) = (\exp a)(\exp 0) = \exp a$$
and that
$$y_1' = \exp(a+t) = y_1 \quad \text{and} \quad y_2' = (\exp a)(\exp t) = y_2$$
Thus both y_1 and y_2 are solutions of the same initial value problem
$$y' = y, \quad y(0) = \exp a$$
So, by Euler's Theorem, $y_1 = y_2$. In particular,
$$\exp(a+b) = y_1(b) = y_2(b) = (\exp a)(\exp b)$$
completing the proof of the theorem.

Using this property we can immediately do a number of calculations.
$$\exp 2 = \exp(1+1) = (\exp 1)(\exp 1) = e^2$$

$$\exp n = \underbrace{\exp(1+1+\ldots+1)}_{n \text{ times}} = \underbrace{(\exp 1)(\exp 1)\ldots(\exp 1)}_{n \text{ times}} = e^n$$

$(\exp(1/2))(\exp(1/2)) = \exp 1 = e$. Therefore, $\exp(1/2) = \sqrt{e} = e^{\frac{1}{2}}$

Using the same kinds of arguments it is easy to show

Theorem 6 *For any rational number r, $\exp(r) = e^r$*

§5. The Exponential Function

Proof:
This proof is a good exercise.

What about irrational numbers? Notice that while it is not entirely clear what e^π means, $\exp(\pi)$ is just as easy to estimate as $\exp 2$ using Euler's Method. Since $\exp(r) = e^r$ for all rational numbers, it is reasonable to define e to any irrational power to be the corresponding value of exp. We thus *define e^x* to be $\exp(x)$ for any x. From now on, we will feel free to use either the notation e^x or $\exp(x)$—they are different names for the same function.

Armed with our new function we are now able to solve many important problems.

Example:

Suppose that some machinery used in a nuclear power plant has been contaminated by a radioactive isotope. The amount of the isotope that is present on January 1, 1989 is 100 grams. One year later, due to radioactive decay, only 95 grams is present. How many grams will be present on January 1, 1999, ten years after the first measurement?

To solve this problem we let t denote time in years after January 1, 1989 and $A(t)$ denote the amount of the isotope that is present after t years. Thus
$$A(0) = 100 \quad \text{and} \quad A(1) = 95.$$

The rate of decay of a radioactive isotope is proportional to the amount present. Thus
$$A' = kA$$
and
$$A(t) = C\exp(kt).$$
We want to determine the two constants C and k.

Using the fact that $A(0) = 100$ we see that
$$100 = A(0) = C\exp(0) = C.$$

Thus $C = 100$ and
$$A(t) = 100\exp(kt).$$

Now using the fact that $A(1) = .95$ we see that
$$95 = 100\exp(k1) = 100\exp k$$

Thus
$$\exp k = 95/100 = 0.95.$$

We can determine the value of k by using the Bisection Method or Newton's Method. The result is roughly

$$k = -.051293,$$

so
$$A(t) = 100 \exp(-.051293t)$$

and the answer to our original question is

$$A(10) = 100 \exp(-.051293(10)) = 59.8737 \text{ grams}.$$

Exercises

*Exercise III.5.7:
The rate of growth of the population of a particular country is proportional to the population. The last two censuses determined that the population in 1980 was 40,000,000 and in 1985 it was 45,000,000. What will the population be in 1995?

Exercise III.5.8:
The rate of growth of the population of a particular country is proportional to the population. The last two censuses determined that the population in 1980 was 20,000,000 and in 1985 it was 25,000,000. What will the population be in 1995?

*Exercise III.5.9:
There are two advertisements in the newspaper for savings accounts in two different banks. The first offers 6% interest compounded quarterly (that is, four times per year). The second offers 5.5% interest compounded continuously. Which account is better? Explain.

---

§6. The Natural Logarithm

In the example at the end of the section 7 we needed to solve the equation
$$\exp(k) = 0.95$$

§6. The Natural Logarithm

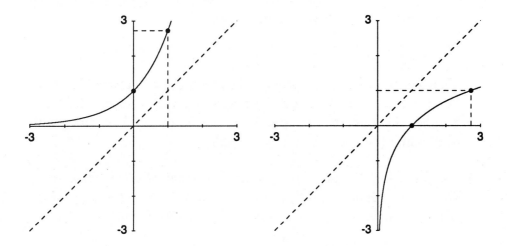

Figure III.43: The function $y = e^x$ and its inverse

Because this kind of equation comes up frequently it is worthwhile to study the inverse of the function $y = \exp x$ or $y = e^x$. In fact, most calculators have the inverse of the function $y = e^x$ built in. It is called the *natural logarithm* and is denoted "ln." Because the natural logarithm is the inverse of the exponential function we can graph it simply by picking up a graph of the exponential function and flipping it over around the diagonal line $y = x$. Figure III.43 shows the graph of the exponential function and its inverse, the natural logarithm, side-by-side.

Because the two functions are inverses of each other, every fact about the exponential function tells us something about the natural logarithm. The basic fact is that the two equations

$$y = e^x \quad \text{and} \quad x = \ln y$$

mean exactly the same thing. The following theorem lists some other facts about the function $y = e^x$ and the corresponding facts about the function $y = \ln x$.

Theorem 7

- $e^0 = 1$ *and, hence,* $\ln 1 = 0$.

    This fact is illustrated in Figure III.43 where you can see the point $(0, 1)$ indicated by a dot on the graph of $y = e^x$ (i.e., the lefthand graph). This point corresponds to the point $(1,0)$ on the graph of $y = \ln x$ (i.e., the right hand graph).

- $e^1 = e$ and, hence, $\ln e = 1$.

  This fact is illustrated in Figure III.43 where you can see the point $(1, e)$ indicated by a dot on the graph of $y = e^x$ (i.e., the lefthand graph). This point corresponds to the point $(e, 1)$ on the graph of $y = \ln x$ (i.e., the right hand graph).

- *The image of the function $y = e^x$ is the set of all positive reals. Hence, the domain of the function $y = \ln x$ is the set of all positive reals.* We haven't actually proved this fact and won't do so here.

- *The domain of the function $y = e^x$ is the set of all reals. Hence, the image of the function $y = \ln x$ is the set of all reals.* Again, we haven't actually proved this fact about the function $y = e^x$ and won't do so here.

- *For any real number x, $\ln(e^x) = x$ since the function $\ln$ is the inverse of the function $\exp$.*

- *For any positive real number x, $e^{\ln x} = x$ since the function $\exp$ is the inverse of the function $\ln$.*

- *For any two reals a and b, $e^{(a+b)} = e^a e^b$. Hence, for any two positive reals u and v, $\ln(uv) = \ln u + \ln v$.*

  Proof: Let $a = \ln u$ and $b = \ln v$. Therefore $u = e^a$ and $v = e^b$.

  Now $uv = e^a e^b = e^{(a+b)}$ and by taking the natural logarithm of both sides we see
  $$\ln(uv) = a + b = \ln u + \ln v.$$

- *If $y = \ln x$ then $y' = 1/x$.*

  Proof: *By the Inverse Function Rule*
  $$\left(\frac{d}{dx}\right)\ln(x) = \frac{1}{\exp(\ln x)} = \frac{1}{x}.$$

§6. The Natural Logarithm

The Functions $y = a^x$

You have been using expressions like 2^3 and $2^{\frac{1}{3}}$ for many years. It seems perfectly natural to use expressions like 2^π or $3^{\sqrt{2}}$. Even though it seems natural, however, we don't really know what these expressions mean. For a positive integer n we know that

$$a^n = \underbrace{a \times a \times a \ldots \times a}_{n \text{ times}}$$

and for a rational number $\frac{p}{q}$ we can deduce that

$$a^{\frac{p}{q}} = \sqrt[q]{a^p}$$

from the observation that

$$\underbrace{a^{\frac{p}{q}} \times a^{\frac{p}{q}} \times \ldots \times a^{\frac{p}{q}}}_{q \text{ times}} = a^p$$

However, we do not know what a^b means if b is not a rational number. We can now fill in this gap by making the following definition.

Definition:

For any positive real number a and any real number b

$$a^b = e^{(b \ln a)}$$

NOTE: This definition only makes sense if a is positive, since we cannot find the natural logarithm of a negative number.

This restriction is not surprising since we cannot find the square root of a negative number.

In order for this definition to make sense we had better check that it agrees with our old definition when the exponent b happens to be a rational number. The following theorem verifies this fact.

Theorem 8 *Using the definition above*

1. $a^0 = 1$

2. $a^1 = a$

3. $a^{(b+c)} = a^b a^c$

4. For any rational number $\frac{p}{q}$,
$$a^{\frac{p}{q}} = \sqrt[q]{a^p}$$

Proof:

1. $a^0 = e^{0 \ln a} = e^0 = 1$

2. $a^1 = e^{1 \ln a} = e^{\ln a} = a$

3. $a^{(b+c)} = e^{(b+c) \ln a} = e^{(b \ln a)+(c \ln a)} = e^{b \ln a} e^{c \ln a} = a^b a^c$

4. This follows from the observation that

$$\underbrace{a^{\frac{p}{q}} \times a^{\frac{p}{q}} \times \ldots \times a^{\frac{p}{q}}}_{q \text{ times}} = a^{\overbrace{\left(\frac{p}{q} + \frac{p}{q} + \ldots + \frac{p}{q}\right)}^{q \text{ times}}} = a^p$$

Thus our new definition of a^b agrees with our old definition and extends it so that it makes sense for any real exponent. As with any new function we need to find the derivative.

Theorem 9 *If $f(x) = a^x$ then $f'(x) = a^x \ln a$*

Proof:
$$f(x) = \exp(x \ln a)$$

So, by the Chain Rule,
$$f'(x) = [\exp(x \ln a)] \ln a = a^x \ln a.$$

Finally, we need to make an observation. We have now defined e^x two different ways. Our first definition was $e^x = \exp x$. The second

§6. The Natural Logarithm

definition is $e^x = \exp(x \ln e)$. Note that these two definitions agree since $\ln e = 1$.

Exercises

*Exercise III.6.1: Find the derivative of the function $y = e^{(\sin x)}$.

Exercise III.6.2: Find the derivative of the function $y = e^{(x^2)}$.

Exercise III.6.3: Find the derivative of the function $y = 2^x$.

Exercise III.6.4: Find the derivative of the function $y = 2^{(\sin x)}$.

Exercise III.6.5: Find the derivative of the function $y = 3^{(\sin x)}$.

Exercise III.6.6: Find the derivative of the function $y = 3^{(x^2)}$.

*Exercise III.6.7: Find the derivative of the function $p = \ln(x^2 - 2x)$.

Exercise III.6.8: Find the derivative of the function $p = \ln \sin x$.

Exercise III.6.9: Find the derivative of the function
$$w = \exp(2x - \sin x).$$

Exercise III.6.10: Find the derivative of the function
$$v = \ln(4 + \exp x).$$

Exercise III.6.11: Find the derivative of the function $T = 7^{x^2 - 1}$.

Exercise III.6.12: Find the derivative of the function
$$y = \exp\left(\frac{1+x}{1-x}\right)$$

The following theorem summarizes some facts about these new functions.

Theorem 10

- For any positive real number a and any real number b,
$$a^{-b} = \frac{1}{a^b}$$

Proof:

$$a^{-b}a^b = a^{-b+b} = a^0 = 1,$$

so

$$a^{-b} = \frac{1}{a^b}.$$

- *For any positive real number a,*

$$\ln\left(\frac{1}{a}\right) = -\ln a.$$

Proof:

$$e^{-\ln a} = \frac{1}{e^{\ln a}} = \frac{1}{a}.$$

- *For any positive real number a and any real number b,*

$$\ln(a^b) = b \ln a$$

Proof: This is immediate from the fact that $a^b = e^{b \ln a}$.

- *For any positive real number a and any real numbers b and c,*

$$\left(a^b\right)^c = a^{bc}.$$

Proof:

$$\left(a^b\right)^c = e^{c \ln(a^b)}.$$

But

$$\ln(a^b) = b \ln a,$$

so

$$\left(a^b\right)^c = e^{c \ln(a^b)} = e^{cb \ln a} = a^{bc}.$$

The fact that

$$\ln(a^b) = b \ln a$$

is the basis for a very useful differentiation trick called *logarithmic differentiation*. We illustrate this trick with an example.

§6. The Natural Logarithm

Example:

Differentiate the function $y = x^x$.

Answer:

Taking the natural logarithm of both sides of this equation we obtain

$$\ln y = x \ln x.$$

Now if we differentiate both sides of this equation with respect to the variable x we get

$$\left(\frac{1}{y}\right)\left(\frac{dy}{dx}\right) = 1 + \ln x.$$

So

$$\frac{dy}{dx} = y(1 + \ln x) = x^x(1 + \ln x).$$

These new functions allow us to express the answers to some earlier questions more clearly. For example, let us return to the example at the end of the last section.

Example:

Suppose that some machinery used in a nuclear power plant has been contaminated by a radioactive isotope. The amount of the isotope that is present on January 1, 1989 is 100 grams. One year later due to radioactive decay only 95 grams is present. How many grams will be present on January 1, 1999 ten years after the first measurement?

To solve this problem we let t denote time in years after January 1, 1989 and $A(t)$ denote the amount of the isotope that is present at time t. Thus

$$A(0) = 100 \quad \text{and} \quad A(1) = 95.$$

The rate of decay of a radioactive isotope is proportional to the amount present. Thus

$$A' = kA$$

and

$$A(t) = C \exp(kt).$$

We want to determine the two constants C and k.

Using the fact that $A(0) = 100$ we see that
$$100 = A(0) = C\exp(0) = C.$$
Thus C= 100 and
$$A(t) = 100\exp(kt).$$
Now using the fact that $A(1) = .95$ we see that
$$95 = 100\exp(k1) = 100\exp(k),$$
so
$$\exp(k) = \frac{95}{100} = 0.95$$
and $k = \ln 0.95$.

Thus,
$$A(t) = 100\exp(t\ln.95) = 100(.95)^t$$
and the answer to our original question is
$$A(10) = 100(.95)^{10} = 59.8737 \quad \text{grams}.$$

By expressing the function $A(t)$ in the form
$$A(t) = 100(0.95)^t$$
instead of
$$100\exp(-.051293t)$$
as we did in the last section, we emphasize the fact that the amount of the isotope falls by 5% each year.

Frequently, a function like the function $A(t)$ can be expressed in several different ways, all of which are mathematically equivalent but which emphasize different properties of the function. For example, people dealing with radiactive decay often talk about the **half-life** of a particular radioactive isotope. This is the amount of time required for half of the original amount to decay. We can rewrite the function $A(t)$ as follows.

$$\begin{aligned}A(t) &= 100\exp(t\ln.95)\\ &= 100\exp\left(\frac{(\ln 0.5)t\ln.95}{\ln 0.5}\right)\end{aligned}$$

§6. The Natural Logarithm

$$= 100 \left(\frac{1}{2}\right)^{\left(\frac{t \ln .95}{\ln 0.5}\right)}$$

$$= 100 \left(\frac{1}{2}\right)^{0.074001t}$$

$$= 100 \left(\frac{1}{2}\right)^{\left(\frac{t}{13.51}\right)}$$

Written in this way we can see that the half-life of this particular isotope is 13.51 years.

Exercises

*Exercise III.6.13: Find the derivative of the function
$$y = \ln\left(\frac{1+x}{1-x}\right)$$

*Exercise III.6.14: Find the derivative of the function $y = (\sin x)^x$

Exercise III.6.15: Find the derivative of the function $z = t^{(\sin t)}$

Exercise III.6.16: Sketch a graph of the function $w = t - e^t$. Does this function have a minimum? a maximum?

*Exercise III.6.17: Sketch a graph of the function $z = xe^x$. Does this function have a minimum? a maximum?

Exercise III.6.18: Sketch a graph of the function $y = x \ln x$ on the interval $(0, \infty)$. Does this function have a maximum on this interval? a minimum?

*Exercise III.6.19: A sealed barrel contains 100 grams of a radioactive isotope in 1980. Five years later it contains 75 grams. Find a function $A(t)$ that describes how many grams it will have t years after 1980. Express your answer in two different ways, one emphasizing the yearly change in $A(t)$ and the other emphasizing the half-life of this isotope.

Exercise III.6.20: The population of a particular country is growing at a rate proportional to its current population. The census in 1980 reported its population was 35,000,000 and the census in 1985 reported its population was 40,000,000. Find a function $p(t)$ that describes the population at time t where t is the number of years after 1980. Express your answer in two different ways, one emphasizing the yearly change in population and the other emphasizing how long it will take for the population to double.

Although the only logarithm function that we will need and use is the natural logarithm, there are others.

Definition:

For each positive real number b, the function $\log_b$ is defined by

$$\log_b(x) = \frac{\ln x}{\ln b}$$

Exercises

*Exercise III.6.21:
  Prove that the function $\log_b$ is the inverse of the function $f(x) = b^x$. Use this fact to answer most of the remaining problems in this set of exercises.

*Exercise III.6.22:
  What is $\log_2(8)$?

*Exercise III.6.23:
  What is $\log_3(9)$?

*Exercise III.6.24:
  What is $\log_3(27)$?

*Exercise III.6.25:
  What is $\log_3(20)$?

*Exercise III.6.26:
  What is $\log_{10}(1/1000)$?

*Exercise III.6.27:
  Find the derivative of the function $y = \log_b x$.

§7. Vertical Asymptotes

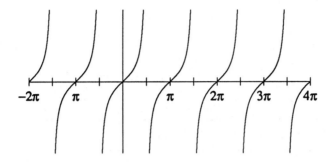

Figure III.44: $y = \tan \theta$

§7. Vertical Asymptotes

Many functions that are used to model practical situations are undefined at some points. For example, the force of gravity between two objects having mass m_1 and m_2 respectively and whose distance from each other is r is given by the function

$$F(r) = \frac{Gm_1m_2}{r^2}$$

where G is a constant called the gravitational constant. This function is undefined when $r = 0$ since the denominator is zero. In this section we are interested in studying functions like $F(r)$ that are undefined at one or more points. Figure III.44 shows another example, the function $\tan \theta$ which is undefined at all the right angles $\theta = \pi/2, -\pi/2, 3\pi/2, -3\pi/2$, and so forth. Figure III.45 shows yet another example, the function $y = \ln x$ which is only defined for x positive.

The first step in sketching the graph of a function like this is to make a diagram showing its domain. For example, Figure III.46 shows the domain of the function $y = \tan x$ and Figure III.47 shows the domain of the function $y = \ln x$.

The next step is to analyze what happens very close to a point at which the function is undefined. Consider the function $y = \tan \theta$ shown in Figure III.44. If you look at this graph near the point $x = \pi/2$ you will see that just to the left of this point $\tan x$ is very large and positive. You can see why this is true from Figure III.48. When θ is very close to $\pi/2$ but slightly less than $\pi/2$ we are looking at a triangle that is

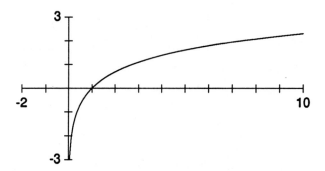

Figure III.45: $y = \ln x$

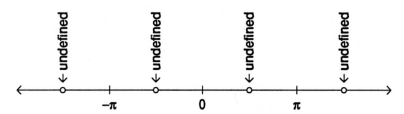

Figure III.46: The domain of $y = \tan \theta$

Figure III.47: The domain of $y = \ln x$

§7. Vertical Asymptotes

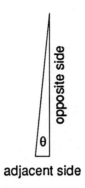

Figure III.48: An angle slightly less than $\pi/2$

very tall and thin. The opposite side of the angle θ will be very long and the adjacent side will be very short. As a result $\tan\theta$ will be very large, since

$$\tan\theta = \frac{\text{opposite side}}{\text{adjacent side}}$$

On the other hand if θ is close to $\pi/2$ but slightly bigger than $\pi/2$ then $\tan\theta$ is very large and negative. This is because we are in the second quadrant where $\tan\theta$ is negative and the length of the opposite side is still very big compared to the length of the adjacent side.

We introduce the following terminology to describe this kind of behavior.

Definition:

We say that the function $y = f(x)$ has a **vertical asymptote** of $+\infty$ as x approaches a from the left and write

$$\lim_{x \to a^-} f(x) = +\infty$$

if for very small positive values of h, $f(a - h)$ is very large and positive.

Definition:

We say that the function $y = f(x)$ has a **vertical asymptote of $-\infty$ as x approaches a from the left** and write

$$\lim_{x \to a^-} f(x) = -\infty$$

if for very small positive values of h, $f(a-h)$ is very large and negative.

Definition:

We say that the function $y = f(x)$ has a **vertical asymptote of $+\infty$ as x approaches a from the right** and write

$$\lim_{x \to a^+} f(x) = +\infty$$

if for very small positive values of h, $f(a+h)$ is very large and positive.

Definition:

We say that the function $y = f(x)$ has a **vertical asymptote of $-\infty$ as x approaches a from the right** and write

$$\lim_{x \to a^+} f(x) = -\infty$$

if for very small positive values of h, $f(a+h)$ is very large and negative.

These definitions capture the idea of a vertical asymptote. As usual, however, for practical problems it is often important to know how close x must be to a in order for $f(x)$ to reach a certain size. The following variant of the first definition captures this idea. Each of the other three definitions has a similar variant.

§7. Vertical Asymptotes

Definition:

We say that the function $y = f(x)$ has a **vertical asymptote** of $+\infty$ as x approaches a from the left and write

$$\lim_{x \to a^-} f(x) = +\infty$$

if for every real number B there is a positive ϵ such that for every $x \in (a - \epsilon, a)$, $f(x) > B$.

Example:

Graph the function

$$f(x) = x + \frac{1}{x}.$$

Answer:

The first step in analyzing this function is noticing that it is undefined when $x = 0$. Notice if x is a very small positive number then $1/x$ is a very large positive number so that $x + 1/x$ will be a very large positive number. That is,

$$\lim_{x \to 0^+} x + \frac{1}{x} = +\infty$$

or, in other words, $f(x)$ has a vertical asymptote of $+\infty$ as x approaches 0 from the right. We can indicate this on a piece of graph paper by an arrow pointing upward just to the right of $x = 0$ as shown in Figure III.49.

On the other hand if x is a very small negative number than $1/x$ will be a very large negative number and $x + 1/x$ will be very large and negative. That is,

$$\lim_{x \to 0^-} x + \frac{1}{x} = -\infty$$

or, in other words, $f(x)$ has a vertical asymptote of $-\infty$ as x approaches 0 from the left. We can add this piece of information to Figure III.49 with an arrow pointing downward just to the left of $x = 0$.

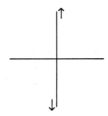

Figure III.49: The vertical asymptotes near $x = 0$

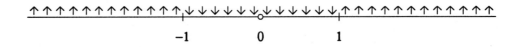

Figure III.50: Where $y = x + 1/x$ is increasing and decreasing

We continue graphing this function using the same techniques as in the past. The next step is to examine the derivative.

$$f'(x) = 1 - \frac{1}{x^2} = \frac{x^2 - 1}{x^2}$$

The derivative is undefined at zero (since the original function was undefined at zero). There are two points $x = 1$ and $x = -1$ at which the derivative is zero. Thus there are three points to worry about.

Kind of point	x-coordinate	y-coordinate
Critical Point	-1	-2
Undefined Point	0	undefined
Critical Point	1	2

We make our usual diagram, Figure III.50, showing where the derivative is positive and negative and, hence, the function is increasing and decreasing.

We see that $f(x)$ is increasing on the interval $(-\infty, -1)$; then there is a local maximum at the point $(-1, -2)$; then the function is decreasing to a vertical asymptote of $-\infty$ to the left of $x = 0$. To the right of

§7. Vertical Asymptotes

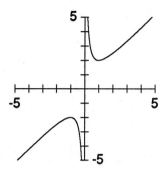

Figure III.51: $y = x + 1/x$

$x = 0$ the function is decreasing from a vertical asymptote of $+\infty$ to a local minimum at the point $(1, 2)$. To the right of $x = 1$ it is increasing.

Finally by looking at the second derivative

$$y'' = \frac{1}{x^3}$$

we see that the graph is concave down to the left of $x = 0$ and concave up to the right of $x = 0$. Putting all this together we get Figure III.51.

Example:

Sketch a graph of the function

$$y = \frac{x^2 - 4}{x^2 - 1}.$$

Answer:

The first step is to note that this function is undefined at $x = 1$ and at $x = -1$. Notice that if x is slightly bigger than 1, $x^2 - 1$ is a small positive number and $x^2 - 4$ is close to -3 thus $(x^2 - 4)/(x^2 - 1)$ is a large negative number. That is,

$$\lim_{x \to 1^+} \left(\frac{x^2 - 4}{x^2 - 1} \right) = -\infty$$

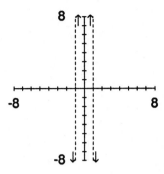

Figure III.52: Vertical asymptotes of $y = (x^2 - 4)/(x^2 - 1)$

and we have a vertical asymptote of $-\infty$ as x approaches 1 from the right. Similar calculations show that

$$\lim_{x \to 1^-} \left(\frac{x^2 - 4}{x^2 - 1} \right) = +\infty$$

$$\lim_{x \to -1^+} \left(\frac{x^2 - 4}{x^2 - 1} \right) = +\infty$$

$$\lim_{x \to -1^-} \left(\frac{x^2 - 4}{x^2 - 1} \right) = -\infty$$

We can indicate this information on a piece of graph paper as shown in III.52.

We continue by analyzing the derivative in the usual way.

$$y = \frac{x^2 - 4}{x^2 - 1}$$

$$= \frac{2x^3 - 2x - (2x^3 - 8x)}{(x^2 - 1)^2}$$

$$= \frac{6x}{(x^2 - 1)^2}$$

Of course, the derivative is undefined at the same two points $x = -1$ and $x = 1$ at which the original function was undefined. In addition

§7. VERTICAL ASYMPTOTES

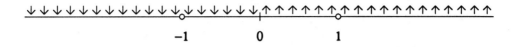

Figure III.53: Where $y = (x^2 - 4)/(x^2 - 1)$ is increasing and decreasing

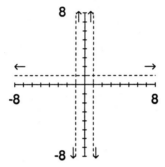

Figure III.54: Asymptotes of $y = (x^2 - 4)/(x^2 - 1)$

there is a critical point at $x = 0$. At this point $y = 4$. We determine the sign of y' and, hence, where y is increasing and decreasing in the usual way. The results are shown in Figure III.53.

The next step is to look for horizontal asymptotes.

$$\lim_{x \to +\infty} \left(\frac{x^2 - 4}{x^2 - 1} \right) = 1$$

and

$$\lim_{x \to -\infty} \left(\frac{x^2 - 4}{x^2 - 1} \right) = 1.$$

We add this information to Figure III.52 to get Figure III.54.

For this function we will not analyze the second derivative. Finally, we use the information we have obtained to draw Figure III.55

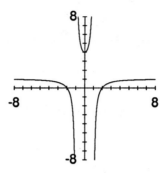

Figure III.55: $y = (x^2 - 4)/(x^2 - 1)$

Exercises

*Exercise III.7.1:
Use the terminology we have developed in this section to describe the vertical asymptotes in Figures III.44 and III.45.

*Exercise III.7.2:
Sketch a graph of the function
$$y = \frac{1}{x(x-3)}.$$
You do not need to analyze the second derivative.

Exercise III.7.3:
Sketch a graph of the function
$$y = \frac{x}{x-3}.$$
You do not need to analyze the second derivative.

Exercise III.7.4:
Sketch a graph of the function
$$y = \frac{x^2}{(x-1)(x-3)}.$$
You do not need to analyze the second derivative.

§7. Vertical Asymptotes

Exercise III.7.5:
Sketch a graph of the function
$$y = \csc x.$$
You do not need to analyze the second derivative.

Exercise III.7.6:
Sketch a graph of the function
$$y = x + \frac{1}{x^2 - 3x + 2}.$$
You do not need to analyze the second derivative.

*Exercise III.7.7:
Sketch a graph of the function
$$y = \frac{1}{x^2(x-1)(x-2)}.$$
You do not need to analyze the second derivative.

*Exercise III.7.8:
Sketch a graph of the function
$$z = t + \frac{1}{t-1} + \frac{2}{t-2}.$$
You do not need to analyze the second derivative.

Exercise III.7.9:
Sketch a graph of the function
$$w = \frac{x^2 - 1}{x^2 - 3x + 2}.$$
You do not need to analyze the second derivative.

*Exercise III.7.10:
A farmer wants to enclose a rectangular field whose area is 100 square feet with the least possible amount of fencing. What should the dimensions of the field be.

*Exercise III.7.11:
A manufacturer wants to design a cyclindrical can that will hold one liter of liquid. (One liter is 1000 cubic centimeters.) The total area of the outside of the can will be $A = 2\pi r^2 + 2\pi rh$ where r is the radius of the cylinder and h is its height. What should the dimensions of the cylinder be to minimize its total area and, hence, the cost of materials.

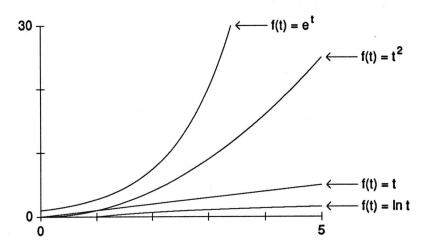

Figure III.56: $f(t) = t, f(t) = t^2, f(t) = \ln t, f(t) = e^t$

§8. L'Hôpital's Rule

How Big is Infinity?

We have seen many examples of functions $f(t)$ that go to infinity as t goes to infinity, for example, the functions: $f(t) = t, f(t) = t^2, f(t) = \ln t$, and $f(t) = e^t$ shown in Figure III.56.

Although all four of these functions go to infinity as t goes to infinity, they do so at very different rates. For example, for the function $f(t) = t$ to exceed 1000, t needs to be greater than 1000 while for the function $f(t) = \ln t$ to exceed 1000, t needs to be greater than 1.97×10^{434}. The function $f(t) = e^t$ will exceed 1000 if t is greater than 6.91 and the function $f(t) = t^2$ will exceed 1000 if t is greater than $\sqrt{1000}$.

Frequently we want to compare two different functions $f(t)$ and $g(t)$ that go to infinity as t goes to infinity. One way to do so is to examine the behavior of the quotient $f(t)/g(t)$ as t goes to infinity.

Example:

Compare the two functions $f(t) = t^2 + 4$ and $g(t) = 2t^2 - 3$ as t goes to infinity.

Answer:

§8. L'Hôpital's Rule

Consider the quotient

$$\frac{f(t)}{g(t)} = \frac{t^2 + 4}{2t^2 - 3}$$

Notice

$$\lim_{t \to +\infty} \frac{t^2 + 4}{2t^2 - 3} = \lim_{t \to +\infty} \frac{1 + \frac{4}{t^2}}{2 - \frac{3}{t^2}}$$
$$= \frac{1}{2}$$

Thus, for very large t, $f(t)$ is roughly half as big as $g(t)$.

This example involved a limit that was easy to determine. Some of the functions that we wish to compare, however, lead to more difficult limits. For example, to compare the functions $f(t) = e^t$ and $f(t) = t^2$ we must determine the limit

$$\lim_{t \to +\infty} \frac{e^t}{t^2}.$$

This limit cannot be evaluated by the same trick that was used in the examples above. In this section we develop a set of techniques, known collectively as **L'Hôpital's Rule**, that can be used to evaluate these and other limits.

L'Hôpital's Rule

Theorem 11 *L'Hôpital's Rule (version I)*
Suppose that

$$\lim_{t \to +\infty} f(t) = \lim_{t \to +\infty} g(t) = \pm\infty$$

and that

$$\lim_{t \to +\infty} \frac{f'(t)}{g'(t)} = L.$$

Then

$$\lim_{t \to +\infty} \frac{f(t)}{g(t)} = L$$

L'Hôpital's Rule is not terribly surprising. It says that if two functions are heading off to infinity and the rate at which one function is increasing is approaching L times the rate at which the other function is increasing then after many years the first function will be roughly L times as large as the second function. We will not prove this version of L'Hôpital's Rule, but we will prove another version later in this section.

Example: Find the limit

$$\lim_{t \to +\infty} \frac{t}{e^t}.$$

The first step in attempting to evaluate any limit involving a quotient is to look at the limits of the numerator and the denominator separately. Notice that

$$\lim_{t \to +\infty} t = +\infty$$

and

$$\lim_{t \to +\infty} e^t = +\infty.$$

So we can use L'Hôpital's Rule. Let $f(t) = t$ denote the numerator and $g(t) = e^t$ denote the denominator. Now $f'(t) = 1$ and $g'(t) = e^t$ so

$$\frac{f'(t)}{g'(t)} = \frac{1}{e^t}$$

and the limit

$$\lim_{t \to +\infty} \frac{1}{e^t}$$

is easy to evaluate since the numerator is going to 1 and the denominator is going to $+\infty$. Thus

$$\lim_{t \to +\infty} \frac{f'(t)}{g'(t)} = \lim_{t \to +\infty} \frac{1}{e^t} = 0$$

and by L'Hôpital's Rule

$$\lim_{t \to +\infty} \frac{f(t)}{g(t)} = \lim_{t \to +\infty} \frac{f'(t)}{g'(t)} = 0.$$

---

§8. L'Hôpital's Rule

Example: Find the limit
$$\lim_{t \to +\infty} \frac{t^2}{e^t}.$$
The first step in attempting to evaluate any limit involving a quotient is to look at the limits of the numerator and the denominator separately. Notice that
$$\lim_{t \to +\infty} t^2 = +\infty$$
and
$$\lim_{t \to +\infty} e^t = +\infty.$$
So we can use L'Hôpital's Rule. Let $f(t) = t^2$ denote the numerator and $g(t) = e^t$ denote the denominator. Now $f'(t) = 2t$ and $g'(t) = e^t$ so
$$\frac{f(t)}{g(t)} = \frac{2t}{e^t}.$$
Unfortunately, the limit
$$\lim_{t \to +\infty} \frac{2t}{e^t}$$
is no easier (at least from the point of view of pre-L'Hôpital days) to evaluate than the original one. Both the numerator and the denominator are going to $+\infty$. Now, however, we can use L'Hôpital's Rule a second time to evaluate this new limit. Taking the derivative of the new numerator we get 2 and taking the derivative of the new denominator we get e^t and now the limit
$$\lim_{t \to +\infty} \frac{2}{e^t}$$
is easy to evaluate since the numerator is going to 2 and the denominator is going to $+\infty$. Thus
$$\lim_{t \to +\infty} \frac{f''(t)}{g''(t)} = \lim_{t \to +\infty} \frac{2}{e^t} = 0$$
and by L'Hôpital's Rule
$$\lim_{t \to +\infty} \frac{f(t)}{g(t)} = \lim_{t \to +\infty} \frac{f'(t)}{g'(t)} = \lim_{t \to +\infty} \frac{f''(t)}{g''(t)} = 0.$$

There are several different variations of L'Hôpital's Rule. The second variation is

Theorem 12 *L'Hôpital's Rule (version II)*
 Suppose that
$$\lim_{t \to +\infty} f(t) = \lim_{t \to +\infty} g(t) = 0$$
and that
$$\lim_{t \to +\infty} \frac{f'(t)}{g'(t)} = L.$$
Then
$$\lim_{t \to +\infty} \frac{f(t)}{g(t)} = L$$

The first two versions of L'Hôpital's Rule involved limits as t approaches $+\infty$. There are eight additional versions of L'Hôpital's Rule for limits as t approaches $-\infty$, a, a from the left, and a from the right. For example,

Theorem 13 *L'Hôpital's Rule (version III)*
 Suppose that
$$\lim_{t \to a^+} f(t) = \lim_{t \to a^+} g(t) = \pm\infty$$
and that
$$\lim_{t \to a^+} \frac{f'(t)}{g'(t)} = L.$$
Then
$$\lim_{t \to a^+} \frac{f(t)}{g(t)} = L$$

The remaining variations are similar. There are ten versions because there are five variations of the limit. Notice that each version requires either that
$$\lim f(t) = \lim g(t) = \pm\infty$$
or
$$\lim f(t) = \lim g(t) = 0.$$

Warning: L'Hôpital's Rule only works if one of these conditions is met. If you try to apply L'Hôpital's Rule when neither of these conditions is met than it may give a wrong answer! As is fitting for a man who gave his name to a theorem with so many variations, L'Hôpital's name has two variations. Sometimes it is spelled "L'Hôpital" and sometimes it is spelled "L'Hospital." It should not be pronounced "hospital." The "s" is silent.

§8. L'Hôpital's Rule

Example: Find the limit

$$\lim_{t \to 0^+} (t \ln t).$$

This limit is difficult to evaluate since it involves the product of two factors, one of which is going to zero and the other of which is going to $+\infty$. At first glance it looks as if L'Hôpital's Rule will be of no help since L'Hôpital's Rule applies to quotients not to products. We can, however, use a sneaky trick. Notice that

$$t \ln t = \frac{\ln t}{\left(\frac{1}{t}\right)}$$

so that our product can be written as a quotient. The first exercise below asks you to finish this example.

To give some idea of why L'Hôpital's Rule is true we give a proof of a weak form of one version of L'Hôpital's Rule.

Theorem 14 *Suppose that $f(x)$, $g(x)$, $f'(x)$ and $g'(x)$ are all continuous at the point $x = a$, that*

$$\lim_{x \to a} f(x) = \lim_{x \to a} g(x) = 0,$$

and that

$$\lim_{x \to a} \frac{f'(x)}{g'(x)} = L.$$

Then

$$\lim_{x \to a} \frac{f(x)}{g(x)} = L.$$

Proof: First notice that since $f(x)$, $g(x)$, $f'(x)$ and $g'(x)$ are all continuous at the point $x = a$,

$$\begin{aligned}
\lim_{x \to a} f(x) &= f(a) = 0 \\
\lim_{x \to a} g(x) &= g(a) = 0 \\
\lim_{x \to a} f'(x) &= f'(a) \\
\lim_{x \to a} g'(x) &= g'(a)
\end{aligned}$$

$$\lim_{x \to a} \frac{f(x)}{g(x)} = \lim_{x \to a} \frac{f(x) - f(a)}{g(x) - g(a)}$$
$$= \lim_{x \to a} \frac{\frac{f(x) - f(a)}{x - a}}{\frac{g(x) - g(a)}{x - a}}$$
$$= \frac{\lim_{x \to a} \frac{f(x) - f(a)}{x - a}}{\lim_{x \to a} \frac{g(x) - g(a)}{x - a}}$$
$$= \frac{f'(a)}{g'(a)}$$
$$= \frac{\lim_{x \to a} f'(x)}{\lim_{x \to a} g'(x)}$$

Exercises

*Exercise III.8.1: Finish the example on the preceding page to find
$$\lim_{t \to 0^+} (t \ln t).$$

Exercise III.8.2: Use L'Hôpital's Rule to find the limit
$$\lim_{x \to +\infty} \frac{2x - 3}{3x + 4}$$

*Exercise III.8.3: Use L'Hôpital's Rule to find the limit
$$\lim_{x \to 1^+} \frac{x - 1}{x^2 - 1}$$

Exercise III.8.4: Find the limit
$$\lim_{t \to 0^+} t^2 \ln t$$

*Exercise III.8.5: Find the limit
$$\lim_{x \to 0^+} (\sqrt{x}) \ln x$$

Exercise III.8.6: Sketch a graph of the function
$$f(x) = \frac{e^x - 1}{x}.$$

Exercise III.8.7: Sketch a graph of the function $y = xe^{-x}$.

Exercise III.8.8: Consider the function $y = xe^{-x}$ from the previous exercise. Notice the region between the curve $y = xe^{-x}$ and the x-axis to the right of the y-axis. Find the area of this region.

Exercise III.8.9: Suppose that the region described in the previous exercise is revolved about the x-axis. Find the volume of the resulting solid.

Exercise III.8.10: Compare the two functions $f(x) = x^n$ where n is a positive integer and $g(x) = e^x$.

Exercise III.8.11: Compare the two functions $f(x) = x^c$ where c is a positive constant and $g(x) = e^x$.

Exercise III.8.12: Compare the two functions $f(x) = x^c$ where c is a positive constant and $g(x) = \ln x$.

---

§9. Summary[+]

The most important single idea in calculus is the idea of a function, a relationship in which one quantity, called the **dependent variable**, is determined by one or more other quantities, called **independent** or **explanatory** variables. In this first semester we deal most frequently with situations in which there is only one independent or explanatory variable. Mathematically, the idea of a function is simple and straightforward—a function is a mechanism that determines the value of the dependent variable that corresponds to each appropriate value of the independent variable. The set of "appropriate" values for the independent variable is called the **domain** of the function.

One of the most important ways of expressing a function is by an initial value problem. This method provides a link between the real world and mathematics. Initial value problems express the relationship between two quantities using immediate information about the quantities and the principles that govern the way they vary. For example, a falling baseball (ignoring air resistance) accelerates at -32 feet per second per second. This physical fact can be immediately expressed as the differential equation $h'' = -32$. As another example, when a new species is first introduced in a habitat with an abundance of food and the other necessities of life its population grows at a constant relative rate. This translates immediately into the differential equation $p' = kp$. A differential equation together with initial data is called an initial value problem.

In this chapter we studied one method (Euler's Method) for computing the values of a function that has been expressed as the solution of an initial value problem. Later in the course we will discuss ways of expressing a function determined by an initial value problem by an algebraic formula. We have seen some examples already in the section on falling bodies. Euler's method, however, is much more general. It can be used to study any initial value problem whereas there are relatively few initial value problems for which one can find algebraic solutions.

Much of this chapter was devoted to expanding our repertoire of functions. In particular, we looked at the exponential function and its inverse, the natural logarithm. These functions are now part of our basic toolkit. We can differentiate them and we can bring all the power of calculus to bear on problems involving these new functions.

We also looked at the second derivative. Like the first derivative the second derivative has many different interpretations. Geometrically it gives us information about the shape of a graph. Its most important physical application is in situations involving motion. If a function $f(t)$ describes the position or location of an object at time t then the derivative $f'(t)$ describes its velocity and the second derivative $f''(t)$ describes its acceleration. This is especially important because many physical problems involve the study of how objects react to forces. Newton's Second Law says that when a force of strength F acts on an object of mass m the resulting acceleration is given by F/m. Thus, the second derivative will be an essential tool for studying the motion of an object subjected to various forces. This will be an important theme later in the course.

We introduced the idea of a vertical asymptote to describe certain features of functions.

Finally, we discussed L'Hôpital's Rule as a powerful new technique for evaluating limits.

Chapter IV

Integration

§1. The Definite Integral

Many problems can be solved by simple multiplication. For example,

- If an object is traveling at the rate of k miles per hour for T hours then the total distance traveled will be kT miles.

- If a rectangle has a height of h feet and a width of w feet then its area will be hw feet.

- Suppose that a hiker is hiking north along a trail that slopes upward at the rate of 0.10 feet for each foot traveled northward. If the hiker goes x feet northward she will climb $0.10x$ feet.

- A cylindrical tank whose cross-sectional area is 25 square feet and whose length is x feet has a volume of $25x$ cubic feet.

- Suppose that a dam allows water to flow out of a lake at the rate of 30,000 gallons per day. In T days, $30,000T$ gallons will flow out of the lake.

All the examples above involve multiplying two constants. Sometimes, however, we need to "multiply" two factors one of which is not constant. For example, we want to be able to solve problems like the following.

- Consider the shaded region in Figure IV.1. Notice that the height of this region varies. Find its area.

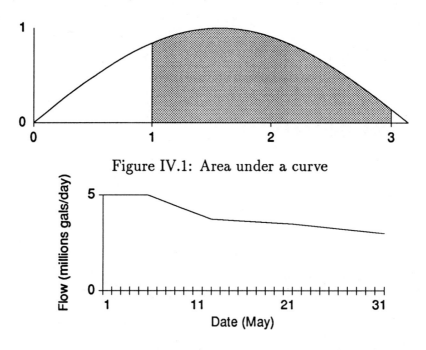

Figure IV.1: Area under a curve

Figure IV.2: Water over the dam

- Suppose the graph shown in Figure IV.2 shows the rate at which water is flowing over a dam. How much water flows over the dam from May 1 through May 15?

- Find the volume of the cone shown in Figure IV.3. Notice that the cross-sectional area of the cone varies depending on where it is sliced.

- Figure IV.4 shows the slope of a particular trail. The x-axis represents the distance in miles from the trailhead measured in a straight horizontal direction. The y-axis represents the slope of the trail. How much altitude will a climber gain hiking from the trailhead to mile marker 5.0? to mile marker 10?

- The velocity of a falling body is given by the function $v = -32t$ where t is time in seconds after it is dropped. How far will it fall during the first three seconds after it is dropped?

In this chapter we develop a tool, the **definite integral**, that can be used to solve a wide variety of problems involving the "multiplication"

§1. The Definite Integral 405

Figure IV.3: Volume of a cone

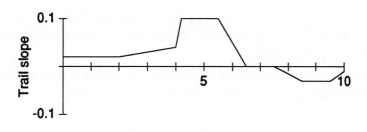

Figure IV.4: Slope of a trail

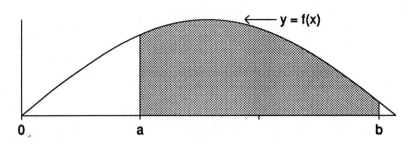

Figure IV.5: Region enclosed by the x-axis, $x = a$, $x = b$, and $y = f(x)$

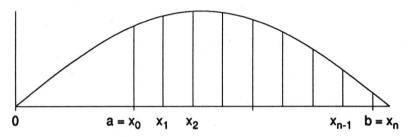

Figure IV.6: Breaking a region into strips

of two numbers, one of which is not constant. This tool can be used to solve all of the problems listed above.

We begin with the first of these problems, finding the area of a region like the one shown in Figure IV.5, because this problem is the easiest to visualize. Our strategy is to *approximate* the area of this region by breaking it up into thin strips and approximating the area of each strip by a rectangle.

We begin by breaking the interval $[a, b]$ into n subintervals each of which has width $h = (b - a)/n$. We use the notation

$$x_i = a + ih,$$

so that the interval $[a, b]$ is broken up into the n subintervals

$$[x_0, x_1], [x_1, x_2], \ldots, [x_{i-1}, x_i], \ldots, [x_{n-1}, x_n]$$

and the original region is broken up into n strips of equal width as shown in Figure IV.6.

We can approximate each strip by a rectangle as shown in Figures IV.7, IV.8 and IV.9. As you can see from these three figures there are

§1. The Definite Integral

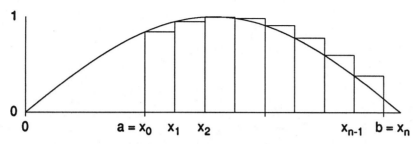

Figure IV.7: Approximating strips by "left" rectangles

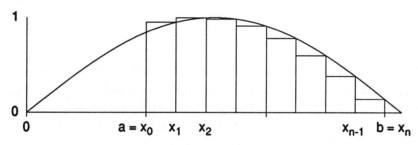

Figure IV.8: Approximating strips by "right" rectangles

several different ways that we might do this. Figure IV.7 approximates each strip by a rectangle whose height is determined by the left edge of the strip, Figure IV.8 approximates each strip by a rectangle whose height is determined by the right edge of the strip, and Figure IV.9 approximates each strip by a rectangle whose height is determined by the midpoint of each strip.

More generally, in each subinterval $[x_{i-1}, x_i]$ we choose a **sample point** s_i and we look at the rectangle whose height $f(s_i)$ is measured at the sample point s_i. For example, in Figure IV.7 the sample point s_i for the ith strip is x_{i-1}, and in Figure IV.8 the sample point s_i for the ith strip is x_i. The area $f(s_i)h$ of the ith rectangle is its height $f(s_i)$ times its width h. The total area of all the rectangles is the sum

$$f(s_1)h + f(s_2)h + \cdots + f(s_n)h.$$

Example:

Estimate the area of the region enclosed by the x-axis, the lines $x = 1$, $x = 3$ and the curve $y = \sin x$.

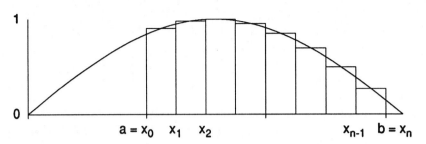

Figure IV.9: Approximating strips by "midpoint" rectangles"

Answer:

We find three different estimates all based on ten strips each of width 0.2. The first estimate is based on rectangles whose height is determined by the left edge of each strip. That is, the sample point s_i for the ith strip is x_{i-1}. The following table shows the calculations.

strip (i)	x_{i-1}	x_i	s_i	height ($f(s_i)$)	width	area
1	1.0000	1.2000	1.0000	.8415	.2000	.1683
2	1.2000	1.4000	1.2000	.9320	.2000	.1864
3	1.4000	1.6000	1.4000	.9854	.2000	.1971
4	1.6000	1.8000	1.6000	.9996	.2000	.1999
5	1.8000	2.0000	1.8000	.9738	.2000	.1948
6	2.0000	2.2000	2.0000	.9093	.2000	.1819
7	2.2000	2.4000	2.2000	.8085	.2000	.1617
8	2.4000	2.6000	2.4000	.6755	.2000	.1351
9	2.6000	2.8000	2.6000	.5155	.2000	.1031
10	2.8000	3.0000	2.8000	.3350	.2000	.0670
Total						1.5952

The second estimate is based on rectangles whose height is determined by the right edge of each strip. That is, the sample point s_i for the ith strip is x_i. The following table shows the calculations.

§1. The Definite Integral

strip (i)	x_{i-1}	x_i	s_i	height ($f(s_i)$)	width	area
1	1.0000	1.2000	1.2000	.9320	.2000	.1864
2	1.2000	1.4000	1.4000	.9854	.2000	.1971
3	1.4000	1.6000	1.6000	.9996	.2000	.1999
4	1.6000	1.8000	1.8000	.9738	.2000	.1948
5	1.8000	2.0000	2.0000	.9093	.2000	.1819
6	2.0000	2.2000	2.2000	.8085	.2000	.1617
7	2.2000	2.4000	2.4000	.6755	.2000	.1351
8	2.4000	2.6000	2.6000	.5155	.2000	.1031
9	2.6000	2.8000	2.8000	.3350	.2000	.0670
10	2.8000	3.0000	3.0000	.1411	.2000	.0282
Total						1.4552

The final estimate is based on rectangles whose height is determined by the midpoint of each strip. That is, the sample point s_i for the ith strip is

$$s_i = \frac{x_{i-1} + x_i}{2}.$$

The following table shows the calculations.

strip (i)	x_{i-1}	x_i	s_i	height ($f(s_i)$)	width	area
1	1.0000	1.2000	1.1000	.8912	.2000	.1782
2	1.2000	1.4000	1.3000	.9636	.2000	.1927
3	1.4000	1.6000	1.5000	.9975	.2000	.1995
4	1.6000	1.8000	1.7000	.9917	.2000	.1983
5	1.8000	2.0000	1.9000	.9463	.2000	.1893
6	2.0000	2.2000	2.1000	.8632	.2000	.1726
7	2.2000	2.4000	2.3000	.7457	.2000	.1491
8	2.4000	2.6000	2.5000	.5985	.2000	.1197
9	2.6000	2.8000	2.7000	.4274	.2000	.0855
10	2.8000	3.0000	2.9000	.2392	.2000	.0478
Total						1.5328

This gives us three different estimates for the area of our region—1.5952, 1.4552, and 1.5328.

Riemann Sums

Sums like those used in our approximations of the area of the region bounded by the curve $y = f(x)$, the x-axis, the line $x = a$, and the line

$x = b$ are called **Riemann sums**. For each n we divide the interval $[a, b]$ into n subintervals, each of width $h = (b-a)/n$, using the notation

$$x_i = a + ih.$$

Next we choose a **sample point** s_i in each subinterval. That is,

$$s_i \in [x_{i-1}, x_i].$$

The sum

$$f(s_1)h + f(s_2)h + f(s_3)h + \cdots + f(s_n)h$$

is called a **Riemann sum**. Notice that each term of the Riemann sum computes the area of a rectangle that approximates the area of one strip of the original region. We shall see later that these same Riemann sums can be used to approximate other quantities that are computed by "multiplication" of two numbers one of which is not constant.

Summation Notation

We want to introduce some notation, called **summation notation** that is very useful for representing sums like Riemann sums. Using summation notation we write the sum

$$f(s_1)h + f(s_2)h + \cdots + f(s_n)h$$

as

$$\sum_{i=1}^{n} f(s_i)h.$$

The symbol "Σ" is the capital Greek letter sigma. For this reason summation notation is sometimes called **sigma notation**. If you are familiar with a programming language like BASIC or Pascal you should think of this notation as being very similar to a FOR...NEXT loop.

```
LET sum = 0                              ! Start with sum = 0.
FOR i = 1 TO n                           ! Start of loop.
    LET sum = sum + function(s(i)) * h   ! Add area of rectangle.
NEXT i                                   ! End of loop.
```

The English equivalent is

§1. The Definite Integral

Sum n terms, each of the form $f(s_i)h$. Start with $i = 1$ and continue through $i = n$.

The best way to learn sigma notation is to see some examples.

- The sum $1 + 4 + 9 + 16 + \cdots + 100$ (i.e. the sum of the first ten squares) can be written using sigma notation as

$$\sum_{i=1}^{10} i^2.$$

- The Riemann sum estimating the area under the function $f(x)$ between the lines $x = a$ and $x = b$ above the x-axis, using the left end of each subinterval as the sample point can be written using sigma notation as

$$\sum_{i=1}^{n} f(x_{i-1})h.$$

- The Riemann sum estimating the area under the function $f(x)$ between the lines $x = a$ and $x = b$ above the x-axis, using the right end of each subinterval as the sample point can be written using sigma notation as

$$\sum_{i=1}^{n} f(x_i)h.$$

- The Riemann sum estimating the area under the function $f(x)$ between the lines $x = a$ and $x = b$ above the x-axis, using the midpoint of each subinterval as the sample point can be written using sigma notation as

$$\sum_{i=1}^{n} f\left(\frac{x_{i-1} + x_i}{2}\right) h.$$

The Definite Integral

Each of the Riemann sums

$$\sum_{i=1}^{n} f(s_i) h$$

computes an estimate for the area of the region bounded by the curve $y = f(x)$, the x-axis, the line $x = a$, and the line $x = b$. Figures IV.10 and IV.11 compare typical estimates with the actual area. In each figure the difference between the estimate and the actual area is shaded. Figure IV.11 uses more strips than Figure IV.10. Because so many strips are used in Figure IV.11 the difference between the exact area and the estimate is quite small. In general it is possible to obtain very good estimates by using very large numbers of strips. In fact, one can obtain as good an estimate as might be required for any given purpose by using sufficiently many strips. Once again we see one of the main themes of calculus. There is no obvious way to find an exact answer but we do have a method for obtaining estimates. This method allows us to obtain as good an estimate as might be necesssary for any given purpose. We can say that

$$\text{Area} = \lim_{n \to \infty} \sum_{i=1}^{n} f(s_i)h.$$

Intuitively this means that by using very many strips the estimate

$$\sum_{i=1}^{n} f(s_i)h$$

will be very close to the area. This description captures the intuitive idea but, as usual, for practical purposes it is important to know how many strips are required to obtain an estimate that is sufficiently close to the exact area. The following definition captures this idea.

Definition:

We say
$$\lim_{n \to \infty} \sum f(s_i)h = L$$
if for any positive tolerance ϵ there is an N such that for every $n \geq N$ and every choice of sample points $s_1, s_2, \ldots, s_n$

$$\left| \sum_{i=1}^{n} f(s_i)h - L \right| < \epsilon.$$

§1. The Definite Integral

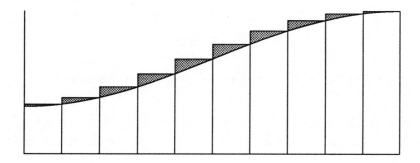

Figure IV.10: Comparing an estimate with the exact area

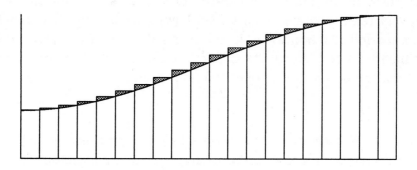

Figure IV.11: Comparing an estimate with the exact area

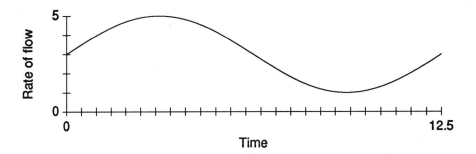

Figure IV.12: The flow of water into the ocean

We introduced Riemann sums in the context of determining area. Exactly the same calculations, however, come up in many different contexts.

Example:

Engineers have been studying the waterflow at a point on a river near its mouth. They have determined that water is flowing into the ocean at the rate of

$$r(t) = 3 + 2\sin\left(\frac{2\pi t}{12.5}\right)$$

where $r(t)$ is measured in millions of gallons per hour and t is measured in hours. This function is shown in Figure IV.12. The rate at which water flows into ocean varies because of the tides. At low tide water flows very rapidly and at high tide it flows more slowly. The function $r(t)$ has period 12.5 hours (i.e. it repeats itself every 12.5 hours) because the time from one high tide to the next is 12.5 hours. How much water flows into the ocean during one 12.5 hour period?

Answer:

We can estimate the total waterflow by breaking the interval $[0, 12.5]$ up into n subintervals each of duration $h = 12.5/n$ and determining the rate of flow $r(s_i)$ at some sample time s_i during each subinterval. We estimate the waterflow on each subinterval by $r(s_i)h$. If n is large than each subinterval is very short and the rate of waterflow during that subinterval does not vary much. As a result $r(s_i)h$ is a reasonable estimate for the waterflow during that subinterval. The total waterflow during the whole 12.5 hour period is estimated by the usual Riemann

§1. The Definite Integral

sum

$$\sum_{i=1}^{n} r(s_i)h.$$

By choosing n very large we obtain very good estimates of the total waterflow. Thus

$$\text{Total waterflow} = \lim_{n \to \infty} \sum r(s_i)h.$$

The table below shows the calculations for an estimate based on 10 subintervals.

strip (i)	x_{i-1}	x_i	s_i	height ($f(s_i)$)	width	area
1	.0000	1.2500	.6250	4.0717	1.2500	5.0896
2	1.2500	2.5000	1.8750	4.8596	1.2500	6.0744
3	2.5000	3.7500	3.1250	4.9372	1.2500	6.1715
4	3.7500	5.0000	4.3750	4.2748	1.2500	5.3435
5	5.0000	6.2500	5.6250	3.1256	1.2500	3.9070
6	6.2500	7.5000	6.8750	1.9283	1.2500	2.4104
7	7.5000	8.7500	8.1250	1.1404	1.2500	1.4255
8	8.7500	10.0000	9.3750	1.0628	1.2500	1.3285
9	10.0000	11.2500	10.6250	1.7252	1.2500	2.1565
10	11.2500	12.5000	11.8750	2.8744	1.2500	3.5931
Total						37.5000

We can draw two very important conclusions from this example.

- We can estimate total waterflow given the rate of waterflow using exactly the same calculations as we used for estimating area.

- The total waterflow is the same as the "area" under the curve $r(t)$ between $t = 0$ and $t = 12.5$ above the t-axis.

We put the word "area" above in quotes because we must be careful about units. Normally, area is measured in square feet, square meters, or some other square units since area is the product of length and width both of which are measured in feet or meters or other units of linear measure. In this case we are multiplying time measured in hours by the rate of waterflow measured in millions of gallons per hour. The result is measured in millions of gallons.

The two conclusions we drew at the end of this example illustrate two extremely important points.

- The solutions to many seemingly very different problems can be solved by an estimation procedure involving the exact same arithmetical computations, Riemann sums.

- The observation above allows us to draw analogies between apparently very different situations.

Because many problems can be solved by determining by the limit of Riemann sums this limit is worth studying in its own right.

Definition:

Suppose that $f(x)$ is any continuous function on the interval $[a, b]$. The **definite integral** of $f(x)$ on $[a, b]$ is defined by

$$\int_a^b f(x)\, dx = \lim_{n \to \infty} \sum_{i=1}^n f(s_i) h.$$

Using this new notation we can write

$$\text{area} = \int_a^b f(x)\, dx$$

or

$$\text{total waterflow} = \int_0^{12.5} r(t)\, dt.$$

Notice that the definite integral is a number. Its definition as a limit of Riemann sums tells us two important things.

- One can estimate

$$\int_a^b f(x)\, dx$$

to whatever degree of precision might be needed for any given purpose by computing a Riemann sum with n sufficiently large.

- The definite integral gives us the exact value of any quantity that can be approximated by Riemann sums.

The definite integral enables us to "multiply" two factors one of which is not constant.

§1. The Definite Integral

Exercises

*Exercise IV.1.1: Using the midpoint method with five subintervals, estimate the volume of a cone whose height is five feet and whose base is a circle of radius one foot. Express the exact volume as an integral.

*Exercise IV.1.2: The rate at which water is flowing over a particular dam during the month of May is given by

$$r(t) = \begin{cases} 5.0, & \text{if } 0 \leq t \leq 5; \\ 5.5 - 0.1t, & \text{if } 5 \leq t \leq 15; \\ 4.0, & \text{if } 15 \leq t \leq 31. \end{cases}$$

where t is measured in days with $t = 0$ midnight the night of April 30–May 1 and $t = 31$ is midnight the night of May 31–June 1. The rate of waterflow $r(t)$ is measured in millions of gallons per day.

Using the midpoint method with 31 subintervals, estimate the total waterflow during the month of May. Express the exact total as an integral. Find the exact total waterflow.

*Exercise IV.1.3: A hiker is hiking along a trail that goes straight north from the trailhead. There are mile markers along the trail giving the distance from the trailhead measured horizontally. The slope of the trail is given by the function

$$s(d) = \begin{cases} 0, & \text{if } 0 \leq d \leq 1; \\ 0.05(d-1), & \text{if } 1 \leq d \leq 3; \\ 0.1, & \text{if } 3 \leq d \leq 5; \\ 0.1 - 0.05(d-5), & \text{if } 5 \leq d \leq 7; \\ 0, & \text{if } 7 \leq d \leq 10. \end{cases}$$

where d is measured in miles from the trailhead and $s(d)$ is the slope measured in miles climbed per horizontal mile.

Using the midpoint method with ten subintervals, estimate the difference in altitude (elevation) between the trailhead and mile marker 10. Express the exact difference as an integral. Determine the exact difference.

*Exercise IV.1.4: The velocity of a falling body is given by the function

$$v(t) = -32t$$

where t is measured in seconds after the body is dropped and $v(t)$ is measured in feet per second.

Using the midpoint method with five subintervals, estimate how far a body will fall during the first two seconds after it is dropped. Express the exact distance a body will fall during the first two second after it is dropped as an integral. Determine the exact distance a body will fall during the first two seconds after it is dropped.

*Exercise IV.1.5:

Engineers have been studying the waterflow near the point where a river empties into the ocean. They have determined that water is flowing into the ocean at the rate of

$$r(t) = 2 + 3\sin\left(\frac{2\pi t}{12.5}\right)$$

where $r(t)$ is measured in millions of gallons per hour and t is measured in hours. Notice that, because of the tides, water is sometimes flowing into the river from the ocean.

The engineers are interested in two things. First, the net flow of water from the river into the ocean during the period $t = 0$ to $t = 25$. Second, they are interested in building a plant to generate electric power. This plant will be able to generate power whichever way the water is flowing. In order to determine how much power the plant might generate they need to know the total water flow *ignoring direction* during the period $t = 0$ to $t = 25$.

What is the net water flow during the time $t = 0$ to $t = 25$? First find an estimate using the midpoint method with ten subintervals, then express the exact answer as an integral.

What is the total water flow *ignoring direction* during the period $t = 0$ to $t = 25$? First find an estimate using ten subintervals, then express the exact answer as an integral.

§2. Properties of the Definite Integral

In the previous section we developed a new tool, the *definite integral*,

$$\int_a^b f(x)\,dx = \lim_{n\to\infty} \sum_{i=1}^n f(s_i)h.$$

§2. PROPERTIES OF THE DEFINITE INTEGRAL

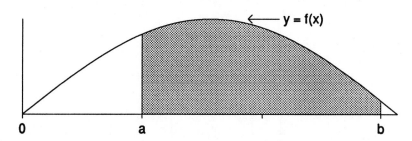

Figure IV.13: The definite integral as an area

In this section we discuss some of its properties. The single most important property is that the definite integral is a real number if $f(x)$ is a continuous function on the interval $[a, b]$.

The second most important property is that we can interpret the definite integral as area. If the function $f(x)$ is nonnegative then the definite integral is the area of the region enclosed by the x-axis, the curve $y = f(x)$, the line $x = a$, and the line $x = b$ as shown in Figure IV.13.

The same picture works when $f(x)$ has both positive and negative values with the proviso that area below the x-axis is counted as being negative. For example,

$$\int_1^4 (x-2)\, dx = 3/2$$

which is the difference obtained by subtracting the area of the triangle below the x-axis from the area of the triangle above the x-axis in Figure IV.14

Exercise

*Exercise IV.2.1: Using the midpoint method with six subintervals, estimate $\int_1^4 (x^2 - 4)\, dx$. Notice, as you do your calculations, the way in which the portion of the region below the x-axis contributes negative area.

The next property of the definite integral is called **linearity**.

Theorem 1 *If $f(x)$ and $g(x)$ are continuous functions on the interval $[a, b]$ and k is any constant then*

$$\int_a^b (f(x) + g(x))\, dx = \int_a^b f(x)\, dx + \int_a^b g(x)\, dx$$

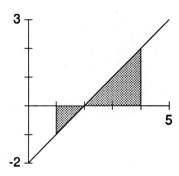

Figure IV.14: Positive and negative area

and
$$\int_a^b kf(x)\,dx = k\int_a^b f(x)\,dx$$

Proof:
Suppose that n is a positive integer and $s_1, s_2, \ldots, s_n$ are sample points. We consider the Riemann sum
$$\sum_{i=1}^n (f+g)(s_i)h$$
approximating
$$\int_a^b (f(x)+g(x))\,dx.$$

$$\sum_{i=1}^n (f+g)(s_i)h = \sum_{i=1}^n (f(s_i)+g(s_i))h$$
$$= \sum_{i=1}^n (f(s_i)h + g(s_i)h)$$
$$= \sum_{i=1}^n f(s_i)h + \sum_{i=1}^n g(s_i)h$$

Now if we take limits as $n \to \infty$ we see
$$\int_a^b (f(x)+g(x))\,dx = \lim_{n\to\infty} \sum_{i=1}^n (f(s_i)+g(s_i))h$$

§2. Properties of the Definite Integral

$$= \lim_{n\to\infty} \left(\sum_{i=1}^{n} f(s_i)h + \sum_{i=1}^{n} g(s_i)h\right)$$

$$= \lim_{n\to\infty} \sum_{i=1}^{n} f(s_i)h + \lim_{n\to\infty} \sum_{i=1}^{n} g(s_i)h$$

$$= \int_a^b f(x)\,dx + \int_a^b g(x)\,dx$$

Similarly,

$$\int_a^b kf(x)\,dx = \lim_{n\to\infty} \sum_{i=1}^{n} kf(s_i)h$$

$$= \lim_{n\to\infty} k\sum_{i=1}^{n} f(s_i)h$$

$$= k \lim_{n\to\infty} \sum_{i=1}^{n} f(s_i)h$$

$$= k\int_a^b f(x)\,dx.$$

The next property of the definite integral is called **positivity**.

Theorem 2 *Suppose $f(x)$ is a continuous function on the interval $[a, b]$ and for every $x \in [a, b]$, $f(x) \geq 0$. Then*

$$\int_a^b f(x)\,dx \geq 0.$$

Proof:
If we look at a typical Riemann sum approximating the integral

$$\sum_{i=1}^{n} f(s_i)h$$

we see that all the terms are nonnegative. Since a limit of nonnegative numbers must also be nonnegative

$$\int_a^b f(x)\,dx \geq 0.$$

As an immediate consequence of these two theorems we have

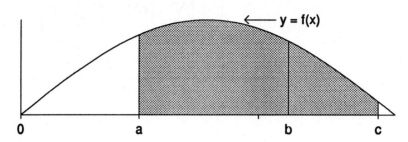

Figure IV.15: The additivity property

Theorem 3 *Suppose $f(x)$ and $g(x)$ are two continuous functions on the interval $[a, b]$ and that for every $x \in [a, b]$, $f(x) \leq g(x)$. Then*

$$\int_a^b f(x)\,dx \leq \int_a^b g(x)\,dx.$$

Proof:
By linearity

$$\int_a^b g(x)\,dx - \int_a^b f(x)\,dx = \int_a^b (g(x) - f(x))\,dx.$$

Since $f(x) \leq g(x)$, $(g(x) - f(x)) \geq 0$. So, by positivity, the integral on the right is nonnegative. Hence

$$\int_a^b g(x)\,dx - \int_a^b f(x)\,dx \geq 0$$
$$\int_a^b g(x)\,dx \geq \int_a^b f(x)\,dx$$

The next property is called **additivity**

Theorem 4 *Suppose that $f(x)$ is continuous on the interval $[a, c]$ and that $b \in [a, c]$. Then*

$$\int_a^c f(x)\,dx = \int_a^b f(x)\,dx + \int_b^c f(x)\,dx$$

Proof:

§2. Properties of the Definite Integral

Our argument is based on the interpretation of the definite integral as area. Look at Figure IV.15. The area of the shaded region is $\int_a^c f(x)\,dx$. The area of the part of this region to the left of the line $x = b$ is $\int_a^b f(x)\,dx$. The area of the part of this region to the right of the line $x = b$ is $\int_b^c f(x)\,dx$. Thus, it is clear from this picture that

$$\int_a^c f(x)\,dx = \int_a^b f(x)\,dx + \int_b^c f(x)\,dx.$$

Next, we want to discuss a technical point. All our integrals so far have been of the form

$$\int_a^b f(x)\,dx$$

where $a < b$. We make the following definition for the definite integral when $a > b$.

Definition:

Suppose that $b < a$ and that $f(x)$ is continuous on the interval $[b, a]$. Then we define

$$\int_a^b f(x)\,dx = -\int_b^a f(x)\,dx.$$

The properties discussed above, particularly the additivity property, are consistent with this new definition.

Finally, we note that

$$\int_a^a f(x)\,dx = 0$$

since for any approximating Riemann sum

$$\sum_{i=1}^n f(s_i)h$$

h will be zero.

Exercises

*Exercise IV.2.2: Find each of the following definite integrals.

$$\int_1^3 (3-x)\,dx$$

$$\int_3^4 (3-x)\,dx$$

$$\int_1^4 (3-x)\,dx$$

$$\int_4^1 (3-x)\,dx$$

*Exercise IV.2.3: Given that

$$\int_0^{\pi/2} \sin x\,dx = 1$$

$$\int_{\pi/2}^{3\pi/4} \sin x\,dx = \frac{1}{\sqrt{2}}$$

find each of the following definite integrals.

$$\int_0^{3\pi/4} \sin x\,dx$$

$$\int_0^{\pi/2} 3\sin x\,dx$$

$$\int_{\pi/2}^{0} 4\sin x\,dx$$

*Exercise IV.2.4: Prove that if $y = f(x)$ is symmetric about the y-axis then

$$\int_{-a}^0 f(x)\,dx = \int_0^a f(x)\,dx.$$

*Exercise IV.2.5: Prove that if $y = f(x)$ is symmetric about the origin then

$$\int_{-a}^0 f(x)\,dx = -\int_0^a f(x)\,dx.$$

§3. The Fundamental Theorem of Calculus

Calculus courses are often divided into two parts—differential and integral calculus. The words "differential" and "integral" refer to two of the grand constellations in the calculus heavens. The first of these constellations is centered around the concept of the derivative. The derivative is the tool we need to study many very diverse situations, for example,

- If a curve is described by a function $y = f(x)$ then the slope of that curve at each point is described by the derivative $f'(x)$.

- If the location of an object at time t is described by a function $y = f(t)$ then its velocity at time t is described by the derivative $f'(t)$.

- If the amount of water in a reservoir at time t is described by the function $w(t)$ then the rate at which water is flowing out of the reservoir is described by the derivative $w'(t)$.

In this chapter we study integration, the second of the grand constellations. This constellation is centered around the concept of the definite integral. Like the derivative, the definite integral is useful in many very diverse situations, for example,

- If a function $f(x)$ describes the height of a region over the point x then
$$\int_a^b f(x)\, dx$$
describes the area of the region between $x = a$ and $x = b$.

- If a function $v(t)$ describes the velocity of an object at time t then
$$\int_a^b f(t)\, dt$$
describes the change in the position of the object between time $t = a$ and $t = b$.

- If the function $f(x)$ describes the slope of a trail at mile marker x then
$$\int_a^b f(x)\,dx$$
describes the change in elevation between $x = a$ and $x = b$.

- If a function $f(t)$ describes the rate at which water is flowing out of a reservoir at time t then
$$\int_a^b f(t)\,dt$$
describes the change in the amount of water in the reservoir between time $t = a$ and $t = b$.

These two constellations are the same! Differentiation and integration talk about exactly the same situations from different perspectives. We begin by considering two examples and then prove the keystone theorem of calculus—The Fundamental Theorem of Calculus.

Example:

Experienced hikers know that the rate at which they walk depends on how much climbing is involved. For example, a hiker who walks four miles per hour on level ground may walk only one mile per hour on a trail that has a slope of 1/5, that is, on a trail that rises one foot for every five feet of horizontal length. When you're planning a hike it is important to know not only distances but slopes and elevations. The map shown in Figure IV.16 is a trail map that shows the elevation (or altitude) along a trail. The x-axis represents points on the trail measured in horizontal miles from the trailhead. The y-axis represents elevation in miles above sea level.

A hiker interested in how easy or difficult different parts of the trail might draw a graph showing the slope of the trail at each point. In mathematical terms she might draw a graph showing the derivative of the elevation function. Figure IV.17 shows such a graph.

These two graphs describe the same situation from different perspectives. Figure IV.16 emphasizes the elevation at each point on the trail. One can immediately read the elevation of Mt. Newton, roughly 0.45 miles, from this graph. Figure IV.17 emphasizes the steepness of the trail. One can immediately see from this graph that the trail begins very steeply with a slope of 0.2 for the first mile. Because the two

§3. The Fundamental Theorem of Calculus

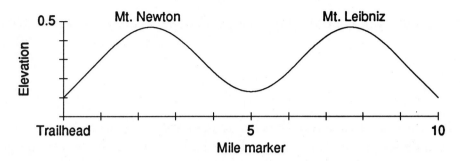

Figure IV.16: Calculus Range Trail (elevation)

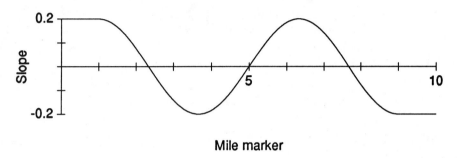

Figure IV.17: Calculus Range Trail (slope)

graphs describe the same situation one can reconstruct either graph from the other one. Figure IV.17 can be constructed from Figure IV.16 by differentiation. If the curve in Figure IV.16 is described by the function $y = A(x)$ then the curve in Figure IV.17 is described by the function $A'(x)$.

We can almost construct Figure IV.16 from Figure IV.17 using the definite integral. Suppose that $S(x)$ describes the steepness of the trail at the point x, that is, $S(x)$ is the function shown in Figure IV.17. We need one additional piece of information, the elevation (0.10 miles above sea level) at the trailhead. With this information we can find $A(x)$ from $S(x)$ as follows.

Suppose we want to estimate the elevation of the trail at some particular point b. We can estimate the change in elevation between the trailhead and the point b by dividing the interval $[0, b]$ up into n subintervals of size $h = b/n$, choosing a sample point s_i in each subinterval, and computing

$$\sum_{i=1}^{n} S(s_i) h.$$

If n is large the slope will be nearly constant on each subinterval. Thus $S(s_i)h$ will be a good estimate of the change in elevation on each subinterval and
$$\sum_{i=1}^{n} S(s_i)h$$
will be a good estimate of the change in elevation between the trailhead and mile marker b. With a very large number n of subintervals we get a very good estimate. Thus the exact change in elevation is
$$\int_0^b S(x)\,dx = \lim_{n \to \infty} \sum_{i=1}^{n} S(s_i)h$$
and the elevation at b is
$$A(b) = 0.10 + \int_0^b S(x)\,dx.$$

This allows us to recover the original function describing the elevation at each point x
$$A(x) = 0.10 + \int_0^x S(x)\,dx.$$

Too Many x's: A Note about Notation

The formula above uses the letter x in two different ways. The most natural letter to use in the first definite integral,
$$\int_0^b S(x)\,dx,$$
is the letter x because we think of the function $S(x)$ as a function of the variable x. But, later when we want to discuss the function $A(x)$ it is natural to use the letter x again because we think of the function $A(x)$ as describing the altitude or elevation at each point x. This leads us to the formula
$$A(x) = 0.10 + \int_0^x S(x)\,dx,$$
in which the variable x is used for two different purposes.

§3. The Fundamental Theorem of Calculus

The author of a calculus book faces a problem at this point. Should he or she use two different letters, writing, for example,

$$A(t) = 0.10 + \int_0^t S(x)\, dx$$

or

$$A(x) = 0.10 + \int_0^x S(t)\, dt$$

to avoid the potential confusion of one letter serving two different purposes, or is it better to use the letter x in both ways and rely on the reader to keep them straight? This author responded to this quandary by including these paragraphs.

Example:

As a practical matter it is often much easier to measure velocity than to measure distance. Velocity can be measured by means of the **Doppler Effect.** You may have noticed the Doppler Effect while waiting for a train to pass. As the train is coming toward you its whistle has a higher pitch than it does as the train recedes. The whistle's pitch is very easy to measure with very high precision. Since the pitch is directly related to the train's velocity, it is very easy to measure the train's velocity with very high precision. This method is very important in astronomy. Stars emit light at certain very specific frequencies. The perceived frequency (color) of emitted light like the perceived frequency (pitch) of sound depends directly on the velocity of the star relative to the observer. Using this idea one can determine the velocity of stars easily and with high precision.

Figure IV.18 shows the graph of the velocity $v(t)$ of a particular object at time t. Given this information we can reconstruct the change in position of the object between two times a and b as follows.

Divide the interval $[a, b]$ up into n subintervals, each of duration $h = (b - a)/n$. Pick a sample point s_i in each subinterval. If n is large, each subinterval will be short and the velocity will be nearly constant on each subinterval. Hence the change in the position of the object during each subinterval will be roughly $v(s_i)h$ and the change in position between time a and time b will be roughly

$$\sum_{i=1}^n v(s_i)h.$$

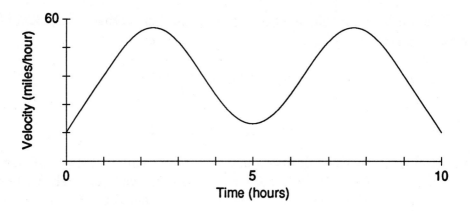

Figure IV.18: Velocity of an object

This method allows us to compute estimates as good as needed for any particular purpose. The exact change in position is

$$\int_a^b v(t)\,dt = \lim_{n\to\infty} \sum_{i=1}^n v(s_i)h.$$

If we know the position of the object at time $t = a$ then we can compute the position at any other time by

$$p(t) = p(a) + \int_a^t v(t)\,dt.$$

The definite integral is exactly the tool needed to reconstruct the position function $p(t)$ given the velocity function $v(t) = p'(t)$.

These two examples illustrate the same fact, that differentiation and integration are opposite sides of the same coin. More precisely, we have the **Fundamental Theorem of Calculus**.

Theorem 5 *(The Fundamental Theorem of Calculus)*

- *Suppose that $f(x)$ is any differentiable function, that $f'(x)$ is continuous, and that a is any point. Then*

$$f(x) = f(a) + \int_a^x f'(x)\,dx.$$

We can also write

$$f(x) = f(a) + \int_a^x f'(t)\,dt$$

§3. The Fundamental Theorem of Calculus

to avoid potential confusion caused by two different uses of the letter x.

- *Suppose that $g(x)$ is any continuous function, a is any point, and that the function $G(x)$ is defined by*

$$G(x) = \int_a^x g(x)\, dx$$

or, to avoid confusion caused by the use of the same letter for two different purposes,

$$G(x) = \int_a^x g(t)\, dt.$$

Then

$$G'(x) = g(x).$$

Proof:

The first assertion follows from the second assertion and Theorem 7 of Section III.4 as follows. Let

$$F(x) = f(a) + \int_a^x f'(t)\, dt.$$

By the second assertion

$$F'(x) = f'(x).$$

By Theorem 7 of Section III.4

$$f(x) = F(x) + c$$

where c is a constant. But

$$F(a) = f(a) + \int_a^a f'(t)\, dt = f(a) + 0 = f(a),$$

so $c = 0$ and $f(x) = F(x)$, which was what we needed to prove.

The second assertion is proved as follows. Given $\epsilon > 0$ we must find $\delta > 0$ such that if $0 < |h| < \delta$ then

$$\left| \frac{G(x+h) - G(x)}{h} - g(x) \right| < \epsilon.$$

But
$$G(x+h) - G(x) = \int_a^{x+h} g(t)\, dt - \int_a^x g(t)\, dt$$
$$= \int_x^{x+h} g(t)\, dt.$$

Since $g(x)$ is continuous, there is a $\delta > 0$ such that for t in the interval $(x - \delta, x + \delta)$
$$g(x) - \epsilon < g(t) < g(x) + \epsilon.$$
Hence for $|h| < \delta$
$$(g(x) - \epsilon)h < \int_x^{x+h} g(t)\, dt < (g(x) + \epsilon)h$$
and
$$g(x) - \epsilon < \frac{\int_x^{x+h} g(t)\, dt}{h} < g(x) + \epsilon.$$
Since
$$G(x+h) - G(x) = \int_x^{x+h} g(t)\, dt$$
this implies
$$\left| \frac{G(x+h) - G(x)}{h} - g(x) \right| < \epsilon$$
completing the proof.

The Fundamental Theorem of Calculus has many important practical applications. We discuss two of them here. We state the first one as a theorem.

Theorem 6 *Suppose that $f(x)$ is a continuous function and that $F(x)$ is any function whose derivative is $f(x)$. Then*
$$\int_a^b f(x)\, dx = F(b) - F(a).$$

Proof:
By the Fundamental Theorem of Calculus
$$F(x) = F(a) + \int_a^x F'(x)\, dx = F(a) + \int_a^x f(x)\, dx$$

§3. The Fundamental Theorem of Calculus

and, in particular,
$$F(b) = F(a) + \int_a^b f(x)\, dx,$$
so
$$\int_a^b f(x)\, dx = F(b) - F(a).$$

This theorem allows us to compute the exact value of an integral
$$\int_a^b f(x)\, dx$$
if we know any function whose derivative is $f(x)$. This leads us to make the following definition.

Definition:

If $F'(x) = f(x)$ then we say that $F(x)$ is an **antiderivative** of $f(x)$. We frequently use the notation
$$[F(x)]_a^b = F(b) - F(a).$$

For example,
$$\left[x^2\right]_2^3 = (3)^2 - (2)^2 = 9 - 4 = 5.$$

Using this notation, we rephrase the conclusion of Theorem 6 as
$$\int_a^b f(x)\, dx = [F(x)]_a^b.$$

Notice that if $F(x)$ is any antiderivative of $f(x)$ then so is $F(x) + c$ where c is any constant. We use the following terminology.

Definition:

If $F(x)$ is any antiderivative of $f(x)$ the family of functions $F(x) + c$ is called the **indefinite integral** of $f(x)$ and we write
$$\int f(x)\, dx = F(x) + c.$$

The constant c is very important. The indefinite integral is the whole family of possible antiderivatives of $f(x)$. See, for example, Figure IV.19.

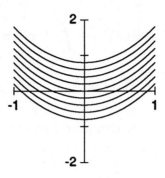

Figure IV.19: Some of the antiderivatives of $f(x) = 2x$

Examples:

- $\int 4 \, dx = 4x + c.$
- $\int x \, dx = x^2/2 + c.$
- $\int \cos x \, dx = \sin x + c.$
- $\int \sin x \, dx = -\cos x + c.$
- $\int x^n \, dx = x^{n+1}/(n+1) + c.$
- $\int e^x \, dx = e^x + c.$

Example:

Find the exact area under the curve $y = \sin x$ between $x = 0$ and $x = \pi$.

Answer:

The area is
$$\int_0^\pi \sin x \, dx.$$
Since $F(x) = -\cos x$ is an antiderivative of of $\sin x$ we have

$$\int_0^\pi \sin x \, dx = [-\cos x]_0^\pi = (-\cos \pi) - (-\cos 0) = -(-1) - (-1) = 2.$$

§3. THE FUNDAMENTAL THEOREM OF CALCULUS

Notice that this is an exact answer not an approximation.

Antidifferentiation is a powerful tool for evaluating definite integrals. When it works it is wonderful since it gives the exact answer. There is, however, a catch. One must know the antiderivative. It is often difficult to find antiderivatives and sometimes impossible. For example, there is no simple formula for the antiderivative of the function $y = \sin(x^2)$. This leads us to the second major application of the Fundamental Theorem of Calculus. It can be used to find antiderivatives. The Fundamental Theorem tells us that

$$F(x) = y_0 + \int_a^x f(x)\, dx$$

is a solution of the initial value problem

$$y' = f(x), \quad y(a) = y_0.$$

This gives us a way to compute antiderivatives. This use of the Fundamental Theorem is especially important in the age of computers since there are many fast and powerful programs for computing definite integrals.

Exercises

*Exercise IV.3.1: Find

$$\int_1^4 (x - 2)\, dx.$$

*Exercise IV.3.2: Find the solution to the initial value problem

$$y' = x - 2, \quad y(0) = 4.$$

*Exercise IV.3.3: Find the solution to the initial value problem

$$y' = x - 2, \quad y(1) = 4.$$

*Exercise IV.3.4: Find the exact area under the curve $y = x(1-x)$ between $x = 0$ and $x = 1$.

*Exercise IV.3.5: Find

$$\int \cos t\, dt.$$

*Exercise IV.3.6: Find
$$\int \cos 2t \, dt.$$

Exercise IV.3.7: Find
$$\int_0^\pi \cos t \, dt.$$

Exercise IV.3.8: Find
$$\int_0^\pi \cos 2t \, dt.$$

*Exercise IV.3.9: Find the solution to the initial value problem
$$y' = t^2, \quad y(0) = 4.$$

Exercise IV.3.10: Find
$$\int_0^2 (t + t^2) \, dt.$$

Exercise IV.3.11: Find
$$\int_0^1 e^x \, dx.$$

Exercise IV.3.12: Find
$$\int (3 - e^t) \, dt.$$

---

§4. Antidifferentiation

In the last section we saw that if $F(x)$ is any antiderivative of $f(x)$ then
$$\int_a^b f(x) \, dx = [F(x)]_a^b = F(b) - F(a).$$

Thus, there are two different ways of evaluating definite integrals and, hence, of solving problems involving "multiplication" of two numbers, one of which is not constant.

§4. Antidifferentiation

- Approximation by Riemann sums.

- If (and that's a big if) we can find an antiderivative $F(x)$ of the function $f(x)$ then

$$\int_a^b f(x)\, dx = [F(x)]_a^b = F(b) - F(a).$$

The latter method has two big advantages. First, it gives an exact answer rather than an estimate. More importantly, however, it allows us to solve problems involving parameters. We will see some examples of this later on. The former method has one big advantage. It always works. There are many functions, for example $\sin x^2$, for which it is impossible to find an antiderivative. Thus the only way to evaluate a definite integral like

$$\int_0^1 \sin x^2\, dx$$

is numerically. In section 7 we develop a very efficient numerical method for approximating a definite integral. First, we look at some methods for finding antiderivatives.

Any differentiation formula can be used to obtain an antidifferentiation formula. Here is a list of some differentiation formulas and corresponding antidifferentiation formulas.

- $$\left(\frac{d}{dx}\right) x^n = nx^{n-1}$$

  If $n \neq -1$ then

  $$\int x^n\, dx = \frac{x^{n+1}}{n+1} + c$$

- $$\left(\frac{d}{dx}\right) \ln |x| = \frac{1}{x}$$

  $$\int \frac{1}{x}\, dx = \ln |x| + c$$

The next two antidifferentiation formulas are frequently referred to together as **linearity**.

- $$\left(\frac{d}{dx}\right)(f(x)+g(x)) = \left(\frac{d}{dx}\right)f(x) + \left(\frac{d}{dx}\right)g(x)$$

  $$\int (f(x)+g(x))\,dx = \int f(x)\,dx + \int g(x)\,dx$$

- $$\left(\frac{d}{dx}\right)kf(x) = k\left(\frac{d}{dx}\right)f(x)$$

  $$\int kf(x)\,dx = k\int f(x)\,dx$$

- $$\left(\frac{d}{dx}\right)\sin x = \cos x$$

  $$\int \cos x\,dx = \sin x + c$$

- $$\left(\frac{d}{dx}\right)\cos x = -\sin x$$

  $$\int \sin x\,dx = -\cos x + c$$

- $$\left(\frac{d}{dx}\right)\tan x = \sec^2 x$$

  $$\int \sec^2 x\,dx = \tan x + c$$

- $$\left(\frac{d}{dx}\right)\cot x = -\csc^2 x$$

  $$\int \csc^2 x\,dx = -\cot x + c$$

§4. Antidifferentiation

- $$\left(\frac{d}{dx}\right)\sec x = \sec x \tan x$$

$$\int \sec x \, \tan x \, dx = \sec x + c$$

- $$\left(\frac{d}{dx}\right)\csc x = -\csc x \cot x$$

$$\int \csc x \cot x \, dx = -\csc x + c$$

- $$\left(\frac{d}{dx}\right)e^x = e^x$$

$$\int e^x \, dx = e^x + c$$

- $$\left(\frac{d}{dx}\right)a^x = a^x \ln a$$

$$\int a^x \, dx = \frac{a^x}{\ln a} + c$$

- $$\left(\frac{d}{dx}\right)\arctan x = \frac{1}{1+x^2}$$

$$\int \frac{1}{1+x^2} \, dx = \arctan x + c$$

- $$\left(\frac{d}{dx}\right)\arcsin x = \frac{1}{\sqrt{1-x^2}}$$

$$\int \frac{1}{\sqrt{1-x^2}} \, dx = \arcsin x + c$$

•
$$\left(\frac{d}{dx}\right) \operatorname{arcsec} x = \frac{1}{|x|\sqrt{x^2-1}}$$

$$\int \frac{1}{|x|\sqrt{x^2-1}}\, dx = \operatorname{arcsec} x + c$$

Using these formulas we can evaluate many definite integrals. Here are some examples.

Example:
$$\int_1^3 (x+x^2)\, dx$$

Answer:

The first step is to notice that $\int x\, dx = \frac{x^2}{2} + c$ and $\int x^2\, dx = \frac{x^3}{3} + c$. By linearity $\int (x+x^2)\, dx = \frac{x^2}{2} + \frac{x^3}{3} + c$. Thus

$$\int_1^3 (x+x^2)\, dx = \left[\frac{x^2}{2} + \frac{x^3}{3}\right]_1^3 = \frac{38}{3}$$

Example:
$$\int_0^{\pi/2} (\sin x + 2\cos x)\, dx$$

Answer:

The first step is to notice that $\int \sin x\, dx = -\cos x + c$ and $\int \cos x\, dx = \sin x + c$. By linearity $\int (\sin x + 2\cos x)\, dx = -\cos x + 2\sin x + c$. Thus

$$\int_0^{\pi/2} (\sin x + 2\cos x)\, dx = [-\cos x + 2\sin x]_0^{\pi/2} = 3$$

§4. Antidifferentiation

Exercises

*Exercise IV.4.1: Evaluate
$$\int_1^3 \left(x + \frac{1}{x}\right) dx.$$

*Exercise IV.4.2: Evaluate
$$\int_0^\pi 3 \sin x \, dx.$$

*Exercise IV.4.3: Evaluate
$$\int_1^3 \left(\frac{x}{2} + \frac{x^2}{3} + 4x^3\right) dx.$$

Exercise IV.4.4: Evaluate
$$\int_0^{\frac{\pi}{2}} 3 \cos x \, dx.$$

*Exercise IV.4.5: Evaluate
$$\int_0^{10} 4x + \sin x \, dx.$$

Exercise IV.4.6: Evaluate
$$\int_0^2 3x + 2e^x \, dx.$$

*Exercise IV.4.7: Evaluate
$$\int_2^5 \left(\frac{x+1}{x}\right) dx.$$

Exercise IV.4.8: Evaluate
$$\int_1^3 x(x+4) \, dx.$$

*Exercise IV.4.9: Evaluate
$$\int_2^6 \sqrt{x} \, dx.$$

Exercise IV.4.10: Evaluate
$$\int_1^2 \frac{1}{x^2}\, dx.$$

*Exercise IV.4.11: Evaluate
$$\int_2^4 x(4-x)\, dx.$$

Exercise IV.4.12: Evaluate
$$\int_{-1}^5 x^2(x+1)\, dx.$$

Exercise IV.4.13: Evaluate
$$\int_3^2 2x + x^2 \, dx.$$

*Exercise IV.4.14: One of the pairs of formulas in our long list of differentiation and antidifferentiation formulas was

$$\left(\frac{d}{dx}\right) \ln|x| = \frac{1}{x}$$

$$\int \frac{1}{x}\, dx = \ln|x| + c.$$

Check that the differentiation formula

$$\left(\frac{d}{dx}\right) \ln|x| = \frac{1}{x}$$

is correct when x is negative by showing that for x negative

$$\left(\frac{d}{dx}\right) \ln|x| = \frac{1}{x}$$

*Exercise IV.4.15: In this exercise we make a distinction between **signed area** and **unsigned area**. Suppose we are interested in the area of the region between the curve $y = \sin x$ and the x-axis between $x = 0$ and $x = 2\pi$. (See Figure IV.20.)

§4. ANTIDIFFERENTIATION

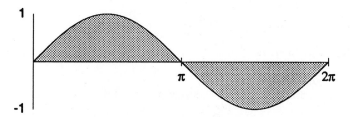

Figure IV.20: Area between the curve $y = \sin x$ and the x-axis

Notice that the region has two pieces. The first piece between $x = 0$ and $x = \pi$ is above the x-axis. Hence if we measure the height of this piece of the region at any point the height will be positive. Thus this piece of the region will have a positive area when we evaluate

$$\int_0^\pi \sin x \, dx.$$

The second piece of this region, between $x = \pi$ and $x = 2\pi$, is below the x-axis. If we measure the height of this piece of the region at any point the height will be negative. Hence when we evaluate the integral

$$\int_\pi^{2\pi} \sin x \, dx$$

we will see that this piece of the region has a negative area. This is not what one usually means by "area." To be precise we will call this the **signed area** of the region. The signed area of a region is negative for regions below the x-axis and positive for regions above the x-axis. If a region has some parts below the x-axis and some parts above the x-axis then the total signed area of the region is the sum of the signed area of the parts above the x-axis and the parts below the x-axis. It might be positive or negative or zero depending on the relative size of the parts below and the parts above the x-axis.

Very frequently one is interested in the usual or **unsigned area** of a region that is partly below and partly above the x-axis. In this case one must be careful to find the signed area of the part above the x-axis (this will be positive) and the signed area of the part below the x-axis (this will be negative) and then add **the absolute values** of these two signed areas. For example, the signed area of the region in Figure IV.20

is zero since
$$\int_0^{2\pi} \sin x \, dx = 0$$
but the unsigned area is
$$\left|\int_0^{\pi} \sin x \, dx\right| + \left|\int_{\pi}^{2\pi} \sin x \, dx\right| = 2 + |-2| = 4$$

Now consider the region between the function $f(x) = x(1-x)$ and the x-axis, and the lines $x = 0$ and $x = 2$.

- Draw a careful graph of this region.
- Find the signed area of the piece above the x-axis.
- Find the signed area of the piece below the x-axis.
- Find the signed area of the whole region.
- Find the unsigned area of the whole region.

Exercise IV.4.16: Consider the region between the curve $f(x) = \cos x$ and the x-axis and the lines $x = 0$ and $x = 3\pi$.

- Draw a careful graph of this region.
- Find the signed area of the piece above the x-axis.
- Find the signed area of the piece below the x-axis.
- Find the signed area of the whole region.
- Find the unsigned area of the whole region.

Exercise IV.4.17: A similar distinction is necesssary when we talk about the motion of an object whose velocity is given by a function $f(t)$. When $f(t)$ is positive the integral
$$\int_a^b f(t) \, dt$$

§5. Integration by Substitution

can be interpreted in two ways. It gives us the change in the location of the object between time $t = a$ and time $t = b$ and it also gives us the distance traveled. Notice that if $f(x)$ is negative then the integral

$$\int_a^b f(x)\,dx$$

will also be negative. Thus the change in location will be negative because the object is traveling backward. If one is interested in the amount of fuel necessary for a trip, however, or in the wear-and-tear on the tires of a car then negative distance doesn't make sense. Driving backward does not put gas into your car's tank. Thus we need to make a distinction between **change in location** or **signed distance traveled** and **unsigned distance traveled**. Informally, one often uses the term "distance traveled" for "unsigned distance traveled." The term **displacement** is often used for signed distance traveled.

Suppose that a baseball is thrown straight upward with an initial velocity of 64 feet per second so that at time t (in seconds after the ball is thrown) its velocity is $v(t) = 64 - 32t$.

- Draw a careful graph of the function $v(t) = 64 - 32t$ for $0 \leq t \leq 3$.

- What is the signed distance traveled while the ball is traveling upward?

- What is the signed distance traveled while the ball is traveling downward? (Note: we are only interested in what happens before $t = 3$).

- What is the total signed distance traveled?

- What is the total unsigned distance traveled?

---

§5. Integration by Substitution

The Chain Rule is one of the most important methods of differentiation. Thus, not surprisingly, it leads to the most important method of antidifferentiation. Consider the following example of differentiation using the Chain Rule.

Example:

Find the derivative of the function.
$$y = \sin(x^2 + 2)$$

We think of this as
$$y = \sin u \quad \text{where} \quad u = x^2 + 2$$

So that the derivative is
$$\frac{dy}{dx} = \left(\frac{dy}{du}\right)\left(\frac{du}{dx}\right) = (\cos u)(2x) = 2x\cos(x^2 + 2).$$

Now if we are immediately confronted with the indefinite integral
$$\int 2x \cos(x^2 + 2)\, dx$$
we will, of course, recognize that this integral is
$$\int 2x \cos(x^2 + 2)\, dx = \sin(x^2 + 2) + c.$$

The purpose of this section is to develop an ability to recognize integrals like the one above even when we have not had the good fortune to have just happened to have found the derivative
$$\left(\frac{d}{dx}\right)\sin(x^2 + 2) = 2x\cos(x^2 + 2).$$

The method that we develop is called **Integration by Substitution**. It is the opposite side of differentiation by the Chain Rule. The basic idea behind integration by substitution is very simple but in practice it can be a little tricky. For example, in the example above one needs to recognize that the integrand
$$2x\cos(x^2 + 2)$$
came from a Chain Rule differentiation in which the intermediate function u was the function $u = x^2 + 2$. This is the entire trick behind the

§5. Integration by Substitution

method of integration by substitution—finding the intermediate function u and breaking the integrand up into two factors, one of which comes from the "dy/du" part of the Chain Rule and the other of which comes from the "du/dx" part of the Chain Rule.

We introduce some notation to help find indefinite integrals by the method of substitution. The notation du is an abbreviation for $u'(x)dx$. Thus we can write

$$\int \cos(x^2 + 2)\, 2x\, dx = \int \cos u\, du = \sin u + c = \sin(x^2 + 2) + c$$

where $u = (x^2 + 2)$.

The following examples illustrate integration by substitution.

Example:

Find the indefinite integral

$$\int \sin x \cos x\, dx.$$

Answer:

The crucial step here is to recognize that the intermediate function is $u = \sin x$ so that $du = \cos x\, dx$ and

$$\int \sin x \cos x\, dx = \int u\, du = \frac{u^2}{2} + c = \frac{\sin^2 x}{2} + c.$$

Example: Find the indefinite integral

$$\int \sqrt{1 + x^2}\, 2x\, dx.$$

Answer:

The crucial step here is to recognize that the intermediate function is $u = 1 + x^2$ so that $du = 2x\, dx$ and

$$\int \sqrt{1 + x^2}\, 2x\, dx = \int \sqrt{u}\, du = \int u^{\frac{1}{2}}\, du = \frac{2u^{\frac{3}{2}}}{3} + c = \frac{2(1 + x^2)^{\frac{3}{2}}}{3} + c.$$

Unfortunately, except in calculus textbooks, integrals that can be found by the method of substitution rarely occur in such a transparent form. Consider the following examples.

Example:

Find the indefinite integral

$$\int x 2^{(x^2)}\, dx.$$

Answer:

The natural guess for the intermediate function u is the function $u = x^2$. This choice takes care of part of the integrand. Now we see

$$\int x 2^{(x^2)} dx = \int x \underbrace{2^{(x^2)}}_{2^u}\, dx = \int \underbrace{2^{(x^2)}}_{2^u} x\, dx$$

If we had started with the problem

$$\int 2x\, 2^{(x^2)}\, dx$$

then we would now have

$$\int \underbrace{2^{(x^2)}}_{2^u} \underbrace{2x\, dx}_{du} = \int 2^u\, du = \frac{2^u}{\ln 2} + c = \frac{2^{(x^2)}}{\ln 2} + c$$

and we would be done. But, alas, the desired 2 is missing. We can save the situation, however, by recalling one of the basic properties of the integral. Recall that for any constant k

$$\int k f(x)\, dx = k \int f(x)\, dx.$$

So we can rewrite our original integral as

$$\int x\, 2^{(x^2)}\, dx = \int \left(\frac{1}{2}\right) 2x\, 2^{(x^2)}\, dx$$
$$= \frac{1}{2} \int 2x\, 2^{(x^2)}\, dx$$

§5. Integration by Substitution

$$= \frac{1}{2} \int \underbrace{2^{(x^2)}}_{2^u} \underbrace{2x\, dx}_{du}$$

$$= \frac{1}{2} \int 2^u\, du$$

$$= \frac{1}{2} \frac{2^u}{\ln 2}$$

$$= \frac{2^u}{2 \ln 2}$$

$$= \frac{2^{(x^2)}}{2 \ln 2}$$

We must emphasize that the formula that enabled us to introduce the factor 2 that made the problem solvable *only works for constants*. You cannot introduce anything other than a constant in this way.

Example:

Find the indefinite integral

$$\int \tan x\, dx$$

Answer:

We begin by rewriting this as

$$\int \tan x\, dx = \int \frac{\sin x}{\cos x}\, dx$$

Now if we let $u = \cos x$ and notice that $du = -\sin x\, dx$ we see that

$$\int \frac{\sin x}{\cos x}\, dx = -\int \frac{-\sin x}{\cos x}\, dx$$

$$= -\int \frac{du}{u}$$

$$= -\ln|u| + c$$

$$= -\ln|\cos x| + c$$

The following example shows how integration by substitution can be used to determine a definite integral.

Example: Evaluate the definite integral

$$\int_0^1 \frac{x}{x^2+1}\,dx$$

Answer:

The first step is to find the indefinite integral

$$\int \frac{x}{x^2+1}\,dx$$

using the substitution $u = x^2 + 1$ as follows.

$$\begin{aligned}
\int \frac{x}{x^2+1}\,dx &= \frac{1}{2}\int \frac{2x}{x^2+1}\,dx \\
&= \frac{1}{2}\int \frac{1}{u}\,du \\
&= \frac{1}{2}\ln|u| \\
&= \frac{1}{2}\ln|x^2+1| + c
\end{aligned}$$

Thus

$$\int_0^1 \frac{x}{x^2+1}\,dx = \left[\frac{\ln|x^2+1|}{2}\right]_0^1 = \frac{\ln 2 - \ln 1}{2} = \frac{\ln 2}{2}$$

There is an alternative way to evaluate the definite integral in the preceding example.

Example: Evaluate the definite integral

$$\int_0^1 \frac{x}{x^2+1}\,dx$$

Answer:

§5. Integration by Substitution

We use the substitution $u(x) = x^2 + 1$ to express the entire problem *including the limits of integration* in terms of u. The integral goes from $x = 0$ to $x = 1$. When $x = 0$ we have

$$u(0) = 0^2 + 1 = 1$$

and when $x = 1$ we have

$$u(1) = 1^2 + 1 = 2$$

so the integral becomes

$$\begin{aligned}
\int_{x=0}^{x=1} \frac{x}{x^2+1}\, dx &= \frac{1}{2}\int_{x=0}^{x=1} \frac{2x}{x^2+1}\, dx \\
&= \frac{1}{2}\int_{u=1}^{u=2} \frac{1}{u}\, du \\
&= \left[\frac{\ln|u|}{2}\right]_1^2 \\
&= \frac{\ln 2}{2} - \frac{\ln 1}{2} \\
&= \frac{\ln 2}{2}
\end{aligned}$$

Notice that either method involves exactly the same computations.

The method illustrated in the preceding example is important enough to be stated as a theorem.

Theorem 7 *Suppose that $u(x)$ is a differentiable function on the interval $[a, b]$ and that $u(a) = p$ and $u(b) = q$. Then*

$$\int_{x=a}^{x=b} f(u(x))u'(x)\, dx = \int_{u=p}^{u=q} f(u)\, du.$$

Notice that the integral on the lefthand side involved only the variable x and the integral on the righthand side involves only the variable u.

Example: Evaluate the definite integral

$$\int_0^4 \sqrt{2x+1}\, dx$$

Answer:

We use the substitution $u(x) = 2x + 1$. Notice that

$$du = u'(x)\, dx = 2\, dx$$

and

$$u(0) = 1$$
$$u(4) = 9$$

so that

$$\int_{x=0}^{x=4} \sqrt{2x+1}\, dx = \frac{1}{2} \int_{x=0}^{x=4} 2\sqrt{2x+1}\, dx$$
$$= \frac{1}{2} \int_{u=1}^{u=9} \sqrt{u}\, du$$
$$= \frac{1}{2} \int_{u=1}^{u=9} u^{1/2}\, du$$
$$= \left[\frac{u^{3/2}}{3} \right]_1^9$$
$$= \frac{27}{3} - \frac{1}{3}$$
$$= \frac{26}{3}$$

Exercises

*Exercise IV.5.1:

$$\int \sqrt{x+1}\, dx$$

Exercise IV.5.2:

$$\int_0^\pi \sin^2 x \cos x\, dx$$

§5. Integration by Substitution

*Exercise IV.5.3:
$$\int_0^\pi \sin^3 x \cos x \, dx$$

*Exercise IV.5.4:
$$\int_1^2 \frac{1}{1+4x^2} \, dx$$

Exercise IV.5.5:
$$\int_1^2 \frac{1}{4+x^2} \, dx$$

*Exercise IV.5.6: Find a general formula for
$$\int \frac{1}{x^2+k^2} \, dx$$
where k is a constant.

Exercise IV.5.7:
$$\int \frac{1}{16+x^2} \, dx.$$

Exercise IV.5.8:
$$\int \frac{x}{16+x^2} \, dx.$$

*Exercise IV.5.9:
$$\int \frac{2+3x}{16+x^2} \, dx.$$

*Exercise IV.5.10:
$$\int \frac{1}{\sqrt{4-x^2}} \, dx$$

*Exercise IV.5.11:
$$\int \frac{x}{\sqrt{4-x^2}} \, dx$$

*Exercise IV.5.12:
$$\int \frac{2+3x}{\sqrt{4-x^2}} \, dx$$

Exercise IV.5.13: Find a general formula for
$$\int \frac{1}{\sqrt{k^2-x^2}} \, dx$$

where k is a constant.

Exercise IV.5.14:
$$\int_0^{\pi/2} \sin 2x \, dx$$

Exercise IV.5.15:
$$\int_0^1 x \cos x^2 \, dx$$

Exercise IV.5.16:
$$\int_0^{\pi/2} \sin^2 2x \cos 2x \, dx$$

*Exercise IV.5.17: Find the integral
$$\int \sin x \cos x \, dx$$
two different ways. First, let $u = \sin x$ and then let $u = \cos x$. You will get two answers that appear to be different. Explain.

*Exercise IV.5.18:
$$\int_2^3 \frac{x+1}{x^2+1} \, dx$$

*Exercise IV.5.19:
$$\int_0^2 \frac{x}{9-x^2} \, dx$$

Exercise IV.5.20:
$$\int_0^{\frac{\pi}{2}} \sin^2 \theta \, d\theta$$

Hint: Use the trigonometric identity
$$\sin^2 \theta = \frac{1-\cos 2\theta}{2}$$

*Exercise IV.5.21:
$$\int_0^{\frac{\pi}{4}} \frac{\sin \theta}{\cos^4 \theta} \, d\theta$$

Exercise IV.5.22:
$$\int e^x \cos e^x \, dx$$

§5. Integration by Substitution

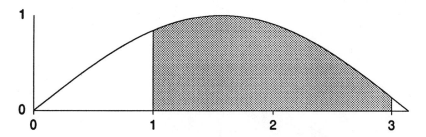

Figure IV.21: Area under a curve

Exercise IV.5.23:
$$\int \frac{\ln x}{x}\, dx$$

*Exercise IV.5.24:
$$\int_1^2 \frac{1}{x^2+4x+5}\, dx$$

*Exercise IV.5.25:
$$\int_1^2 \frac{1}{x^2+4x+13}\, dx$$

Exercise IV.5.26:
$$\int_0^1 \frac{e^x}{e^{2x}+1}\, dx$$

Area by Cross Sections

We can determine the area of a region by "multiplying" its height by its width. If the region is a rectangle then we don't need the quotation marks; we simply multiply two numbers. If the region is not a rectangle, however, the "multiplication" is really integration. We have already seen how to find the area of a region like the one shown in Figure IV.21. Now we look at some variations on this theme. We begin with an example.

Example:

Find the area of the region trapped between the two curves $y = x^2$ and $y = 2x$. Figure IV.22 shows this region.

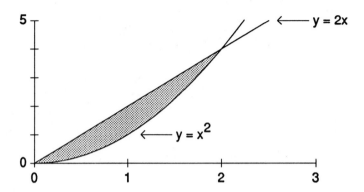

Figure IV.22: Region trapped between $y = x^2$ and $y = x$.

Answer:

The first step in finding the area of this region is determining the two points at which the two curves intersect. We can do this by solving the equation

$$x^2 = 2x$$
$$x^2 - 2x = 0$$
$$x(x - 2) = 0$$

Thus $x = 0$ and $x = 2$ are the two points at which the two curves intersect. Hence we want to find the definite integral

$$\int_0^2 \text{height } dx.$$

Figure IV.23 shows a typical cross section. Notice the height of this cross section is $2x - x^2$. So the integral above becomes.

$$\int_0^2 2x - x^2 \, dx = \left[x^2 - \frac{x^3}{3}\right]_0^2 = 4 - \frac{8}{3} = \frac{4}{3}$$

As before, we need to be careful about the distinction between unsigned area and signed area. Consider the following example.

§5. Integration by Substitution

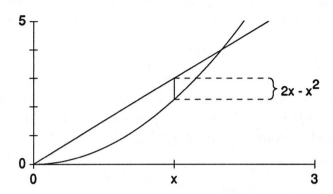

Figure IV.23: Height of a typical cross section

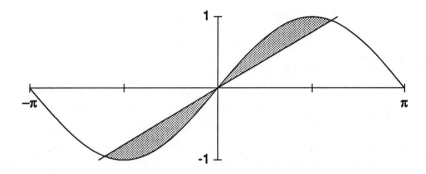

Figure IV.24: Region between $y = x/2$ and $y = \sin x$

Example:

Find the area of the region between the curve $y = \sin x$ and the line $y = x/2$. This region is shown in Figure IV.24.

Answer:

The first step in determining the area of this region is finding the points at which the curves $y = \sin x$ and $y = x/2$ intersect. As you can see from Figure IV.24 these two curves intersect in three places. One of these places is easy to find, namely $x = 0$. The other two are more difficult. One needs to use a numerical method like the Bisection Method or Newton's Method. Either of these methods can be used to find the other two points, $x = 1.895494$ and $x = -1.895494$. Now we want to determine the area of the indicated region by finding the integral.

$$\int_{-1.895494}^{1.895494} \text{height } dx.$$

Some care, however, is necessary. For negative values of x notice that the curve $y = x/2$ is above the curve $y = \sin x$. So for this part of the region we need to evaluate the integral

$$\int_{-1.895495}^{0} \left(\frac{x}{2} - \sin x\right) dx.$$

For positive values of x the curve $y = \sin x$ is above the line $y = x/2$. Thus for this part of the region we need to evaluate the integral

$$\int_{0}^{1.895494} \left(\sin x - \frac{x}{2}\right) dx.$$

The final result will be the sum

$$\int_{-1.895495}^{0} \left(\frac{x}{2} - \sin x\right) dx + \int_{0}^{1.895494} \left(\sin x - \frac{x}{2}\right) dx.$$

This is a fairly common situation. We must draw careful graphs for these problems so that we can break the integral up into pieces if necessary as in this example.

Exercises

*Exercise IV.5.27: Find the area of the region enclosed by the curves $y = 2x + 1$ and $y = x^2$.

Exercise IV.5.28: Find the area of the region enclosed by the curves $y = \sin 2x$ and $y = x$.

*Exercise IV.5.29: Find the area of the region enclosed by the curves $y = x$ and $y = x^3$.

Exercise IV.5.30: Find the area of the region enclosed by the curves $y = xe^{x^2}$ and $y = 2x$.

*Exercise IV.5.31: A particular country is just able to produce enough food to feed its current population. Unfortunately its population is growing at a rate of 5% per year but its food production is

§6. Numerical Integration, I

growing only at the rate of 2% per year. Let the constant C denote the current rate of food consumption and food supply in units of calories per year. Let t denote time in years with $t = 0$ being right now. Let $f(t)$ denote the rate of food consumption at time t and let $s(t)$ denote the rate of food production at time t. Notice that

$$f(t) = Ce^{.05t} \quad \text{and} \quad s(t) = Ce^{.02t}.$$

- What will be the total food needs of this country for the next ten years?

- What will be the total food production of this country for the next ten years?

- What will be the total amount of food that this country will need to import over the next ten years to feed its population?

- Draw a picture interpreting the quantities above as areas.

Exercise IV.5.32: A particular country is able to produce 110% of the food that it needs to feed its current population. Unfortunately its population is growing at the rate of 4% per year but its food production is growing only at the rate of 3% per year. Let C denote the current rate of food consumption in units of calories per year.

- In how many years will the country be just exactly able to feed its population?

- How many excess calories will it have produced between now and the time you found above?

---

§6. Numerical Integration, I

The purpose of this section is to explore the accuracy of two moderately good methods of numerical integration. Both of these methods are motivated by the interpretation of the definite integral

$$\int_a^b f(x)\, dx$$

as the area of the region bounded by the curve $y = f(x)$, the x-axis, the line $x = a$, and the line $x = b$.

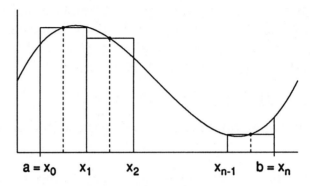

Figure IV.25: The midpoint method

The Midpoint Method

The **midpoint method** is based on Figure IV.25. It approximates the desired region by n rectangles of width $h = (b-a)/n$. If we let x_i denote the point

$$x_i = a + ih$$

then the ith rectangle has its left edge at x_{i-1} and its right edge at x_i. The height of the ith rectangle is taken at the midpoint

$$s_i = \frac{x_{i-1} + x_i}{2} = a + \left(i + \frac{1}{2}\right)h.$$

Thus the area of the ith rectangle is $f(s_i)h$ and the integral is approximated by

$$\sum_{i=1}^{n} f(s_i)h.$$

The Trapezoid Method

The **trapezoid method** is based on Figure IV.26. It approximates the desired region by n trapezoids of width $h = (b-a)/n$. If we let x_i denote the point

$$x_i = a + ih$$

then the ith trapezoid has its left edge at x_{i-1} and its right edge at x_i. Thus the area of the ith trapezoid is

$$\left(\frac{f(x_{i-1}) + f(x_i)}{2}\right)h$$

§6. Numerical Integration, I

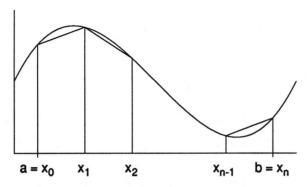

Figure IV.26: The trapezoid method

and the integral is approximated by

$$\sum_{i=1}^{n} \left(\frac{f(x_{i-1}) + f(x_i)}{2} \right) h.$$

Exercises

For each of the following integrals use either a computer or a calculator to

- Estimate the integral using the midpoint method and 20 rectangles.

- Estimate the integral using the trapezoid method and 20 rectangles.

Then

- Find the exact answer using antidifferentiation.

- Compute the error for the midpoint method and for the trapezoid method.

Exercise IV.6.1:

$$\int_0^{\pi/2} \sin x \, dx.$$

Exercise IV.6.2:

$$\int_0^{\pi/2} \sin 2x \, dx.$$

Exercise IV.6.3:

$$\int_0^{\pi} (2 - \sin(x/2)) \, dx.$$

Exercise IV.6.4:

$$\int_0^1 e^x \, dx.$$

Exercise IV.6.5: Any integral of your choice.

Exercise IV.6.6: Another integral of your choice.

Exercise IV.6.7: Make a table showing your results from the preceding exercises. Write down any observations you may have.

*Exercise IV.6.8: Show that the trapezoid method can be written

$$\text{Estimate} = (f(a) + f(b))h/2 + \sum_{i=1}^{n-1} f(x_i)h.$$

If you compare two computer programs for the trapezoid method, one based on the original formula

$$\text{estimate} = \sum_{i=1}^{n} \left(\frac{f(x_{i-1}) + f(x_i)}{2} \right) h$$

and the other on this new formula, the second program will run approximately twice as fast since it needs to evaluate the function $f(x)$ roughly half as many times.

```
!    Approximate an integral by the Trapezoid Method
!    using the original formula.
!
DEF f(x) = sin(x)    ! Function to be integrated.
PRINT "Enter the limits of integration: ";
INPUT a, b
PRINT "Enter the number of subintervals: ";
INPUT n
```

```
LET h = (b - a)/n
LET sum = 0
FOR i = 1 TO n
    LET left = a + (i - 1)*h
    LET right = left + h
    LET sum = sum + h * (f(left) + f(right))/2
NEXT i
PRINT USING "###.######": sum
END
```

```
!   Approximate an integral by the Trapezoid Method
!   using the new formula.
!
DEF f(x) = sin(x)   ! Function to be integrated.
PRINT "Enter the limits of integration: ";
INPUT a, b
PRINT "Enter the number of subintervals: ";
INPUT n
LET h = (b - a)/n
LET sum = h * (f(a) + f(b))/2
FOR i = 1 TO (n - 1)
    LET x = a + i * h
    LET sum = sum + h * f(x)
NEXT i
PRINT USING "###.######": sum
END
```

---

§7. Numerical Integration II: Simpson's Rule

The following table compares the results of using the trapezoid method and the midpoint method with the exact answer for the integrals

$$\int_0^{\pi/2} \sin x \, dx$$

and

$$\int_0^1 e^x \, dx.$$

The two numerical methods were applied with $n = 20$.

Exact	Midpoint	Error	Trapezoid	Error
1.000000000	1.000257067	0.000257067	0.999485905	−0.000514095
1.718281828	1.718102854	−0.000178974	1.718639789	0.000357961

There is something very striking about these results. Notice, first, that the errors for the trapezoid method are almost exactly twice as big as the errors for the midpoint method. Second, notice that the errors for the trapezoid method have the opposite signs of the errors for the midpoint method. If you look back at your results from the last lab you will see the same pattern. Thus we have roughly the picture shown in Figure IV.27.

Figure IV.27: Comparing the trapezoid and midpoint methods

These observations suggest that a very good numerical estimate can be obtained by choosing the number between the midpoint estimate and the trapezoid estimate that is twice as far from the trapezoid estimate as the midpoint estimate. This number is

$$\frac{\text{Midpoint Estimate} + \text{Midpoint Estimate} + \text{Trapezoid Estimate}}{3}.$$

This formula is called **Simpson's Rule** and yields extremely good estimates for the value of a definite integral. You may use a program based on this method to evaluate any definite integral that comes up in this course. A numerical method like Simpson's Rule is often the only way to evaluate such an integral because it may be impossible to find the necessary antiderivative for an exact answer.

In this section we developed Simpson's Rule on the basis of experimentation, observing some examples of errors for the trapezoid method and the midpoint method. It is possible to put Simpson's Method on a more solid theoretical footing but we will not do so here.

Exercises

The following exercises involve the same integrals as the exercises for the midpoint and trapezoid methods in the previous section. You have already done most of the work for these exercises in that section. For each integral

§7. Numerical Integration II: Simpson's Rule

- Estimate the integral using the midpoint method and 20 rectangles (i.e. $n = 20$).

- Estimate the integral using the trapezoid method and 20 rectangles (i.e. $n = 20$).

- Estimate the integral using Simpson's Rule and $n = 10$. Notice that by using $n = 10$ we ask the computer to do the same amount of work as if it were using either the midpoint method or the trapezoid method with $n = 20$.

- Find the exact answer using antidifferentiation.

- Compute the error of the midpoint method, the trapezoid method, and Simpson's Rule.

Exercise IV.7.1: See Exercise IV.6.1

$$\int_0^{\pi/2} \sin x \, dx.$$

Exercise IV.7.2: See Exercise IV.6.2

$$\int_0^{\pi/2} \sin 2x \, dx.$$

Exercise IV.7.3: See Exercise IV.6.3.

$$\int_0^{\pi/2} (2 - \sin(x/2)) \, dx.$$

Exercise IV.7.4: See Exercise IV.6.4.

$$\int_0^1 e^x \, dx.$$

Exercise IV.7.5: Any integral of your choice. See Exercise IV.6.5.

Exercise IV.7.6: Another integral of your choice. See Exercise IV.6.6.

§8. Computer-Based Antidifferentiation

Several computer programs, including *Mathematica*, Maple, DERIVE and MathCad, that run on personal computers can do extensive symbolic antidifferentiation. Although some of these programs are quite expensive and require fairly expensive personal computers, prices are dropping both for hardware and software. DERIVE is now available on a hand-held IBM compatible made by Hewlett-Packard for a total price of approximately $700.

The following *Mathematica* session does *all* the exercises in section 5 at the press of a few keys.

```
(* Exercise IV.5.1 *)   Integrate[Sqrt[x + 1], x]

  2    2 x
(- + ---) Sqrt[1 + x]
  3    3


(* Exercise IV.5.2 *)   Integrate[Sin[x]^2 Cos[x], {x, 0, Pi}]

0


(* Exercise IV.5.3 *)   Integrate[Sin[x]^3 Cos[x], {x, 0, Pi}]

0


(* Exercise IV.5.4 *)   Integrate[1/(1 + 4 x^2), {x, 1, 2}]

-ArcTan[2]    ArcTan[4]
---------- + ---------
    2            2


(* Exercise IV.5.5 *)   Integrate[1/(4 + x^2), {x, 1, 2}]

-Pi   ArcTan[2]
--- + ---------
 8        2


(* Exercise IV.5.6 *)   Integrate[1/(x^2 + k^2),x]

        x
ArcTan[-]
        k
---------
    k
```

§8. Computer-Based Antidifferentiation

```
(* Exercise IV.5.7 *)   Integrate[1/(16 + x^2), x]

        4
-ArcTan[-]
        x
----------
    4

(* Exercise IV.5.8 *)   Integrate[x/(16 + x^2), x]

         2
Log[16 + x ]
------------
     2

(* Exercise IV.5.9 *)   Integrate[(2 + 3x)/(16 + x^2), x]

        4
-ArcTan[-]                2
        x     3 Log[16 + x ]
----------  + ---------------
    2               2

(* Exercise IV.5.10 *) Integrate[1/Sqrt[4 - x^2], x]

       x
ArcSin[-]
       2

(* Exercise IV.5.11 *) Integrate[x/Sqrt[4 - x^2], x]

         2
-Sqrt[4 - x ]

(* Exercise IV.5.12 *) Integrate [(2 + 3x)/Sqrt[4 - x^2], x]

            2          x
-3 Sqrt[4 - x ] + 2 ArcSin[-]
                           2

(* Exercise IV.5.13 *) Integrate[1/Sqrt[k^2 - x^2], x]

            x
ArcSin[--------]
             2
       Sqrt[k ]
```

```
(* Exercise IV.5.14 *) Integrate[Sin[2x], {x, 0, Pi/2}]

1

(* Exercise IV.5.15 *) Integrate[x Cos[x^2], {x, 0, 1}]

Sin[1]
------
  2

(* Exercise IV.5.16 *) Integrate[Sin[2 x]^2 Cos[2 x], {x, 0, Pi/2}]

0

(* Exercise IV.5.17 *) Integrate[Sin[x] Cos[x], x]

         2
 -Cos[x]
 --------
    2

(* Exercise IV.5.18 *) Integrate[(x + 1)/(x^2 + 1), {x, 2, 3}]

                         Log[5]   Log[10]
-ArcTan[2] + ArcTan[3] - ------ + -------
                           2         2

(* Exercise IV.5.19 *) Integrate[x/(9 - x^2), {x, 0, 2}]

-(I Pi + Log[5])   I Pi + Log[9]
---------------- + -------------
       2                 2

(* Exercise IV.5.20 *) Integrate[Sin[x]^2, {x, 0, Pi/2}]

Pi
--
4

(* Exercise IV.5.21 *) Integrate[Sin[x]/Cos[x]^4, {x, 0, Pi/4}]

         3/2
   1    2
 -(-) + ----
   3    3
```

§8. Computer-Based Antidifferentiation

```
(* Exercise IV.5.22 *) Integrate[Exp[x] Cos[Exp[x]], x]

      x
Sin[E ]

(* Exercise IV.5.23 *) Integrate[Log[x]/x, x]

      2
Log[x]
--------
   2

(* Exercise IV.5.24 *) Integrate[1/(x^2 + 4x + 5), {x, 1, 2}]

-ArcTan[3] + ArcTan[4]

(* Exercise IV.5.25 *) Integrate[1/(x^2 + 4x + 13),{x, 1, 2}]

             3
      ArcTan[-]
Pi           4
-- - ----------
12        3

(* Exercise IV.5.26 *) Integrate[Exp[x]/(Exp[2 x] + 1), {x, 0, 1}]

-Pi
--- + ArcTan[E]
 4
```

Now that programs like *Mathematica* can do symbolic antidifferentiation quicker and better than most calculus students (and instructors) we do not have to invest as much time and energy learning antidifferentiation techniques as we did in the recent past. Nonetheless we do need to learn and understand the basic techniques of antidifferentiation for several reasons.

- In order to use programs like *Mathematica* intelligently we need to understand something about how they work.

- Any program is limited to the tasks that it was written to do. Computer algbera systems like *Mathematica* are no exception to

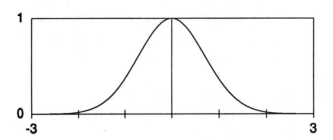

Figure IV.28: $y = e^{-x^2}$

this rule. Even though they are extremely flexible and powerful, they do have limitations. See the short story *The Profession*[1] by Isaac Asimov for further reading on this subject.

- Sometimes the answers produced by a computer algebra system like *Mathematica* are correct yet in a form that is not as easy to understand as the answer a human being might discover. In order to understand the answers produced by a computer algebra system we need to have a good understanding of the process of antidifferentiation and an ability to do ourselves the kinds of algebraic manipulations involved in antidifferentiation.

Below and in the next chapter we include some comments about computer algebra systems like *Mathematica* that can help you to use them more effectively.

Example:

There are many functions that do not have antiderivatives that can be expressed as simple functions. One of the most important such functions is the function
$$y = e^{-x^2}.$$
The graph of this function is a "bell-shaped" curve. See Figure IV.28. One standard exercise for beginning calculus students is

Exercise Find
$$\int e^{-x^2}\, dx.$$

---

[1] *Other Worlds of Isaac Asimov*, Edited by Martin H. Greenberg, Avenel Books, New York (1987).

§8. Computer-Based Antidifferentiation

This is in some ways a nasty exercise because it is impossible. The whole point of this exercise is that it is impossible. There is no simple well-known function whose derivative is e^{-x^2}. But look what happens if we give this problem to *Mathematica*.

```
Integrate[Exp[-x^2], x]
```

```
Sqrt[Pi] Erf[x]
----------------
       2
```

How did *Mathematica* solve this impossible problem? *Mathematica* doesn't have the ability to solve impossible problems. The seeming solution is really a sneaky trick. The function erf(x) is defined by

Definition:

$$\operatorname{erf}(x) = \int_0^x \frac{2}{\sqrt{\pi}} e^{-x^2} \, dx$$

or, to avoid using the same letter for two different purposes,

$$\operatorname{erf}(x) = \int_0^x \frac{2}{\sqrt{\pi}} e^{-t^2} \, dt.$$

In other words the function erf(x) is defined as the integral of a function that is very similar to the function e^{-x^2} that we are trying to integrate. The function erf(x) is extremely important in probability and statistics. It can be computed numerically on the basis of the definition above. Thus, it is a perfectly fine function. However, to say that we have found the integral

$$\int_0^x \frac{2}{\sqrt{\pi}} e^{-x^2/2} \, dx$$

by saying the answer is erf(x) is misleading. Try, for example, on your next exam to answer a question like

Question Find
$$\int \frac{x}{1-3x^2}\, dx$$
by

Answer:

Define
$$f(x) = \int_0^x \frac{x}{1-3x^2}\, dx.$$

The answer is $f(x) + c$.

Technically the answer is correct, but only in the most narrow sense. It's like saying $2+2 = 2+2$. It's true but not very helpful.

Mathematica's response is not entirely useless. It does show that the integral
$$\int e^{-x^2}\, dx$$
can be expressed in terms of the function $\text{erf}(x)$. This is useful information, but not nearly as useful as you might expect if you didn't know how $\text{erf}(x)$ was defined.

Exercise

Exercise IV.8.1: Show that *Mathematica*'s solution of
$$\int e^{-x^2}\, dx$$
is correct.

Example:

One use of a computer algebra system like *Mathematica* is to check answers that have been found using other methods. Sometimes, however, this is more difficult than you might expect. For example, we can find the integral
$$\int \frac{1}{4+x^2}\, dx$$

§8. Computer-Based Antidifferentiation

as follows by using the substitution $u = x/2$.

$$\int \frac{1}{4+x^2}\, dx = \frac{1}{4}\int \frac{1}{1+\left(\frac{x}{2}\right)^2}\, dx$$

$$= \frac{1}{2}\int \frac{\frac{1}{2}\, dx}{1+\left(\frac{x}{2}\right)^2}\, dx$$

$$= \frac{1}{2}\int \frac{du}{1+u^2}$$

$$= \frac{1}{2}\arctan u$$

$$= \frac{1}{2}\arctan(x/2) + c.$$

Here is the answer obtained using *Mathematica*.

```
Integrate[1/(4 + x^2),x]

          2
-ArcTan[-]
        x
----------
     2
```

This answer looks very different than the answer we found above. The two answers, however, are the same since

$$\tan(\pi/2 - x) = 1/\tan(x),$$

so that

$$\arctan(1/x) = \pi/2 - \arctan(x).$$

Thus, the difference between *Mathematica*'s answer and our answer is a constant. Since the indefinite integral always has a built in constant of ambiguity the two answers are the same.

This is a fairly common situation. Even two humans may get answers that appear, at first, to be quite different. The problem is even more pronounced with computer algebra systems like *Mathematica*.

§9. Summary[+]

Many simple problems can be solved by the multiplication of two constants. More complicated variations of these problems often call for multiplication of two factors one of which is a constant and the other of which is a function. For example, the volume of a cylinder is the area of its cross-section multiplied by its length. In order to compute the volume of a cone, however, one must take into account the fact that the area of a cross-section depends on where that cross-section is taken. We began this chapter by developing a tool, the **definite integral**, for solving problems of this kind.

The solution of a problem involving this kind of multiplication can be approximated by Riemann sums of the form

$$\sum_{i=1}^{n} f(s_i) h.$$

The definite integral is the limit of such Riemann sums. Although the definite integral can be used to solve a variety of problems, it can always be thought of as computing area. This leads to a number of numerical methods for approximating the definite integral. We developed one fairly sophisticated numerical method **Simpson's Rule**. This method can be used to estimate any definite integral. Simpson's Rule or some other computer or calculator-based method for numerical estimation of definite integrals should be a regular part of your mathematical tool kit.

The most important theorem in this chapter was the Fundamental Theorem of Calculus which says that integration and differentiation are two sides of the same coin. One consequence of this theorem is that if $F(x)$ is any antiderivative of a continuous function $f(x)$ then

$$\int_a^b f(x)\, dx = [F(x)]_a^b.$$

In view of this theorem, antidifferentiation is an important tool for finding definite integrals. It is important to remember, however, that finding an antiderivative can be quite difficult or even impossible. Thus, numerical estimation of definite integrals remains a necessary tool. We began the study of antidifferentiation in this chapter, developing some fundamental tools in section 4 and then the single most

§9. Summary[+]

important technique of antidifferentiation, integration by substitution, in section 5. We develop more techniques for antidifferentiation in the next chapter.

Finally, we discussed computer algebra systems like *Mathematica*, Maple, DERIVE and Mathcad that can do extensive symbolic antidifferentiation. To some extent these systems relieve us of the burden of studying antidifferentaiation exhaustively. It is still, however, important to study enough antidifferentiation to be able to understand how these systems work and to be able to interpret their results.

The following exercises can help you review this chapter.

Exercises

For each of the following integrals

- *If possible* find the exact answer via antidifferentiation.

- Estimate the answer using Simpson's Rule or another calculator or computer-based numerical method.

- If you have access to a computer algebra system like *Mathematica,*, Maple, DERIVE, or Mathcad find the answer using that system.

- Check that your answers agree.

*Exercise IV.9.1:
$$\int_0^\pi \sin 3x \, dx$$

*Exercise IV.9.2:
$$\int_0^4 \frac{x^2 + 3x - 4}{x} \, dx$$

*Exercise IV.9.3:
$$\int_1^4 x + \sqrt{x} \, dx$$

*Exercise IV.9.4:
$$\int_0^{\pi/2} \sin x^2 \, dx$$

*Exercise IV.9.5:
$$\int_0^{\pi/2} x \sin x^2 \, dx$$

*Exercise IV.9.6:
$$\int_0^{\pi/2} x^2 \sin x^2 \, dx$$

*Exercise IV.9.7:
$$\int_0^{\pi} \sin x/2 \, dx$$

*Exercise IV.9.8:
$$\int_0^3 \sqrt{x+1} \, dx$$

*Exercise IV.9.9:
$$\int_0^3 x\sqrt{x+1} \, dx$$

*Exercise IV.9.10:
$$\int_0^1 \frac{x}{4x^2+9} \, dx$$

*Exercise IV.9.11:
$$\int_0^1 \frac{1}{4x^2+9} \, dx$$

*Exercise IV.9.12:
$$\int_0^1 \frac{3x+5}{4x^2+9} \, dx$$

Chapter V

Methods and Applications of Integration

§1. Integration by Parts

We begin this chapter with an integration technique, **integration by parts**, that is like the hyperspace button on some video games. When the aliens are closing in on you, when you've tried outrunning them and outsmarting them, you're in a corner and your demise is imminent, then you can try the hyperspace button. The hyperspace button will get you out of the immediate danger, out of the frying pan. Unfortunately, however, as often as not you escape the frying pan only to land in the fire. You will eventually develop some feel for integration by parts but in the beginning it seems more like a hyperspace button than like the more conventional techniques of integration.

Integration by parts comes from the product rule for differentiation. Suppose that we have two functions, $u(x)$ and $v(x)$. The product rule says that

$$\left(\frac{d}{dx}\right) u(x)v(x) = u(x)v'(x) + v(x)u'(x).$$

Taking the antiderivative of both sides we see that

$$u(x)v(x) = \int u(x)v'(x)\, dx + \int v(x)u'(x)\, dx.$$

Using the notation $du = u'(x)dx$ and $dv = v'(x)dx$ we can rewrite this

as
$$uv = \int u\, dv + \int v\, du$$
or
$$\int u\, dv = uv - \int v\, du.$$
This last formula is the **integration by parts formula**. To see how integration by parts works we look at several examples.

Example: Find the integral
$$\int x \sin x\, dx.$$
To integrate this by parts we will let $u(x)$ be the function $u(x) = x$ and dv or $v'(x)\, dx$ be $dv = \sin x\, dx$. Thus
$$du = u'(x)\, dx = 1\, dx = dx$$
and
$$v(x) = \int v'(x)\, dx = \int \sin x\, dx = -\cos x$$
Plugging these functions into the integration by parts formula we see that
$$\int x \sin\, dx = \int u\, dv = uv - \int v\, du$$
$$= -x \cos x - \int -\cos x\, dx$$
$$= -x \cos x + \sin x.$$

So that
$$\int x \sin x\, dx = -x \cos x + \sin x + c.$$

This example shows integration by parts in its most attractive guise. We started with an integral that cannot be found by any of our earlier methods. We broke the integrand $x \sin x\, dx$ up into two factors x and $\sin x\, dx$ corresponding to the u and dv parts of the integration by parts formula. Then we calculated du and v and substituted u, v, du and dv into the integration by parts formula. This allowed us to express our old and difficult integral in terms of a new and easier integral. The

§1. Integration by Parts

hyperspace button worked. It extracted us from the immediate danger posed by the alien warships and plunked us down into a nice safe place.

Unfortunately, integration by parts doesn't always work quite so nicely. The following example shows what can happen. This example looks at exactly the same problem as the previous example but makes a less felicitous choice of u and dv.

Example: Find the integral

$$\int x \sin x \, dx.$$

This time we will let $u(x)$ be the function $u(x) = \sin x$ and dv or $v'(x)dx$ be $x \, dx$. Thus

$$du = \cos x \, dx$$

and

$$v = \int v'(x) \, dx = \int x \, dx = \frac{x^2}{2}$$

Plugging these functions into the integration by parts formula we see that

$$\int x \sin x \, dx = \int u \, dv = uv - \int v \, du = \left(\frac{x^2}{2}\right)\sin x - \int \left(\frac{x^2}{2}\right)\cos x \, dx$$

and this time the integral on the right hand side is worse than the original one.

Integration by parts isn't quite as unpredictable as the hyperspace button. With practice one can learn to anticipate its effects. Look at the integration by parts formula.

$$\int u \, dv = uv - \int v \, du$$

The key step in applying this formula is breaking the integrand up into two factors. One factor will be the u and the other factor will be the dv in the left hand side of the formula. The effect of the integration by parts formula is to leave us with the integral on the right hand side. An application of the formula will be successful if the integral on the right hand side is more tractable than the original integral on the left hand side. In our first example, that is exactly what happened. We started with a tough integral

$$\int x \sin x \, dx$$

and after applying integration by parts we had the easier integral

$$\int \cos x \, dx.$$

The effect of integration by parts is to replace u by $du = u'(x)dx$ and $dv = v'(x)dx$ by

$$v(x) = \int v'(x) \, dx.$$

When we chose $u = x$ and $dv = \sin x$ in the first example this worked well because $du = dx$ was simpler than $u = x$ and $v = -\cos x$ was no worse than $dv = \sin x \, dx$. In the second example integration by parts made things worse since in that example we chose $dv = x$ and $v = x^2/2$ was worse than v. The key idea behind the successful application of integration by parts is to choose dv so that v will be better or no worse than dv and to choose u so that du will be better or no worse than u. In the remainder of this section we look at a number of examples that illustrate successful applications of integration by parts.

Example: Find

$$\int \ln x \, dx.$$

The problem here is the function $\ln x$, which we do not know how to integrate. So we choose $u(x) = \ln x$. This leaves us with $dv = dx$. Now

$$v = \int dx = x,$$

$$du = u'(x)dx = \left(\frac{1}{x}\right) dx,$$

and integration by parts gives us

$$\int \ln x \, dx = \int u \, dv = uv - \int v \, du$$

$$= x \ln x - \int x \left(\frac{1}{x}\right) dx$$

$$= x \ln x - \int 1 \, dx$$

$$= x \ln x - x$$

§1. Integration by Parts

Thus
$$\int \ln x \, dx = x \ln x - x + c$$

Example: Find
$$\int x^2 \sin x \, dx.$$
We let $u = x^2$ and $dv = \sin x$ so that
$$du = 2x \, dx$$
and
$$v = -\cos x.$$
Applying integration by parts we get
$$\int x^2 \sin x \, dx = -x^2 \cos x - \int -\cos x \, 2x \, dx$$
$$= -x^2 \cos x + 2 \int x \cos x \, dx.$$

This doesn't solve our problem because we still have a difficult integral on the right but at least it is not as bad as the original integral.

Now we use integration by parts to find
$$\int x \cos x \, dx$$
by letting $u = x$ and $dv = \cos x \, dx$ so that
$$du = dx$$
and
$$v = \sin x$$
which leads to
$$\int x \cos x \, dx = x \sin x - \int \sin x \, dx = x \sin x + \cos x.$$

Now we put this together with the results of our first integration by parts to get
$$\int x^2 \sin x \, dx = -x^2 \cos x + 2 \int x \cos x \, dx = -x^2 \cos x + 2(x \sin x + \cos x).$$

So, finally,
$$\int x^2 \sin x \, dx = -x^2 \cos x + 2x \sin x + 2 \cos x + c.$$

Example: Find
$$\int e^x \cos x \, dx.$$

We let $u = e^x$ and $dv = \cos x \, dx$ so that
$$du = e^x dx$$

and
$$v = \sin x$$

leading to
$$\int e^x \cos x \, dx = e^x \sin x - \int e^x \sin x \, dx.$$

This does not look very promising but after staring at it for a while and failing to find any other way of attacking this problem we might try to apply integration by parts one more time working with the right hand integral
$$\int e^x \sin x \, dx.$$

This time we will let $u = e^x$ and $dv = \sin x \, dx$ so that
$$du = e^x dx$$

and
$$v = -\cos x$$

leading to
$$\int e^x \sin x \, dx = -e^x \cos x - \int -e^x \cos x \, dx = -e^x \cos x + \int e^x \cos x \, dx.$$

Now putting this together with our first integration by parts we get
$$\int e^x \cos x \, dx = e^x \sin x - \int e^x \sin x \, dx$$
$$= e^x \sin x - (-e^x \cos x + \int e^x \cos x \, dx)$$

§1. Integration by Parts

$$= e^x \sin x + e^x \cos x - \int e^x \cos x \, dx$$

$$2 \int e^x \cos x \, dx = e^x \sin x + e^x \cos x$$

$$\int e^x \cos x \, dx = \frac{e^x \sin x + e^x \cos x}{2}$$

$$\int e^x \cos x \, dx = \frac{e^x \sin x + e^x \cos x}{2} + c$$

Integration by parts often leads to formulas, called **reduction formulas** that can be used to solve a difficult integral by a series of simplifying steps. The following example shows both how such a reduction formula is found and how it is used.

Example:

Consider an integral of the form

$$\int x^n e^x \, dx.$$

If we apply integration by parts to this integral with $u(x) = x^n$ and $dv = e^x \, dx$ we see that

$$\int x^n e^x \, dx = x^n e^x - \int n x^{n-1} e^x \, dx$$

or

$$\int x^n e^x \, dx = x^n e^x - n \int x^{n-1} e^x \, dx.$$

This is called a **reduction formula** because it it reduces the difficult integral $\int x^n e^x \, dx$ to the simpler integral $\int x^{n-1} e^x \, dx$.

This reduction formula can be used, for example, to determine $\int x^2 e^x \, dx$ as follows.

$$\int x^2 e^x \, dx = x^2 e^x - 2 \int x e^x \, dx.$$

But now, using the same formula

$$\int x e^x \, dx = x e^x - \int e^x \, dx \quad \text{since } x^0 = 1$$
$$= x e^x - e^x$$

So,

$$\int x^2 e^x \, dx = x^2 e^x - 2 \int x e^x \, dx$$
$$= x^2 e^x - 2(x e^x - e^x)$$

Thus,

$$\int x^2 e^x \, dx = x^2 e^x - 2x e^x + 2e^x + c.$$

Although reduction formulas like the one that we developed above can be tedious to apply, they are extremely powerful antidifferentiation tools. They are an important part of the antidifferentiation routines built in to computer algebra systems like *Mathematica*, Maple, DERIVE, or Mathcad. For example, the following *Mathematica* statements show how the reduction formula that we developed above can be written in *Mathematica*.

```
MyIntegral[Exp[var_]] := Exp[var]

MyIntegral[Exp[var_] var_] := Exp[var] var - Exp[var]

MyIntegral[Exp[var_] var_^n_] :=
    var^n Exp[var] - n MyIntegral[Exp[var] var^(n-1)]
```

These three statements tell *Mathematica* how to integrate any integrand of the form

$$e^v v^n$$

where v is any variable. We need three statements because *Mathematica* is unable[1] to recognize without further instruction that 1 can be written as x^0 or that x can be written as x^1. This is one of the difficulties of writing a program like *Mathmetica*. Things like x^1 and x that a human being immediately recognizes as being identical look different to a computer. One reason that programs like *Mathematica* require

---

[1] The designers of *Mathematica* were aware of the problems inherent in the fact that x^0 and x^1 look different than 1 and x, respectively, and they built into *Mathematica* the ability to recognize that these expressions are identical. Because we don't want to get into the details of *Mathematica* here we do not discuss this capability in detail.

§1. Integration by Parts

large computers, lots of memory, and longish running times is that they must painstakingly check for this kind of situation.

One way that *Mathematica* deals with the problems inherent in the possibility of writing the same expression in several different ways is by using standard forms for expressions whenever possible. This is the reason why we work with

`Exp[x] x^n`

which is *Mathematica's* standard form rather than

`x^n Exp[x]}`

which is closer to the way most humans would write $x^n e^x$.

The following *Mathematica* session shows how this new function can be used to evaluate integrals of the form $\int e^x x^n \, dx$

`MyIntegral[Exp[v]]`

```
 v
E
```

`MyIntegral[Exp[v] v]`

```
   v     v
-E  + E  v
```

`MyIntegral[Exp[v] v^3]`

```
 v  3       v  2        v     v
E  v  - 3 (E  v  - 2 (-E  + E  v))
```

Exercises

*Exercise V.1.1: Find
$$\int x 2^x \, dx$$

*Exercise V.1.2: Find
$$\int \arcsin x \, dx$$

*Exercise V.1.3: Find
$$\int \arctan x \, dx$$

Exercise V.1.4: Find
$$\int x^3 e^x \, dx$$

*Exercise V.1.5: Find
$$\int x^2 e^{-x} \, dx$$

*Exercise V.1.6: Find a reduction formula for
$$\int x^n e^{-x} \, dx$$

If *Mathematica* is available to you, write a procedure **MyIntegral** that implements this reduction formula to integrate any integral of the form
$$\int e^{-x} x^n \, dx.$$

Exercise V.1.7: Find
$$\int x^3 e^{-x} \, dx$$

*Exercise V.1.8: Find
$$\int x \ln x \, dx$$

Exercise V.1.9: Find a general formula for
$$\int x^n \ln x \, dx$$

*Exercise V.1.10: Find
$$\int \sin^2 x \, dx$$

by writing it as
$$\int (\sin x)(\sin x) \, dx$$

and applying integration by parts and then using the fact that
$$\cos^2 x = 1 - \sin^2 x$$

*Exercise V.1.11: Prove that
$$\int \sin^n x \, dx = -\frac{\sin^{n-1} x \cos x}{n} + \frac{n-1}{n} \int \sin^{n-2} x \, dx.$$

Exercise V.1.12: Find a reduction formula that expresses $\int x^n \sin x\, dx$ in terms of $\int x^{n-1} \cos x\, dx$.

Exercise V.1.13: Find a reduction formula that expresses $\int x^n \cos x\, dx$ in terms of $\int x^{n-1} \sin x\, dx$.

Exercise V.1.14: Use the formulas from the last two exercises to determine
$$\int x^3 \sin x\, dx.$$

*Exercise V.1.15:** If *Mathematica* is available to you write a procedure `MyIntegral` that uses the reduction formulas developed above to integrate any integral of the form
$$\int (\sin x) x^n\, dx \quad \text{or}$$
$$\int (\cos x) x^n\, dx.$$

*Exercise V.1.16:** Explain how any integral of the form
$$\int \sin^m x \cos^n x\, dx$$
can be evaluated.

*Exercise V.1.17:** Suppose that α and β are *different* constants. Find
$$\int \sin \alpha x \cos \beta x\, dx$$

*Exercise V.1.18:** Suppose that α is a constant. Find
$$\int \sin^2 \alpha x\, dx$$

---

§2. Solids of Revolution

In this section we study solids that can be constructed by taking a region like the one shown in Figure V.1 and revolving it around the x-axis. As it revolves around the x-axis such a region will sweep out a solid as shown in Figure V.2. Such a solid is called a **Solid of Revolution**.

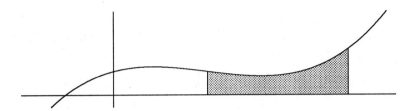

Figure V.1: Region about to sweep out a solid

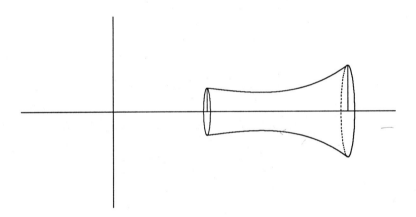

Figure V.2: Solid of revolution

Many interesting solids can be made in this way. For example, suppose that one would like to construct a cone whose base is a circle with radius 2 units and whose height is 6 units. One can construct this cone by revolving a triangle like the one shown in Figure V.3 about the x-axis as shown in Figure V.4. Similarly, one can construct a sphere of radius R by revolving a semicircle of radius R around the x-axis as shown in Figures V.5 and V.6.

We are interested in finding the volume of a solid of revolution. Suppose that we start with a region enclosed by the x-axis, the curve $y = f(x)$, and the lines $x = a$ and $x = b$ as shown in Figure V.7 and we revolve this region about the x-axis to get a solid like the one shown in Figure V.8.

For simple solids like cylinders and rectangular prisms whose cross

§2. Solids of Revolution

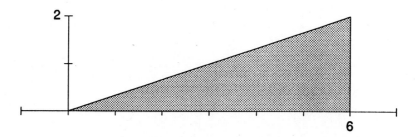

Figure V.3: Triangle about to sweep out a cone

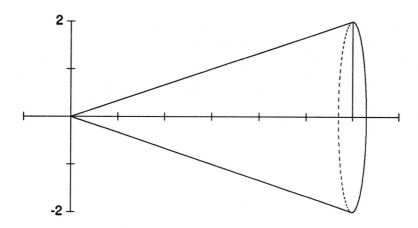

Figure V.4: Cone born of a triangle

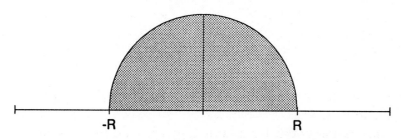

Figure V.5: Semicircle about to sweep out a sphere

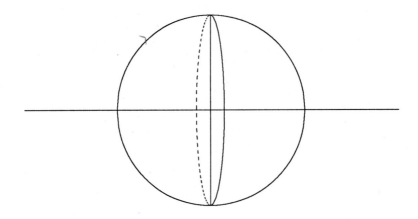

Figure V.6: Sphere born of a semicircle

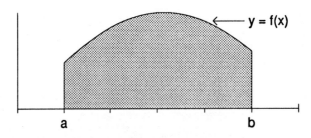

Figure V.7: Region about to sweep out a solid

§2. Solids of Revolution

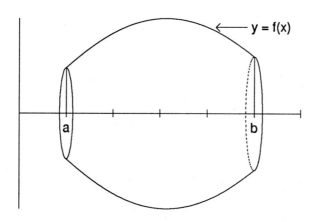

Figure V.8: Solid of revolution

sections are all the same we determine the volume by simple multiplication, multiplying the area of each cross section by the height of the solid. Solids like the ones shown in Figures V.2, V.4, V.6 and V.8, however, do not have constant cross sectional area. For this reason we need to use integration rather than simple multiplication. Figure V.9 shows a typical section of a typical solid of revolution.

This section is swept out by the line which goes from the x-axis at the point $(x, 0)$ to the point $(x, f(x))$ on the curve $y = f(x)$. As the line sweeps around the x-axis it traces out a circle of radius $f(x)$. The area of this circle is $\pi f(x)^2$. Integration multiplies this (nonconstant) area by the length of the solid of revolution to give us the volume.

$$V = \int_a^b \pi f(x)^2 \, dx = \pi \int_a^b f(x)^2 \, dx$$

This formula enables us to find the volume of many interesting solids.

Example:

Find the volume of a cone whose radius is 2 units and whose height is 6 units.

We picture this cone as shown in Figure V.4 with its base perpendicular to the x-axis and its height measured along the x-axis. It would be better to use the word length instead of height with the cone placed

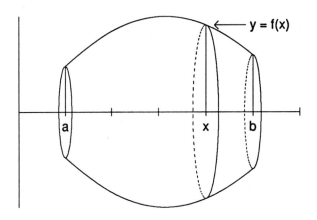

Figure V.9: Typical section

this way. This cone is thought of as the solid of revolution produced by revolving the region shown in Figure V.3 around the x-axis. The top edge of this region is the line

$$y = \frac{x}{3}.$$

Thus the volume of the solid is

$$\pi \int_0^6 \left(\frac{x}{3}\right)^2 dx = \pi \int_0^6 \frac{x^2}{9} dx = \pi \left[\frac{x^3}{27}\right]_0^6 = 8\pi$$

Example:

Find the volume of a sphere of radius R.

We picture this sphere as shown in Figure V.6. It is the result of revolving the semicircle shown in Figure V.5 around the x-axis. The top edge of this semicircle is described by the function

$$f(x) = \sqrt{R^2 - x^2}\,.$$

Thus the volume of the sphere is

$$\pi \int_{-R}^{R} \left(\sqrt{R^2 - x^2}\right)^2 dx \;=\; \pi \int_{-R}^{R} (R^2 - x^2)\, dx$$

§2. SOLIDS OF REVOLUTION

$$= \pi \left[R^2 x - \frac{x^3}{3} \right]_{-R}^{R}$$

$$= \pi \left[\left(R^3 - \frac{R^3}{3} \right) - \left(-R^3 + \frac{R^3}{3} \right) \right]$$

$$= \frac{4\pi R^3}{3}$$

Exercises

Exercise V.2.1: Find the volume of a cone whose base has radius 2 and whose height (length) is 4.

Exercise V.2.2: Find the volume of a cone whose base has radius R and whose height (length) is H.

Exercise V.2.3: Suppose that the region bounded by the curve $y = \sin x$, the x-axis, and the line $x = \pi/2$ is revolved about the x-axis. Find the volume of the resulting solid.

Exercise V.2.4: Suppose the semiellipse

$$y = b\sqrt{1 - \frac{x^2}{a^2}}$$

is revolved about the x-axis. The resulting solid is called an **ellipsoid**. Find its volume.

Exercise V.2.5: Suppose the region bounded by the x-axis, the curve $y = xe^{-x}$, and the line $x = 5$ is revolved about the x-axis. What is the volume of the resulting solid?

Exercise V.2.6: Suppose that the region bounded by the x-axis and the curve $y = x\sqrt{1 - x^2}$ between $x = 0$ and $x = 1$ is revolved about the x-axis. What is the volume of the resulting solid?

Exercise V.2.7: Find the volume of a pyramid with a square base 50 feet by 50 feet and a height of 100 feet. Note that the top of this pyramid is directly above the center of the base and that the sides are flat planes. See Figure V.10.

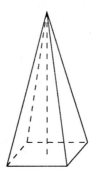

Figure V.10: Pyramid

*Exercise V.2.8: Find a general formula for the volume of a pyramid with a square base x by x and a height h.

*Exercise V.2.9: Find the volume of a sphere whose radius is two feet from which a cylinder with a radius of one foot has been drilled out of the center.

Exercise V.2.10: Find the volume of a cone whose original height was six feet and whose base has a radius of two feet from which a cylinder of radius one foot has been drilled out of the center.

---

§3. Black Holes and Escape Velocity

" 'The devil take you,' grumbled Master Andry Musnier.

" 'Master Andry,' replied Jehan, still hanging from the capital (of the pillar), 'Shut up or I'll drop on your head.'

"Master Andry raised his eyes, seemed to measure the height of the pillar, the weight of the rascal, multiplied mentally the weight by the square of the velocity and was quiet.[2]"

In this section we examine two questions about space and space travel. The first question concerns escaping from the earth—is it possible to launch a projectile from the surface of the earth with sufficient initial velocity so that it can escape without ever being pulled back

---

[2] *Notre Dame de Paris,* by Victor Hugo.

§3. Black Holes and Escape Velocity

by gravity? We will be able to answer this question completely. The second question is more ambitious and more difficult. We want to investigate "black holes." Because a complete understanding of black holes requires some knowledge of relativity, a subject beyond the scope of this book, we will be able to make only a small beginning on the study of black holes.

Mass, Work, Energy and Force

We begin by clarifying four ideas—mass, work, energy, and force. Our discussion starts with a simple observation—if an object is at rest then unless some force acts on that object it will remain at rest. We know from experience that "larger" objects resist our efforts to move them more than "smaller" objects do. This resistance to acceleration is called **inertia** and is determined by the **mass** of the object. Mass is measured in kilograms.

In ordinary everyday English people often confuse two concepts—mass and weight. They are, however, quite different. Mass is an intrinsic property of an object. It is a measure of how difficult it is to make the object move. An object's weight, however, is not an intrinsic property of the object alone. It depends on both its mass and its location. An object whose mass is one kilogram weighs 2.2 pounds on the surface of the earth but only 0.37 pounds on the surface of the moon.

The next remark is anything but natural—if an object is moving then unless some force acts on that object it will continue moving in a straight line with the same velocity. This contradicts our everyday experience because our everyday world is filled with friction, a force that acts on moving objects in such a way as to slow them down and eventually bring them to a halt.

When a force acts on an object it produces an acceleration given by the equation

$$\text{acceleration} = \left(\frac{1}{m}\right) F$$

where F denotes the force and m is the mass of the object.

We measure force in units of **newtons** or kilogram meters per second^2. A force of one newton will cause an object whose mass is one kilogram to accelerate at the rate of one meter per second^2. Looking at the units involved in the equation

$$\text{acceleration} = \left(\frac{1}{m}\right) F$$

we see

$$\frac{\text{meters}}{\text{second}^2} = \left(\frac{1}{\text{kilograms}}\right) \left(\frac{\text{kilogram meters}}{\text{second}^2}\right).$$

If an object is on the earth's surface then gravity pulls it downward toward the center of the earth with a force of $-9.80m$ newtons where m is the mass of the object and the minus sign indicates that the force of gravity is pulling the object downward. Thus the object will accelerate at the rate

$$\text{acceleration} = -\left(\frac{1}{m}\right) 9.80m = -9.80 \text{ meters per second}^2$$

where the minus sign indicates that the acceleration is downward.

If we denote the height of an object near the surface of the earth by y this gives us the differential equation

$$y'' = -9.80 \text{ meters per second}^2.$$

You may have seen this differential equation before expressed in English units.

$$y'' = -32 \text{ feet per second}^2.$$

In order to move an object against a force, for example, to lift an object against the force of gravity, we must do some work. Work and energy are two sides of the same coin. Energy is the capacity to do **work**. Thus work and energy are measured using the same units, **newton meters** or kilogram meter^2 per second^2. One newton meter is the amount of work performed or the amount of energy used to move an object one meter against a force of one newton. Thus near the surface of the earth the amount of work performed or the amount of energy required to lift an object whose mass is m kilograms h meters is

$$\text{work} = 9.80mh \text{ kilogram meter}^2 \text{ per second}^2.$$

Notice that the work required to lift an object is positive because the force of gravity is pushing against the force lifting the object. Notice also that the amount of work to lift a given object is proportional to

§3. Black Holes and Escape Velocity

the distance it is lifted. This matches our everyday experience—it is twice as hard to lift a given object two feet as to lift it one foot.

Perhaps the single most important principle of physics is the law of conservation of energy, you don't get something for nothing. Although energy can be converted from one form[3] to another, it cannot be created out of nothing nor may it be destroyed. Conservation laws are extraordinarily important for our understanding of the physical world and are important in the social sciences as well. For example, economists have a saying that "There is no such thing as a free lunch." This is a conservation law.

Energy comes in many forms. In this section we are concerned with two forms—the **kinetic energy** associated with a moving object and the **potential energy** associated with an object at a given height. When you catch a baseball or get hit by a truck you feel the effects of the kinetic energy associated with a moving object. When an object is lifted, the energy expended to lift it does not disappear. Instead it is stored in the form of **potential energy**. For example, if a painter lifts a bucket of paint from the ground to the top of her ladder and then drops it, the potential energy stored while she lifted the bucket will be converted to kinetic energy as the bucket drops onto any unlucky passerby. This is a dramatic and often colorful demonstration of the law of conservation of energy. The total amount of energy in the system remains constant. As the bucket falls, its potential energy is converted to kinetic energy.

Since we know how much work is required to lift on object h meters, we know the amount of potential energy acquired by an object of mass m kilograms lifted h meters, $9.80mh$. We can use this fact together with the law of conservation of energy to determine the amount of kinetic energy associated with an object of mass m moving at v meters per second.

Theorem 1 *The kinetic energy associated with an object of mass m moving at velocity v is*

$$E = \frac{mv^2}{2}.$$

---

[3] One of Einstein's contributions to our understanding of our physical world was the realization that matter is a form of energy. His famous equation $E = mc^2$ expresses the way in which energy in the form of matter can be converted to other forms of energy.

Proof:

Suppose that the object is dropped from a height of h meters. The change in its potential energy as it drops is $-9.80mh$ newton meters. If its height at time t is $y(t)$, measured in meters where t is time in seconds after the object is released, then $y(t)$ is the solution of the initial value problem
$$y'' = -9.80, \qquad y(0) = h.$$
The solution of this initial value problem is
$$y = h - \frac{9.80t^2}{2}.$$
The object will reach ground level at time
$$t = \sqrt{\frac{2h}{9.80}}$$
at which time its velocity will be
$$v = -9.80\sqrt{\frac{2h}{9.80}}.$$
So
$$v^2 = 9.80(2h)$$
and
$$\frac{v^2}{2} = 9.80h.$$

At this time the potential energy $9.80mh$ has been converted to kinetic energy. Let E denote this kinetic energy, which was formerly potential energy.

$$E = 9.80mh$$
$$= \frac{1}{2}mv^2,$$

as advertised.

When a projectile of mass m is launched from the surface of the earth with an initial velocity v_0 it has kinetic energy $mv_0^2/2$. As the

§3. Black Holes and Escape Velocity

projectile rises it slows down as this kinetic energy is converted into potential energy. At the top of its flight the projectile will have zero velocity and, hence, zero kinetic energy. If h denotes the height of the projectile at the top of its trajectory, the initial kinetic energy $mv_0^2/2$ will have been completely converted into the potential energy $9.80mh$.

This gives us the equation

$$\frac{mv_0^2}{2} = 9.80mh$$

or

$$v_0^2 = 19.60h,$$

which allows us to calculate the highest altitude reached by a projectile if we know its initial velocity or to calculate the initial velocity required to reach a given altitude. Notice that the mass of the object makes no difference.

Example:

How fast would you have to throw a baseball if you wanted it to rise 30 meters?

Answer:

$$v_0^2 = 19.60(30)$$

$$v_0^2 = 588$$

$$v_0 = \sqrt{588}$$

$$= 24.25 \text{ meters per second.}$$

Exercises

Exercise V.3.1:

How fast would you have to throw a baseball if you wanted it to rise 15 meters?

*Exercise V.3.2:
How fast would you have to throw a baseball if you wanted it to rise 50 feet?

Exercise V.3.3:
How fast would you have to throw a baseball if you wanted it to rise 100 feet?

*Exercise V.3.4:
Suppose a baseball is thrown straight upward with an initial velocity of 15 meters per second. How high will it rise?

Exercise V.3.5:
Suppose a baseball is thrown straight upward with an initial velocity of 30 meters per second. How high will it rise?

*Exercise V.3.6:
Suppose a baseball is thrown straight upward with an initial velocity of 40 feet per second. How high will it rise?

Exercise V.3.7:
Suppose a baseball is thrown straight upward with an initial velocity of 80 feet per second. How high will it rise?

Escape Velocity

In this section we want to determine whether it is possible to launch a projectile from the surface of the earth with sufficient velocity to escape from the earth and its gravitational pull. If it is possible then we call the necessary initial velocity the **escape velocity**. Our discussion so far ignores air resistance. We will continue to ignore air resistance. There is another factor, however, that we cannot ignore. The force exerted by gravity is stronger when two bodies are close together than when they are far apart. In general the strength of the force exerted by gravity is given by the formula

$$F = \frac{Gm_1 m_2}{r^2}$$

where G is a constant called the **gravitational constant**, m_1 and m_2 are the masses of the two bodies, and r is the distance between them

§3. BLACK HOLES AND ESCAPE VELOCITY

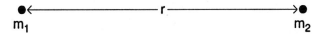

Figure V.11: Gravitational attraction between two bodies

as shown in Figure V.11. Notice when r is very large this force will be very small because of the denominator r^2. Symbolically,

$$\lim_{r \to +\infty} F = 0,$$

which corresponds to our intuitive feeling that we really don't feel any noticeable gravitational effect from objects that are very, very far away from us. In this case "very, very far" is quite a distance. Even though the moon is quite far away, roughly 400,000 kilometers, we feel a considerable gravitational effect. The moon's gravitational effect causes tides. Even the sun, which is much farther away, roughly 150,000,000 kilometers, but also much more massive than the moon, exerts a considerable gravitational effect and has a strong impact on tides.

The distance r is the distance between the centers of gravity of the two objects. This distance will be measured in meters. The constant G is 6.67×10^{-11} newton meter^2 per kilogram^2. For example, the distance from the surface of the earth to the center of the earth is 6,380 kilometers or 6.38×10^6 meters and the mass of the earth is 5.98×10^{24} kilograms. Thus the force exerted by gravity on an object whose mass is m kilograms at the surface of the earth is

$$\begin{aligned}\frac{G m_{earth} m}{r^2} &= \frac{(6.67 \times 10^{-11})(5.98 \times 10^{24})m}{(6.38 \times 10^6)^2} \\ &= 9.80m\end{aligned}$$

which is the same value used earlier although we usually write $-9.80m$ because the force is directed downward.

Exercises

*Exercise V.3.8:
The earth is not a perfect sphere. Its diameter measured across the equator is 12,760 kilometers but its diameter measured from the north pole to the south pole is 12,720 kilometers. What is the force exerted by gravity on an object at the north pole whose mass is m kilograms?

*Exercise V.3.9:
The mass of the moon is 7.4×10^{22} kilograms and its radius is 1,738 kilometers. What is the force exerted by gravity on an object whose mass is m kilograms near the surface of the moon?

Exercise V.3.10:
How fast would you have to throw a baseball if you wanted it to rise 15 meters from the surface of the moon?

*Exercise V.3.11:
How fast would you have to throw a baseball if you wanted it to rise 50 feet from the surface of the moon?

Exercise V.3.12:
How fast would you have to throw a baseball if you wanted it to rise 100 feet from the surface of the moon?

*Exercise V.3.13:
Suppose a baseball is thrown straight upward with an initial velocity of 15 meters per second from the surface of the moon. How high will it rise?

Exercise V.3.14:
Suppose a baseball is thrown straight upward with an initial velocity of 30 meters per second from the surface of the moon. How high will it rise?

*Exercise V.3.15:
Suppose a baseball is thrown straight upward with an initial velocity of 40 feet per second from the surface of the moon. How high will it rise?

Exercise V.3.16:
Suppose a baseball is thrown straight upward with an initial velocity of 80 feet per second from the surface of the moon. How high will it rise?

§3. Black Holes and Escape Velocity

To determine, for example, the amount of work required to lift an object of mass m from the earth's surface to an altitude of 1,000 kilometers above the earth's surface one cannot simply multiply the force of gravity times the distance (1,000 kilometers) that the object will rise because the force of gravity is not constant. This is another job for integration. The work required will be

$$\int_{6,380,000}^{7,380,000} \frac{Gm_{earth}m}{r^2} \, dr = \int_{6,380,000}^{7,380,000} \frac{(6.67 \times 10^{-11})(5.98 \times 10^{24})m}{r^2} \, dr$$

$$= \int_{6,380,000}^{7,380,000} \frac{3.99 \times 10^{14} m}{r^2} \, dr$$

$$= \left[\frac{-m(3.99 \times 10^{14})}{r} \right]_{6,380,000}^{7,380,000}$$

$$= 8.47 \times 10^6 m$$

Notice the limits of integration are 6,380,000 meters, the distance from the object to the center of the earth when the object is on the surface of the earth, and 7,380,000 meters, the distance from the object to the center of the earth when the object is at an altitude of 1,000 kilometers above the surface of the earth.

Now suppose we want to know the initial velocity required for a projectile to reach an altitude of 1000 kilometers before dropping back to the earth. The kinetic energy for an object with an initial velocity of v kilometers per second would be $mv^2/2$ and, since this would all be expended at the top of the trajectory, we want

$$\frac{mv^2}{2} = 8.47 \times 10^6 m$$

or

$$v^2 = 16.94 \times 10^6.$$

So

$$v = \sqrt{16.94 \times 10^6} = 4,116 \text{ meters per second.}$$

Exercises

*Exercise V.3.17: What initial velocity would be required for a projectile to reach an altitude of 2,000 kilometers above the surface of the earth?

Exercise V.3.18: What initial velocity would be required to reach an altitude of 5,000 kilometers above the surface of the earth?

*Exercise V.3.19: Find a general formula for the initial velocity required for a projectile to reach an altitude H above the surface of the earth?

*Exercise V.3.20: Find the limit as H goes to infinity of the answer you obtained to the previous exercise.

Now we would like to ask how much energy would be required for a projectile to escape completely from the earth's gravitational pull. Intuitively we need to find the amount of work that would be done to raise a projectile of mass m from the surface of the earth to an altitude of "infinity." Symbolically we want to find

$$\text{Escape Energy} = \int_{6,380,000}^{+\infty} \frac{Gm_{earth}m}{r^2}\, dr$$

$$= Gm_{earth}m \int_{6,380,000}^{+\infty} \frac{1}{r^2}\, dr$$

$$= 3.99 \times 10^{14} m \int_{6,380,000}^{+\infty} \frac{1}{r^2}\, dr$$

Once we have computed this **escape energy** we can compute the necessary initial velocity.

We can estimate the escape energy by computing the energy required to lift a projectile to a very high altitude, for example,

$$\text{Estimated Escape Energy} = \int_{6,380,000}^{1,000,000,000,000} \frac{Gm_{earth}m}{r^2}\, dr$$

This leads us to the following definition.

§3. Black Holes and Escape Velocity

Improper Integrals

Definition:

The improper integral
$$\int_a^{+\infty} f(x)\, dx$$
is defined by
$$\int_a^{+\infty} f(x)\, dx = \lim_{b \to +\infty} \int_a^b f(x)\, dx.$$

Example:

To solve the problem that motivated this definition we need to evaluate
$$\int_{6,380,000}^{+\infty} \frac{1}{r^2}\, dr$$

$$\begin{aligned}
\int_{6,380,000}^{+\infty} \frac{1}{r^2}\, dr &= \lim_{b \to +\infty} \int_{6,380,000}^{b} \frac{1}{r^2}\, dr \\
&= \lim_{b \to +\infty} \left[-\frac{1}{r} \right]_{6,380,000}^{b} \\
&= \lim_{b \to +\infty} \left(-\frac{1}{b} + \frac{1}{6,380,000} \right) \\
&= \frac{1}{6,380,000} \\
&= 1.57 \times 10^{-7}
\end{aligned}$$

Hence the escape energy is
$$E = (3.99 \times 10^{14} m)(1.57 \times 10^{-7}) = 6.26 \times 10^7 m$$

and we can find the escape velocity by solving the equation

$$\frac{mv^2}{2} = 6.26 \times 10^7 m$$

$$v^2 = 1.25 \times 10^8$$

$$\begin{aligned}v &= \sqrt{1.25 \times 10^8} = 1.12 \times 10^4 \text{meters per second} \\ &= 11.2 \text{ kilometers per second,}\end{aligned}$$

and we see that it is possible to launch a projectile with sufficient initial velocity to escape from the earth completely and that the escape velocity is roughly eleven kilometers per second.

This is only one example of situations where an integral with an infinite limit makes physical and mathematical sense. The following exercises look at some other examples.

Exercises

*Exercise V.3.21: Find the area of the region between the curve

$$y = \frac{1}{x^2}$$

and the x-axis to the right of the line $x = 1$.

*Exercise V.3.22: Find the area of the region between the curve

$$y = \frac{1}{x}$$

and the x-axis to the right of the line $x = 1$.

*Exercise V.3.23: Suppose the region in the preceding exercise is revolved about the x-axis. Find the volume of the resulting solid.

A Hint About "Black Holes"

A complete understanding about black holes requires some knowledge about the Theory of Relativity. Relativity Theory is beyond this course. The work in this section, however, provides a hint of why black holes

§3. Black Holes and Escape Velocity

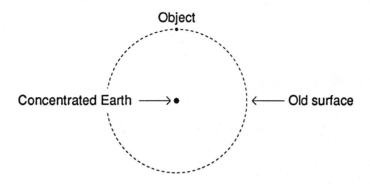

Figure V.12: Concentrated earth

exist. I would like to recommend one of my favorite books, *The Collapsing Universe* by Isaac Asimov, if you would like to pursue the subject of black holes.

Suppose that the earth's entire mass were concentrated at a single point at the center of the original earth and that we were to drop an object from the present surface of the earth (i.e. a point 6,380 kilometers above the imaginary point at which the earth's mass is hypothetically concentrated) as shown in Figure V.12. We would like to know how fast the object would be going when it reached the imaginary point-earth. Our strategy will be the same as before. As the object falls the force of gravity is doing work. That work is converted into kinetic energy. We compute the amount of work done by the force of gravity. This is just the integral

$$\int_0^{6,380,000} \frac{3.99 \times 10^{14} m}{r^2} \, dr = 3.99 \times 10^{14} m \int_0^{6,380,000} \frac{1}{r^2} \, dr.$$

Actually this integral computes the amount of work that would be done raising the object from the point-earth to the surface of the original earth. The same amount of work is done by gravity pulling the object down. We need to evaluate the integral

$$\int_0^{6,380,000} \frac{1}{r^2} \, dr$$

and there is a problem. The function $1/r^2$ is not defined when $r = 0$. So, a priori, this integral doesn't makes sense. We can, however, estimate the answer to our question by evaluating

$$\int_a^{6,380,000} \frac{3.99 \times 10^{14} m}{r^2} \, dr = 3.99 \times 10^{14} m \int_a^{6,380,000} \frac{1}{r^2} \, dr.$$

for a very small. This leads to the following definition.

Improper Integrals, An Encore

Definition:

Suppose that $f(x)$ is defined and continuous at every x in the interval $[a, b]$ except for the point a. Then we define the **Improper Integral**

$$\int_a^b f(x) \, dx$$

by

$$\int_a^b f(x) \, dx = \lim_{p \to a^+} \int_p^b f(x) \, dx$$

Example:

We want to evaluate the integral we ran into above

$$\begin{aligned}
\int_0^{6,380,000} \frac{1}{r^2} \, dr &= \lim_{p \to 0^+} \int_p^{6,380,000} \frac{1}{r^2} \, dr \\
&= \lim_{p \to 0^+} \left[\frac{-1}{r} \right]_p^{6,380,000} \\
&= \lim_{p \to 0^+} \left(\frac{-1}{6,380,000} - \frac{-1}{p} \right) \\
&= \lim_{p \to 0^+} \left(\frac{-1}{6,380,000} + \frac{1}{p} \right) \\
&= +\infty
\end{aligned}$$

Thus if our object were to fall all the way to our point-earth it would acquire an infinite amount of kinetic energy and, hence, an infinite velocity — TILT!!! This is impossible. The flip side of this observation is that an object that was on surface of our point-earth could not escape. In fact, it could not reach the former surface of the earth because to do so would require an infinite amount of energy. This is essentially why "black holes" are black. Nothing, not even light, can escape. When one considers relativistic effects one can show that because no object can travel faster than the speed of light black holes have a nonzero radius.

As with our first improper integral our new improper integrals have more mundane uses.

Exercises

*Exercise V.3.24: Find the area of the region between the curve $y = 1/x$, the y-axis, the x-axis, and the line $x = 1$.

*Exercise V.3.25: Find the area of the region between the curve $y = 1/x^2$, the y-axis, the x-axis, and the line $x = 1$.

*Exercise V.3.26: Find the area of the region between the curve $y = 1/\sqrt{x}$, the y-axis, the x-axis, and the line $x = 1$.

*Exercise V.3.27: Suppose the region in the preceding problem is revolved about the x-axis. Find the volume of the resulting solid.

§4. Implicit Differentiation

A relationship between two variables x and y can often be expressed as a function $y = f(x)$ in which the value of y is determined by the value of x. Sometimes, however, the relationship is more complicated. For example, if x and y are the two coordinates of a point whose distance from the origin is 5 then $x^2 + y^2 = 25$. Although this equation expresses a relationship between x and y, y is not a function of x; knowing the value of x does not determine the value of y. For example, if $x = 4$ then y could be either 3 or -3.

To check whether a particular point (a, b) is on the curve described by an equation, we substitute the values a for x and b for y into the equation and see if it is satisfied.

510 Chapter V. Methods and Applications of Integration

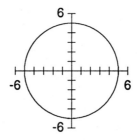

Figure V.13: $x^2 + y^2 = 25$

Example:

Is the point $(3, 4)$ on the curve described by the equation

$$x^2 + y^2 = 25?$$

Answer:

Substituting 3 for x and 4 for y we get

$$\begin{aligned} (3)^2 + (4)^2 &= 25? \\ 9 + 16 &= 25? \end{aligned}$$

and we see that $(3, 4)$ is on the curve.

In this section we study curves like those shown in Figures V.13-V.16 in which two variables are related by an equation.

Frequently, as in these examples, the graphs of such equations are smooth curves and if we look at a particular point on such a curve it makes sense to talk about the tangent to the curve at that point. Figure V.17 indicates several such tangents on the curve described by the equation

$$(x^2 + y^2 - 25)(y^2 - x^2 - 4) = 0.$$

If we look through a microscope at a curve like the one shown in Figure V.18 near a point at which the curve is smooth we see that the curve looks like the graph of a function. Since the slope of a smooth

§4. Implicit Differentiation

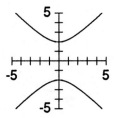

Figure V.14: $y^2 - x^2 = 4$

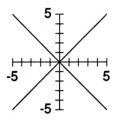

Figure V.15: $y^2 - x^2 = 0$

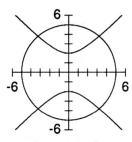

Figure V.16: $(x^2 + y^2 - 25)(y^2 - x^2 - 4) = 0$

512 CHAPTER V. METHODS AND APPLICATIONS OF INTEGRATION

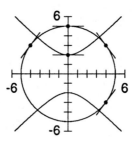

Figure V.17: Tangents to the curve $(x^2 + y^2 - 25)(y^2 - x^2 - 4) = 0$.

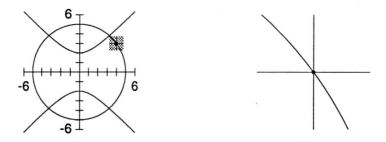

Figure V.18: Microscopic view near a point

curve at a point depends only on the part of the curve in the immediate neighborhood of the point we can work as if the equation did define y as a function of x. We illustrate the method by considering the equation $x^2 + y^2 = 25$.

Example:

Find the slope of the tangent to the curve $x^2 + y^2 = 25$ at the point (x, y). See Figure V.13.

Answer:

Since this curve looks like the graph of a function near any point except the points $(-5, 0)$ and $(5, 0)$ we work as if y were a function of x. We differentiate both sides of the equation $x^2 + y^2 = 25$ with respect to the variable x.

$$\left(\frac{d}{dx}\right)(x^2 + y^2) = \left(\frac{d}{dx}\right)(25)$$

§4. Implicit Differentiation

$$2x + 2y\left(\frac{dy}{dx}\right) = 0$$

Notice that the right side of the original equation is simply the constant 25. The derivative of any constant is 0. The left side of the equation contains the term y^2. This term is differentiated using the chain rule since y can be treated as a function of x. We continue the calculations begun above as follows.

$$2x + 2y\left(\frac{dy}{dx}\right) = 0$$
$$2y\left(\frac{dy}{dx}\right) = -2x$$
$$\frac{dy}{dx} = -\frac{x}{y}$$

This gives us a formula for the slope of the tangent to the curve $x^2 + y^2 = 25$ at almost any point on the curve. The only exceptions are the points $(-5, 0)$ and $(5, 0)$ at which the denominator of the formula $dy/dx = -x/y$ is equal to zero and at which we expected trouble from the start.

Notice that this formula involves both the variable x and the variable y. This is not surprising since both variables are needed to describe a point on the curve. For a function $y = f(x)$ the value of x determines the value of y but in this situation there are two possible values $y = \sqrt{25 - x^2}$ and $y = -\sqrt{25 - x^2}$ for y for each value of x between -5 and 5. This particular equation determines two functions, one whose graph is the upper semicircle and the other whose graph is the lower semicircle. If we were interested in one of these, for example, the upper semicircle given by $y = \sqrt{25 - x^2}$ then we could write

$$\frac{dy}{dx} = -\frac{x}{y} = -\frac{x}{\sqrt{25 - x^2}}.$$

This example ilustrates the technique of **implicit differentiation**. Implicit differentiation can be used to find the slope of the tangent to a curve defined by an equation at a point at which the curve looks like a

514 CHAPTER V. METHODS AND APPLICATIONS OF INTEGRATION

smooth function under a microscope. The essence of the method is to differentiate both sides of the equation using the chain rule whenever the variable y occurs.

Example:

Find the slope of the tangent to the curve $y^2 - x^2 = 4$ at the point (x, y).

Answer:

$$y^2 - x^2 = 4$$
$$2y\left(\frac{dy}{dx}\right) - 2x = 0$$
$$2y\left(\frac{dy}{dx}\right) = 2x$$
$$\frac{dy}{dx} = \frac{x}{y}$$

Example:

Find the slope of the tangent to the curve $(x^2+y^2-25)(y^2-x^2-4) = 0$ at the point (x, y).

Answer:

$$(x^2 + y^2 - 25)(y^2 - x^2 - 4) = 0$$
$$(x^2 + y^2 - 25)(2yy' - 2x) + (2x + 2yy')(y^2 - x^2 - 4) = 0$$
$$2x^2yy' + 2y^3y' - 50yy' - 2x^3 - 2xy^2 + 50x + 2xy^2 - 2x^3 - 8x + 2y^3y' - 2x^2yy' - 8yy' = 0$$
$$4y^3y' - 58yy' - 4x^3 + 42x = 0$$
$$2y^3y' - 29yy' - 2x^3 + 21x = 0$$
$$(2y^3 - 29y)y' = 2x^3 - 21x$$
$$y' = \frac{2x^3 - 21x}{2y^3 - 29y}$$
$$y' = \frac{x(2x^2 - 21)}{y(2y^2 - 29)}$$

§4. Implicit Differentiation   515

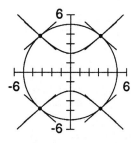

Figure V.19: Four weird points

This is the formula we seek. You might still, however, be left with some questions about this example. Looking at Figure V.19, the graph of the equation
$$(x^2 + y^2 - 25)(y^2 - x^2 - 4) = 0,$$
we see that there are four points at which two curves cross. At these four points one might expect something to go wrong since there are two possible tangents at each of the four points.

These four points can be found by using the fact that they lie on the intersection of the curves defined by
$$x^2 + y^2 - 25 = 0$$
and
$$y^2 - x^2 - 4 = 0.$$
Solving these two equations simultaneously we obtain
$$\begin{aligned} x^2 + y^2 - 25 &= 0 \\ -x^2 + y^2 - 4 &= 0. \end{aligned}$$

Adding these two equations we obtain
$$\begin{aligned} 2y^2 - 29 &= 0 \\ y^2 &= \frac{29}{2} \\ y &= \pm\sqrt{\frac{29}{2}}. \end{aligned}$$

Substituting these possible values for y back into the equation

$$y^2 - x^2 - 4 = 0$$

we get

$$\frac{29}{2} - x^2 - 4 = 0$$

$$\frac{21}{2} - x^2 = 0$$

$$x^2 = \frac{21}{2}$$

$$x = \pm\sqrt{\frac{21}{2}}.$$

Thus the four points are

$$\left(\pm\sqrt{\frac{21}{2}}, \pm\sqrt{\frac{29}{2}}\right).$$

If we substitute any of these four points into the formula obtained earlier for dy/dx

$$\frac{dy}{dx} = \frac{x(2x^2 - 21)}{y(2y^2 - 29)}$$

we get

$$\frac{dy}{dx} = \frac{0}{0}$$

which is not surprising since at these points the idea of a tangent does not make sense.

Exercises

*Exercise V.4.1: Is the point $(3,2)$ on the curve

$$xy = 5?$$

Exercise V.4.2: Is the point $(0,3)$ on the curve

$$\frac{x^2}{4} + \frac{y^2}{9} = 1?$$

§4. Implicit Differentiation

*Exercise V.4.3: Use implicit differentiation to find a formula for dy/dx at any point (x, y) on the curve described by the equation

$$xy = 1.$$

Exercise V.4.4: Use implicit differentiation to find a formula for dy/dx at any point (x, y) on the curve described by the equation

$$(x + y)^2 = 1.$$

Exercise V.4.5: Use implicit differentiation to find a formula for dy/dx at any point (x, y) on the curve described by the equation

$$x^2 + x + y + xy = 4.$$

*Exercise V.4.6: Find the equation of the tangent to the curve

$$\frac{x^2}{4} + \frac{y^2}{9} = 1$$

at the point $(0, 3)$.

Exercise V.4.7: Find the equation of the tangent to the curve $xy = 6$ at the point $(2, 3)$.

*Exercise V.4.8: The graph for the equation

$$(x^2 + y^2 - 25)(y^2 - x^2 - 4) = 0$$

shown in Figures V.16, V.17, V.18, and V.19 consists of two parts—the circle

$$x^2 + y^2 - 25 = 0$$

shown in Figure V.13 and the part

$$y^2 - x^2 - 4 = 0$$

shown in Figure V.14. These two parts correspond to the two factors

$$(x^2 + y^2 - 25)$$

and

$$(y^2 - x^2 - 4)$$

in the product
$$(x^2 + y^2 - 25)(y^2 - x^2 - 4).$$
You would expect that if a point (x, y) that satisfies the equation
$$(x^2 + y^2 - 25)(y^2 - x^2 - 4) = 0$$
lies on the first part
$$x^2 + y^2 - 25 = 0$$
then at that point
$$\frac{dy}{dx} = \frac{x(2x^2 - 21)}{y(2y^2 - 29)}$$
would be $-x/y$, the formula obtained by applying implicit differentiation to the equation
$$x^2 + y^2 - 25 = 0.$$
Verify that this is true.

Similarly, if a point lies on the second part then at that point dy/dx would be x/y, the formula obtained by applying implicit differentiation to the equation
$$y^2 - x^2 - 4 = 0.$$

*Exercise V.4.9:   More generally, the graph of the equation
$$f(x, y)g(x, y) = 0$$
consists of two parts—the graph of
$$f(x, y) = 0$$
and the graph of
$$g(x, y) = 0.$$
Show that applying implicit differentiation to the equation
$$f(x, y)g(x, y) = 0$$
yields the same results for points on the curve
$$f(x, y) = 0$$

§5. Separation of Variables

as applying implicit differentiation to that curve, and yields the same results for points on the curve

$$g(x,y) = 0$$

as applying implicit differentiation to that curve.

*Exercise V.4.10: There are some exceptions to the results described in the last two exercises. If a point lies on both parts then you would expect problems. See Figure V.19. What happens at these points?

§5. Separation of Variables

If we apply implicit differentiation to the equation

$$y^2 = x^3$$

we obtain

$$2yy' = 3x^2.$$

More generally, if we start with an equation of the form

$$f(y) = g(x)$$

and apply implicit differentiation we obtain

$$f'(y)y' = g'(x)$$

or

$$\left(\frac{df}{dy}\right)\left(\frac{dy}{dx}\right) = \frac{dg}{dx}.$$

Any differentiation formula can be read backward to yield an integration formula. This formula leads us to a particularly nice technique for solving differential equations, called **separation of variables**. We illustrate this method with an example.

Example:

Solve the initial value problem

$$y' = xy, \quad y(0) = 2.$$

Answer:

The first step is algebraic

$$y' = xy$$
$$\left(\frac{1}{y}\right) y' = x$$

This equation is of the form

$$f'(y)y' = g'(x)$$

and, hence, comes from

$$f(y) = g(x)$$

by implicit differentiation. Since $f'(y) = 1/y$,

$$f(y) = \int \frac{1}{y} \, dy = \ln|y| + c_1,$$

and, since $g'(x) = x$,

$$g(x) = \int x \, dx = \frac{x^2}{2} + c_2.$$

Thus

$$\ln|y| + c_1 = \frac{x^2}{2} + c_2$$

$$\ln|y| = \frac{x^2}{2} + c_2 - c_1$$

$$\ln|y| = \frac{x^2}{2} + c.$$

where the last line is obtained by using the notation c for the constant $c_1 + c_2$.

This equation describes a relationship between the two variables x and y. This is often the best that we can do. In this case, however, if we apply the exponential function to both sides of this equation we see that

$$|y| = e^{\frac{x^2}{2} + c} = e^{\frac{x^2}{2}} e^c = C e^{\frac{x^2}{2}}$$

§5. Separation of Variables

where $C = e^c$. Thus,
$$y = \pm Ce^{\frac{x^2}{2}}.$$

There is a technical point here that obscures the poetry of our presentation. The constant $C = e^c$ must be positive, since e^c is always positive. The absolute value $|y|$ that came from the fact that
$$\int \frac{dy}{y} = \ln|y|,$$
however, means that we eventually wind up with the equation
$$y = Ce^{\frac{x^2}{2}}$$
where the constant C can be positive or negative.

Finally, we can determine the value of the constant C by substituting the initial condition $y(0) = 2$ into the equation
$$y = Ce^{\frac{x^2}{2}}$$
to get
$$2 = Ce^0 = C,$$
so that the solution of this initial value problem is
$$y = 2e^{\frac{x^2}{2}}.$$

Example:

Solve the initial value problem
$$y' = -\frac{x}{y}, \quad y(3) = 4.$$

Answer:

The first step is algebraic
$$y' = -\frac{x}{y}$$
$$yy' = -x$$

Chapter V. Methods and Applications of Integration

Now that we have rewritten the equation in the form

$$f'(y)y' = g'(x)$$

we find $f(y)$ and $g(x)$ by integrating as follows

$$f(y) = \int f'(y)\, dy$$

$$= \int y\, dy$$

$$= \frac{y^2}{2} + c_1$$

$$g(x) = \int g'(x)\, dx$$

$$= \int -x\, dx$$

$$= -\frac{x^2}{2} + c_2.$$

This leads to the equation

$$\frac{y^2}{2} + c_1 = -\frac{x^2}{2} + c_2$$

$$y^2 + x^2 = 2c_2 - 2c_2.$$

We can write this equation more simply as

$$y^2 + x^2 = c.$$

Using the initial condition $y(3) = 4$ we see that

$$y^2 + x^2 = c$$
$$(4)^2 + (3)^2 = c$$
$$25 = c$$

§5. Separation of Variables

leading to the equation
$$y^2 + x^2 = 25.$$

In this case we can find an explicit formula for y as a function of x.

$$\begin{aligned} y^2 + x^2 &= 25 \\ y^2 &= 25 - x^2 \\ y &= \pm\sqrt{25 - x^2} \end{aligned}$$

giving us two possibilities. The initial condition $y(3) = 4$ implies that we should choose the positive square root. Thus

$$y = \sqrt{25 - x^2}.$$

The method of separation of variables begins by expressing an equation like
$$\frac{dy}{dx} = xy$$
in the form
$$f(y)\frac{dy}{dx} = g(x),$$
for example,
$$\frac{1}{y}\frac{dy}{dx} = x.$$

We often write this in the form
$$f(y)\, dy = g(x)\, dx,$$
for example,
$$\frac{dy}{y} = x\, dx.$$

When we use this notation we think of dy as an abbreviation for $\frac{dy}{dx}dx$. This notation is suggestive of the next step

$$\int f(y)\, dy = \int g(x)\, dx.$$

Exercises

Exercise V.5.1: Use the method of separation of variables to find a function $p(t)$ such that
$$\frac{dp}{dt} = kp, \quad p(0) = 1{,}000, p(1) = 800$$
where k is a constant.

*Exercise V.5.2: Solve the initial value problem
$$\frac{dy}{dx} = \frac{y}{x}, \quad y(3) = 5.$$

Exercise V.5.3: Solve the initial value problem
$$\frac{dy}{dx} = y^2, \quad y(0) = 1.$$

Exercise V.5.4: Solve the initial value problem
$$\frac{dy}{dx} = y \cos x, \quad y(0) = 1.$$

*Exercise V.5.5: Experimentation has shown that when a hot object is cooling in a room of constant temperature A then its temperature $T(t)$ changes according to the differential equation
$$\frac{dT}{dt} = k(A - T)$$
where k is a positive constant. This experimental observation is called **Newton's Law of Cooling**. Suppose that a cup of coffee is cooling in a room whose temperature is 70 degrees Fahrenheit. Suppose that at time $t = 0$ the temperature of the coffee is 212 degrees Fahrenheit and that ten minutes later its temperature is 180 degrees Fahrenheit. What will its temperature be at time $t = 25$? Find
$$\lim_{t \to +\infty} T(t).$$
Does your answer match your intuition?

§6. The Logistic Equation and Partial Fractions

In chapter I we studied population dynamics for a species like temperate zone insects that had discrete generations. We described the population for such a species by a sequence of numbers $p_1, p_2, p_3, \ldots$ where p_n was the population of the nth generation. Most species, however, do not have discrete generations. We describe the population of such a species by a function $p(t)$ rather than by a sequence.

The same distinction can be made in economics. In chapter I we studied changing prices for a product called Byties that had discrete generations in the same mathematical sense as temperate zone insects—Byties were baked fresh each morning and had to be sold the same day. Many other products, however, do not have discrete generations. We describe the price of such a product by a function rather than by a sequence.

Models involving continuously changing functions are called **continuous** models. In a continuous model change is described by a differential equation or **continuous dynamical system**

$$\frac{dp}{dt} = f(p)$$

rather than by a discrete dynamical system

$$p_{n+1} = f(p_n).$$

Continuous Exponential Models

The simplest continuous model of population growth assumes that population grows at a constant percentage rate. For example, if we measure time in years then the differential equation

$$\frac{dp}{dt} = 0.02p$$

says that p is growing at the constant rate of 2% per year.[4] More generally, a **continuous exponential model** is an initial value problem of the form

$$\frac{dp}{dt} = kp, \quad p(t_0) = p_0$$

---

[4] The *effective* rate of growth would be greater than 2%. See section III.7.

where k is a constant called the **relative growth rate** and p_0 is the **initial value** of p at time t_0.

Using the method of separation of variables we can solve such a differential equation as follows:

$$p' = kp$$
$$\left(\frac{1}{p}\right)p' = k$$
$$\ln|p| = kt + c$$
$$|p| = e^{kt+c}$$
$$p = Ce^{kt}$$

where the constant C is $\pm e^c$. For population models C is never negative since negative population makes no sense.

These models behave very much like the discrete exponential models discussed in chapter I. If the constant k is positive then

$$\lim_{t \to \infty} Ce^{kt} = +\infty$$

and if the constant k is negative then

$$\lim_{t \to \infty} Ce^{kt} = 0.$$

For population models neither of these situations is very interesting. The first is uninteresting because it is unrealistic. Populations exist in finite habitats and cannot rise without bound. The second is uninteresting because it describes a population that will eventually die out. There are very few such populations around.

Exercises

A population whose natural relative growth rate in a particular habitat is negative will eventually die out unless some outside factor intervenes. One possible outside factor is immigration. We can modify the exponential model above by adding immigration to obtain the differential equation

$$p' = kp + m.$$

The positive constant m is called the **rate of immigration**. This model describes a situation in which individuals are immigrating at a constant rate.

§6. The Logistic Equation and Partial Fractions

Exercise V.6.1: Suppose a particular population can be described by the initial value problem

$$p' = -0.2p + 100, \quad p(0) = 100.$$

- Find the function $p(t)$ by solving this initial value problem.

- Find the longterm behavior of this population by determining

$$\lim_{t \to \infty} p(t).$$

Exercise V.6.2: Suppose a particular population can be described by the initial value problem

$$p' = -0.2p + 100, \quad p(0) = 1000.$$

- Find the function $p(t)$ by solving this initial value problem.

- Find the longterm behavior of this population by determining

$$\lim_{t \to \infty} p(t).$$

Exercise V.6.3: Suppose a particular population can be described by the initial value problem

$$p' = 0.2p + 100, \quad p(0) = 100.$$

- Find the function $p(t)$ by solving this initial value problem.

- Find the longterm behavior of this population by determining

$$\lim_{t \to \infty} p(t).$$

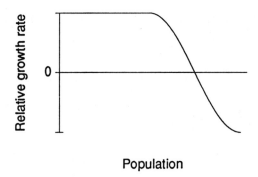

Figure V.20: Possible relative growth rate

Continuous Logistic Models

Exponential models are not very realistic because they assume that the relative rate of growth is constant. In any realistic situation the habitat is finite with a finite amount of food, water, and the other necessities of life. If the population is low then there will be plenty of these resources for all—the species will be healthy, birth rates high, and death rates low. As a result the relative population growth rate will be high. When the population is high, however, there will not be enough resources to go around—health will suffer, the birth rate will be lower and the death rate higher. As a result the relative population growth rate will be low. When the population is very high the relative population growth rate will be negative because of extreme shortages of food, water, and shelter. Thus the relative population growth rate is usually a function $k(p)$ roughly like the one shown in Figure V.20.

Although this kind of relative growth rate is the most common, there are other possibilities. For example, the relative population growth rate for a species that hunts in packs might look roughly like the function shown in Figure V.21. In this situation the growth rate is negative when the population is very low because the population is not large enough to form efficient hunting packs. When the population is larger the relative growth rate is positive because the population can form efficient hunting packs and there is plenty of food, water, and shelter to go around, but when the population is extremely large the relative growth rate is negative because there is no longer enough food, water, and shelter.

In this section we are interested in the simplest relative growth rate

§6. The Logistic Equation and Partial Fractions

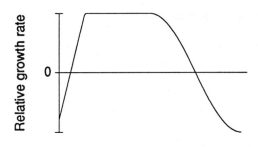

Figure V.21: Relative growth rate for a pack-hunting species

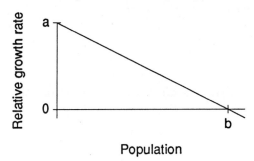

Figure V.22: $k(p) = a(1 - p/b)$

functions, functions of the form

$$k(p) = a(1 - p/b)$$

shown in Figure V.22. The constant a represents the theoretical highest possible relative growth. When the population is extremely low then $k(p)$ is close to a. The constant b is sometimes called the **carrying capacity** of the habitat. When $p = b$ the relative growth rate is zero. This is the population level that the habitat can support in equilibrium. There is exactly enough food, water, and shelter so that at this population level the birth rate and death rate are in perfect balance.

Models based on this relative growth rate function are called **continuous logistic models** and are described by the differential equation

$$\frac{dp}{dt} = a(1 - p/b)p.$$

This equation is sometimes called the **logistic equation**.

530 CHAPTER V. METHODS AND APPLICATIONS OF INTEGRATION

When we studied discrete dynamical systems

$$p_{n+1} = f(p_n)$$

one of the most important concepts was the concept of an equilibrium point, a solution of the equation

$$p = f(p),$$

because if p_n is at an equilibrium then p_{n+1} and all subsequent terms of the sequence will remain there. We have a similar situation here. Given a continuous dynamical system

$$\frac{dp}{dt} = f(p)$$

an **equilibrium point** is a solution of the equation

$$f(p) = 0$$

because if $p(t_0)$ is at such a point then $p(t)$ will remain there forever.

The logistic equation

$$\frac{dp}{dt} = a(1 - p/b)p$$

has two equilibrium points, $p = 0$ and $p = b$, corresponding to zero population and a population equal to the carrying capacity.

The method of separation of variables together with a new technique called the method of partial fractions enables us to solve initial value problems involving the logistic equation. We start as follows.

$$\frac{dp}{dt} = a(1 - p/b)p$$

$$\left(\frac{1}{p(1 - p/b)}\right) dp = a\, dt$$

The right side of this equation is easy to integrate. The left side can be integrated using the method of partial fractions.

§6. The Logistic Equation and Partial Fractions

The Method of Partial Fractions

The **method of partial fractions** can be used to find the integral of any rational function, that is, any function of the form

$$\frac{f(x)}{g(x)}$$

where $f(x)$ and $g(x)$ are polynomials. The best way to learn this method is from examples.

Example:

Find the integral

$$\int \left(\frac{1}{(p-1)(p-2)}\right) dp.$$

Answer:

We already know how to find the integrals

$$\int \left(\frac{1}{p-1}\right) dp$$

and

$$\int \left(\frac{1}{p-2}\right) dp.$$

The idea behind the method of partial fractions is to break the original fraction

$$\frac{1}{(p-1)(p-2)}$$

into a sum of simpler fractions

$$\frac{1}{(p-1)(p-2)} = \frac{A}{p-1} + \frac{B}{p-2}$$

where A and B are constants. We work backward from the equation above. We want to find A and B so that

$$\frac{1}{(p-1)(p-2)} = \frac{A}{p-1} + \frac{B}{p-2}$$

$$\frac{1}{(p-1)(p-2)} = \frac{A(p-2) + B(p-1)}{(p-1)(p-2)}$$

$$1 = A(p-2) + B(p-1)$$

$$1 = (A+B)p - 2A - B$$

This last equation must be true for all values of p. Thus

$$A + B = 0$$

and

$$-2A - B = 1.$$

From the first equation we see that $A = -B$ and substituting $A = -B$ into the second equation we get

$$-2(-B) - B = 1$$
$$B = 1$$

and since $A = -B$

$$A = -1.$$

Thus

$$\frac{1}{(p-1)(p-2)} = -\frac{1}{p-1} + \frac{1}{p-2}$$

and

$$\int \frac{1}{(p-1)(p-2)} \, dp = \int \left(-\frac{1}{p-1} + \frac{1}{p-2} \right) dp$$

$$= -\ln|p-1| + \ln|p-2| + c$$

§6. The Logistic Equation and Partial Fractions

We found the constants A and B such that
$$1 = A(p-2) + B(p-1)$$
by doing some algebra that lead to the pair of equations
$$A + B = 0$$
$$-2A - B = 1.$$

There is, however, a shortcut. If the equation
$$1 = A(p-2) + B(p-1)$$
is true for all values of p then, in particular, it is true for $p = 1$. Substituting $p = 1$ into this equation we obtain
$$1 = A(1-2) = -A$$
so that $A = -1$. Similarly, substituting $p = 2$ into the same equation we obtain
$$1 = B(2-1) = B$$
so that $B = 1$. We chose to substitute $p = 1$ and $p = 2$ into this equation because these values of p simplify the right hand side of the equation.

Example:

Consider the rational function
$$\frac{x^2 + 7x - 3}{(x-1)^2(x+2)}.$$
We want to express this fraction as a sum of the form
$$\frac{x^2 + 7x - 3}{(x-1)^2(x+2)} = \frac{A}{x-1} + \frac{B}{(x-1)^2} + \frac{C}{x+2}$$
The first step is to write the right side as a single fraction.
$$\frac{x^2 + 7x - 3}{(x-1)^2(x+2)} = \frac{A}{x-1} + \frac{B}{(x-1)^2} + \frac{C}{x+2}$$
$$= \frac{A(x-1)(x+2) + B(x+2) + C(x-1)^2}{(x-1)^2(x+2)}$$

Thus
$$x^2 + 7x - 3 = A(x-1)(x+2) + B(x+2) + C(x-1)^2.$$

If we substitute $x = 1$ into this equation we get
$$5 = 3B.$$

So
$$B = \frac{5}{3}.$$

If we substitute $x = -2$ into the same equation we get
$$-13 = 9C.$$

So
$$C = -\frac{13}{9}.$$

Now,
$$\begin{aligned}x^2 + 7x - 3 &= A(x-1)(x+2) + B(x+2) + C(x-1)^2 \\ &= A(x^2 + x - 2) + Bx + 2B + C(x^2 - 2x + 1) \\ &= (A+C)x^2 + (A+B-2C)x + (-2A+2B+C).\end{aligned}$$

So that
$$A + C = 1$$
and, since $C = -13/9$,
$$A - \frac{13}{9} = 1$$
and
$$A = \frac{22}{9}.$$

Thus
$$\frac{x^2 + 7x - 3}{(x-1)^2(x+2)} = \left(\frac{22}{9}\right)\left(\frac{1}{x-1}\right) + \left(\frac{5}{3}\right)\left(\frac{1}{(x-1)^2}\right) - \left(\frac{13}{9}\right)\left(\frac{1}{x+2}\right).$$

§6. The Logistic Equation and Partial Fractions

Example: Consider the rational function

$$\frac{x^2+4}{(x^2+x+1)(x-1)}$$

We want to express fraction this as a sum of the form

$$\frac{x^2+4}{(x^2+x+1)(x-1)} = \frac{Ax+B}{x^2+x+1} + \frac{C}{x-1}$$

The first step is to write the right side as a single fraction.

$$\frac{x^2+4}{(x^2+x+1)(x-1)} = \frac{Ax+B}{x^2+x+1} + \frac{C}{x-1}$$

$$= \frac{(Ax+B)(x-1) + C(x^2+x+1)}{(x-1)(x^2+x+1)}.$$

Thus,

$$x^2 + 4 = (Ax+B)(x-1) + C(x^2+x+1).$$

If we substitute $x = 1$ into this equation we see that

$$5 = 3C.$$

So that

$$C = \frac{5}{3}.$$

Now,

$$x^2 + 4 = (Ax+B)(x-1) + C(x^2+x+1)$$

$$= (A+C)x^2 + (B-A+C)x + (-B+C).$$

So that

$$A + C = 1$$
$$B - A + C = 0$$
$$-B + C = 4.$$

Now since
$$A + C = 1$$
we see that
$$A + \frac{5}{3} = 1$$
and
$$A = -\frac{2}{3}.$$
Finally since
$$-B + C = 4$$
we see that
$$-B + \frac{5}{3} = 4$$
and
$$B = -\frac{7}{3}.$$
So that finally
$$\frac{x^2 + 4}{(x^2 + x + 1)(x - 1)} = \frac{-\left(\frac{2}{3}\right)x - \left(\frac{7}{3}\right)}{x^2 + x + 1} + \left(\frac{5}{3}\right)\left(\frac{1}{x - 1}\right).$$

Exercises

*Exercise V.6.4:
Factor the denominator of the function
$$\frac{x - 2}{x^2 + 3x + 2}$$
and then write the function as a sum of two appropriate terms

*Exercise V.6.5:
Factor the denominator of the function
$$\frac{x^2 + 2x + 3}{x^3 + x}$$
and then write the function as a sum of two appropriate terms.

Exercise V.6.6:
Express the function
$$\frac{x^2 - 2x - 4}{(x - 2)^3}$$
as a sum of three appropriate terms.

§6. The Logistic Equation and Partial Fractions

Integrating the Simple Terms

The ideas discussed in the preceding section can be used to reduce the problem of integrating a rational function to several problems involving terms of the form

$$\frac{q}{bx+c},$$

$$\frac{q}{(bx+c)^n},$$

$$\frac{px+q}{ax^2+bx+c},$$

or

$$\frac{px+q}{(ax^2+bx+c)^n}.$$

In this section we discuss integrating terms in which the denominator is either of the form $(bx+c)$ or $(bx+c)^n$. These integrals can be evaluated using the substitution $u = bx + c$. Here are two examples.

Example:

Evaluate the integral

$$\int \frac{2}{3x-1}\, dx$$

Let $u = 3x - 1$ and notice that $du = 3dx$ so

$$\int \frac{2}{3x-1}\, dx = \left(\frac{2}{3}\right)\int \frac{3}{3x-1}\, dx$$

$$= \left(\frac{2}{3}\right)\int \frac{1}{u}\, du$$

$$= \left(\frac{2}{3}\right)\ln|u|$$

$$= \left(\frac{2}{3}\right)\ln|3x-1| + c$$

Example:

Evaluate the integral

$$\int \frac{4}{(-x+6)^3}\, dx$$

Let $u = -x + 6$ and notice that $du = -dx$ so that

$$\int \frac{4}{(-x+6)^3}\, dx = -4 \int \frac{-1}{(-x+6)^3}\, dx$$

$$= -4 \int u^{-3}\, du$$

$$= -\frac{4u^{-2}}{-2}$$

$$= 2(-x+6)^{-2} + c.$$

Exercises

Find the following integrals

*Exercise V.6.7:

$$\int \frac{2x-3}{x^2 - 5x + 4}\, dx$$

Exercise V.6.8:

$$\int \frac{x^2 - 2x + 1}{x^3 + x^2 - x - 1}\, dx$$

Exercise V.6.9:

$$\int \frac{x^3 - 3x + 2}{(x^2 - 4)(x^2 - 9)}\, dx$$

The method of partial fractions is exactly what is needed to solve logistic equations.

§6. The Logistic Equation and Partial Fractions

Example:

Solve the initial value problem

$$p' = p(2-p), \quad p(0) = 1.$$

First we write this as

$$\left(\frac{1}{p(2-p)}\right) p' = 1.$$

Integrating the right side is easy

$$\int 1 \, dt = t + c.$$

To integrate the left side

$$\int \left(\frac{1}{p(2-p)}\right) p' \, dt = \int \left(\frac{1}{p(2-p)}\right) dp$$

we use the method of partial fractions.

$$\frac{1}{p(2-p)} = \frac{A}{p} + \frac{B}{2-p}$$

$$\frac{1}{p(2-p)} = \frac{A(2-p) + Bp}{p(2-p)}$$

$$1 = A(2-p) + Bp$$

Substituting $p = 0$ in this last equation we see

$$1 = 2A$$
$$A = 1/2$$

and substituting $p = 2$ in the same equation,

$$1 = B2$$
$$B = 1/2.$$

Thus
$$\frac{1}{p(2-p)} = \frac{1}{2}\left(\frac{1}{p}\right) + \frac{1}{2}\left(\frac{1}{2-p}\right)$$
and
$$\int \frac{1}{p(2-p)}\, dp = \int \frac{1}{2}\left(\frac{1}{p}\right) dp + \int \frac{1}{2}\left(\frac{1}{2-p}\right) dp$$
$$= \frac{1}{2}\ln|p| - \frac{1}{2}\ln|2-p|$$
$$= \ln\left|\frac{p}{2-p}\right|^{\frac{1}{2}}$$

Thus
$$\ln\left|\frac{p}{2-p}\right|^{\frac{1}{2}} = t + c$$
and applying the function exp to both sides
$$\left|\frac{p}{2-p}\right|^{\frac{1}{2}} = e^{t+c}$$
$$\left|\frac{p}{2-p}\right| = e^{2t+2c}$$
$$\frac{p}{2-p} = Ce^{2t}$$
$$p = 2Ce^{2t} - pCe^{2t}$$
$$p + pCe^{2t} = 2Ce^{2t}$$
$$p = \frac{2Ce^{2t}}{1 + Ce^{2t}}$$

From the initial condition $p(0) = 1$ we see that
$$1 = \frac{2Ce^0}{1 + Ce^0}$$

§6. The Logistic Equation and Partial Fractions

$$1 = \frac{2C}{1+C}$$

$$1 + C = 2C$$

$$1 = C$$

and finally

$$p = \frac{2e^{2t}}{1 + e^{2t}}.$$

It is interesting to look at the longterm behavior of this solution using L'Hôpital's Rule

$$\lim_{t \to \infty} \frac{2e^{2t}}{1 + e^{2t}} = \lim_{t \to \infty} \frac{4e^{2t}}{2e^{2t}} = 2.$$

Exercises

*Exercise V.6.10:

- Find the solution of the initial value problem

$$\frac{dp}{dt} = 2(1 - .001p)p, \quad p(0) = 100.$$

- Find

$$\lim_{t \to \infty} p(t).$$

- Sketch the slope field for this differential equation and check that your answer above agrees with what would be expected on the basis of the slope field.

Exercise V.6.11:

- Find the solution of the initial value problem

$$\frac{dp}{dt} = 2(1 - .001p)p, \quad p(0) = 2000.$$

- Find
$$\lim_{t\to\infty} p(t).$$

- Sketch the slope field for this differential equation and check that your answer above agrees with what would be expected on the basis of the slope field.

Exercise V.6.12:

- Find the solution of the initial value problem
$$\frac{dp}{dt} = 2(1 - .001p)p, \quad p(0) = 1000.$$

- Find
$$\lim_{t\to\infty} p(t).$$

- Sketch the slope field for this differential equation and check that your answer above agrees with what would be expected on the basis of the slope field.

Exercise V.6.13:

- Find the solution of the initial value problem
$$\frac{dp}{dt} = 3(1 - .002p)p, \quad p(0) = 100.$$

- Find
$$\lim_{t\to\infty} p(t).$$

- Sketch the slope field for this differential equation and check that your answer above agrees with what would be expected on the basis of the slope field.

Exercise V.6.14:

- Find the solution of the initial value problem
$$\frac{dp}{dt} = 3(1 - .002p)p, \quad p(0) = 1000.$$

§6. The Logistic Equation and Partial Fractions

- Find
$$\lim_{t\to\infty} p(t).$$

- Sketch the slope field for this differential equation and check that your answer above agrees with what would be expected on the basis of the slope field.

Exercise V.6.15:

- Find the solution of the initial value problem
$$\frac{dp}{dt} = 3(1 - .002p)p, \quad p(0) = 500.$$

- Find
$$\lim_{t\to\infty} p(t).$$

- Sketch the slope field for this differential equation and check that your answer above agrees with what would be expected on the basis of the slope field.

*Exercise V.6.16:

For a discrete logistic dynamical system
$$p_{n+1} = a(1 - p_n)p_n$$
the positive constant a has a very strong impact on the behavior of the system. Find the solution of the differential equation
$$\frac{dp}{dt} = a(1 - p)p$$
and compare the effect of the constant a on the behavior of a continuous logistic model with the effect of the constant a on the behavior of a discrete logistic model.

Mathematica and Computer Algebra Systems

Computer algebra systems like *Mathematica* can be powerful tools for studying problems like the ones we studied in this section. First, because the decomposition of a rational function into a sum of simple terms is an important tool in other situations besides integration *Mathmeatica* has a procedure `Apart` that will do this. The following *Mathematica* session illustrates this with using examples from earlier in this section.

```
Apart[1/((p-1)(p-2))]
```

$$\frac{1}{-2+p} - \frac{1}{-1+p}$$

```
Apart[(x^2 + 7x - 3)/((x-1)^2(x+2))]
```

$$\frac{5}{3(-1+x)^2} + \frac{22}{9(-1+x)} - \frac{13}{9(2+x)}$$

```
Apart[(x^2 + 4)/((x^2 + x + 1)(x - 1))]
```

$$\frac{5}{3(-1+x)} - \frac{7+2x}{3(1+x+x^2)}$$

You can also use the ability of *Mathematica* to perform algebraic manipulations to help with the calculations used to solve a problem like the preceding initial value problem. Refer back to that example as you look at the following lines of *Mathematica*. The first step is finding the integral

$$\int \left(\frac{1}{p(2-p)}\right) dp$$

which *Mathematica* does in a single line.

```
Integrate[1/(p(2 - p)),p]
```

$$\frac{-\text{Log}[-2+p]}{2} + \frac{\text{Log}[p]}{2}$$

§6. THE LOGISTIC EQUATION AND PARTIAL FRACTIONS

This gives us the equation

$$\frac{-\log(-2+p)}{2} + \frac{\log(p)}{2} = t + c$$

which leads to

$$\exp(\frac{-\log(-2+p)}{2} + \frac{\log(p)}{2}) = C \exp t.$$

We want to solve this equation for p. *Mathematica* can do this in a single step[5].

```
Solve[Exp[%] == C Exp[t],p]
```

```
                 2
{{p -> 2 - ------------}}
              2   2 t
            1 - C  E
```

Replacing the constant C^2 by the constant C, we think of this as

$$p = 2 - \frac{2}{1 - C\exp(2t)}.$$

Using the initial condition $p(0) = 1$, we want to solve the equation

$$1 = 2 - \frac{2}{1 - C}$$

for C. Again, this is a simple single step for *Mathematica*.

```
Solve[1 == 2 - 2/(1 - C),C]
```

```
{{C -> -1}}
```

Thus, the function $p(t)$ is

$$p(t) = 2 - \frac{2}{e^{2t}}.$$

We can use *Mathematica* to draw a graph of this function as shown below.

```
p[t_] := 2 - 2/(1 Exp[2 t])

Plot[p[t], {t, 0, 5}]
```

---

[5]Note the use of the symbol % to refer to the result of the previous *Mathematica* computation.

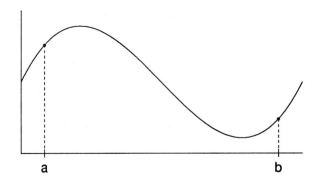

Figure V.23: Curve given by $y = f(x)$

§7. Arclength

The Arclength of a Curve Given by $y = f(x)$

In this section we are interested in the length of a curve. We begin with curves that can be described by a function like $y = f(x)$. We want to find the length of a piece of such a curve between the lines $x = a$ and $x = b$ as shown in Figure V.23.

We can approximate a curve like the one in Figure V.23 by a curve made up of straight line segments as shown in Figure V.24. Such an approximation to a curve is called a **piecewise linear** approximation. By using a large number of very short straight line segments we can obtain a very good approximation of the original curve as shown in Figure V.25.

We obtain a sequence of estimates for the length of the original arc as follows. For each n divide the interval $[a, b]$ up into n equal pieces each of which has length $h = (b-a)/n$. Let $x_0 = a$, $x_1 = a + h, \ldots x_n = b$. Figure V.24 illustrates this for $n = 5$.

The length of the ith straight line segment can be calculated by the Pythagorean Theorem. Refer to Figure V.26.

$$\text{Length} = \sqrt{((x_i - x_{i-1})^2 + (f(x_i) - f(x_{i-1}))^2}$$

$$= \sqrt{1 + \left(\frac{f(x_i) - f(x_{i-1})}{x_i - x_{i-1}}\right)^2} (x_i - x_{i-1})$$

§7. Arclength

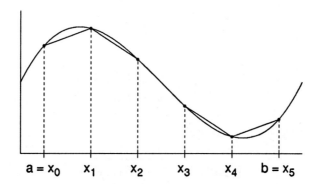

Figure V.24: Piecewise linear approximation

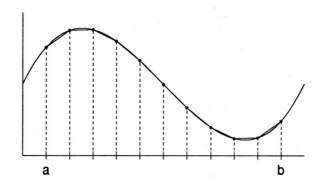

Figure V.25: Better piecewise linear approximation

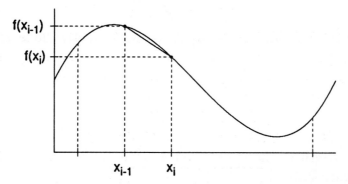

Figure V.26: Finding the length of one segment

Now, by the Mean Value Theorem, there is a point s_i in the interval $[x_{i-1}, x_i]$ such that

$$f'(s_i) = \left(\frac{f(x_i) - f(x_{i-1})}{x_i - x_{i-1}}\right).$$

and, since $h = x_i - x_{i-1}$, we get

$$\text{Length} = \sqrt{1 + f'(s_i)^2}\, h$$

as the length of the ith segment. To find the length of the entire piecewise approximation we calculate the sum

$$\sum_{i=1}^{n} \sqrt{1 + f'(s_i)^2}\, h.$$

Notice that this is a Riemann sum for the function

$$g(x) = \sqrt{1 + f'(x)^2}$$

with sample points s_i. Taking the limit we see that the length of the original curve is

$$\int_a^b \sqrt{1 + f'(x)^2}\, dx$$

or

$$\int_a^b \sqrt{1 + \left(\frac{dy}{dx}\right)^2}\, dx.$$

In theory this formula enables us to compute exactly the length of a curve described as shown in Figure V.23. In practice the integrals that arise from arclength problems are usually very difficult or impossible to evaluate except numerically.

Example:

Find the length of the curve $f(x) = x^2$ from $x = 0$ to $x = 1$ as shown in Figure V.27.

Answer:

§7. Arclength

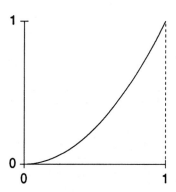

Figure V.27: $y = x^2$

Notice that
$$\frac{dy}{dx} = 2x$$
so the integral that we need to evaluate is
$$\int_0^1 \sqrt{1 + 4x^2}\, dx.$$

This integral can be evaluated numerically. The result is 1.478943. It is also one of the very few integrals that arises from an arclength problem that can be evaluated via antidifferentiation. A technique for finding the antiderivative of
$$\sqrt{1 + 4x^2}$$
is discussed in the next section.

Arclength of a Curve Given by $x(t)$, $y(t)$

Many curves can be described by a pair of functions $x(t)$ and $y(t)$. For example, the trajectory of a moving object can be described by functions $x(t)$ and $y(t)$ that describe its location at each time t. As another example, a circle of radius R is described by the pair of functions

$$\begin{aligned} x(t) &= R\cos t \\ y(t) &= R\sin t \end{aligned}$$

CHAPTER V. METHODS AND APPLICATIONS OF INTEGRATION

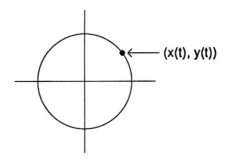

Figure V.28: Circle $x(t) = R\cos t$, $y(t) = R\sin t$

as t runs from $t = 0$ to $t = 2\pi$. See Figure V.28.

The formula for the arclength of a curve that is described by a pair of functions $x(t)$ and $y(t)$ for $t = a$ to $t = b$ is

$$\int_a^b \sqrt{x'(t)^2 + y'(t)^2}\, dt.$$

Using this formula we can find the length of a circle of radius R using the description above as follows.

Example:

Find the length of the circle of radius R described by

$$x(t) = R\cos t$$
$$y(t) = R\sin t$$

as t runs from $t = 0$ to $t = 2\pi$ as shown in Figure V.28.

Answer:

Notice that

$$x'(t) = -R\sin t$$
$$y'(t) = R\cos t.$$

§7. Arclength

So
$$\int_0^{2\pi} \sqrt{x'(t)^2 + y'(t)^2}\, dx = \int_0^{2\pi} \sqrt{R^2 \sin^2 t + R^2 \cos^2 t}\, dx$$
$$= \int_0^{2\pi} R\, dx$$
$$= 2\pi R,$$

the familiar formula.

Exercises

For many of the exercises below you will have to evaluate integrals numerically using either a computer or a calculator.

*Exercise V.7.1: Find the arclength of the curve $y = \sin x$ between $x = 0$ and $x = 2\pi$.

*Exercise V.7.2: Find the arclength of the curve $y = 2x$ between $x = 0$ and $x = 1$ two different ways, by using the formula we developed above and by elementary geometry.

*Exercise V.7.3: Find the arclength of the curve $y = mx$ between $x = a$ and $x = b$ two different ways, by using the formula we developed above and by elementary geometry.

*Exercise V.7.4: Find the arclength of the ellipse described by

$$x(t) = \cos t$$
$$y(t) = 2\sin t$$

for $t = 0$ to $t = 2\pi$.

Exercise V.7.5: Find the arclength of the curve $y = \ln \cos x$ between $x = 0$ and $x = 1$. The main claim to fame of this particular curve is that it is one of the few which lead to an integral that can be evaluated exactly.

Exercise V.7.6: Find the arclength of the spiral described by

$$x(t) = t\cos 2\pi t$$
$$y(t) = t\sin 2\pi t$$

552 CHAPTER V. METHODS AND APPLICATIONS OF INTEGRATION

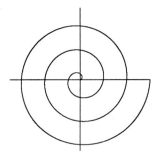

Figure V.29: Spiral

for $t = 0$ to $t = 3$. See Figure V.29.

*Exercise V.7.7:   Using elementary geometry show that the length of the straight line described by the linear function $y = mx + k$ between $x = a$ and $x = b$ is $(b-a)\sqrt{1+m^2}$.

*Exercise V.7.8:   Using the result of the preceding exercise explain the arclength formula

$$\int_a^b \sqrt{1 + f'(x)^2}\, dx$$

as an application of integration rather than multiplication.

*Exercise V.7.9:   Consider a curve described by the pair of functions $x(t)$ and $y(t)$ for t between a and b. Suppose that $f(s)$ is any increasing function such that $f(p) = a$ and $f(q) = b$. Then the curve described by the pair of functions $x(f(s))$ and $y(f(s))$ for s between p and q is physically the same as the original curve. Thus, you would expect this curve to have the same arclength as the original curve. Prove that this is true.

§8. Integration Miscellany

In this section we discuss very briefly a number of integration techniques. We discuss them briefly because programs like *Mathematica*, Maple, MathCad, and DERIVE are able to solve integration problems involving these techniques so easily that, although it is important to

§8. Integration Miscellany

understand how these techniques could be used, it is no longer essential for us to do all the sometimes tedious calculations.

Integrals of the Form $\int \sin^n \theta \cos^m \theta \, d\theta$

The simplest of these integrals to evaluate are those in which either one or both of the two exponents, m and n, is odd. The following example is typical.

Example: Find the integral

$$\int \sin^8 \theta \cos^5 \theta \, d\theta.$$

Answer:

The first step is to rewrite $\cos^5 \theta$ as follows

$$\int \sin^8 \theta \cos^5 \theta \, d\theta = \int \sin^8 \theta \cos^4 \theta \cos \theta \, d\theta$$
$$= \int \sin^8 \theta \, (\cos^2 \theta)^2 \cos \theta \, d\theta$$
$$= \int \sin^8 \theta \, (1 - \sin^2 \theta)^2 \cos \theta \, d\theta$$

Now we make the substitution $x = \sin \theta$ so that $dx = \cos \theta \, d\theta$ and the integral becomes

$$\int \sin^8 \theta \, (1 - \sin^2 \theta)^2 \cos \theta \, d\theta = \int x^8 (1 - x^2)^2 \, dx,$$

which is easy to evaluate.

This same method can be used to integrate the secant.

Example: Find

$$\int \sec \theta \, d\theta.$$

Answer:

CHAPTER V. METHODS AND APPLICATIONS OF INTEGRATION

$$\int \sec\theta \, d\theta = \int \frac{1}{\cos\theta} \, d\theta$$

$$= \int \frac{\cos\theta}{\cos^2\theta} \, d\theta$$

$$= \int \frac{1}{1 - \sin^2\theta} \cos\theta \, d\theta$$

Now we make the substitution $x = \sin\theta$ so that $dx = \cos\theta \, d\theta$ and the integral becomes.

$$\int \frac{1}{1 - \sin^2\theta} \cos\theta \, d\theta = \int \frac{1}{1 - x^2} \, dx$$

which is easily evaluated by the method of partial fractions.

$$\int \frac{dx}{1 - x^2} = \frac{1}{2} \ln\left|\frac{1 + x}{1 - x}\right| + c$$

and since $x = \sin\theta$

$$\int \sec\theta \, d\theta = \frac{1}{2} \ln\left|\frac{1 + x}{1 - x}\right| + c$$

$$= \frac{1}{2} \ln\left|\frac{1 + \sin\theta}{1 - \sin\theta}\right| + c.$$

The absolute value is unnecessary since neither the numerator nor the denominator can be negative. So

$$\int \sec\theta \, d\theta = \frac{1}{2} \ln\left(\frac{1 + \sin\theta}{1 - \sin\theta}\right) + c$$

The method above can be applied when one of the two exponents is odd. If both exponents are even then the problem is more difficult. In this case we use the identities

$$\sin^2\theta = \frac{1 - \cos 2\theta}{2} \quad \text{and} \quad \cos^2\theta = \frac{1 + \cos 2\theta}{2}$$

§8. Integration Miscellany

as shown in the following example.

Example: Find the integral

$$\int \sin^4 \theta \cos^2 \theta \, d\theta.$$

Answer:

Using the identities above we get

$$\int \sin^4 \theta \cos^2 \theta \, d\theta = \int \left(\frac{1 - \cos 2\theta}{2}\right)^2 \left(\frac{1 + \cos 2\theta}{2}\right) d\theta$$

$$= \frac{1}{8} \int (1 - \cos 2\theta - \cos^2 2\theta + \cos^3 2\theta) \, d\theta.$$

Each of the terms in the last line can be easily integrated. The only term that is at all difficult is

$$\int \cos^2 2\theta \, d\theta.$$

This term is handled using the fact that

$$\cos^2 2\theta = \frac{1 + \cos 4\theta}{2}.$$

Exercises

Evaluate the following integrals.

*Exercise V.8.1:

$$\int_0^{\frac{\pi}{2}} \cos^3 \theta \sin^3 \theta \, d\theta$$

Exercise V.8.2:

$$\int_0^{\frac{\pi}{2}} \cos^2 \theta \sin^3 \theta \, d\theta$$

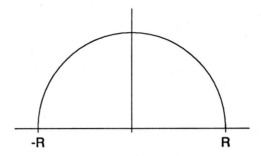

Figure V.30: Semicircle of radius R

*Exercise V.8.3:

$$\int_0^{\frac{\pi}{2}} \cos^2\theta \sin^2\theta \, d\theta$$

*Exercise V.8.4:

$$\int \sec^3\theta \, d\theta$$

Exercise V.8.5:

$$\int_0^{\frac{\pi}{4}} \frac{\sin^5\theta}{\cos^9\theta} \, d\theta$$

Trigonometric Substitution

Consider the following problem.

Example:

Find the area of a circle of radius R.

Answer:

The top of a semicircle of radius R is described by the function

$$y = \sqrt{R^2 - x^2}$$

§8. Integration Miscellany

as shown in Figure V.30. Thus the area of a semicircle of radius R is

$$\int_{-R}^{R} \sqrt{R^2 - x^2}\, dx$$

and the area of a circle of radius R is just twice the area of a semicircle of radius R or

$$2\int_{-R}^{R} \sqrt{R^2 - x^2}\, dx.$$

The only problem is evaluating the integral

$$\int_{-R}^{R} \sqrt{R^2 - x^2}\, dx.$$

A technique, called **trigonometric substitution**, enables us to evaluate this integral as well as a number of other common integrals. Trigonometric substitution is based on the fact that the trigonometric functions satisfy certain very useful identities.

$$\sin^2 \theta + \cos^2 \theta = 1$$
$$1 + \tan^2 \theta = \sec^2 \theta$$
$$\sec^2 \theta - 1 = \tan^2 \theta$$

Because of these identities, functions involving the expressions

$$x^2 + a^2$$
$$a^2 - x^2$$
$$x^2 - a^2$$

where a is a constant can often be successfully integrated by an appropriate substitution involving trigonometric functions. For example, we can finish the problem above as follows.

Example:

Find

$$\int_{-R}^{R} \sqrt{R^2 - x^2}\, dx.$$

Answer:

558 Chapter V. Methods and Applications of Integration

We make the substitution $x = R\sin\theta$. Thus

$$dx = \left(\frac{dx}{d\theta}\right) d\theta = R\cos\theta \; d\theta$$

and

$$\begin{aligned}
\int \sqrt{R^2 - x^2} \; dx &= \int \sqrt{R^2 - R^2 \sin^2\theta} \; R\cos\theta \; d\theta \\
&= \int R\sqrt{1 - \sin^2\theta} \; R\cos\theta \; d\theta \\
&= \int R\sqrt{\cos^2\theta} \; R\cos\theta \; d\theta \\
&= R^2 \int \cos^2\theta \; d\theta
\end{aligned}$$

This last integral

$$\int \cos^2\theta \; d\theta$$

can be integrated either by using the trigonometric identity

$$\cos^2\theta = \frac{1 + \cos 2\theta}{2}$$

or by integration by parts. Either method yields

$$\int \cos^2\theta \; d\theta = \frac{\theta}{2} + \frac{\sin 2\theta}{4} + c.$$

Thus

$$R^2 \int \cos^2\theta \; d\theta = R^2 \left(\frac{\theta}{2} + \frac{\sin 2\theta}{4}\right) + c.$$

Now we have two choices for how to proceed. One possibility is to express this integral in terms of the original variable x. Our original substitution was

$$x = R\sin\theta.$$

So

$$\theta = \arcsin\left(\frac{x}{R}\right).$$

Now

$$\int \sqrt{R^2 - x^2} \; dx = R^2 \int \cos^2\theta \; d\theta$$

§8. Integration Miscellany

$$= R^2 \left(\frac{\theta}{2} + \frac{\sin 2\theta}{4} \right) + c$$

$$= R^2 \left(\frac{\arcsin(\frac{x}{R})}{2} + \frac{\sin 2 \arcsin(\frac{x}{R})}{4} \right) + c$$

and

$$\int_{-R}^{R} \sqrt{R^2 - x^2}\, dx = R^2 \left[\frac{\arcsin(\frac{x}{R})}{2} + \frac{\sin 2 \arcsin(\frac{x}{R})}{4} \right]_{-R}^{R}$$

$$= \frac{\pi R^2}{2}.$$

Thus the area of the circle is

$$2 \int_{-R}^{R} \sqrt{R^2 - x^2}\, dx = \pi R^2,$$

the familiar formula.

There is a second possibility. When we make the substitution

$$x = R \sin \theta$$

we can at the same time express the limits of integration in terms of the variable θ. The lower limit of the integral is $x = -R$ and this corresponds to

$$\theta = \arcsin\left(\frac{x}{R}\right) = \arcsin\left(\frac{-R}{R}\right) = \arcsin(-1) = -\frac{\pi}{2}.$$

The upper limit of the integral is $x = R$ and this corresponds to

$$\theta = \arcsin\left(\frac{x}{R}\right) = \arcsin\left(\frac{R}{R}\right) = \arcsin(1) = \frac{\pi}{2}.$$

So the integral becomes

$$\int_{-R}^{R} \sqrt{R^2 - x^2}\, dx = \int_{-\frac{\pi}{2}}^{\frac{\pi}{2}} \sqrt{R^2 - R^2 \sin^2 \theta}\, R \cos \theta\, d\theta,$$

which leads us (following the same calculations as above) to

$$\int_{-\frac{\pi}{2}}^{\frac{\pi}{2}} \sqrt{R^2 - R^2 \sin^2 \theta}\, R \cos \theta\, d\theta = R^2 \int_{-\frac{\pi}{2}}^{\frac{\pi}{2}} \cos^2 \theta\, d\theta$$

$$= R^2 \left[\frac{\theta}{2} + \frac{\sin 2\theta}{4}\right]_{-\frac{\pi}{2}}^{\frac{\pi}{2}}$$

$$= \frac{R^2 \pi}{2}.$$

The difficulty of the original problem stemmed from the expression $\sqrt{R^2 - x^2}$. Our success above was due to the trigonometric identity

$$1 - \sin^2 \theta = \cos^2 \theta$$

that allowed the troublesome expression to be simplified. As a general rule, the following substitutions often can simplify integrals involving the given expressions. They are worth trying if other, simpler methods do not work.

For the expression	Try
$a^2 - x^2$	$x = a \sin \theta$
$x^2 - a^2$	$x = a \sec \theta$
$x^2 + a^2$	$x = a \tan \theta$

Example:

Find the integral

$$\int \frac{dx}{(a^2 + x^2)^2}.$$

Answer:

We make the substitution $x = a \tan \theta$. Thus $dx = a \sec^2 \theta \, d\theta$ and the integral becomes

$$\int \frac{dx}{(a^2 + x^2)^2} = \int \frac{a \sec^2 \theta \, d\theta}{(a^2 + a^2 \tan^2 \theta)^2}$$

$$= \frac{1}{a^3} \int \frac{\sec^2 \theta}{(1 + \tan^2 \theta)^2} \, d\theta$$

$$= \frac{1}{a^3} \int \frac{\sec^2 \theta}{(\sec^2 \theta)^2} \, d\theta$$

§8. Integration Miscellany

$$= \frac{1}{a^3} \int \frac{1}{\sec^2 \theta} \, d\theta$$
$$= \frac{1}{a^3} \int \cos^2 \theta \, d\theta$$

The final integral is one of our old friends.

Example:

Find the integral
$$\int_{-1}^{1} \sqrt{x^2 - 1} \, dx.$$

Answer:

We make the substitution $x = \sec \theta$ so that $dx = \sec \theta \tan \theta \, d\theta$. When $x = -1$, $\theta = -\pi/2$ and when $x = 1$, $\theta = \pi/2$ so the integral becomes

$$\int_{-1}^{1} \sqrt{x^2 - 1} \, dx = \int_{-\frac{\pi}{2}}^{\frac{\pi}{2}} \sqrt{\sec^2 \theta - 1} \sec \theta \tan \theta \, d\theta$$
$$= \int_{-\frac{\pi}{2}}^{\frac{\pi}{2}} \sqrt{\tan^2 \theta} \sec \theta \tan \theta \, d\theta$$
$$= \int_{-\frac{\pi}{2}}^{\frac{\pi}{2}} \tan^2 \theta \sec \theta \, d\theta$$
$$= \int_{-\frac{\pi}{2}}^{\frac{\pi}{2}} (\sec^2 \theta - 1) \sec \theta \, d\theta$$
$$= \int_{-\frac{\pi}{2}}^{\frac{\pi}{2}} \sec^3 \theta - \sec \theta \, d\theta$$

The final integral can be evaluated using the techniques of the last section.

Exercises

Evaluate the following integrals.

*Exercise V.8.6:
$$\int_{0}^{5} \sqrt{25 - x^2} \, dx$$

Exercise V.8.7:
$$\int_1^2 \frac{x^2}{\sqrt{25-x^2}}\,dx$$

Exercise V.8.8:
$$\int_2^3 \frac{x^2}{\sqrt{x^2-25}}\,dx$$

*Exercise V.8.9:
$$\int_0^{100} \frac{dx}{1+4x^2}$$

Exercise V.8.10:
$$\int_0^{+\infty} \frac{dx}{1+4x^2}$$

Exercise V.8.11:
$$\int_0^{100} \frac{dx}{(1+4x^2)^2}$$

Rational Functions

A **rational function** is a function of the form

$$\frac{p(x)}{q(x)}$$

where $p(x)$ and $q(x)$ are polynomials. There is a straightforward but tedious recipe for integrating any rational function. Its ingredients are

- **Long Division for Polynomials.** When we divide one integer by another we often express the answer as a quotient with a remainder or as an integer plus a fraction that is less than one. For example,

$$\frac{47}{7} = 6 \text{ with remainder of } 5 \quad \text{or}$$
$$= 6 + \frac{5}{7}$$

§8. Integration Miscellany

We can do exactly the same thing when one polynomial is divided by another polynomial. We can express the answer as a polynomial plus a fraction in which the numerator is a polynomial of lower degree than the denominator. For example,

$$\frac{2x^3 - x^2 + 2x - 1}{x^2 + 2} = (2x - 1) + \left(\frac{-2x + 1}{x^2 + 2}\right)$$

This first step allows us to reduce any integration problem involving a rational function to two integration problems one of which involves a polynomial and the other of which involves a rational function in which the numerator is a polynomial of lower degree than the denominator. For example,

$$\int \frac{2x^3 - x^2 + 2x - 1}{x^2 + 2} \, dx = \int (2x - 1) \, dx + \int \frac{-2x + 1}{x^2 + 2} \, dx.$$

The first integral involving a polynomial is simple. We need only worry about the second integral.

- Any polynomial can be factored into a product of factors each of which has degree one or two. For example,

$$x^3 - x^2 + 2x - 2 = (x - 1)(x^2 + 2).$$

This second step allows us to factor the denominator of the second integral left from the previous step.

- Once the denominator has been factored the method of partial fractions enables us to break the second integral up into a sum of simpler integrals. For example,

$$\int \frac{2x^2 - 2x + 3}{(x^2 + 1)(x - 1)} \, dx = \int \frac{x - 1}{x^2 + 1} \, dx + \int \frac{2}{x - 1} \, dx$$

- Each of the simpler integrals remaining after the previous step is of one of the forms

$$\frac{q}{bx + c}$$

$$\text{or} \quad \frac{q}{(bx+c)^n}$$

$$\text{or} \quad \frac{px+q}{ax^2+bx+c}$$

$$\text{or} \quad \frac{px+q}{(ax^2+bx+c)^n}$$

In an earlier section we discussed integrating the first two forms. In this section we discuss the last two forms.

Completing the Square

We begin with a technique that is of use in many situations. Suppose that we have the expression

$$ax^2 + bx + c.$$

Notice that

$$ax^2 + bx + c = a\left(x^2 + \frac{bx}{a} + \frac{c}{a}\right)$$

$$= a\left(x^2 + \frac{bx}{a} + \frac{b^2}{4a^2} - \frac{b^2}{4a^2} + \frac{c}{a}\right)$$

$$= a\left(x + \frac{b}{2a}\right)^2 - a\left(\frac{b^2}{4a^2} - \frac{c}{a}\right)$$

$$= a\left(x + \frac{b}{2a}\right)^2 - a\left(\frac{b^2}{4a^2} - \frac{4ac}{4a^2}\right)$$

$$= a\left(x + \frac{b}{2a}\right)^2 - \frac{b^2 - 4ac}{4a}$$

This bit of algebra allows us to rewrite the original expression as a square plus a constant. This is exactly the trick that is used to find the quadratic formula, i.e., the formula for the roots of a quadratic

§8. Integration Miscellany

equation.

$$ax^2 + bx + c = 0$$

$$a\left(x + \frac{b}{2a}\right)^2 - \frac{b^2 - 4ac}{4a} = 0$$

$$a\left(x + \frac{b}{2a}\right)^2 = \frac{b^2 - 4ac}{4a}$$

$$\left(x + \frac{b}{2a}\right)^2 = \frac{b^2 - 4ac}{4a^2}$$

$$x + \frac{b}{2a} = \frac{\pm\sqrt{b^2 - 4ac}}{2a}$$

$$x = \frac{-b \pm \sqrt{b^2 - 4ac}}{2a}$$

The same trick is useful in many contexts. We use it below to integrate the third of the basic forms listed above.

The Denominator $ax^2 + bx + c$

We develop two formulas, one for the integral

$$\int \frac{dx}{ax^2 + bx + c}$$

and the other for the integral

$$\int \frac{x\,dx}{ax^2 + bx + c}.$$

First, notice that we may assume that the denominator $ax^2 + bx + c$ cannot be factored. If it could be factored then using the method of partial fractions the integrals above could be reduced to simpler integrals. Since a quadratic expression $ax^2 + bx + c$ can be factored if and only if the equation $ax^2 + bx + c = 0$ has real roots, we know that this equation does not have real roots. Since the roots of the equation

566 CHAPTER V. METHODS AND APPLICATIONS OF INTEGRATION

$ax^2 + bx + c$ are given by the Quadratic Formula

$$\frac{-b \pm \sqrt{b^2 - 4ac}}{2a},$$

we know that $b^2 - 4ac$ is negative and $4ac - b^2$ is positive. Now

$$\int \frac{dx}{ax^2 + bx + c} = \frac{1}{a} \int \frac{dx}{x^2 + \frac{b}{a}x + \frac{c}{a}}$$
$$= \frac{1}{a} \int \frac{dx}{\left(x + \frac{b}{2a}\right)^2 + \left(\frac{4ac-b^2}{4a^2}\right)}.$$

Next we make the substitution

$$u = x + \frac{b}{2a}.$$

So that $du = dx$ and

$$x = u - \frac{b}{2a}.$$

Let k denote the constant

$$k = \sqrt{\frac{4ac - b^2}{4a^2}} = \frac{\sqrt{4ac - b^2}}{2a}.$$

We can take the square root since we know that $4ac - b^2$ is positive.
With these substitutions our integral becomes

$$\frac{1}{a} \int \frac{du}{u^2 + k^2}$$

which can easily be evaluated.

$$\frac{1}{a} \int \frac{du}{u^2 + k^2} = \frac{1}{ka} \arctan\left(\frac{u}{k}\right)$$

Hence

$$\int \frac{dx}{ax^2 + bx + c} = \frac{1}{a} \int \frac{dx}{\left(x + \frac{b}{2a}\right)^2 + \left(\frac{4ac-b^2}{4a^2}\right)}$$

§8. Integration Miscellany

$$= \frac{1}{a} \int \frac{du}{u^2 + k^2}$$

$$= \frac{1}{a} \left(\frac{1}{k}\right) \arctan\left(\frac{u}{k}\right)$$

$$= \frac{1}{a} \left(\frac{2a}{\sqrt{4ac - b^2}}\right) \arctan\left(\frac{2au}{\sqrt{4ac - b^2}}\right)$$

$$= \left(\frac{2}{\sqrt{4ac - b^2}}\right) \arctan\left(\frac{2a\left(x + \frac{b}{2a}\right)}{\sqrt{4ac - b^2}}\right)$$

$$= \left(\frac{2}{\sqrt{4ac - b^2}}\right) \arctan\left(\frac{2ax + b}{\sqrt{4ac - b^2}}\right).$$

Similar calculations yield the following formula

$$\int \frac{x \, dx}{ax^2 + bx + c} = \frac{1}{2a} \ln|ax^2 + bx + c| - \frac{b}{2a} \int \frac{dx}{ax^2 + bx + c}$$

Using these two formulas one can evaluate any integral of the form

$$\int \frac{px + q}{ax^2 + bx + c} \, dx$$

Exercises

*Exercise V.8.12: Find the integral

$$\int \frac{x + 1}{x^2 - 4x + 20} \, dx$$

Exercise V.8.13: Find the integral

$$\int \frac{2x - 5}{4x^2 - 24x + 40}$$

The Denominator $(ax^2 + bx + c)^n$

Finally consider the last of the four basic forms

$$\int \frac{px + q}{(ax^2 + bx + c)^n} \, dx.$$

The techniques of completing the square and an appropriate substitution combine to reduce the problem to two integrals of the form

$$\int \frac{u}{(k^2 + u^2)^n} \, dx \quad \text{and} \quad \int \frac{1}{(k^2 + u^2)^n} \, dx$$

where k is an appropriate constant.

The first of these integrals can be evaluated by the substitution $v = k^2 + u^2$ and the second can be evaluated by a trigonometric substitution. In practice one uses the following formulas.

$$\int \frac{x \, dx}{(ax^2 + bx + c)^{n+1}} = -\frac{1}{2an(ax^2 + bx + c)^n} -$$

$$\frac{b}{2a} \int \frac{dx}{(ax^2 + bx + c)^{n+1}}$$

$$\int \frac{dx}{(ax^2 + bx + c)^{n+1}} = \frac{2ax + b}{n(4ac - b^2)(ax^2 + bx + c)^n} +$$

$$\frac{2(2n - 1)a}{n(4ac - b^2)} \int \frac{dx}{(ax^2 + bx + c)^n}$$

§9. Summary[+]

Techniques of Integration

The ideal way to evaluate a definite integral

$$\int_a^b f(x) \, dx$$

is via antidifferentiation using the Fundamental Theorem of Calculus. Unfortunately, however, antidifferentiation is often difficult or even impossible. In most calculus textbooks, the exercises involving integration

§9. Summary†

are carefully chosen so that the integrals can be evaluated via antidifferentiation. In this book and in the real world, many integrals can only be evaluated using a numerical method like, for example, Simpson's Rule.

In this chapter and chapter IV we studied various methods of antidifferentiation one-by-one. At the end of each section the exercises required the methods just studied. The location of these exercises was an important clue to which method could be successfully used. The exercises below, like practical problems, do not come with such an implicit clue. Often the hardest part of each exercise is determining which method to use.

Exercises

For each of the following exercises find the indicated integral. Whenever possible, find the exact answer using antidifferentiation. For some of these exercises, however, you may need to find an approximation using numerical methods rather than an exact answer using antidifferentiation.

*Exercise V.9.1:
$$\int_0^1 \sin^2 x \; dx.$$

*Exercise V.9.2:
$$\int_0^1 \sin x^2 \; dx.$$

*Exercise V.9.3:
$$\int_0^1 x \sin x \; dx.$$

*Exercise V.9.4:
$$\int_0^1 x^2 \sin x \; dx.$$

*Exercise V.9.5:
$$\int_0^1 x \sin^2 x \; dx.$$

Chapter V. Methods and Applications of Integration

*Exercise V.9.6:
$$\int_0^1 x \sin x^2 \, dx.$$

*Exercise V.9.7:
$$\int_0^1 x^2 \sin x^2 \, dx.$$

*Exercise V.9.8:
$$\int_0^1 x^3 \sin x^2 \, dx.$$

*Exercise V.9.9:
$$\int_{-1}^1 \frac{1}{4-x^2} \, dx.$$

*Exercise V.9.10:
$$\int_{-1}^1 \frac{1}{\sqrt{4-x^2}} \, dx.$$

*Exercise V.9.11:
$$\int_0^1 \tan x \sec^3 x \, dx.$$

*Exercise V.9.12:
$$\int_0^1 \sqrt{1+4x^2} \, dx.$$

*Exercise V.9.13:
$$\int_0^1 x\sqrt{1+4x^2} \, dx.$$

*Exercise V.9.14:
$$\int_0^1 x^2\sqrt{1+4x^2} \, dx.$$

§9. Summary[+]

*Exercise V.9.15:
$$\int_0^2 \sqrt{4-x^2}\, dx.$$

*Exercise V.9.16:
$$\int_0^2 x\sqrt{4-x^2}\, dx.$$

*Exercise V.9.17:
$$\int_0^2 x^2\sqrt{4-x^2}\, dx.$$

*Exercise V.9.18:
$$\int_1^4 e^x \ln x\, dx.$$

*Exercise V.9.19:
$$\int_0^{\pi/2} \sin^3 x \cos x\, dx.$$

*Exercise V.9.20:
$$\int_0^{\pi/2} \sin^3 x \cos^2 x\, dx.$$

*Exercise V.9.21:
$$\int_0^{\pi/2} \sin^2 x \cos^2 x\, dx.$$

*Exercise V.9.22:
$$\int_0^{\pi/2} \tan x\, dx.$$

*Exercise V.9.23:
$$\int_1^{+\infty} \ln x\, dx.$$

Exercise V.9.24:
$$\int_0^1 \ln x \, dx.$$

Exercise V.9.25:
$$\int_0^1 e^{3x} \cos x \, dx.$$

Exercise V.9.26:
$$\int_0^3 \frac{1}{x^2 + 2x + 5} \, dx.$$

Exercise V.9.27:
$$\int_0^3 \frac{x}{x^2 + 2x + 5} \, dx.$$

Exercise V.9.28:
$$\int_0^3 \frac{2x + 7}{x^2 + 2x + 5} \, dx.$$

Exercise V.9.29:
$$\int_1^3 \frac{x^4 - x^3 + 2x + 1}{x^3 + x} \, dx.$$

Sketch a graph of each of the following functions

Exercise V.9.30:
$$F(x) = \int_0^x \cos t \, dt.$$

Exercise V.9.31:
$$F(x) = \int_1^x \cos t \, dt.$$

§9. Summary†

*Exercise V.9.32:
$$F(x) = \int_x^0 \cos t \, dt.$$

*Exercise V.9.33:
$$F(x) = \int_x^1 \cos t \, dt.$$

*Exercise V.9.34:
$$F(x) = \int_0^x e^{-t^2/2} \, dt.$$

*Exercise V.9.35:
$$F(x) = \frac{1}{\sqrt{2\pi}} \int_{-\infty}^x e^{-t^2/2} \, dt.$$

Solve each of the following initial value problems.

*Exercise V.9.36:
$$y'' = 2; \quad y(0) = 1, y'(0) = 2.$$

*Exercise V.9.37:
$$y'' = 2; \quad y(1) = 1, y'(1) = 2.$$

*Exercise V.9.38:
$$y'' = 2; \quad y(0) = 1, y'(1) = 2.$$

*Exercise V.9.39:
$$p' = 2p(2-p); \quad p(0) = 1.$$

*Exercise V.9.40:
$$y' = 2y/x; \quad y(1) = 2.$$

*Exercise V.9.41:

$$y' = 3x/y; \qquad y(1) = 2.$$

Applications of Integration

We covered a number of applications of integration in this chapter and chapter IV. You should review those applications. The real purpose of this chapter, however, is not to learn a list of particular applications but rather to develop a general understanding of how integration is used to solve real problems, so that one can solve not only those problems that we have studied but also other problems involving the same ideas in different settings. In this section, instead of reviewing applications already covered, we look at some new applications and see how the ideas we studied in this chapter can be used in new contexts.

The basic tool we use in all these applications is the Riemann sum

$$\sum_{i=1}^{n} f(s_i) h$$

together with the definition of the definite integral as the limit of Riemann sums

$$\int_a^b f(x)\, dx = \lim_{n \to \infty} \sum_{i=1}^{n} f(s_i) h = \int_a^b f(x)\, dx.$$

Riemann sums can be used to estimate many different quantities as precisely as might be required for any particular purpose. If a particular quantity Q can be estimated in this way then the exact value of Q is given by the integral, since

$$Q = \lim_{n \to \infty} \sum_{i=1}^{n} f(s_i) h = \int_a^b f(x)\, dx.$$

For example, the area of the region under the curve $y = f(x)$ between the lines $x = a$ and $x = b$ shown in Figure V.31 can be estimated by a Riemann sum of the form

$$\sum_{i=1}^{n} f(s_i) h$$

§9. Summary[†]

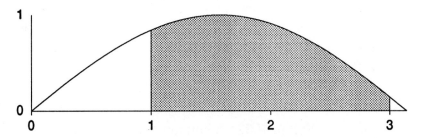

Figure V.31: Area under a curve

where, for example, s_i is the midpoint of the interval $[x_{i-1}, x_i]$. Thus, the exact value of this area is given by the integral

$$\int_a^b f(x)\, dx.$$

In many applications, the definite integral can be thought of as "multiplication" with one factor being nonconstant. For example, we can think of the area above as the product of the width $(b - a)$ by the (nonconstant) height $f(x)$.

Levers and Centroids

We use many machines to increase the effectiveness of our efforts. Perhaps the most common of these machines are levers. We even use the word "leverage" to describe the way in which our efforts can be multiplied. For example, when a corporate raider purchases a company using $100,000,000 of her own money together with $400,000,000 borrowed from a consortium of banks, we describe this as a "leveraged buyout."

Wouldbe corporate raiders honing their skills on seesaws like the one shown in Figure V.32 know that a 40 pound weight placed ten feet from the fulcrum or pivot point of a seesaw can be "balanced" by an eighty pound weight placed only five feet from the fulcrum on the opposite side.

To understand levers we need to study the notion of **torque**. Torque can measured in units of pound-feet and is analogous to the notion of force. Force is used to move an object by causing it to accelerate. Torque is used to rotate an object about a pivot point by causing it to accelerate about that point. When we study the vertical motion of an object we usually set up problems so that the positive direction

576 CHAPTER V. METHODS AND APPLICATIONS OF INTEGRATION

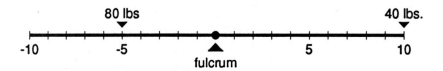

Figure V.32: A simple seesaw

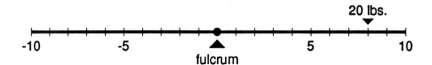

Figure V.33: This seesaw will rotate clockwise

is upward and the negative direction is downward. Thus a positive force will accelerate an object in the upward direction. For problems involving levers and torque we measure motion around a pivot point so that clockwise is positive and counter-clockwise is negative. Consider, for example, the set-ups shown in Figures V.33 and V.34. The weight or force on the right side of the seesaw in Figure V.33 will pull that side down, accelerating the seesaw in a clockwise direction. The weight or force on the left side of the seesaw in Figure V.34 will pull that side down, accelerating the seesaw in a counterclockwise direction.

In all these figures the x-axis is placed so that the fulcrum of the lever is at the point $x = 0$. When a force or weight F is applied pushing

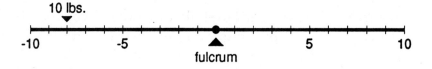

Figure V.34: This seesaw will rotate clockwise

§9. Summary†

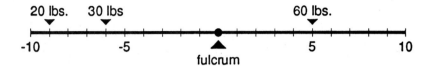

Figure V.35: Which way will this seesaw accelerate?

downward at the point x the resulting torque is given by the formula

$$\text{Torque} = Fx,$$

where x is negative if the force is applied at a point to the left of the fulcrum and x is positive if the force is applied at a point to the right of the fulcrum. Because we are usually interested in weights, we will measure force so that the downward direction is positive. Thus in Figure V.33 see that

$$\text{Torque} = (20)(8) = 160,$$

which is positive and causes the seesaw to acccelerate in the clockwise direction. Similarly, in Figure V.34

$$\text{Torque} = (10)(-8) = -80,$$

which is negative and causes the seesaw to accelerate in the counterclockwise direction. In Figure V.32 we have two weights or forces and the resulting torque is

$$\text{Torque} = (80)(-5) + (40)(10) = 0,$$

which causes no acceleration about the fulcrum. This represents mathematically the fact that the combination of these two forces or weights are in "balance."

Exercises

*Exercise V.9.42:   What is the total torque for the seesaw shown in Figure V.35? Will it accelerate in the clockwise or the counterclockwise direction?

578 Chapter V. Methods and Applications of Integration

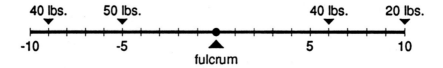

Figure V.36: Which way will this seesaw accelerate?

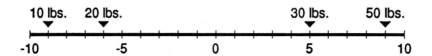

Figure V.37: Where should the fulcrum be placed for balance?

*Exercise V.9.43: What is the total torque for the seesaw shown in Figure V.36? Will it accelerate in the clockwise or the counterclockwise direction?

*Exercise V.9.44: Where should the fulcrum be placed in Figure V.37 so that the seesaw will be balanced?

*Exercise V.9.45: Where should the fulcrum be placed in Figure V.38 so that the seesaw will be balanced?

Now suppose that instead of a simple seesaw with a few weights we have a thin sheet of metal or wood, for example, the triangle shown in Figure V.39. Picture a rod running through this piece of metal

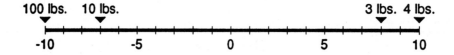

Figure V.38: Where should the fulcrum be placed for balance?

§9. Summary[+]

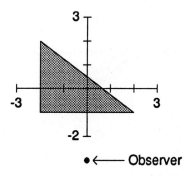

Figure V.39: A piece of metal

along the y-axis in such a way that the metal can pivot around the rod. Hold this setup so that it is horizontal. We are interested in the torque generated by the piece of metal around the rod, and in whether the piece of metal will accelerate in the clockwise or counter-clockwise direction as viewed by an observer whose eye is placed along the y-axis at a point where y is negative. Figure V.39 shows a picture of this setup as viewed from above.

Suppose that the piece of metal is of uniform thickness so that each square unit weighs 0.1 pound. We can *estimate* the torque exerted by gravity acting on this setup by approximating the triangle by rectangles as shown in Figures V.40 and V.41. We do this by first dividing the interval $[-2, 2]$ in the usual way into n subintervals, each of width $h = 4/n$. We choose the midpoint of each interval as a sample point $s_i = (x_i + x_{i-1})/2$. Notice that the top edge of the triangle is described by the equation

$$y = -\frac{3}{4}x + \frac{1}{2}$$

and the bottom edge is just

$$y = -1,$$

so that the area of each rectangle is given by

$$\text{area of typical rectangle} = \left(-\frac{3}{4}s_i + \frac{3}{2}\right)h$$

and, since the metal weighs 0.1 pound per square unit, its weight is

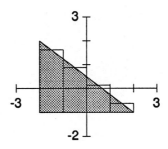

Figure V.40: Approximating by rectangles

given by

$$\text{weight of typical rectangle} = 0.1\left(-\frac{3}{4}s_i + \frac{3}{2}\right)h.$$

The distance of this weight from the rod along the y-axis is approximated by s_i. This leads to the estimate

$$\text{Torque} \approx \sum_{i=1}^{n} 0.1 s_i \left(-\frac{3}{4}s_i + \frac{3}{2}\right)h$$

and, taking the limit of these estimates,

$$\text{Torque} = \int_{-2}^{2} 0.1 x \left(-\frac{3}{4}x + \frac{3}{2}\right) dx = -0.4.$$

Notice that the piece of metal will rotate counterclockwise about the y-axis, that is, the left side will fall and the right side will rise, as is obvious from the picture.

Exercises

*Exercise V.9.46: Find the torque around the line $x = 1$ of the piece of metal shown in Figure V.39.

*Exercise V.9.47: Find a general formula for the torque around the line $x = x_0$ of the piece of metal shown in Figure V.39.

*Exercise V.9.48: If you wanted to support the piece of metal shown in Figure V.39 by a single rod parallel to the y-axis in such a way that the piece of metal would not rotate about the rod, where should the rod be placed?

§9. Summary†

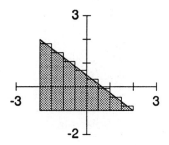

Figure V.41: Approximating by more rectangles

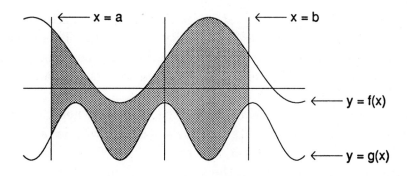

Figure V.42: A piece of metal

*Exercise V.9.49: Find the balance point for the piece of metal shown in Figure V.39.

*Exercise V.9.50: Find a general formula for the torque around the y-axis of the thin metal plate whose shape is shown in Figure V.42. Suppose that the plate has a uniform thickness and that it weighs k pounds per square unit.

*Exercise V.9.51: Find a general formula for the torque around the line $x = x_0$ of the thin metal plate whose shape is shown in Figure V.42. Suppose that the plate has a uniform thickness and that it weighs k pounds per square unit.

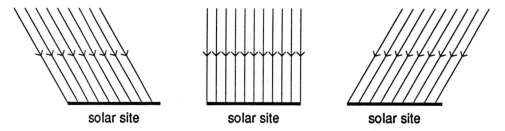

Figure V.43: Three different times of day

Solar Power

Designers of solar power generating facilities must take into account the variation during the day of the intensity of sunlight. This variation is caused by the fact that the relative position of the sun changes during the day as the earth rotates. In this section we examine a simple case to see some of the ideas involved. Consider a solar generating facility located at a site on the equator on the day of either the vernal or autumnal equinox. At this site on this day the sun will rise at 6:00 AM[6], pass directly overhead at noon, and then set at 6:00 PM. Suppose that at noon, energy is arriving on the site at a rate of 1,000 units per hour. Thus, if the intensity of light were constant for 12 hours, the site would collect 12,000 units of energy. Unfortunately, however, energy form the sun does not arrive at a constant rate. Except at noon, sunlight falls on the site at an angle as shown in Figure V.43

The angle at which sunlight strikes the solar site affects the rate at which energy arrives in several different ways. The most important effect is that sunlight is "thinned out," that is, the same amount of energy is spread over a larger area. This is the only effect that we consider in this section. Another, smaller effect, is due to the fact that light arriving at different angles must pass through different thicknesses of the atmosphere. Because this effect is much smaller than the first effect, we ignore it.

---

[6]Due the vagaries of time zones, the sun might rise anytime from 5:30 AM to 6:30 AM with corresponding changes in the time at which it passes directly overhead and sets. The important thing is that it passes directly overhead exactly six hours after it rises and then sets six hours later.

§9. Summary⁺

Exercises

*Exercise V.9.52: At what rate is solar energy arriving at the solar site at 9:00 AM?

*Exercise V.9.53: At what rate is solar energy arriving at the solar site at 3:00 PM?

*Exercise V.9.54: At what rate is solar energy arriving at the solar site at 6:00 PM?

*Exercise V.9.55: Find the total amount of energy received at the solar site from 6:00 AM to 6:00 PM.

Chapter VI

Real and Complex Numbers

§1. Polar Coordinates

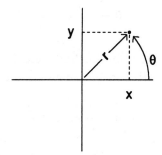

Figure VI.1: One point, two sets of coordinates

In this section we introduce a new coordinate system for describing the location of a point in the plane. This system, called **polar coordinates**, describes the location of a point using a pair, (r, θ), of real numbers. The first coordinate, r, gives the distance of the point from the origin. The second coordinate, θ, describes the direction of the point from the origin by measuring the angle from the x-axis to the line going from the origin to the given point. Figure VI.1 shows a typical point with its usual Cartesian coordinates, (x, y), and its polar coordinates (r, θ).

It is easy to convert from polar coordinates to Cartesian coordinates

$$\begin{aligned} x &= r\cos\theta \\ y &= r\sin\theta \end{aligned}$$

and almost as easy to convert from Cartesian coordinates to polar coordinates. The basic idea is

$$r = \sqrt{x^2 + y^2}$$

$$\theta = \arctan \frac{y}{x}.$$

This last equation, however, must be used intelligently. The arctangent of θ is always a number between $-\pi/2$ and $\pi/2$ but the angle θ is not restricted to this range. The equation

$$\theta = \arctan \frac{y}{x}$$

can be used as is for points in the first and fourth quadrants (i.e., for points for which $x > 0$). If $x < 0$, however, then

$$\theta = \pi + \arctan \frac{y}{x}.$$

If $x = 0$ and $y > 0$ then $\theta = \pi/2$. If $x = 0$ and $y < 0$ then $\theta = -\pi/2$. Finally, if both x and y are zero then θ is irrelevant.

Example:

Find the polar coordinates of the point whose Cartesian coordinates are $(-3, 4)$.

Answer:

First,
$$r = \sqrt{(-3)^2 + 4^2} = \sqrt{25} = 5.$$

Next, since $x < 0$,

$$\theta = \arctan\left(\frac{4}{-3}\right) + \pi = -0.927 + \pi = 2.214.$$

So this point has polar coordinates $(5, 2.214)$.

§1. Polar Coordinates

Example:

Find the Cartesian coordinates of the point whose polar coordinates are $(2, \pi)$.

Answer:

$$x = 2 \cos \pi = -2$$
$$y = 2 \sin \pi = 0.$$

So this point has Cartesian coordinates $(-2, 0)$.

Exercises

*__Exercise VI.1.1:__ Find the polar coordinates of each of the following points given in Cartesian coordinates: $(1, 1)$, $(1, -1)$, $(-1, 1)$, and $(-1, -1)$.

Exercise VI.1.2: Find the polar coordinates of each of the following points given in Cartesian coordinates: $(4, 3)$, $(-4, 3)$, $(4, -3)$, $(-4, -3)$, $(0, 4)$, $(0, -4)$, $(3, 0)$, $(-3, 0)$.

*__Exercise VI.1.3:__ Find the Cartesian coordinates of the point whose polar coordiantes are $(4, \pi)$.

Exercise VI.1.4: Find the Cartesian coordinates of each the following points given in polar coordinates: $(5, \pi/4)$, $(5, 3\pi/4)$, $(5, 5\pi/4)$, $(5, 7\pi/4)$, $(5, 9\pi/4)$, $(5, -\pi/4)$, $(5, -3\pi/4)$, $(5, -5\pi/4)$.

Polar coordinates are amibiguous because angles are ambiguous. For any given angle θ,

$$\theta, \ (\theta + 2\pi), \ (\theta + 4\pi), \ldots, \ (\theta - 2\pi), \ (\theta - 4\pi), \ldots$$

all denote the same angle. Thus a point can be described using polar coordinates in many different ways. For example,

$$\left(1, \frac{\pi}{4}\right) \quad \text{and} \quad \left(1, \frac{-7\pi}{4}\right)$$

denote the same point. See Figure VI.2.

588   CHAPTER VI. REAL AND COMPLEX NUMBERS.

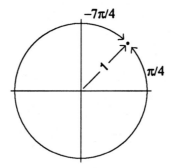

Figure VI.2: Two names for one point

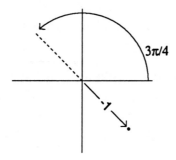

Figure VI.3: The point $(-1, 3\pi/4)$

§1. Polar Coordinates

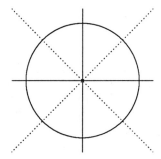

Figure VI.4: Sensitivity of an omnidirectional microphone

There is another source of ambiguity in polar coordinates. Although usually r is positive, it can be negative. To find the point $(-1, 3\pi/4)$, for example, one would first measure the angle $3\pi/4$ and then set off in the indicated direction a "distance" of -1 units. See Figure VI.3.

Graphs in polar coordinates are particularly useful for describing phenomena in which the angle is a signficant factor. For example, one of the most important characteristics of a microphone is its sensitivity to sound from various directions. Some microphones, called "omnidirectional" microphones, are designed to pick up sound from all directions. If we let $f(\theta)$ denote the sensitivity of a microscope to sound from the direction θ then the graph of the function $r = f(\theta)$ for a perfect omnidirectional microphone would look like Figure VI.4.

Omnidirectional microphones are appropriate if one wants to pick up all the sound in a particular situation. In many situations, however, one wants to pick up some particular sound, for example, a singer, without background noise. In such situations a microphone called a "cardioid" microphone is frequently used. The sensitivity pattern for a cardioid microphone is described by a function like

$$r = 1 + \cos(\theta)$$

(Figure VI.5). You can see how cardioid microphones get their name. Notice that a cardioid microphone is very sensitive to sound coming from the angle $\theta = 0$. This angle is toward the "front" of the microphone. A cardioid microphone is much less sensitive to sound that is off to the sides ($\theta = \pi/2$ or $3\pi/2$) and has little or no sensitivity to sounds coming from the back of the microphone.

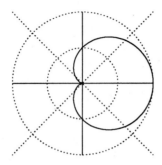

Figure VI.5: The function $r = 1 + \cos\theta$

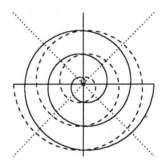

Figure VI.6: The spiral $r = \theta$

The ambiguity of polar coordinates can cause some difficulty but it also can have some very nice consequences. For example, it is very easy to describe a spiral using polar coordinates. Figure VI.6 shows the spiral $r = \theta$. The solid part of the graph corresponds to positive values of θ and, hence, r. The dashed part corresponds to negative values of θ and, hence, r.

Graphs like Figures VI.4, VI.5, and VI.6 of functions expressed in polar coordinates can be sketched using the same ideas used for Cartesian coordinates. The derivative $y' = dy/dx$, of a function $y = f(x)$ in Cartesian coordinates tells us whether y is increasing or decreasing as x increases. In polar coordinates the derivative $r' = dr/d\theta$ tells us whether r is increasing or decreasing as θ increases. In polar coordinates, however, $dr/d\theta$ is somewhat difficult to interpret because its

§1. Polar Coordinates

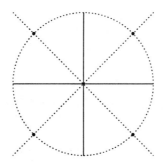

Figure VI.7: The four critical points for $r = \sin 2\theta$

interpretation depends on the sign of r. If $r > 0$ then when $dr/d\theta$ is positive the curve $r = f(\theta)$ is moving away from the origin (as you follow it in the counterclockwise direction) and if $dr/d\theta$ is negative the curve is moving toward the origin. If $r < 0$, however, the situation is reversed.

To graph a function in either Cartesian and polar coordinates we begin by finding the critical points and then analyze the sign of the derivative.

Example:

Graph the function $r = \sin 2\theta$ on the interval $[0, 2\pi]$.

Answer:

First notice that $dr/d\theta = 2\cos 2\theta$, which is zero when

$$\theta = \frac{\pi}{4}, \frac{3\pi}{4}, \frac{5\pi}{4}, \text{ and } \frac{7\pi}{4}.$$

Hence there are four critical points,

$$\left(1, \frac{\pi}{4}\right), \left(-1, \frac{3\pi}{4}\right), \left(1, \frac{5\pi}{4}\right), \text{ and } \left(-1, \frac{7\pi}{4}\right),$$

indicated in Figure VI.7.

The next step is to make a chart like the one shown in Figure VI.8 showing where the derivative is positive and where it is negative. This chart is constructed in exactly the same way as we did when we were

Figure VI.8: Analysis of the derivative of the function $r = \sin 2\theta$

working with graphs in Cartesian coordinates. Figure VI.9(a) shows a graph of the function $r = \sin 2\theta$ for θ running from 0 to $\pi/2$. Notice that as θ increases from 0 to $\pi/4$, r increases from 0 to 1. As θ increases from $\pi/4$ to $\pi/2$, r decreases from 1 to 0.

Figure VI.9(b) continues this same graph. It shows the same function for θ running from 0 to π. Notice that for θ between $\pi/2$ and π, r is negative. Figure VI.9(c) shows the same function for θ running from 0 to $3\pi/2$. Notice that for θ between π and $3\pi/2$, r is positive. Figure VI.9(d) the same function for θ running from 0 to 2π. Notice that for θ between $3\pi/2$ and 2π, r is negative.

Exercises

*Exercise VI.1.5: Sketch a graph in polar coordinates of the function $r = 2 + \sin \theta$.

Exercise VI.1.6: Sketch a graph in polar coordinates of the function $r = 1 + \sin 2\theta$.

*Exercise VI.1.7: Sketch a graph in polar coordinates of the function $r = \sin 3\theta$.

Exercise VI.1.8: Sketch a graph in polar coordinates of the function $r = \sin 4\theta$.

Exercise VI.1.9: Sketch a graph in polar coordinates of the function $r = \tan \theta$.

Exercise VI.1.10: Sketch a graph in polar coordinates of the function $r = 1/\theta$ for $\theta > 0$.

§1. Polar Coordinates

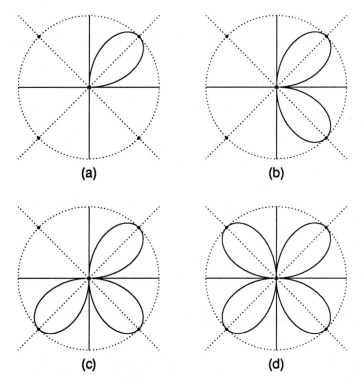

Figure VI.9: $r = \sin 2\theta$

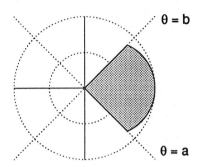

Figure VI.10: Region described in polar coordinates

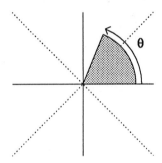

Figure VI.11: Sector of a circle

Area in Polar Coordinates

Many interesting curves and, hence, many interesting regions can be described most naturally using polar coordinates. In this section we describe how to find the area of a region like the one shown in Figure VI.10 that is enclosed by the lines $\theta = a$, $\theta = b$, and the curve $r = f(\theta)$.

First, consider a sector (or pie-shaped piece) of a circle like the one shown in Figure VI.11. An entire circle of radius R has an area of πR^2. An entire circle would be the sector formed by an angle of 2π radians. Thus a sector formed by an angle of θ radians would have an area of

$$\left(\frac{\theta}{2\pi}\right)\pi R^2 = \left(\frac{R^2}{2}\right)\theta.$$

For a region like that shown in Figure VI.10 the radius is not con-

§1. Polar Coordinates

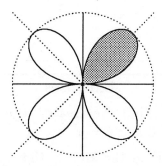

Figure VI.12: Four petals of $r = \sin 2\theta$

stant, so we use integration rather than simple multiplication. The formula for the area of a region like the one shown in Figure VI.10 is

$$\int_a^b \frac{f(\theta)^2}{2}\, d\theta.$$

Notice that if $a < b$ this formula yields a nonnegative number since the function being integrated is $f(\theta)^2$. Thus this formula computes unsigned area. For example, a circle of radius -2 would have an area of $\pi(-2)^2 = 4\pi$.

Example:

The graph of the function $r = \sin 2\theta$ looks like a flower with four petals. See Figure VI.12. Find the area of the shaded petal.

Answer:

This petal is the portion of the graph between $\theta = 0$ and $\theta = \pi/2$. So we need to evaluate the integral

$$\int_0^{\frac{\pi}{2}} \frac{\sin^2 2\theta}{2}\, d\theta = \frac{1}{2} \int_0^{\frac{\pi}{2}} \sin^2 2\theta\, d\theta$$

$$= \frac{1}{2} \int_0^{\frac{\pi}{2}} \frac{1 - \cos 4\theta}{2}\, d\theta$$

$$= \frac{1}{4} \int_0^{\frac{\pi}{2}} d\theta - \frac{1}{4} \int_0^{\frac{\pi}{2}} \cos 4\theta\, d\theta$$

$$= \frac{\pi}{8}$$

Exercises

*Exercise VI.1.11: Find the area of region formed by the spiral $r = \theta$ and the lines $\theta = 0$ and $\theta = \pi$.

Exercise VI.1.12: Find the area of the region inside the cardioid $r = 1 + \cos\theta$.

Exercise VI.1.13: Find the area of one petal of the region inside the curve $r = \sin 6\theta$.

---

§2. Complex Numbers

In this section we study another way of looking at the plane. We use the notation $z = a + bi$ to denote the point (a, b). Thought of in this way z is called a **complex number**. We call a the **real part** of z and b the **imaginary part** of z. A real number a can be thought of as the complex number $a + 0i$.

Two complex numbers $z = a + bi$ and $w = c + di$ can be added or subtracted as follows.

$$\begin{aligned} z + w &= (a + bi) + (c + di) \\ &= a + c + bi + di \\ &= (a + c) + (b + d)i \\ z - w &= (a + bi) - (c + di) \\ &= a - c + bi - di \\ &= (a - c) + (b - d)i \end{aligned}$$

Complex numbers behave very much like the real numbers. For example, for any three complex numbers u, v and w

$$\begin{aligned} u + v &= v + u \\ u + (v + w) &= (u + v) + w \end{aligned}$$

§2. Complex Numbers

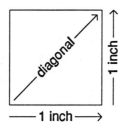

Figure VI.13: One inch by one inch square

We are engaged in one of the long-running tasks of mathematics, inventing new numbers to cope with new situations. The first numbers that mathematicians invented were the positive integers: $1, 2, 3, \ldots$. The positive integers were invented (or discovered depending on your philosophy) to enable us to count, to keep track of the "number" of things. The next step in this long line of invention was a single number, the number zero. Although the number zero is all too familiar to those of us with empty wallets, it was actually a very bold intellectual step to invent a symbol for nothing. The next step was the negative integers: $-1, -2, -3, \ldots$. At this point mathematicians had invented all the whole numbers or **integers**.

The set of integers is sufficient if all one wants to do is to keep track of items, of things. The integers, however, have a flaw. One cannot always divide. A mathematician armed solely with the integers, a single candy bar, and one friend is at a loss if he or she tries to share the candy bar. Enter the rational numbers. Each person will receive 1/2 candy bar.

Our work, however, is still incomplete. In order to describe the real world one must be able to measure lengths but there are some lengths that cannot be measured by rational numbers. Figure VI.13 shows a square, each of whose sides is one inch long. By the Pythagorean Theorem the diagonal of this square is $\sqrt{2}$ inches long but there is no rational number x such that $x^2 = 2$. This discovery eventually lead to the invention of the real numbers. But the real number system is still lacking—negative real numbers do not have real square roots. The complex numbers fill this lack. We define multiplication for complex

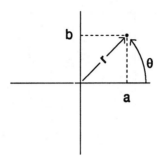

Figure VI.14: Polar coordinates of a complex number

numbers by defining i^2 to be -1 so that for $z = a + bi$ and $w = c + di$

$$\begin{aligned} zw &= (a+bi)(c+di) \\ &= ac + adi + bci + bdi^2 \\ &= ac + adi + bci - bd \quad (\text{since } i^2 = -1) \\ &= (ac - bd) + (ad + bc)i \end{aligned}$$

The complex numbers with the three operations defined so far behave very much like the real numbers. For example, for any three complex numbers, u, v and w,

$$\begin{aligned} uv &= vu \\ u(vw) &= (uv)w \\ u(v+w) &= uv + uw \end{aligned}$$

It is often helpful to visualize a complex number in terms of its polar coordinates as shown in Figure VI.14. Notice that $a = r \cos \theta$ and that $b = r \sin \theta$ so that we can write

$$z = a + bi = r \cos \theta + ir \sin \theta = r(\cos \theta + i \sin \theta).$$

The number

$$r = \sqrt{a^2 + b^2}$$

§2. Complex Numbers

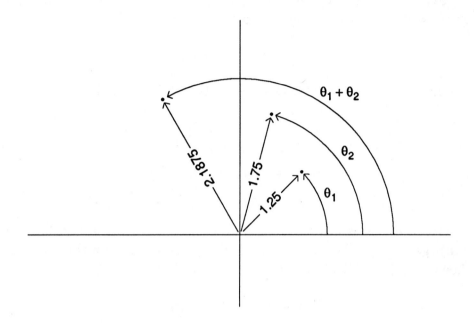

Figure VI.15: Multiplying two complex numbers

is called the **magnitude** of the complex number z. We use the notation
$$|z| = \sqrt{a^2 + b^2}$$
for the magnitude of z. Notice that if $z = a$ is a real number then
$$|z| = \sqrt{a^2} = |a|.$$

Thinking of complex numbers in polar coordinates leads to a particularly nice visualization of multiplication. If we multiply two complex numbers $z_1 = r_1(\cos\theta_1 + i\sin\theta_1)$ and $z_2 = r_2(\cos\theta_2 + i\sin\theta_2)$ as follows

$$\begin{aligned} z_1 z_2 &= r_1 r_2 (\cos\theta_1 + i\sin\theta_1)(\cos\theta_2 + i\sin\theta_2) \\ &= r_1 r_2 [(\cos\theta_1 \cos\theta_2 - \sin\theta_1 \sin\theta_2) + i(\sin\theta_1 \cos\theta_2 + \sin\theta_2 \cos\theta_1)] \\ &= r_1 r_2 [\cos(\theta_1 + \theta_2) + i\sin(\theta_1 + \theta_2)], \end{aligned}$$

we see that the result is to add their angles and multiply their magnitudes. See Figure VI.15.

Because the expression $\cos\theta + i\sin\theta$ occurs frequently when one works with complex numbers we often use the notation

$$\operatorname{cis}\theta = \cos\theta + i\sin\theta.$$

Using this notation we can express multiplication as follows.

$$\begin{aligned} z_1 &= r_1 \operatorname{cis}\theta_1 \\ z_2 &= r_2 \operatorname{cis}\theta_2 \\ z_1 z_2 &= r_1 r_2 \operatorname{cis}(\theta_1 + \theta_2) \end{aligned}$$

If $z = a + bi$ is any complex number then $\bar{z} = a - bi$ is called the **conjugate** of z. Geometrically, $\bar{z}$ is the reflection of z in the x-axis as shown in Figure VI.16. In polar coordinates we see that

$$z = r(\cos\theta + i\sin\theta)$$

and

$$\bar{z} = r(\cos\theta - i\sin\theta) = r[\cos(-\theta) + i\sin(-\theta)].$$

So the product $z\bar{z}$ will have an angle $\theta + (-\theta) = 0$. It will lie on the x-axis and be a real number. We can also see this algebraically.

$$\begin{aligned} z\bar{z} &= (a+bi)(a-bi) \\ &= a^2 - abi + abi - b^2 i^2 \\ &= (a^2 + b^2) \\ &= |z|^2 \end{aligned}$$

Now notice that if we let

$$z^{-1} = \left(\frac{1}{|z|^2}\right)\bar{z}$$

then

$$zz^{-1} = 1.$$

So that the number z^{-1} deserves its notation. It acts exactly like the number x^{-1} when x is real, namely, it is the multiplicative inverse of z.

§2. Complex Numbers

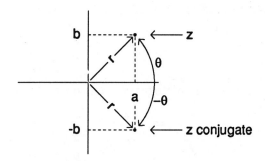

Figure VI.16: A complex number z and its conjugate

Now that we know how to find the multiplicative inverse of a complex number we can divide one complex number by another

$$\frac{z}{w} = zw^{-1}$$

and we can apply all four basic arithmetic operations for real numbers—addition, subtraction, multiplication and division—to complex numbers as well.

Exercises

For each of the following pairs of complex numbers z and w compute $|z|, |w|, \bar{z}, \bar{w}, z+w, z-w, zw$, and z/w.

*Exercise VI.2.1: $z = 3 + 4i$, $w = 12 + 5i$.

Exercise VI.2.2: $z = 1 + 2i$, $w = 2 + 3i$.

Exercise VI.2.3: $z = 1 - 2i$, $w = 2 + 3i$.

Exercise VI.2.4: $z = 2(\cos \pi/4 + i \sin \pi/4)$, $w = 3(\cos \pi/3 + i \sin \pi/3)$.

The geometric picture of multiplication of complex numbers is particularly informative when we consider squaring a complex number.

$$\begin{aligned} z &= r \text{ cis } \theta \\ z^2 &= r^2 \text{ cis } 2\theta \end{aligned}$$

Thus to find the square root of a complex number we need to take the square root of the magnitude and divide the angle by 2. That is, if $z = r \text{ cis } \theta$ then
$$\sqrt{z} = \sqrt{r} \text{ cis } \left(\frac{\theta}{2}\right).$$
In particular, if z is a negative real number then
$$\begin{aligned} z &= r \text{ cis } \pi \\ \sqrt{z} &= \sqrt{r} \text{ cis } \left(\frac{\pi}{2}\right) \\ &= \sqrt{r} \left[\cos\left(\frac{\pi}{2}\right) + i \sin\left(\frac{\pi}{2}\right)\right] \\ &= i\sqrt{r} \end{aligned}$$

or, in other words,
$$\sqrt{-r} = r\sqrt{-1} = i\sqrt{r}$$
exactly as we would expect.

There is an important catch to all this. Recall the basic ambiguity of angles. Because of this ambiguity we can express the same complex number many different ways.
$$\begin{aligned} z &= r \text{ cis } \theta \\ &= r \text{ cis } (\theta + 2\pi) \\ &= r \text{ cis } (\theta + 4\pi) \\ &= r \text{ cis } (\theta - 2\pi) \\ &\vdots \end{aligned}$$

Thus when we take the square root we get many different possibilities.
$$\begin{aligned} \sqrt{z} &= \sqrt{r} \text{ cis } (\theta/2) \\ &= \sqrt{r} \text{ cis } (\theta/2 + \pi) \\ &= \sqrt{r} \text{ cis } (\theta/2 + 2\pi) \\ &= \sqrt{r} \text{ cis } (\theta/2 - \pi) \\ &\vdots \end{aligned}$$

Notice that among all these possibilities there are really only two distinct square roots.
$$\sqrt{z} = \sqrt{r} \text{ cis } (\theta/2)$$
and
$$\sqrt{z} = \sqrt{r} \text{ cis } (\theta/2 + \pi) = -\sqrt{r} \text{ cis } (\theta/2).$$

Exercises

Exercise VI.2.5: Find a formula for z^n where n is any positive integer. Use polar coordinates.

*Exercise VI.2.6: Find all the cube roots of the number 1.

Exercise VI.2.7: Find a formula for the cube root of z. Use polar coordinates. How many distinct cube roots does z have?

Exercise VI.2.8: Find all the fourth roots of 1.

Exercise VI.2.9: Find all the fifth roots of 1.

Exercise VI.2.10: Find a formula for the nth root of z. Use polar coordinates. How many distinct nth roots does z have?

One of the main goals of this chapter is to show that many (but not all) of the functions already defined for real numbers can be defined *in some sense* on the basis of the four basic operations—addition, subtraction, multiplication, and division. For example, the exponential function can be thought of as an "infinite polynomial"

$$e^x = 1 + x + \frac{x^2}{2} + \frac{x^3}{6} + \frac{x^4}{12} + \cdots.$$

Not only will this enable us to define e^z when z is complex but we get a bonus. By expressing a function like e^x in terms of the four basic operations, we show how it can be computed by a computer because computers work with these same operations.

§3. Lotteries and Geometric Series

As this section is being written the siren cry of "no new taxes" has taken over the land. Despite their desire for good public services such as schools, roads, and police and fire protection, voters often rebel when asked to pay the taxes that buy those services. There is an exception to voters' reluctance to raise money for public services—the lottery. By playing the lottery people willingly contribute money that is sorely needed for essential public services. The author is skeptical about the lottery for a variety of reasons. Initially, it was argued that people gambled in any case, that the lottery exploited for a good cause an existing tendancy to gamble. But now the lottery is advertised very heavily and seeks to stimulate new gambling. In addition, there is a question of who plays the lottery. Lottery players subsidize public services for those of us who do not choose to buy lottery tickets. That is very nice of them. We should ask, however, whether this is fair. Who is subsidizing whom?

The lottery advertises large prizes, prizes of a million dollars or even more, but the chances of winning are small. To get some idea of whether it is worth playing the lottery simply as a lottery and not as a form of donating money to the state we need to examine two numbers. The first number is called the **payoff**—the amount of money paid out by the lottery as prizes. The second number is called the **handle**—the total amount of money that is bet on the lottery. If the two numbers were equal then all the money bet would be returned to the players in the form of prizes. Of course, the payoff is normally considerably below the handle. The difference between the two goes for two things. Some of it goes for the costs of running the lottery and some of it goes to the state. From the point of view of the lottery-player if the ratio payoff/handle is close to one then the lottery is a reasonable bet but if this ratio is considerably less than one then the lottery is not a reasonable bet.

Computing the payoff is complicated by fine print. The biggest prizes almost invariably have a footnote. For example, most lotteries have a million dollar[1] prize. Look at that footnote. The author would certainly not turn down $50,000 per year for twenty years. That is not, however, the same thing as one million dollars. If someone were to give the author one million dollars he could invest it in a relatively safe investment that would earn interest at the rate of eight or nine percent. Suppose he invested it conservatively at the rate of eight percent per year. Then each year he would receive $80,000 *forever*. Thus one million dollars is clearly more than $50,000 per year for twenty years.

---

[1]$50,000 per year for twenty years.

§3. Lotteries and Geometric Series

The rest of this section is mathematical. It deals with the mathematics of computing the real value of footnoted prizes.

Suppose that we want to compute the value of a prize of $150,000 paid in three installments of $50,000 each, the first installment paid immediately, the second installment next year and the third installment the following year. We want to find the value of this prize as compared to a prize paid right now. For the sake of argument let us assume that there is a very safe investment available that earns 8% interest (compounded yearly). Thus if you have x dollars available today you can invest it and receive $1.08x$ dollars in one year. Conversely, to receive y dollars next year you need to invest $y/1.08$ dollars this year. Similarly, if you were to invest x dollars this year you could withdraw $(1.08)^2 x$ dollars in two years, and, conversely, to receive y dollars in two years you need to invest $y/(1.08)^2$ dollars this year. Thus the three payments of the prize are worth

$$\$50,000 + \frac{\$50,000}{1.08} + \frac{\$50,000}{(1.08)^2} = \$139,163.24$$

and the prize described above is the equivalent of a lump sum payment of $139.163.24 right now. Economists call this sum the **present value** of the three installments.

To compute the present value of the footnoted lottery prize we would need to compute a sum like

$$\$50,000 + \frac{\$50,000}{1.08} + \frac{\$50,000}{(1.08)^2} + \cdots + \frac{\$50,000}{(1.08)^{19}}$$

or, using sigma notation,

$$\sum_{k=0}^{19} \$50,000 r^k$$

where

$$r = \frac{1}{1.08}$$

The next part of this section examines sums like this.

Geometric Series

A **geometric series** is a sum of the form

$$a + ar + ar^2 + ar^3 + \cdots + ar^n = \sum_{k=0}^{n} ar^k.$$

The following theorem allows us to compute the value of a geometric series.

Theorem 1
$$\sum_{k=0}^{n} ar^k = \frac{a - ar^{n+1}}{1-r}$$

Proof:
Let
$$S = \sum_{k=0}^{n} ar^k = a + ar + ar^2 + ar^3 + \cdots + ar^n$$

Then

$$\begin{aligned} S &= a + ar + ar^2 + ar^3 + \cdots + ar^n \\ rS &= ar + ar^2 + ar^3 + ar^4 + \cdots + ar^{n+1} \\ S - rS &= a - ar^{n+1} \end{aligned}$$

$$S = \frac{a - ar^{n+1}}{1-r}.$$

This theorem can be used to answer many practical questions.

Exercises

*Exercise VI.3.1: What is the present value of the "footnoted" lottery prize—$50,000 per year for twenty years. Assume that there is a safe investment available that pays 8% interest per year.

Exercise VI.3.2: What is the present value of the "footnoted" lottery prize—$50,000 per year for twenty years. Assume that there is a safe investment available that pays 10% interest per year.

Exercise VI.3.3: What is the present value of a lottery prize that pays $50,000 per year for thirty years. Assume that there is a safe investment available that pays 8% interest per year.

Exercise VI.3.4: What is the present value of a lottery prize that pays $50,000 per year for forty years. Assume that there is a safe investment available that pays 8% interest per year.

§3. Lotteries and Geometric Series

Exercise VI.3.5: What is the present value of a lottery prize that pays $50,000 per year for fifty years. Assume that there is a safe investment available that pays 8% interest per year.

*Exercise VI.3.6: Suppose that you want to save enough money to make a down payment on a house in five years. Suppose that the down payment is $30,000. You plan to invest a fixed sum of money starting today and then to make five additional payments at one year intervals. Thus after five years you will have made a total of six payments although the sixth payment will not have had time to earn any interest. How much money do you need to invest each year to accumulate $30,000 after five years? Assume that your investment pays 8% per year.

Exercise VI.3.7: Suppose that you want to save enough money to make a down payment on a house in five years. Suppose that the down payment is $30,000. You plan to invest a fixed sum of money starting today and then to make five additional payments at one year intervals. Thus after five years you will have made a total of six payments although the sixth payment will not have had time to earn any interest. How much money do you need to invest each year to accumulate $30,000 after five years? Assume that your investment pays 6% per year.

*Exercise VI.3.8: Suppose that a country is currently able to build 20,000 houses per year and that each year its capacity to build houses increases by 10%. Find the total number of houses that it can build in twenty years.

Exercise VI.3.9: Suppose that a country is currently able to build 20,000 houses per year and that each year its capacity to build houses increases by 20%. Find the total number of houses that it can build in twenty years.

Most big lottery prizes seem to be paid in twenty yearly installments. It is possible, however, to imagine a prize like the infinite[2] prize. This sounds like a really big prize. We would like to compute its present value. In order to do this we need to be able to compute an infinite geometric sum

$$\sum_{k=0}^{\infty} ar^k.$$

---

[2] $50,000 per year forever.

To estimate this "infinite" sum we would compute

$$\sum_{k=0}^{n} ar^k$$

for very large values of n. We can get better and better estimates by computing the sum for larger and larger values of n. Thus we make the following definition.

Definition:

$$\sum_{k=0}^{\infty} ar^k = \lim_{n \to \infty} \sum_{k=0}^{n} ar^k$$

Theorem 2 *If $|r| < 1$ then*

$$\sum_{k=0}^{\infty} ar^k = \frac{a}{1-r}$$

Proof:
For any n we know that

$$\sum_{k=0}^{n} ar^k = \frac{a - ar^{n+1}}{1-r}.$$

If $|r| < 1$ then

$$\lim_{n \to \infty} r^{n+1} = 0.$$

So

$$\sum_{k=0}^{\infty} ar^k = \lim_{n \to \infty} \sum_{k=0}^{n} ar^k = \lim_{n \to \infty} \frac{a - ar^{n+1}}{1-r} = \frac{a}{1-r}.$$

Exercises

*Exercise VI.3.10: Find the present value of $50,000 per year forever. Assume that a safe investment paying 8% per year is available. Can you think of a more direct way of answering this question without the theorem above?

Exercise VI.3.11: Find the present value of $50,000 per year forever. Assume that a safe investment paying 10% per year is available. Compare your answer to the present value of $50,000 per year for twenty years.

Exercise VI.3.12: Compare the present value of $50,000 per year forever with the present value of $50,000 per year for twenty years. Does the investment rate make any difference?

Exercise VI.3.13: What can you say about

$$\sum_{k=0}^{\infty} ar^k$$

if $|r| \geq 1$?

§4. Infinite Series

In the last section we found the present value of a lottery prize of $50,000 per year for life by computing an infinite sum. Infinite sums are exactly the tool that we need in order to express functions like the exponential function, the logarithm and the trigonometric functions in terms of the four basic operations of addition, subtraction, multiplication and division. In the last section we approached an infinite sum as the limit of finite sums. We use the same idea in this section to make sense more generally of the notion of an infinite sum. Suppose that we have a sequence of numbers: $a_0, a_1, a_2, \ldots, a_n, \ldots$ and we wish to find the sum

$$\sum_{k=0}^{\infty} a_k.$$

We can "estimate" this sum by adding up a large number of the numbers. For example, we could compute

$$\sum_{k=0}^{100} a_k$$

or

$$\sum_{k=0}^{1000} a_k$$

or even
$$\sum_{k=0}^{1,000,000} a_k.$$

Each of these sums is called a **partial sum**. The individual numbers $a_0, a_1, a_2, \ldots$ are called the **terms** and the entire infinite sum is called an **infinite series**. Intuitively, by adding up a large number of the terms we should get a good approximation for the entire infinite series. Thus we would like to say that

$$\sum_{k=0}^{\infty} a_k = \lim_{n \to \infty} \sum_{k=0}^{n} a_k.$$

There is, however, a catch. The limit might not exist. In fact, the infinite sum might not make any sense at all. Consider the following infinite series

$$1 - 1 + 1 - 1 + 1 - 1 + 1 - 1 + 1 - 1 \cdots$$

Notice that in this infinite series all the even-numbered terms (i.e. terms $a_0, a_2, a_4, \ldots$) are $+1$ and all the odd-numbered terms (i.e., terms $a_1, a_3, a_5, \ldots$) are -1. Symbolically,

$$a_k = (-1)^k.$$

Notice that

$$\sum_{k=0}^{0} = 1$$

$$\sum_{k=0}^{1} = 1 - 1 = 0$$

$$\sum_{k=0}^{2} = 1 - 1 + 1 = 1$$

$$\sum_{k=0}^{3} = 1 - 1 + 1 - 1 = 0$$

$$\sum_{k=0}^{4} = 1 - 1 + 1 - 1 + 1 = 1$$

$$\vdots$$

§4. INFINITE SERIES

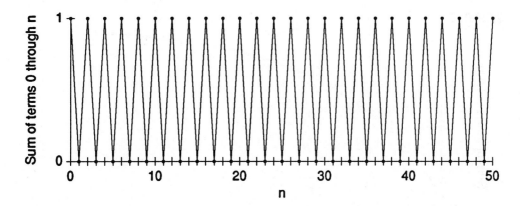

Figure VI.17: Partial sums of the series $1 - 1 + 1 - 1 + \cdots$

Figure VI.17 is a graph of the sequence of partial sums. Notice that the sequence of partial sums is not approaching a limit. It just keeps bouncing back and forth between 0 and 1. This is not an uncommon situation. Frequently, when we write down an infinite sum it doesn't make sense. That is, the limit of the partial sums does not exist. Fortunately, most of the infinite sums that arise from practical problems do make sense. For example, consider the infinite sum that came up in the last section.

$$\$50,000 + \frac{\$50,000}{1.08} + \frac{\$50,000}{(1.08)^2} + \frac{\$50,000}{(1.08)^3} + \cdots$$

or, more compactly,

$$\sum_{k=0}^{\infty} \frac{\$50,000}{(1.08)^k}.$$

Figure VI.18 shows a graph of the partial sums for this infinite series. Notice that for this infinite series the partial sums are approaching a limit as shown in the preceding section. This leads us to make the following definition.

Definition:

The infinite series

$$\sum_{k=0}^{\infty} a_k$$

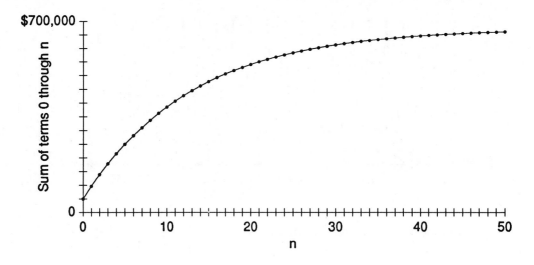

Figure VI.18: Partial sums of $\$50,000 + \frac{\$50,000}{1.08} + \frac{\$50,000}{(1.08)^2} + \frac{\$50,000}{(1.08)^3} + \cdots$

is said to **converge** if the sequence of partial sums has a limit. That is, if

$$\lim_{n \to \infty} \sum_{k=0}^{n} a_k = L.$$

In this case we say that L is the value of the infinite sum and we write

$$\sum_{k=0}^{\infty} a_k = L.$$

If

$$\lim_{n \to \infty} \sum_{k=0}^{n} a_k = +\infty$$

or

$$\lim_{n \to \infty} \sum_{k=0}^{n} a_k = -\infty$$

or the limit doesn't exist at all then we say that the sequence **diverges**.

We can rephrase this definition using the definition of limit.

§4. Infinite Series

Definition:

We say that
$$\sum_{k=0}^{\infty} a_n = L$$
if for every positive tolerance ϵ there is an N such that if $n \geq N$ then
$$\left| \sum_{k=0}^{n} a_k - L \right| < \epsilon.$$

Notice that this definition captures the idea that one can estimate the value
$$\sum_{k=0}^{\infty} a_k$$
to any desired degree of precision by summing sufficiently many terms.

The Harmonic Series

Sometimes it can be very difficult to determine whether or not a particular series converges. One of the nastiest examples is the **harmonic series**,
$$\sum_{k=1}^{\infty} \frac{1}{k} = 1 + \frac{1}{2} + \frac{1}{3} + \frac{1}{4} + \cdots.$$
Notice that we have numbered the terms of this series starting with $k = 1$ rather than with $k = 0$ as we have done with all our previous series. It is often convenient to number the first term $k = 1$. If we calculate a few partial sums we see that

$$\sum_{k=1}^{1} \frac{1}{k} = 1$$

$$\sum_{k=1}^{2} \frac{1}{k} = 1 + \frac{1}{2} = 1.5$$

$$\sum_{k=1}^{3} \frac{1}{k} = 1 + \frac{1}{2} + \frac{1}{3} = 1.8333$$

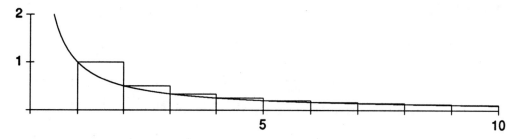

Figure VI.19: Does the harmonic series converge?

$$\sum_{k=1}^{4} \frac{1}{k} = 1 + \frac{1}{2} + \frac{1}{3} + \frac{1}{4} = 2.0833$$

$$\vdots$$

$$\sum_{k=1}^{100} \frac{1}{k} = 5.187377517639621$$

$$\vdots$$

$$\sum_{k=1}^{1000} \frac{1}{k} = 7.485470860550343$$

$$\vdots$$

$$\sum_{k=1}^{2000} \frac{1}{k} = 8.178368103610284$$

It certainly *appears* from these calculations that the harmonic series is convergent. This appearance, however, is deceiving. Consider Figure VI.19. The curved line in Figure VI.19 is the graph of the function $y = 1/x$. Each of the boxes in this figure has width 1. The first box has height 1, the second box has height $1/2$, the third box has height $1/3$ and so forth. Thus the total area of all the boxes is

$$\sum_{k=1}^{\infty} \frac{1}{k}.$$

That is, the total area of the boxes is given by the harmonic series. Next notice that the total area between the curved line $y = 1/x$ and

§4. INFINITE SERIES

the x-axis to the right of the line $x = 1$ is

$$\int_1^\infty \frac{dx}{x} = +\infty.$$

But now notice that the region enclosed in the boxes contains the region between the curved line $y = 1/x$ and the x-axis to the right of the line $x = 1$. Thus

$$\sum_{k=1}^\infty \frac{1}{k} \geq \int_1^\infty \frac{dx}{x} = +\infty$$

and, hence, the harmonic series diverges.

The harmonic series is a good example to keep in mind as we study infinite series. It diverges, but it diverges very, very slowly. In fact, it diverges so slowly that if one were to rely only on computer evidence like the calculations we did above then one would probably guess incorrectly that it converged.

The One-Way Convergence Test

The first question one must ask about any infinite series is whether it converges or diverges. We begin our study of convergence with an important but often misunderstood theorem. This is a one-way theorem. It allows us to make deductions in one direction but not in the other.

Theorem 3 (One-Way Convergence Test) *If the infinite series*

$$\sum_{k=0}^\infty a_k$$

converges then

$$\lim_{k \to \infty} a_k = 0.$$

Proof:

The reason that this theorem is true is clear from Figure VI.20. If you look at the sequence of partial sums for the series

$$\sum_{k=0}^\infty a_k$$

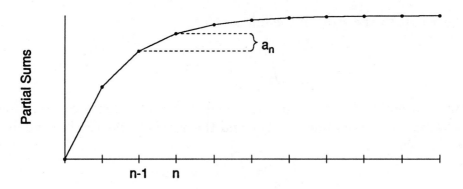

Figure VI.20: The difference between successive partial sums

then the difference between the partial sum

$$\sum_{k=0}^{n-1} a_k$$

and the next partial sum

$$\sum_{k=0}^{n} a_k$$

is just

$$\sum_{k=0}^{n} a_k - \sum_{k=0}^{n-1} a_k = a_n.$$

Since the sequence of partial sums is converging the partial sum themselves are getting closer to the limit and, hence, closer to each other and thus the individual terms a_n must be getting closer to zero.

More formally, we prove this theorem as follows. Suppose that ϵ is positive. We must find a K such that for every $k \geq K$, $|a_k| < \epsilon$. Since the original series converges we know

$$\lim_{n \to \infty} \sum_{k=0}^{n} a_k = L.$$

Hence there is an N such that for every $n \geq N$

$$\left| \sum_{k=0}^{n} a_k - L \right| < \frac{\epsilon}{2}$$

§4. INFINITE SERIES

Thus

$$\left|\sum_{k=0}^{n+1} a_k - L\right| < \frac{\epsilon}{2}$$

$$\left|\sum_{k=0}^{n} a_k - L\right| < \frac{\epsilon}{2}$$

Now

$$\begin{aligned}
|a_{n+1}| &= \left|\sum_{k=0}^{n+1} a_k - \sum_{k=0}^{n} a_k\right| \\
&= \left|\sum_{k=0}^{n+1} a_k - L + L - \sum_{k=0}^{n} a_k\right| \\
&\leq \left|\sum_{k=0}^{n+1} a_k - L\right| + \left|\sum_{k=0}^{n} a_k - L\right| \\
&< \frac{\epsilon}{2} + \frac{\epsilon}{2} \\
&= \epsilon
\end{aligned}$$

and we have shown that if $n \geq N$ then $|a_{n+1}| < \epsilon$. Hence if $k \geq N+1$ then $|a_k| < \epsilon$ which is what we had to show.

This theorem gives us the first and most important test for convergence. If an infinite series

$$\sum_{k=0}^{\infty} a_k$$

converges then

$$\lim_{k \to \infty} a_k = 0.$$

but it is important to remember that this is a one-way theorem. The reverse is not necessarily true. There are many infinite series for which

$$\lim_{k \to \infty} a_k = 0$$

and yet the series does not converge. The harmonic series is an example.

The Integral Test for Convergence

In this section we develop one of the most useful tests for convergence. This test, called the **Integral Test**, is based on the same picture, Figure VI.19, that we used to show that the harmonic series diverged. Before proving this theorem we need some terminology and some more facts about infinite series.

Definition:

An infinite series
$$\sum_{k=0}^{\infty} a_k$$
is said to be **nonnegative** if all of its terms are nonnegative.

Definition:

An infinite series
$$\sum_{k=0}^{\infty} a_k$$
is said to be **bounded** if the sequence of partial sums is bounded. That is, if there is some number B such that for every n,
$$\left| \sum_{k=0}^{n} a_k \right| \leq B.$$

It is possible for a series to be bounded and not converge. For example, the infinite series
$$1 - 1 + 1 - 1 + 1 - 1 + 1 - \cdots$$
is bounded and does not converge. If a *nonnegative* series is bounded, however, then it does converge. We state this fact as a theorem but omit its proof.

Theorem 4 *If the series*
$$\sum_{k=0}^{\infty} a_k$$
is nonnegative and bounded then it converges.

§4. INFINITE SERIES

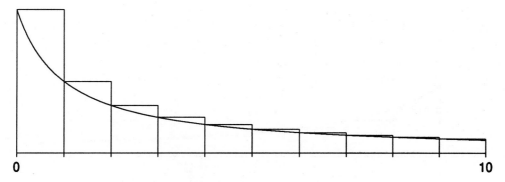

Figure VI.21: The integral test, I

We can now prove the **Integral Test** for convergence.

Theorem 5 (The Integral Test) *Suppose that*

$$\sum_{k=0}^{\infty} a_k$$

is an infinite series whose terms can be described by

$$a_k = f(k)$$

where $f(x)$ is a nonnegative, decreasing function. Then the series converges if and only if

$$\int_0^{\infty} f(x)\, dx$$

is finite.

Proof:

The proof is based on Figures VI.21 and VI.22. Each of the rectangles in Figures VI.21 and VI.22 has a width of one unit. In Figure VI.21 the leftmost rectangle has height $a_0 = f(0)$; the next rectangle has height $a_1 = f(1)$ and so forth. Thus the total area of all the rectangles is

$$\sum_{k=0}^{\infty} a_k.$$

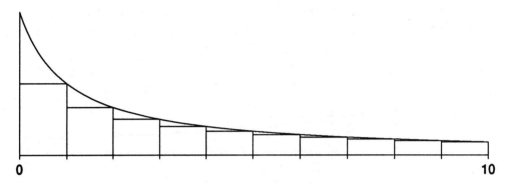

Figure VI.22: The integral test, II

Notice that together these rectangles contain the region bounded by the curve $y = f(x)$ and the x-axis to the right of the y-axis. Thus

$$\int_0^\infty f(x)\, dx \leq \sum_{k=0}^\infty a_k$$

and if the sum converges then the integral is finite. This proves half of the theorem.

For the other half look at Figure VI.22. In this figure the leftmost rectangle has height $a_1 = f(1)$; the next rectangle has height $a_2 = f(2)$, and so forth. Thus the total area of all the rectangles is

$$\sum_{k=1}^\infty a_k$$

Notice that the rectangles are all contained in the region bounded by the curve $y = f(x)$ and the x-axis to the height of the y-axis. Thus, for every n

$$\sum_{k=1}^n a_k \leq \int_0^n f(x)\, dx$$

and

$$\sum_{k=0}^n a_k \leq a_0 + \int_0^n f(x)\, dx \leq a_0 + \int_0^\infty f(x)\, dx$$

That is, the sequence of partial sums is bounded by

$$a_0 + \int_0^\infty f(x)\, dx$$

§4. INFINITE SERIES

which is finite. Hence by Theorem 5 the series is convergent.

Notice that we are writing all of our theorems in terms of series that begin with $k = 0$. The same results apply, however, to series that begin with some other value of k. For example, if $a_k = f(k)$ where $f(x)$ is a nonegative decreasing function then the infinite series $\sum_{k=1}^{\infty} a_k$ converges if and only if the integral $\int_1^{\infty} f(x)\,dx$ is finite.

Example:

Is the series
$$\sum_{k=1}^{\infty} \frac{1}{k^2}$$
convergent?

Answer:

We can check this using the integral test. Let $f(x) = 1/x^2$ and notice that $a_k = f(k)$. Now
$$\int_1^{\infty} \frac{1}{x^2}\,dx = 1$$
is finite. So, by the integral test, the series converges.

The Comparison Test

Often we can determine whether a complicated infinite series converges by comparing it to a simpler one using the following theorem.

Theorem 6 (The Comparison Test) *Suppose that*
$$\sum_{k=0}^{\infty} a_k \quad \text{and} \quad \sum_{k=0}^{\infty} b_k$$
are two nonegative infinite series and that for each k
$$a_k \leq b_k$$
then if
$$\sum_{k=0}^{\infty} b_k$$
converges so does
$$\sum_{k=0}^{\infty} a_k.$$

Proof:

For each of the partial sums

$$\sum_{k=0}^{n} a_k$$

we have

$$\sum_{k=0}^{n} a_k \leq \sum_{k=0}^{n} b_k \leq \sum_{k=0}^{\infty} b_k$$

and, since the right hand side is finite, this shows that the series

$$\sum_{k=0}^{\infty} a_k$$

is bounded and, hence, by Theorem 5, convergent.

Example:

Does the series

$$\sum_{k=1}^{\infty} \left(\frac{1}{k 2^k} \right)$$

converge?

Answer:

We can check this using the comparison test. Consider the series

$$\sum_{k=1}^{\infty} \left(\frac{1}{2^k} \right).$$

This series is a geometric series with $r = 1/2$ and, hence, is convergent. Comparing the original series with this series term-by-term we see that for each k

$$\frac{1}{k 2^k} \leq \frac{1}{2^k}.$$

So, by the comparison test, the original series converges as well.

---

§4. Infinite Series

More Theorems

So far most of our theorems have concerned nonnegative series. The next definition and theorem allow us to apply results about nonnegative series to series in which the terms are real or complex numbers. We omit the proof of the theorem.

Definition:

Suppose that
$$\sum_{k=0}^{\infty}$$
is an infinite series. If
$$\sum_{k=0}^{\infty} |a_k|$$
converges then we say the original series **converges absolutely**.

Theorem 7 *If the series*
$$\sum_{k=0}^{\infty} a_k$$
converges absolutely then it converges.

The next three theorems are not surprising. Their proofs are omitted. They apply to series in which the terms are real or complex numbers.

Theorem 8 *Suppose that the series*
$$\sum_{k=0}^{\infty} a_k \quad \text{and} \quad \sum_{k=0}^{\infty} b_k$$
both converge. Then so does the series
$$\sum_{k=0}^{\infty} (a_k + b_k)$$
and
$$\sum_{k=0}^{\infty} (a_k + b_k) = \sum_{k=0}^{\infty} a_k + \sum_{k=0}^{\infty} b_k.$$

Theorem 9 *Suppose that the series*

$$\sum_{k=0}^{\infty} a_k \quad \text{and} \quad \sum_{k=0}^{\infty} b_k$$

both converge. Then so does the series

$$\sum_{k=0}^{\infty} (a_k - b_k)$$

and

$$\sum_{k=0}^{\infty} (a_k - b_k) = \sum_{k=0}^{\infty} a_k - \sum_{k=0}^{\infty} b_k.$$

Theorem 10 *Suppose that the series*

$$\sum_{k=0}^{\infty} a_k$$

converges and that c is any real or complex number. Then the sequence

$$\sum_{k=0}^{\infty} c a_k$$

also converges and

$$\sum_{k=0}^{\infty} c a_k = c \sum_{k=0}^{\infty} a_k.$$

Exercises

*Exercise VI.4.1: Does the series

$$\sum_{k=1}^{\infty} \frac{1}{\sqrt{k}}$$

converge?

*Exercise VI.4.2: Does the series

$$\sum_{k=1}^{\infty} k^2$$

converge?

§4. INFINITE SERIES

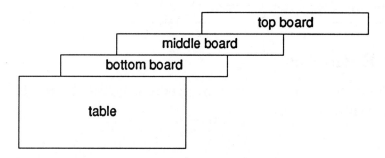

Figure VI.23: Wood pile on table

*Exercise VI.4.3: For what values of p does the series

$$\sum_{k=1}^{\infty} k^p$$

converge?

For what values of p does this series diverge?

Exercise VI.4.4: Does the series

$$\sum_{k=0}^{\infty} \frac{1}{1+k^2}$$

converge?

Exercise VI.4.5: Suppose that you have an infinite supply of boards that are one foot long. You want to make a pile like the one shown in Figure VI.23 on the edge of a table. You must be careful to arrange the pile so that it won't fall over. For example, the top board cannot extend more than six inches beyond the edge of the board just below it.

(a) If your pile is two boards high how far can the end of the top piece extend beyond the edge of the table?

(b) If your pile is three boards high how far can the end of the top piece extend beyond the edge of the table?

(c) If your pile is four boards high how far can the end of the top piece extend beyond the edge of the table?

(d) If your pile is n boards high how far can the end of the top piece extend beyond the edge of the table?

(e) If your pile is an infinite number boards high how far can the end of the top piece extend beyond the edge of the table?

The Ratio Test

One of the most useful tests for convergence is called the **Ratio Test**. This test involves the ratios
$$\frac{a_{k+1}}{a_k}$$
between successive terms of the series. The ratio test applies to series in which the terms are real or complex numbers.

Theorem 11 (The Ratio Test) *Suppose that*
$$\sum_{k=0}^{\infty} a_k$$
is an infinite series such that
$$\lim_{k \to \infty} \left| \frac{a_{k+1}}{a_k} \right| = L.$$

Then

1. *If $|L| < 1$ then the series converges.*

2. *If $|L| > 1$ then the series diverges.*

3. *If $|L| = 1$ then this theorem does not settle the question of convergence either way.*

Notice that this theorem only settles the question of convergence if the limit exists and the limit is neither 1 or -1.

Proof:
Let
$$\lim_{k \to \infty} \left| \frac{a_{k+1}}{a_k} \right| = L.$$
First suppose that $|L| < 1$. Let $\epsilon = 1 - L$ and notice that $\epsilon > 0$. Let
$$r = 1 - \epsilon/2 = \frac{L+1}{2}.$$

§4. Infinite Series

Since
$$\lim_{k \to \infty} \left| \frac{a_{k+1}}{a_k} \right| = L$$
there is an N such that for every $k > N$
$$\left| \left| \frac{a_{k+1}}{a_k} \right| - L \right| < \epsilon/2$$
and, hence,
$$\left| \frac{a_{k+1}}{a_k} \right| < L + \epsilon/2 = \frac{L+1}{2} = r$$

Hence
$$\begin{aligned} |a_{N+2}| &< |a_{N+1}|r \\ |a_{N+3}| &< |a_{N+2}|r < a_{N+1}r^2 \\ |a_{N+4}| &< |a_{N+3}|r < a_{N+1}r^3 \\ &\vdots \end{aligned}$$

Thus
$$|a_{N+n}| < |a_{N+1}|r^{n-1}$$
and by the comparison test comparing
$$\sum_{k=N+1}^{\infty} |a_k|$$
to the geometric series
$$\sum_{k=N+1}^{\infty} |a_{N+1}|r^{k-N-1}$$
we see that the former converges. Hence
$$\sum_{k=0}^{\infty} |a_k| = \sum_{k=0}^{N} |a_k| + \sum_{k=N+1}^{\infty} |a_k|$$
converges and we have shown that if $L < 1$ then the original series converges absolutely and, hence, converges.

Now suppose that $L > 1$. Then

$$\lim_{k \to \infty} a_k \text{ is not } 0$$

and, hence, by the one-way convergence test the series does not converge.

This completes the proof of the theorem.

Exercises

*Exercise VI.4.6: What does the ratio test tell us about the harmonic series?

*Exercise VI.4.7: Does the series

$$\sum_{k=1}^{\infty} \frac{k}{2^k}$$

converge?

Exercise VI.4.8: Does the series

$$\sum_{k=1}^{\infty} \frac{k^2}{2^k}$$

converge?

§5. Some Technical Tools

This section develops some technical tools and notation.

Factorials

Definition:

The notation $n!$ which is read "n factorial" means

$$1 \times 2 \times 3 \times 4 \times \cdots \times n.$$

For example,

$$3! = 1 \times 2 \times 3 = 6 \quad \text{and}$$

$$10! = 1 \times 2 \times 3 \times 4 \times 5 \times 6 \times 7 \times 8 \times 9 \times 10 = 3,628,800.$$

§5. Some Technical Tools

Most scientific calculators have a key to compute $n!$.

Exercises

*Exercise VI.5.1: Compute $1!, 2!, 3!, 4!, 5!, 6!, 7!, 8!$ and $9!$.

Exercise VI.5.2: Use your calculator to compute $20!$, $30!$ and $40!$. Notice how large these numbers are. Find out the largest number whose factorial is within the range of your calculator.

*Exercise VI.5.3: Does the series

$$\sum_{k=1}^{\infty} \frac{2^k}{k!}$$

converge?

Notation for Higher Derivatives

The most commonly used derivatives of a function $y(t)$ are the first derivative $y'(t)$ and the second derivative $y''(t)$, which have immediate physical interpretations, as velocity and acceleration, respectively. In this and the following sections we need higher derivatives as well—the third derivative $y'''(t)$, the fourth derivative $y''''(t)$ and so forth. Because this notation can get tiresome and difficult to read we sometimes use the notation $y^{(n)}(t)$ for the nth derivative of the function $y(t)$.

Extended Initial Value Tables

In the next section we work with initial value problems like

$$y' = y; \quad y(0) = 1$$

or

$$y'' = -y; \quad y(0) = 1, y'(0) = 2.$$

Each of these initial value problems gives general information about a function $y(t)$ in the form of a differential equation and also some very specific information about the value(s) of the function and its derivative in the form of initial condition(s). The following examples show how one can find additional specific information about a function $y(t)$ described by an initial value problem like the ones above.

Example:

Consider the initial value problem.
$$y' = y; \quad y(0) = 1.$$

We would like to make a table of values for $y(0), y'(0), y^{(2)}(0), y^{(3)}(0), \ldots$ like the table below with the blanks filled in. Such a table is called an **extended initial value table.**

$y(0)$	
$y'(0)$	
$y^{(2)}(0)$	
$\vdots$	$\vdots$
$y^{(n)}(0)$	
	$\vdots$

The first entry in this table is easy. The initial condition tells us that $y(0) = 1$.

$y(0)$	1
$y'(0)$	
$y^{(2)}(0)$	
$\vdots$	$\vdots$
$y^{(n)}(0)$	
	$\vdots$

We can find the second entry by looking at the differential equation itself
$$y' = y.$$
So, in particular,
$$y'(0) = y(0)$$
and since we already know that $y(0) = 1$
$$y'(0) = y(0) = 1$$
giving us

§5. Some Technical Tools

$y(0)$	1
$y'(0)$	1
$y^{(2)}(0)$	
$\vdots$	$\vdots$
$y^{(n)}(0)$	
$\vdots$	$\vdots$

Now if we differentiate both sides of the original differential equation

$$y' = y$$

we get

$$y'' = y'$$

and, in particular, we note that

$$y''(0) = y'(0)$$

which we have just seen is 1. Thus

$$y''(0) = y'(0) = 1$$

giving us

$y(0)$	1
$y'(0)$	1
$y^{(2)}(0)$	1
$\vdots$	$\vdots$
$y^{(n)}(0)$	
$\vdots$	$\vdots$

Continuing in this way we see that

$$y''' = y''$$

and, in particular, that

$$y'''(0) = y''(0) = 1.$$

The same argument shows that for all n, $y^{(n)}(0) = 1$. Thus the completed table is

$y(0)$	1
$y'(0)$	1
$y^{(2)}(0)$	1
$\vdots$	$\vdots$
$y^{(n)}(0)$	1
$\vdots$	$\vdots$

Example:

Construct an extended initial value table for the initial value problem
$$y' = 0.5y; \qquad y(0) = 1.$$

Answer:

This begins exactly as the previous table

$y(0)$	1
$y'(0)$	
$y^{(2)}(0)$	
$\vdots$	$\vdots$
$y^{(n)}(0)$	
$\vdots$	$\vdots$

but now notice that
$$y' = 0.5y$$
so that
$$y'(0) = 0.5y(0) = 0.5$$
and the table continues

$y(0)$	1
$y'(0)$	0.5
$y^{(2)}(0)$	
$\vdots$	$\vdots$
$y^{(n)}(0)$	
$\vdots$	$\vdots$

§5. Some Technical Tools

Continuing in this way we construct the table

$y(0)$	1
$y'(0)$	0.5
$y^{(2)}(0)$	0.25
$\vdots$	$\vdots$
$y^{(n)}(0)$	$\left(\frac{1}{2}\right)^n$
$\vdots$	$\vdots$

Example:

Construct an extended initial value table for the initial value problem

$$y'' = -y; \qquad y(0) = 1, y'(0) = 2.$$

Answer:

Notice that the two initial conditions enable us to fill in the first two lines of the table.

$y(0)$	1
$y'(0)$	2
$y^{(2)}(0)$	
$\vdots$	$\vdots$
$y^{(n)}(0)$	
$\vdots$	$\vdots$

Looking at the original differential equation we see that

$$y'' = -y'$$

and, in particular, that

$$y''(0) = -y(0) = -1$$

enabling us to fill in the next line of the table.

$y(0)$	1
$y'(0)$	2
$y^{(2)}(0)$	-1
$\vdots$	$\vdots$
$y^{(n)}(0)$	
$\vdots$	$\vdots$

Next by differentiating both sides of the original differential equation we see that

$$y''' = -y'$$

and, in particular,

$$y'''(0) = -y'(0) = -2$$

allowing us to fill in the next line of the table

$y(0)$	1
$y'(0)$	2
$y^{(2)}(0)$	-1
$y^{(3)}(0)$	-2
$\vdots$	$\vdots$
$y^{(n)}(0)$	
$\vdots$	$\vdots$

Continuing in this way we obtain

$y(0)$	1
$y'(0)$	2
$y^{(2)}(0)$	-1
$y^{(3)}(0)$	-2
$y^{(4)}(0)$	1
$y^{(5)}(0)$	2
$y^{(6)}(0)$	-1
$y^{(7)}(0)$	-2
$\vdots$	$\vdots$

Exercises

*Exercise VI.5.4: Construct an extended initial value table for the initial value problem

$$y' = -y; \qquad y(0) = 1.$$

Exercise VI.5.5: Construct an extended initial value table for the initial value problem

$$y'' = -y; \qquad y(0) = 1, y'(0) = 0$$

Do enough lines of the table so that you see a pattern.

Exercise VI.5.6: Construct an extended initial value table for the initial value problem

$$y'' = -y; \qquad y(0) = 0, y'(0) = 1$$

Do enough lines of the table so that you see a pattern.

§6. Power Series

Perhaps the single most important initial value problem is the initial value problem

$$y' = y, \quad y(0) = 1.$$

We first studied this initial value problem in chapter III. At that time we noticed that none of the functions that we had met so far satisfied the differential equation $y' = y$. For example, no polynomial,

$$p(x) = a_n x^n + a_{n-1} x^{n-1} + \cdots + a_1 x + a_0,$$

can satisfy the differential equation $y' = y$ because the highest power of x in the derivative

$$p'(x) = n x^{n-1} + (n-1) x^{n-2} + \cdot + x_1$$

is less than the highest power of x in $p(x)$.

We eventually solved this initial value problem by using Euler's Theorem on the existance of solutions to initial value problems to conclude that it had a solution and Euler's Method to estimate this solution. Then we gave the solution a name, the exponential function.

In this section we show that the exponential function is a kind of "infinite polynomial" or **power series**,

$$e^x = 1 + x + \frac{x^2}{2!} + \frac{x^3}{3!} + \cdots + \frac{x^n}{n!} + \cdots = \sum_{n=0}^{\infty} \frac{x^n}{n!}.$$

For any particular value of x a power series like this one requires us to evaluate an infinite sum. Consider, for example, the power series

$$T(x) = 1 + x + x^2 + x^3 + \cdots = \sum_{n=0}^{\infty} x^n.$$

To determine $T(1/2)$ we must compute

$$T(1/2) = 1 + \frac{1}{2} + \frac{1}{4} + \frac{1}{8} + \cdots$$

This is a geometric series whose sum is 2. Thus, $T(1/2) = 2$.

In fact, for any value of x, $T(x)$ is a geometric series and by our work in section 3

$$T(x) = \begin{cases} 1/(1-x), & \text{if } |x| < 1; \\ \text{diverges}, & \text{if } |x| \geq 1. \end{cases}$$

This example is typical of power series. Often a power series converges for some values of x and diverges for others.

Exercises

Consider the power series

$$E(x) = 1 + x + \frac{x^2}{2!} + \frac{x^3}{3!} + \cdots + \frac{x^n}{n!} + \cdots = \sum_{n=0}^{\infty} \frac{x^n}{n!}.$$

For each of the following values of x estimate $E(x)$ by computing

$$\sum_{n=0}^{N} \frac{x^n}{n!}$$

for some large values of N. Compare your estimate with e^x.

§6. Power Series

*Exercise VI.6.1: $x = 0$.

Exercise VI.6.2: $x = 1$.

*Exercise VI.6.3: $x = -1$.

Exercise VI.6.4: $x = 10$.

Exercise VI.6.5: $x = 100$.

In the remainder of this section we develop the theory of power series.

Definition:

A **power series centered at zero** is a formal sum of the form
$$p(z) = \sum_{n=0}^{\infty} a_n z^n.$$

This definition needs some explanation. First, we use the letter "z" to emphasize that we can work with either real or complex numbers. We use the phrase "formal" sum because a power series really tells us how to *try* to compute the value $p(z)$ at each point z. Writing down the rules for a computation doesn't automatically guarantee that it is possible to carry out the computation.

Definition:

The **domain** of a power series centered at zero
$$\sum_{n=0}^{\infty} a_n z^n$$
is the set of all complex numbers z such that the series
$$\sum_{n=0}^{\infty} a_n z^n$$
converges *absolutely*.

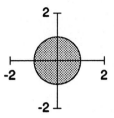

Figure VI.24: Domain of $1 + z + z^2 + z^3 + \cdots$

Example:

Consider the power series

$$p(z) = \sum_{k=0}^{\infty} z^k = 1 + z + z^2 + z^3 + z^4 + \cdots$$

For any particular value of z this is a geometric series that converges absolutely if $|z| < 1$ and diverges if $|z| \geq 1$. Thus the domain of $p(z)$ is the set of all z such that $|z| < 1$. This set is pictured in Figure VI.24.

The first thing to do when confronted with a power series is to ask what its domain is. There is one easy theorem about this.

Theorem 12 *The point 0 is in the domain of any power series centered at zero.*

Proof:

Let $p(z)$ be the power series

$$p(z) = \sum_{n=0}^{\infty} a_n z^n = a_0 + a_1 z + a_2 z^2 + a_3 z^3 + \cdots$$

Notice that

$$\sum_{n=0}^{\infty} |a_n(0)^n)| = |a_0| + 0 + 0 + 0 + \cdots = |a_0|.$$

So the infinite sum $p(0)$ is absolutely convergent.

As you might have guessed from the name, there is a slight variation on the basic idea of a power series centered at zero.

§6. POWER SERIES

Definition:

Suppose that a is any complex number. A **power series centered at** a is a formal sum of the form

$$p(z) = \sum_{n=0}^{\infty} a_n(z-a)^n.$$

The same ideas that we developed for a power series centered at zero carry over to this slightly more general kind of power series.

Definition:

The domain of the power series centered at a

$$p(z) = \sum_{n=0}^{\infty} a_n(z-a)^n$$

is the set of all z for which the series

$$\sum_{n=0}^{\infty} a_n(z-a)^n$$

converges *absolutely*.

Example:

Let a be a complex number. Consider the power series centered at a

$$\sum_{n=0}^{\infty}(z-a)^n = 1 + (z-a) + (z-a)^2 + (z-a)^3 + \cdots.$$

Notice that for any particular value of z this series is a geometric series with $r = (z-a)$. It converges absolutely if $|z-a| < 1$ and diverges if $|z-a| \geq 1$. Thus the domain of this power series is the set of all complex numbers z such that $|z-a| < 1$. This set is pictured in Figure VI.25.

Notice that for both of our examples so far the domain has been a disk. The two examples are actually very similar. So you might think this is a fluke. Surprisingly it isn't. The following theorem describes the possibilities for the domain of a power series centered at a.

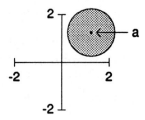

Figure VI.25: Domain of $1 + (z-a) + (z-a)^2 + (z-a)^3 + \cdots$

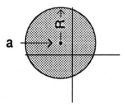

Figure VI.26: First possibility for the domain of $p(z)$.

Theorem 13 *Suppose that*

$$p(z) = \sum_{n=0}^{\infty} a_n(z-a)^n$$

is a power series centered at a. Then there are three possibilities.

1. There is a positive real number R called the **radius of convergence** such that

    - If $|z-a| < R$ then z is in the domain of $p(z)$.
    - If $|z-a| > R$ then z is not in the domain of $p(z)$.
    - If $|z-a| = R$ then z may be in or not in the domain of $p(z)$.

    In this case the domain of $p(z)$ looks like Figure VI.26.

2. The domain of $p(z)$ is just the single point a. In this case we say that **the radius of convergence of $p(z)$ is zero**.

§6. Power Series

3. The domain of $p(z)$ is the set of all complex numbers. In this case we say that **the radius of convergence of $p(z)$ is ∞**.

Proof:
The basic idea behind the proof is the fact that if u is in the domain of $p(z)$ and v is any complex number with $|v - a| < |u - a|$ then for each n

$$|v - a|^n \leq |u - a|^n$$

so

$$|a_n(v-a)^n| = |a_n||v-a|^n \leq |a_n||u-a|^n = |a_n(u-a)^n|$$

and, for each N,

$$\sum_{n=0}^{N} |a_n(v-a)^n| \leq \sum_{n=0}^{N} |a_n(u-a)^n| \leq \sum_{n=0}^{\infty} |a_n(u-a)^n|.$$

The sum on the right converges since u is in the domain of $p(z)$. Thus the series

$$\sum_{n=0}^{\infty} |a_n(v-a)^n|$$

is bounded and converges by Theorem 5. Hence,

$$\sum_{n=0}^{\infty} a_n(v-a)^n$$

converges absolutely which is what we wanted to show.

This basic idea implies that if $p(z)$ converges absolutely at any point u it also converges at any other point that is closer to a. This, in turn, implies the theorem.

As with other functions, we want to differentiate functions that are defined by power series. One is tempted to do this term-by-term exactly as if a power series were just a very long polynomial. The following two theorems state that this is legitimate. Theorem 14 says that the new power series obtained in this way has the same radius of convergence as the original one and thus the function obtained by differentiating the original power series term-by-term has the same domain as the original function. Theorem 15 states that this new function is the derivative of the original one. Both proofs are omitted.

Thus far we have only talked about derivatives for functions of a *real* variable. One can, however, develop an elegant and powerful theory of differentiation for functions of complex variables. Theorem 15 is true for functions of complex variables.

Theorem 14 *Suppose that*

$$p(z) = \sum_{n=0}^{\infty} a_n(z-a)^n = a_0 + a_1(z-a) + a_2(z-a)^2 + a_3(z-a)^3 + \cdots$$

is a power series centered at a. Let

$$q(z) = \sum_{n=1}^{\infty} n a_{n-1}(z-a)^{n-1} = a_1 + 2a_2(z-a) + 3a_3(z-a)^2 + \cdots.$$

Then $p(z)$ and $q(z)$ have the same radius of convergence.

Theorem 15 *Suppose that*

$$\sum_{n=0}^{\infty} a_n(z-a)^n = a_0 + a_1(z-a) + a_2(z-a)^2 + a_3(z-a)^3 + \cdots$$

is a power series centered at a whose radius of convergence is R and that the function $f(z)$ is defined by

$$f(z) = \sum_{n=0}^{\infty} a_n(z-a)^n$$

for every z such that $|z-a| < R$.
Then $f(z)$ is a differentiable function and the derivative of $f(z)$ is given by

$$f'(z) = \sum_{n=1}^{\infty} n a_n(z-a)^{n-1} = a_1 + 2a_2(z-a) + 3a_3(z-a)^2 + \cdots.$$

The proof of Theorem 15 is somewhat difficult. It is beyond the scope of this book. One difficult part of the proof is seeing why the proof is difficult. At first glance you might expect a very easy proof along the following lines.
Let

$$f_k(z) = a_0 + a_1(z-a) + a_2(z-a)^2 + a_3(z-a)^3 + \cdots + a_k(z-a)^k$$

§6. Power Series

Notice that $f_k(z)$ is just an ordinary polynomial so
$$f'_k(z) = a_1 + 2a_2(z-a) + 3a_3(z-a)^2 + \cdots + ka_k(z-a)^{k-1}$$
Notice that
$$f(z) = \lim_{k \to \infty} f_k(z).$$
Now we would like to say
$$f'(z) = \lim_{k \to \infty} f'_k(z).$$
If we could say this we would be done since
$$\lim_{k \to \infty} f'_k(z) = \sum_{n1}^{\infty} na_n(z-a)^{n-1}.$$

Unfortunately, there is a huge flaw in this argument. It is not necessarily true that if we have functions $g(z), g_0(z), g_1(z), g_2(z), \ldots$ such that
$$\lim_{k \to \infty} g_k(z) = g(z)$$
then
$$g'(z) = \lim_{k \to \infty} g'_k(z).$$

Example:

Let
$$g_k(x) = \left(\frac{1}{2^k}\right) \sin(2^k x).$$
Notice that
$$g'_k(x) = \cos(2^k x).$$

Figure VI.27 shows $g_1(x)$ and $g'_1(x)$. Figure VI.28 shows $g_2(x)$ and $g'_2(x)$. Figure VI.29 shows $g_3(x)$ and $g'_3(x)$. Figure VI.30 shows $g_4(x)$ and $g'_4(x)$. Notice that
$$\lim_{k \to \infty} g_k(x) = 0$$
but that
$$\lim_{k \to \infty} g'_k(x)$$
does not even exist.

This example shows that the "very easy proof" of Theorem 15 does not work.

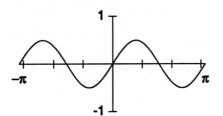 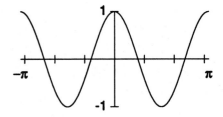

Figure VI.27: $g_1(x)$ and $g_1'(x)$

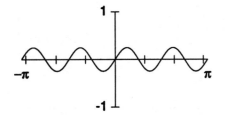 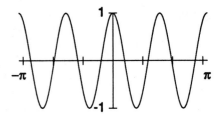

Figure VI.28: $g_2(x)$ and $g_2'(x)$

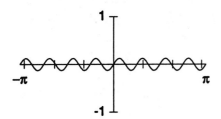 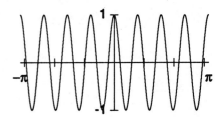

Figure VI.29: $g_3(x)$ and $g_3'(x)$

§6. Power Series

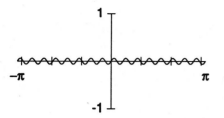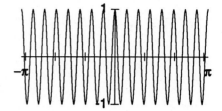

Figure VI.30: $g_4(x)$ and $g_4'(x)$

The Taylor Series for a Function

Suppose that we have a function $f(x)$ that can be described by a power series centered at the point a. That is, we can write

$$f(x) = \sum_{n=0}^{\infty} a_n(x-a)^n.$$

We will call this power series the **Taylor Series for $f(x)$ centered at a**. The following theorem tells us how to find the coefficients $a_0, a_1, a_2, \ldots$.

Theorem 16 *The coefficients a_n of the Taylor Series for $f(x)$ centered at a are given by*

$$\begin{aligned} a_0 &= f(a) \\ a_1 &= f'(a) \\ a_2 &= \frac{f^{(2)}(a)}{2} \\ &\vdots \\ a_n &= \frac{f^{(n)}(a)}{n!} \end{aligned}$$

Proof:

$$f(x) = a_0 + a_1(x-a) + a_2(x-a)^2 + a_3(x-a)^3 + \cdots + a_n(x-a)^n + \cdots.$$

So

$$f(a) = a_0.$$

$$f'(x) = a_1 + 2a_2(x-a) + 3a_3(x-a)^2 + \cdots + na_n(x-a)^{n-1} + \cdots.$$

So
$$f'(a) = a_1.$$

$$f^{(2)}(x) = 2a_2 + 6a_3(x-a) + \cdots + n(n-1)a_n(x-a)^{n-2} + \cdots.$$

So
$$f^{(2)}(a) = 2a_2$$

and
$$a_2 = \frac{f^{(2)}(a)}{2}.$$

$$f^{(n)}(x) = n!a_n + \cdots.$$

So
$$f^{(n)}(a) = n!a_n$$

and
$$a_n = \frac{f^{(n)}(a)}{n!}.$$

In particular, for a Taylor Series centered at 0 we have
$$a_n = \frac{f^{(n)}(0)}{n!}.$$

Many important functions can be described by power series in this way.

Example:

Express the function $y = \sin x$ as a Taylor series centered at 0.

Answer:

The first step is to calculate the derivatives of the function $f(x) = \sin x$.

$$\begin{aligned} f(x) &= \sin x \\ f'(x) &= \cos x \end{aligned}$$

§6. Power Series

$$f^{(2)}(x) = -\sin x$$
$$f^{(3)}(x) = -\cos x$$
$$f^{(4)}(x) = \sin x$$
$$f^{(5)}(x) = \cos x$$
$$\vdots$$

The second step is to evaluate the function and its derivatives at the point 0 to get a table like the one below. Notice how similar this table is to the extended initial value tables with which we worked in the previous section.

$f(0)$	0
$f'(0)$	1
$f^{(2)}(0)$	0
$f^{(3)}(0)$	-1
$f^{(4)}(0)$	0
$f^{(5)}(0)$	1
$\vdots$	$\vdots$
$\vdots$	$\vdots$

Finally, using the previous theorem

$$f(x) = x - \frac{x^3}{3!} + \frac{x^5}{5!} - \frac{x^7}{7!} + \frac{x^9}{9!} + \cdots.$$

This is the Taylor series for the function $f(x) = \sin x$. This is one way that computers and calculators can compute the sine function. Of course, they would not use the entire infinite series. The first few terms are sufficient to get a very accurate approximation of the sine function. In the next section we discuss the question of how many terms are necessary to attain an approximation with a desired tolerance.

Figure VI.31 compares the sine function with the approximation

$$x - \frac{x^3}{3!}.$$

Figure VI.32 compares the sine function with the approximation

$$x - \frac{x^3}{3!} + \frac{x^5}{5!}.$$

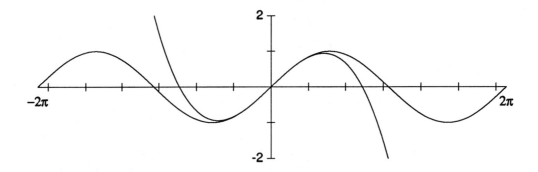

Figure VI.31: Comparison of $y = \sin x$ and $y = x - x^3/6$

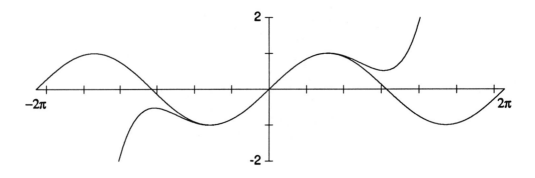

Figure VI.32: Comparison of $y = \sin x$ and $y = x - x^3/6 + x^5/120$

Figure VI.33 compares the sine function with the approximation

$$x - \frac{x^3}{3!} + \frac{x^5}{5!} - \frac{x^7}{7!}.$$

Notice that even with a relatively small number of terms we get a very good approximation of the sine function.

We can find the radius of convergence for this power series by using the ratio test. Looking at the ratio of two successive terms of the series

$$x - \frac{x^3}{3!} - \frac{x^5}{5!} + \cdots \pm \frac{x^{2n-1}}{(2n-1)!} \pm \frac{x^{2n+1}}{(2n+1)!} \pm \cdots$$

§6. Power Series

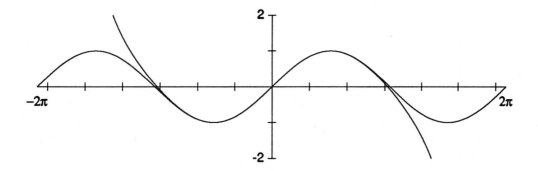

Figure VI.33: $y = \sin x$ and $y = x - x^3/6 + x^5/120 - x^7/5040$

we obtain
$$\frac{\left(\frac{x^{2n+1}}{(2n+1)!}\right)}{\left(\frac{x^{2n-1}}{(2n-1)!}\right)} = \frac{x^2}{2n(2n+1)}$$

and taking the limit
$$\lim_{n \to \infty} \frac{x^2}{2n(2n+1)} = 0$$

for any x. (Notice, as we take the limit x stays fixed.) Thus this series converges for every value of x and the radius of convergence is infinity.

Example:

Next consider the function $f(x) = \log x$. Notice that this function is not defined at the point $x = 0$. Hence, we cannot hope to find a Taylor Series centered at zero for this function. Instead we look for a power series centered at 1. That is, we would like to write $\log x$ as

$$\log(x) = a_0 + a_1(x-1) + a_2(x-1)^2 + a_3(x-1)^3 + \cdots$$

We begin by computing the first few derivatives

$$\begin{aligned} f(x) &= \log x \\ f'(x) &= \frac{1}{x} = x^{-1} \\ f^{(2)}(x) &= -x^{-2} \\ f^{(3)}(x) &= 2x^{-3} \end{aligned}$$

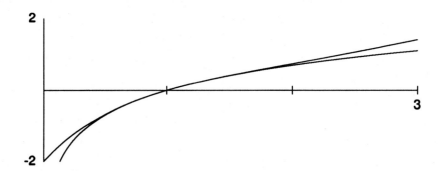

Figure VI.34: $y = \log x$ and $y = (x-1) - (x-1)^2/2 + \cdots + (x-1)^{11}/11$

$$f^{(4)}(x) = -6x^{-4}$$
$$\vdots$$
$$f^{(n)}(x) = (-1)^{(n+1)}(n-1)!x^{-n}$$
$$\vdots$$

This leads to the following table

$f(1)$	0
$f'(1)$	1
$f^{(2)}(1)$	-1
$f^{(3)}(1)$	2
$f^{(4)}(1)$	-6
$\vdots$	$\vdots$
$f^{(n)}(1)$	$(-1)^{(n+1)}(n-1)!$
$\vdots$	$\vdots$

and, hence, to the following Taylor Series

$$\log(x) = (x-1) - \frac{(x-1)^2}{2} + \frac{(x-1)^3}{3} - \frac{(x-1)^4}{4} + \cdots.$$

Figure VI.34 compares the log function with the approximation

$$(x-1) - \frac{(x-1)^2}{2} + \cdots + \frac{(x-1)^{11}}{11}$$

§6. Power Series

The Taylor series (centered at 1) for this function converges much more slowly than then Taylor series for the function $y = \sin x$. The table below compares the values of the function $y = \log x$ for $x = 0.2, 0.4, \ldots, 1.8$ with approximations obtained using the partial sums of this series

$$p_n(x) = (x-1) - \frac{(x-1)^2}{2} \cdots \pm \frac{(x-1)^n}{n}$$

for $n = 3, 5, 9, 20$ and 40.

x	$\log x$	$p_3(x)$	$p_5(x)$	$p_9(x)$	$p_{20}(x)$	$p_{40}(x)$
.2	-1.60944	-1.29067	-1.45860	-1.56814	-1.60755	-1.60943
.4	-0.91629	-0.85200	-0.89995	-0.91495	-0.91629	-0.91629
.6	-0.51083	-0.50133	-0.50978	-0.51081	-0.51083	-0.51083
.8	-0.22314	-0.22267	-0.22313	-0.22314	-0.22314	-0.22314
1.2	0.18232	0.18267	0.18233	0.18232	0.18232	0.18232
1.4	0.33647	0.34133	0.33698	0.33648	0.33647	0.33647
1.6	0.47000	0.49200	0.47515	0.47040	0.47000	0.47000
1.8	0.58779	0.65067	0.61380	0.59401	0.58754	0.58779

Notice how many terms seem to be necessary to get a good approximation for $\log(0.2)$. If we look at points closer to zero then more terms are required. For example,

$$\log(0.1) = -2.302585$$

and

$$\begin{aligned} p_{40}(0.1) &= -2.08388 \\ p_{80}(0.1) &= -2.26335 \\ p_{200}(0.1) &= -2.29985 \end{aligned}$$

Since the function $\log x$ is undefined when $x = 0$ we would not expect the power series to converge for $x = 0$. In fact, for $x = 0$ we get

$$\begin{aligned} \log(0) &= (0-1) - \frac{(0-1)^2}{2} + \frac{(0-1)^3}{3} - \frac{(0-1)^4}{4} + \cdots \\ &= -1 - 1/2 - 1/3 - 1/4 - \cdots \\ &= -(1 + 1/2 + 1/3 + 1/4 + \cdots) \end{aligned}$$

which is just minus the harmonic series and as expected does not converge. Thus the radius of convergence will be less than or equal to 1. We can determine the radius of convergence by using the ratio test. If we look at the ratio of successive terms of the sequence

$$\log(x) = (x-1) - \frac{(x-1)^2}{2} + \frac{(x-1)^3}{3} - \cdots \pm \frac{(x-1)^n}{n} \pm \frac{(x-1)^{n+1}}{n+1} \pm \cdots$$

we obtain

$$\frac{\left(\frac{(x-1)^{n+1}}{n+1}\right)}{\left(\frac{(x-1)^n}{n}\right)} = \frac{n(x-1)}{n+1}$$

and taking the limit

$$\lim_{n \to \infty} \left| \frac{n(x-1)}{n+1} \right| = |x-1|.$$

Hence, by the ratio test, the series will converge if $|x-1| < 1$. Thus the radius of convergence for this series is 1.

Exercises

*Exercise VI.6.6:
Find the Taylor series centered at 0 for the function $f(x) = \cos x$. Find the radius of convergence for this series.

Exercise VI.6.7:
Find the Taylor series centered at 0 for the function $f(x) = e^x$. Find the radius of convergence for this series.

Exercise VI.6.8:
The Taylor series that we have found so far can be used to compute $\sin z$, $\cos z$ and e^z for any complex number z. Using these Taylor series show that for any real number y,

$$e^{iy} = \cos y + i \sin y.$$

To find the coefficients $a_0, a_1, a_2, \ldots$ for a function

$$f(x) = \sum_{n=0}^{\infty} a_n (x-a)^n$$

§6. Power Series

we need to know $f(a), f'(a), f^{(2)}(a), \ldots$ in the last section we saw how to determine these values from an initial value problem by constructing an extended initial value table. Thus we can find power series solutions for initial value problems as shown in the following examples.

Example: Find a power series solution of the initial value problem

$$y' = y, \quad y(0) = 1.$$

(Note: Exercise V.6.7 asks you to find the power series for the function e^x. Since this function is the solution of this initial value problem your answer should agree with the result here.)

Answer:

The extended initial value table for this initial value problem is

$y(0)$	1
$y'(0)$	1
$y^{(2)}(0)$	1
$\vdots$	$\vdots$
$y^{(n)}(0)$	1
$\vdots$	$\vdots$

Thus, for every n, $y^{(n)}(0) = 1$ and $a_n = 1/n!$. leading to the power series

$$\sum_{n=0}^{\infty} \frac{x^n}{n!}.$$

Example:

Find a power series solution of the initial value problem

$$y'' = -y; \quad y(0) = 1, y'(0) = 0.$$

Answer:

The extended initial value table for this initial value problem is

$y(0)$	1
$y'(0)$	0
$y^{(2)}(0)$	-1
$y^{(3)}(0)$	0
$y^{(4)}(0)$	1
$y^{(5)}(0)$	0
$y^{(6)}(0)$	-1
$\vdots$	$\vdots$

Thus

$$a_0 = 1$$
$$a_1 = 0$$
$$a_2 = -1/2$$
$$a_3 = 0$$
$$a_4 = 1/4! = 1/24$$
$$\vdots$$

leading to the poower series

$$1 - \frac{x^2}{2} + \frac{x^4}{4!} + \cdots.$$

Exercises

*Exercise VI.6.9: Find a power series solution to the initial value problem

$$y' = -y, \quad y(0) = 1.$$

*Exercise VI.6.10: Find a power series solution **centered at 1** to the initial value problem

$$y' = y, \quad y(1) = 2.$$

*Exercise VI.6.11: Find a power series solution to the initial value problem

$$y'' = -y, \quad y(0) = 1, y'(0) = 2.$$

§7. Taylor's Theorem

In this section we develop one of the most important theorems of calculus, **Taylor's Theorem**. Although Taylor's Theorem is true both for functions of a real variable and for functions of a complex variable, we give the proof only for functions of a real variable for simplicity. We begin our development of Taylor's Theorem with an example.

Example:

Suppose that a particular object is traveling with a velocity that varies between 10 feet per second and 30 feet per second. How far will it travel in 40 seconds?

Answer:

With this data we cannot determine exactly how far the object will travel but we can say that it will travel at least 400 feet and no more than 1200 feet. The lower figure is the lowest possible velocity (10 feet per second) multiplied by 40 seconds. The higher figure is the highest possible velocity (30 feet per second) multiplied by 40 seconds.

We can generalize this example to a theorem as follows.

Theorem 17 *Suppose that $f(x)$ is a function and that for all x between a and $(a+h)$*
$$L \leq f'(x) \leq U.$$
Then
$$Lh \leq f(a+h) - f(a) \leq Uh.$$

Proof: Since
$$L \leq f'(x) \leq U$$
for all x between a and $a+h$
$$\int_a^{a+h} L \, dx \leq \int_a^{a+h} f'(x) \, dx \leq \int_a^{a+h} U \, dx$$
so that
$$hL \leq f(a+h) - f(a) \leq hU$$
which was what we wanted to show.

This theorem is not surprising. Given bounds on an object's velocity, we can deduce bounds on distance traveled. In the same way, if an object starts at rest and we are given bounds on its acceleration we can deduce bounds on its distance traveled.

Theorem 18 *Suppose that $f(x)$ is a function and that $f'(a) = 0$. Suppose also that for all x between a and $(a+h)$*

$$L \le f''(x) \le U.$$

Then

$$\frac{Lh^2}{2} \le f(a+h) - f(a) \le \frac{Uh^2}{2}.$$

Proof: Since

$$L \le f''(x) \le U,$$

between a and $(a+h)$, we have for every s between a and $(a+h)$

$$\int_a^s L\, dx \ \le\ \int_a^s f''(x)\, dx \ \le\ \int_a^s U\, dx$$

$$L(s-a) \ \le\ f'(s) - f'(a) \ \le\ U(s-a)$$

and, since $f'(a) = 0$,

$$L(s-a) \le f'(s) \le U(s-a).$$

Hence,

$$\int_a^{a+h} L(s-a)\, ds \ \le\ \int_a^{a+h} f'(s)\, ds \ \le\ \int_a^{a+h} U(s-a)\, ds$$

$$Lh^2/2 \ \le\ f(a+h) - f(a) \ \le\ Uh^2/2$$

which was what we wanted to prove.

These two theorems have immediate physical interpretations. They allow us to get bounds on how far an object might travel if we are given bounds either on its velocity or on its acceleration. Our next theorem is very similar to the previous two theorems. It give us important information about how much a function can change between two points, exactly what is needed to determine how quickly Taylor Series converge.

§7. Taylor's Theorem

Theorem 19 *Suppose that $f(x)$ is a function and that the first n derivatives of f are zero at $x = a$. That is, suppose that*

$$f'(a) = 0$$
$$f^{(2)}(a) = 0$$
$$f^{(3)}(a) = 0$$
$$\vdots$$
$$f^{(n)}(a) = 0.$$

Suppose also that for every x between a and $(a + h)$ that

$$L \leq f^{(n+1)}(x) \leq U.$$

Then

$$\frac{Lh^{n+1}}{(n+1)!} \leq f(a+h) - f(a) \leq \frac{Uh^{n+1}}{(n+1)!}.$$

Proof: The proof is identical to the proof of the last theorem except that one must integrate $(n + 1)$ times.

As an immediate consequence of this theorem and the Intermediate Value Theorem we have the following.

Theorem 20 *Suppose that $f(x)$ is a function and that the first n derivatives of f are zero at $x = a$. That is, suppose that*

$$f'(a) = 0$$
$$f^{(2)}(a) = 0$$
$$f^{(3)}(a) = 0$$
$$\vdots$$
$$f^{(n)}(a) = 0.$$

Suppose also that $f^{(n+1)}(x)$ is a continuous function between a and $(a + h)$. Then there is a point ξ between a and $a + h$ such that

$$f(a+h) - f(a) = \frac{f^{(n+1)}(\xi)h^{n+1}}{(n+1)!}.$$

Proof:
Let L denote the minimum value of $f^{(n+1)}(x)$ between a and $(a+h)$ and let U denote the maximum value of $f^{(n+1)}(x)$ between a and $(a+h)$. By the preceding theorem

$$\frac{Lh^{n+1}}{(n+1)!} \leq f(a+h) - f(a) \leq \frac{Uh^{n+1}}{(n+1)!}.$$

So

$$L \leq (f(a+h) - f(a))\left(\frac{(n+1)!}{h^{n+1}}\right) \leq U$$

and by the Intermediate Value Theorem there is a point ξ between a and $(a+h)$ such that

$$f^{(n+1)}(\xi) = [f(a+h) - f(a)]\left(\frac{(n+1)!}{h^{n+1}}\right).$$

Hence,

$$\frac{f^{(n+1)}(\xi)h^{n+1}}{(n+1)!} = f(a+h) - f(a).$$

This is the result we need to prove Taylor's Theorem[3]. Before proving this theorem we make some remarks about notation. Taylor series are often written using either of the following two formulas.

$$p(x) = \sum_{k=0}^{\infty} a_k(x-a)^k$$

$$p(a+h) = \sum_{k=0}^{\infty} a_k h^k$$

The two different notations say exactly the same thing. By introducing the variable $h = (x - a)$ the second notation focuses attention on the difference, h, between x and a.

Theorem 21 (Taylor's Theorem) *Suppose that $f(x)$ is a differentiable function and that all the deriviatives of $f(x)$ are also differentiable. Let $p(x)$ denote the Taylor series for $f(x)$ centered at the point a. That is,*

---
[3]See the note at the end of this section.

§7. Taylor's Theorem

$$p(x) = \sum_{k=0}^{\infty} a_k(x-a)^k = f(a) + f'(a)(x-a) + \left(\frac{f''(a)}{2}\right)(x-a)^2 + \cdots$$

Let $p_n(x)$ denote the nth Taylor Polynomial

$$p_n(x) = \sum_{k=0}^{n} a_k(x-a)^k.$$

Then for each h there is a point ξ between a and $a+h$ such that

$$f(a+h) - p_n(a+h) = \frac{f^{(n+1)}(\xi)h^{n+1}}{(n+1)!}.$$

Notice that the left hand side of the equation above is the difference between the exact value $f(a+h)$ and the approximation $p_n(a+h)$ given by the nth Taylor polynomial. Thus if we know something about $f^{(n+1)}(x)$ we can estimate the size of this difference.

Proof:

Consider the function

$$g(x) = f(x) - p_n(x).$$

It is easy to check from the definition of the Taylor Series that

$$\begin{aligned} g(a) &= 0 \\ g'(a) &= 0 \\ g^{(2)}(a) &= 0 \\ &\vdots \\ g^{(n)}(a) &= 0 \end{aligned}$$

and that

$$g^{(n+1)}(x) = f^{(n+1)}(x).$$

Now, by the preceding theorem, there is a point ξ between a and $(a+h)$ such that

$$\begin{aligned} g(a+h) - g(a) &= \frac{g^{(n+1)}(\xi)h^{n+1}}{(n+1)!} \\ &= \frac{f^{(n+1)}(\xi)h^{n+1}}{(n+1)!}, \end{aligned}$$

but
$$g(a+h)-g(a) = f(a+h)-f(a)-[p_n(a+h)-p_n(a)] = f(a+h)-p_n(a+h).$$

So
$$f(a+h) - p_n(a+h) = \frac{f^{(n+1)}(\xi)h^{n+1}}{(n+1)!}$$

completing the proof.

This theorem has a large number of useful applications. First, it allows us to determine how quickly a Taylor series converges. This application is illustrated in the following example.

Example:

Suppose that you need to write a computer program to estimate $\sin x$ correct to ± 0.000001 for values of x in the interval $[-\pi, \pi]$. Notice that $\sin x$ for other values of x can be determined from your program using the fact that the sine function is periodic.

You will write your program using the Taylor series centered at zero
$$\sin x = x - \frac{x^3}{3!} + \frac{x^5}{5!} - \frac{x^7}{7!} + \cdots.$$

How many terms of this Taylor series must be used to obtain the desired accuracy of ± 0.000001?

Answer:

Let n be even and let $p_n(x)$ denote the Taylor polynomial
$$p_n(x) = x - \frac{x^3}{3!} + \cdots \pm \frac{x^{n-1}}{(n-1)!}.$$

Since n is even the term involving x^n has a coefficient of zero and does not appear in p_n.

By Taylor's theorem for each h there is a point ξ between 0 and h such that
$$\sin h - p_n(h) = \frac{\sin^{(n+1)}(\xi)h^{n+1}}{(n+1)!}.$$

Hence,
$$|\sin h - p_n(h)| = \left|\frac{\sin^{(n+1)}(\xi)h^{n+1}}{(n+1)!}\right|.$$

§7. Taylor's Theorem

Now all the derivatives of the sine function are $\pm \sin x$ and $\pm \cos x$ and these functions are bounded by ± 1. Hence, $|\sin^{(n+1)}(\xi)| \leq 1$ and

$$|\sin h - p_n(h)| = \left|\frac{\sin^{(n+1)}(\xi)h^{n+1}}{(n+1)!}\right|$$

$$\leq \frac{|h|^{n+1}}{(n+1)!}$$

But, now, in the interval $[-\pi, \pi]$, $|h| \leq \pi$. So

$$|\sin h - p_n(h)| \leq \frac{\pi^{n+1}}{(n+1)!}$$

and making some calculations for the right hand side of this inequality we get

n	$\frac{\pi^{n+1}}{(n+1)!}$
2	5.1677128
4	2.5501640
6	0.5992645
8	0.0821459
10	0.0073704
12	0.0004663
14	0.0000219
16	0.0000008

So we see that the Taylor polynomial

$$p_{16}(x) = x - \frac{x^3}{3!} + \frac{x^5}{5!} - \cdots - \frac{x^{15}}{15!}$$

will give us the desired accuracy. Notice that in this particular case the Taylor polynomial $p_{16}(x)$ only has eight nonzero terms.

Example:

Find a Taylor polynomial centered at $a = 1$ for the function $y = \log x$ that can be used to estimate $\log x$ with an error less than $.0001$ on the interval $[0.5, 1.5]$.

Answer:

In the last section we saw that the Taylor Series centered at 1 for $\log x$ is

$$\log x = (x-1) - \frac{(x-1)^2}{2} + \frac{(x-1)^3}{3} - \cdots \pm \frac{(x-1)^n}{n} \pm \cdots.$$

Let $p_n(x)$ be the Taylor polynomial

$$p_n(x) = (x-1) - \frac{(x-1)^2}{2} + \frac{(x-1)^3}{3} - \cdots \pm \frac{(x-1)^n}{n}.$$

Notice that

$$\begin{aligned} f(x) &= \log x \\ f'(x) &= x^{-1} \\ f''(x) &= -x^{-2} \\ f^{(3)}(x) &= 2x^{-3} \\ &\vdots \\ f^{(n)} &= \pm(n-1)!x^{-n} \end{aligned}$$

By Taylor's Theorem, for each point x in the interval $[0.5, 1.5]$ there is a point ξ between 1 and x such that

$$f(x) - p_n(x) = \frac{f^{(n+1)}(\xi)(x-1)^{n+1}}{(n+1)!}.$$

Thus,

$$\begin{aligned} |f(x) - p_n(x)| &= \left| \frac{f^{(n+1)}(\xi)(x-1)^{n+1}}{(n+1)!} \right| \\ &= \left| \frac{n!\xi^{-(n+1)}(x-1)^{n+1}}{(n+1)!} \right| \\ &\leq \frac{1}{n+1} \end{aligned}$$

where the last line follows from the fact that $\xi \geq 0.5$ and $|x-1| \leq 0.5$ for x in the interval $[0.5, 1.5]$.

§7. Taylor's Theorem

Thus, the polynomial $p_{10,001}$ will yield the required approximation.

Exercises

For each of the following functions find a Taylor polynomial that approximates the given function on the given interval with the given accuracy ϵ.

*Exercise VI.7.1:
$$f(x) = \sin x, \qquad [-\pi, \pi], \qquad \epsilon = 0.0001$$

Exercise VI.7.2:
$$f(x) = \sin x, \qquad [-\pi, \pi], \qquad \epsilon = 0.00000001$$

Exercise VI.7.3:
$$f(x) = \cos x, \qquad [-\pi, \pi], \qquad \epsilon = 0.0001$$

Exercise VI.7.4:
$$f(x) = \cos x, \qquad [-\pi, \pi], \qquad \epsilon = 0.000001$$

*Exercise VI.7.5:
$$f(x) = e^x, \qquad [-1, 1], \qquad \epsilon = 0.000001$$

*Exercise VI.7.6:
$$f(x) = \log x, \qquad [0.5, 1.5], \qquad \epsilon = 0.01$$

Exercise VI.7.7:
$$f(x) = \log x, \qquad [0.5, 1.5], \qquad \epsilon = 0.001$$

*Exercise VI.7.8:
$$f(x) = \log x, \qquad [0.1, 1.9], \qquad \epsilon = 0.01$$

Taylor's theorem is often the key step in determining the accuracy of a numerical method for estimating some quantity. We illustrate how useful it is by looking at methods for estimating derivatives numerically. The most straightforward method of estimating a derivative comes directly from the original definition

$$f'(x) = \lim_{h \to 0} \frac{f(x+h) - f(x)}{h}.$$

This leads to the estimate

$$f'(x) \approx \frac{f(x+h) - f(x)}{h}$$

where the symbol $\approx$ is used to mean "approximately equal to."

In theory one can obtain as good an estimate as might be needed for any particular purpose by choosing h sufficiently small, but there is a practical catch. Frequently one is dealing with a function $f(x)$ for which we have incomplete information. For example, the function $f(x)$ might be the altitude of a rocket ship a time x and the only records we have give the altitude at certain particular times. For example, we might be given a record like the following table.

Time(seconds)	Altitude(feet)
0.0	116.00
0.1	119.36
0.2	123.04
0.3	127.04
0.4	131.36
0.5	136.00
0.6	140.96
0.7	146.24
0.8	151.84
0.9	157.76
1.0	164.00

Suppose that we want to estimate the velocity of this rocket at time $x = 0.5$. There are two obvious estimates that we might make, with $h = 0.1$ and $h = -0.1$.

§7. Taylor's Theorem

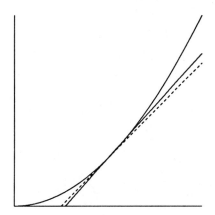

Figure VI.35: Estimation of $f'(1.0)$ using a "righthand chord."

$$f'(0.5) \approx \frac{f(0.6) - f(0.5)}{0.1} = 49.60$$

and

$$f'(0.5) \approx \frac{f(0.4) - f(0.5)}{-0.1} = 46.40$$

Notice how far apart these two estimates are. Because of our limited data we can't simply use a smaller value of h. One possible solution to this problem is suggested by Figures VI.35 - VI.37. These figures show three different estimates of the derivative of a function $f(x)$ at the point $x = 1$. These estimates are based on the following table of values

x	$f(x)$
0.9	0.81
1.0	1.00
1.1	1.21

All three figures show the actual tangent at the point $x = 1$ as a dotted line. The solid line in each figure is a chord estimating the tangent. In Figure VI.35 the chord is a "righthand" chord going through the points $(1.0, 1.0)$ and $(1.1, 1.21)$. Notice that this chord is considerably steeper than the actual tangent. In Figure VI.36 the chord is

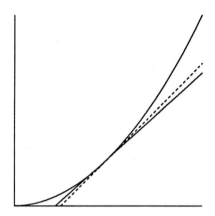

Figure VI.36: Estimation of $f'(1.0)$ using a "lefthand chord."

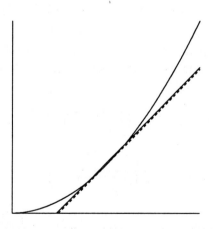

Figure VI.37: Estimation of $f'(1.0)$ using a "two-sided chord."

§7. Taylor's Theorem

a "lefthand chord" going through the point $(0.9, 0.81)$ and $(1.0, 1.0)$. Notice that this chord is considerably less steep than the tangent. In Figure VI.37 the chord is a "two-sided chord" going through the points $(0.9, 0.81)$ and $(1.1, 1.21)$. Notice that this chord appears to be almost parallel to the actual tangent. In these figures it appears that the two-sided chord gives a much better estimate than either of the one-sided chords. We can quantify this observation using Taylor's theorem.

Theorem 22 *This theorem concerns two estimates for the derivative of a function $y = f(x)$ at a point x. The first estimate is called the* **one-sided** *estimate and is given by*

$$E_1 = \frac{f(x+h) - f(x)}{h}.$$

The second-estimate is called the two-sided or the **symmetric difference** *estimate and is given by*

$$E_2 = \frac{f(x+h) - f(x-h)}{2h}.$$

There are points ξ_1 and ξ_2 in the interval $[x-h, x+h]$ such that

$$|f'(x) - E_1| = \left| \frac{f''(\xi_1) h}{2} \right|$$

$$|f'(x) - E_2| = \left| \frac{f^{(3)}(\xi_2) h^2}{3} \right|$$

Proof: By Taylor's theorem with $n = 1$ there is a point ξ_1 between x and $x + h$ such that

$$f(x+h) = f(x) + f'(x)h + \frac{f''(\xi_1) h^2}{2}$$

$$f(x+h) - f(x) = f'(x)h + \frac{f''(\xi_1) h^2}{2}$$

$$\frac{f(x+h) - f(x)}{h} = f'(x) + \frac{f''(\xi_1) h}{2}$$

$$\frac{f(x+h)-f(x)}{h} - f'(x) = \frac{f''(\xi_1)h}{2}$$

$$\left|\frac{f(x+h)-f(x)}{h} - f'(x)\right| = \left|\frac{f''(\xi_1)h}{2}\right|$$

which gives us the error for the one-sided estimate of $f'(x)$.

By Taylor's Theorem with $n=2$ there is a point ξ_3 between x and $x+h$ such that

$$f(x+h) = f(x) + f'(x)h + \frac{f''(x)h^2}{2} + \frac{f^{(3)}(\xi_3)}{6}$$

and there is a point ξ_4 between x and $x-h$ such that

$$f(x-h) = f(x) + f'(x)(-h) + \frac{f''(x)(-h)^2}{2} + \frac{f^{(3)}(\xi_4)(-h)^3}{6}$$

$$= f(x) - f'(x)h + \frac{f''(x)h^2}{2} - \frac{f^{(3)}(\xi_4)h^3}{6}$$

Thus we get

$$f(x+h) - f(x-h) = 2hf'(x) + \frac{f^{(3)}(\xi_3)h^3}{6} + \frac{f^{(3)}(\xi_4)h^3}{6}$$

$$\frac{f(x+h)-f(x-h)}{2h} = f'(x) + \left(\frac{f^{(3)}(\xi_3) + f^{(3)}(\xi_4)}{2}\right)\left(\frac{h^2}{6}\right)$$

$$\frac{f(x+h)-f(x-h)}{2h} - f'(x) = \left(\frac{f^{(3)}(\xi_3) + f^{(3)}(\xi_4)}{2}\right)\left(\frac{h^2}{6}\right)$$

$$\left|\frac{f(x+h)-f(x-h)}{2h} - f'(x)\right| = \left|\left(\frac{f^{(3)}(\xi_3) + f^{(3)}(\xi_4)}{2}\right)\left(\frac{h^2}{6}\right)\right|$$

Notice that

$$\frac{f^{(3)}(\xi_3) + f^{(3)}(\xi_4)}{2}$$

is just the average of $f^{(3)}(x)$ at ξ_3 and ξ_4. Hence, by the Intermediate value Theorem there is a point ξ_2 between ξ_3 and ξ_4 such that

$$f^{(3)}(\xi_2) = \frac{f^{(3)}(\xi_3) + f^{(3)}(\xi_4)}{2}$$

§7. Taylor's Theorem

and, hence,

$$\left|\frac{f(x+h)-f(x-h)}{2h} - f'(x)\right| = \left|\frac{f^{(3)}(\xi_2)h^2}{6}\right|$$

giving us the error for the two-sided or symmetric difference estimate of $f'(x)$.

Now compare the two errors.

$$|f'(x) - E_1| = \left|\frac{f''(\xi_1)h}{2}\right|$$

$$|f'(x) - E_2| = \left|\frac{f^{(3)}(\xi_2)h^2}{6}\right|$$

Usually the numbers $f''(\xi_1)$ and $f^{(3)}(\xi_2)$ are relatively modest numbers and the important factors determining these size of errors are the factors $h/2$ for E_1 and $h^2/6$ for E_2. If h is relatively small (for example, $h = 0.01$) then h^2 will be much smaller (for example, 0.0001) and thus E_2 will be much closer to the actual value of $f'(x)$ then E_1. Roughly speaking the error for the estimate E_1 is "in the same ballpark" as h and the error for the estimate E_2 is "in the same ballpark" as h^2.

Exercises

For each of the following problems compute the actual value of $f'(x)$ at the indicated point x and compute the two one-sided estimates and the symmetric difference estimate for $f'(x)$ using the indicated value of h. Discuss the accuracy of these estimates comparing the predictions given by the theorem above and the actual errors.

*Exercise VI.7.9:

$$f(x) = \sin x, \qquad x = 1.0, \qquad h = 0.01$$

Exercise VI.7.10:

$$f(x) = \sin x, \qquad x = 1.0, \qquad h = 0.001$$

Exercise VI.7.11:

$$f(x) = \sin x, \quad x = 1.0, \quad h = 0.0001$$

Exercise VI.7.12:

$$f(x) = e^x, \quad x = 1.0, \quad h = 0.01$$

Exercise VI.7.13:

$$f(x) = e^x, \quad x = 1.0, \quad h = 0.0001$$

Exercise VI.7.14:

$$f(x) = \log x, \quad x = 1.0, \quad h = 0.01$$

Exercise VI.7.15:

$$f(x) = \log x, \quad x = 1.0, \quad h = 0.0001$$

Exercise VI.7.16:
Another way to approach the problem of finding a better estimate for $f'(x)$ than either of the two one-sided estimates would be to average the two one-sided estimates. Show that this approach leads to the same estimate as the symmetric difference estimate.

Exercise VI.7.17:
Try to show that

$$\frac{f(x+h) - 2f(x) + f(x-h)}{h^2}$$

is a good estimate for $f''(x)$. How good is it? (That is, find a formula for the error.)

Note

The proofs of the theorems leading up to Taylor's Theorem assumed that h was positive because they used the fact that if $f(x) \leq g(x)$ then

$$\int_a^b f(x)\,dx \leq \int_a^b g(x)\,dx$$

§8. Spring-and-Mass Systems

which is true only when $a < b$. If $b < a$ then

$$\int_b^a f(x)\,dx \leq \int_b^a g(x)\,dx,$$

so that

$$\int_a^b f(x)\,dx \geq \int_a^b g(x)\,dx.$$

Nonetheless by going over this series of theorems when h is negative you can show that the final result just before Taylor's Theorem holds in this case as well.

§8. Spring-and-Mass Systems

Differential Equations Revisited

In this section we return to the study of phenomena that can be described or modeled by differential equations. Recall that a differential equation is an equation like

$$y'' = -32$$

or

$$y' = y + 12t$$

that expresses a relationship between a function $y(t)$ and one or more of its derivatives. One of the important features of a differential equation is the order of the derivatives involved in the equation. Differential equations are often classified by this feature.

Definition:

> The **order** of a differential equation is the order of the highest derivative involved in the equation. For example, $y' = y$ is a first order differential equation and $y'' = y' + y$ is a second order differential equation.

> The simplest first order differential equations are those of the form
> $$y'(t) = f(t).$$

These are easily solved by integration

$$y(t) = F(t) + c$$

where $F(t)$ is any antiderivative of the function $f(t)$. Notice that there is one arbitrary constant. This is typical of first order differential equations. The solution always involves one arbitrary constant. Initial value problems with a first order differential equation include one piece of initial data, $y(t_0) = y_0$, that determines the value of the arbitrary constant.

Similarly, the solution of a second order differential equation involves two arbitrary constants and initial value problems involving a second order differential equation include two pieces of initial data that determine the values of the two arbitrary constants. For example, the solution of a typical falling body problem

$$y'' = -32; \quad y(0) = 50, \; y'(0) = 0$$

is

$$y = -16t^2 + c_1 t + c_2,$$

which involves two constants c_1 and c_2 whose values are determined from the initial height $y(0)$ and the initial velocity $y'(0)$. This is exactly what one would expect from the physics—to describe a falling body completely one needs to know its acceleration (given by the differential equation) and its initial height and velocity (given by the two initial conditions).

One of the most important techniques used to study the solutions of an equation involving numbers like

$$x^2 + 4x + 2 = 14$$

is checking a possible solution to see if it is an actual solution. For example, suppose that one suspects that $x = 2$ is a solution of the equation above. We can check this by simply substituting $x = 2$ in this equation to get

$$
\begin{aligned}
(2)^2 + 4(2) + 2 &=? \; 14 \\
4 + 8 + 2 &=? \; 14 \\
14 &=? \; 14
\end{aligned}
$$

§8. Spring-and-Mass Systems

and we see that $x = 2$ is a solution of this equation.

The same idea works for differential equations like

$$y'' = -2y.$$

Suppose that we suspect that $y = \cos 2t$ is a solution of this differential equation. We can check as follows. First we compute the derivatives

$$\begin{aligned} y &= \cos 2t \\ y' &= -2 \sin 2t \\ y'' &= -4 \cos 2t \end{aligned}$$

and then substitute into the original differential equation

$$\begin{aligned} y'' &=? \; -2y \\ -4 \cos 2t &=? \; -2 \cos 2t \end{aligned}$$

and we see that this particular function is not a solution of the differential equation.

Exercises

*Exercise VI.8.1:
Is the function $y(t) = \cos 2t$ a solution of the differential equation

$$y'' = -4y?$$

Exercise VI.8.2:
Is the function $y(t) = e^t$ a solution of the differential equation

$$y'' - 2y' + y = 0?$$

Exercise VI.8.3:
Is the function $y(t) = te^t$ a solution of the differential equation

$$y'' - 2y' + y = 0?$$

674　　　　　Chapter VI. Real and Complex Numbers.

Figure VI.38: Spring at rest

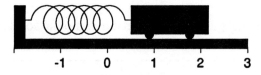

Figure VI.39: Stretched spring

Figure VI.40: Compressed spring

§8. Spring-and-Mass Systems

Spring-and-Mass Systems

We are interested in the motion of a spring-and-mass system like the one shown in Figures VI.38-VI.40.

Spring-and-mass systems have three components.

- A mass that is free to move back-and-forth along a track. In Figures VI.38-VI.40 this mass is shown with wheels to indicate that it is free to move.

- A spring with one end fixed in a solid wall and the other end attached to the movable mass.

- Friction that works against the motion of the movable mass.

Figure VI.38 shows the system at rest in a state of equilibrium. The spring is neither stretched nor compressed and is not exerting any force on the mass. Figure VI.39 shows the system with the spring stretched. It is pulling the mass back toward the equilibrium position. Figure VI.40 shows the system with the spring compressed. It is pushing the mass back toward the equilibrium position. The position or **displacement** of the mass (its left end) is measured in centimeters using coordinates chosen so that the equilibrium position is at zero and points to the right of the rest position have positive coordinates. The mass of the mass is measured in grams. Force is measured in dynes or centimeter-grams per second per second, that is, a force of one dyne will accelerate a mass of one gram at the rate of one centimeter per second per second.

First we consider a spring-and-mass system with very little friction. Suppose that one pulls the mass to the right to position $x = 1$ and then releases it. One would expect the mass to be pulled back toward its equilibrium position by the spring. By the time the mass reaches its equilibrium position it will be moving quite rapidly and will continue past the equilibrium position. Once it is past the equilibrium position the push of the spring will act to slow it down. It will eventually come to an infinitely brief stop at a point to the left of the equilibrium position with the spring compressed. Now the spring will push the mass back toward its equilibrium position. When it reaches the equilibrium position for a second time it will again be moving quite rapidly but this time to the right and it will continue past the equilibrium position.

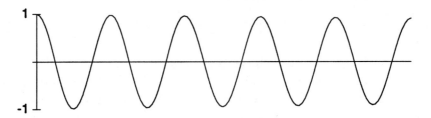

Figure VI.41: Travels of a low friction spring-and-mass system

Once past the equilibrium position the pull of the spring will act to slow the mass down and it will eventually come to an infinitely brief stop at a point to the right of the equilibrium position with the spring stretched. The stretched spring will now pull the mass back toward its equilibrium position and the cycle will repeat. If the system has very little friction then this cycle will continue with each "trip" of the spring slightly shorter than the preceding "trip." Figure VI.41 shows an example of such a situation.

There are many questions that one might ask about this system—for example, is the interval between successive passages of the mass through its equilibrium position always the same?—or how does the behavior of the system depend on how far the mass is pulled to the right before being released?

The mathematical models that describe spring-and-mass systems also describe other important systems involving oscillations, for example, the electrical circuits that are at the heart of television and radio tuners. Thus our work in this section is useful in a wide variety of settings.

Hooke's Law

From personal experience we know that when a spring is stretched it pulls back and the more it is stretched, the harder it pulls back. Similarly, a spring that is compressed pushes back and the more it is compressed, the harder it pushes back. Careful experiments have shown that if a spring is stretched or compressed by a moderate amount than the force with which it pulls or pushes is proportional to the amount by which it has been stretched or compressed. This is generally known

§8. Spring-and-Mass Systems

as **Hooke's**[4] **Law**. Mathematically, Hooke's Law can be expressed in the form

$$F = -kx$$

where x is the displacement of the mass as described above and k is a positive constant called the **spring constant** that represents the strength of the spring. The constant k is measured in dynes per centimeter. Thus, when multiplied by the displacement of the spring measured in centimeters, the result is a force measured in dynes. When the spring is stretched, the displacement x is positive and the force is negative because the spring is pulling the mass to the left. When the spring is compressed, the displacement x is negative and the force is positive because the spring is pushing the mass to the right. Time is measured in seconds and the displacement of the spring at time t will be denoted $x(t)$. As usual acceleration is given by the differential equation

$$x'' = \left(\frac{1}{m}\right) F = -\left(\frac{k}{m}\right) x.$$

The physical description of a spring-and-mass system involves two pieces of initial data—the initial displacement and the initial velocity of the mass. Mathematically, this is reflected by the fact that the differential equation above is a second order differential equation whose general solution involves two constants.

We begin our study of spring-and-mass systems by considering a system in which the constant $k/m = 1$ so that the equation above is

$$x'' = -x.$$

We already know two functions whose second derivative is equal to the negative of the original function, namely,

$$x_1(t) = \sin t$$
$$x_2(t) = \cos t$$

---

[4] Robert Hooke first published this law as an anagram "ceiiinosssttuv" in 1676. For those who had difficulty with word games he later (1678) published it is "ut tensio sic vis." Fortunately, modern scientists rarely publish their work in the form of Latin anagrams.

Exercises

*Exercise VI.8.4:
Show that any function of the form $x(t) = c_1 \cos t + c_2 \sin t$ is a solution of the differential equation
$$x'' = -x.$$

*Exercise VI.8.5:
Find the solution of the initial value problem
$$x'' = -x; \qquad x(0) = 1,\ x'(0) = 0.$$

Exercise VI.8.6:
Find the solution of the initial value problem
$$x'' = -x; \qquad x(0) = 0,\ x'(0) = 1.$$

Exercise VI.8.7:
Find the solution of the initial value problem
$$x'' = -x; \qquad x(0) = 1,\ x'(0) = 2.$$

Now we return to the more general situation
$$x'' = -\left(\frac{k}{m}\right)x.$$

In Exercise V.8.5 above we saw that the solution of the initial value problem
$$x'' = -x; \qquad x(0) = 1,\ x'(0) = 0$$
is $x(t) = \cos t$. This function describes the motion of a spring-and-mass system in which $k = m$. If we think of another system in which $k = 2m$, that is, in which the spring is twice as strong as in the first system, we might expect that everything would happen twice as fast, that is, we might guess that the solution of the initial value problem
$$x'' = -2x; \qquad x(0) = 1,\ x'(0) = 0$$
would be
$$x(t) = \cos 2t.$$

§8. Spring-and-Mass Systems

If we try this guess we see that

$$\begin{aligned} x(t) &= \cos 2t \\ x'(t) &= -2\sin 2t \\ x''(t) &= -4\cos 2t \\ &= -4x(t) \end{aligned}$$

which is not quite what we hoped for. This near miss, however, shows us how to solve this problem. We guess that

$$x(t) = \cos \omega t$$

where ω is some constant. Trying this guess we see that

$$\begin{aligned} x(t) &= \cos \omega t \\ x'(t) &= -\omega \sin \omega t \\ x''(t) &= -\omega^2 \cos \omega t \\ &= -\omega^2 x(t) \end{aligned}$$

Since we want x'' to be equal to $-2x$ we see that ω should be $\sqrt{2}$ and one solution of the differential equation $x'' = -2x$ is $x(t) = \cos \sqrt{2}t$.

Exercises

*Exercise VI.8.8:
Show that any function of the form

$$x(t) = c_1 \cos \omega t + c_2 \sin \omega t$$

where $\omega = \sqrt{k/m}$ is a solution of the differential equation

$$x'' = -\left(\frac{k}{m}\right)x.$$

*Exercise VI.8.9:
Find the solution of the initial value problem

$$x'' = -9x; \qquad x(0) = 1,\ x'(0) = 0.$$

Exercise VI.8.10:
Find the solution of the initial value problem
$$x'' = -9x; \qquad x(0) = 0, \; x'(0) = 2.$$

Exercise VI.8.11:
Find the solution of the initial value problem
$$x'' = -9x; \qquad x(0) = 1, \; x'(0) = 2.$$

In a realistic situation there will always be some friction. The force exerted by friction is proportional to velocity[5] and is always trying to slow down a moving object. Thus this force is described by

$$F = -px'$$

where p is a positive constant. If we add the force of friction to our spring-and-mass model we get

$$\begin{aligned} x'' &= \left(\frac{1}{m}\right)(-kx - px') \\ &= -\left(\frac{k}{m}\right)x - \left(\frac{p}{m}\right)x'. \end{aligned}$$

If we let c denote k/m and b denote p/m this leads to the differential equation

$$x'' = -cx - bx'$$

or

$$x'' + bx' + cx = 0.$$

Our next goal is to find the solutions of this differential equation or any differential equation of the form

$$ax'' + bx' + cx = 0$$

where a, b, and c are constants.

---

[5]Friction is a complicated subject. The force exerted by friction is often proportional to velocity as in this model. There are, however, many different sources of friction and some of them produce friction that is not proportional to velocity.

§8. Spring-and-Mass Systems

The Differential Equation $ax'' + bx' + cx = 0$

We can solve the differential equation

$$ax'' + bx' + cx = 0$$

with the aid of a clever guess—that a solution is of the form $x = e^{rt}$ for some constant r. We try this guess out as follows

$$\begin{aligned} x &= e^{rt} \\ x' &= re^{rt} \\ x'' &= r^2 e^{rt} \end{aligned}$$

Substituting into the left side of differential equation we obtain

$$\begin{aligned} ax'' + bx' + cx &= ar^2 e^{rt} + bre^{rt} + ce^{rt} \\ &= ar^2 x + brx + cx \\ &= (ar^2 + br + c)x. \end{aligned}$$

So we want to choose r so that

$$ar^2 + br + c = 0.$$

This quadratic equation, called the **characteristic equation**, is easy to solve using the quadratic formula

$$r = \frac{-b \pm \sqrt{b^2 - 4ac}}{2a}.$$

The calculations above prove the first part of the following theorem.

Theorem 23

- If r is a solution of the characteristic equation

$$ar^2 + br + c = 0$$

then $x = e^{rt}$ is a solution of the differential equation

$$ax'' + bx' + c = 0.$$

- If $b^2 - 4ac \neq 0$ the characteristic equation has two distinct roots r_1 and r_2 and any function of the form

$$x = c_1 e^{r_1 t} + c_2 e^{r_2 t}$$

is a solution of the differential equation.

- If $b^2 - 4ac = 0$ then the characteristic equation has one root $r = -b/2a$ and any function of the form

$$x = c_1 e^{rt} + c_2 t e^{rt}$$

is a solution of the differential equation.

Proof:

We have already proved the first statement. The proof of the remaining two statements is an exercise.

Exercises

*Exercise VI.8.12: Prove the remaining two statements in the previous theorem.

*Exercise VI.8.13: Find the solution of the initial value problem

$$x'' - 4x = 0; \quad x(0) = 1, \ x'(0) = 2.$$

Exercise VI.8.14: Find the solution of the initial value problem

$$x'' - 4x' + 4x = 0; \quad x(0) = 1, \ x'(0) = 2.$$

Exercise VI.8.15: Find the solution of the initial value problem

$$x'' - 4x' = 0; \quad x(0) = 1, \ x'(0) = 2.$$

*Exercise VI.8.16: Find the solution of the initial value problem

$$x'' + x = 0; \quad x(0) = 1, \ x'(0) = 0.$$

Notice that although solving this problem involves complex numbers, the final result does not involve complex numbers.

§8. Spring-and-Mass Systems

The characteristic equation often has complex roots leading to solutions of the form $x = e^{rt}$ where r is complex. We have already seen that e^z can be computed when z is complex using the Taylor Series

$$e^z = 1 + z + \frac{z^2}{2!} + \frac{z^3}{3!} + \cdots + \frac{z^n}{n!} + \cdots$$

So the solutions to the differential equation

$$ax'' + bx' + cx = 0$$

are easily computable complex-valued functions. In physics this is extremely important. There are many situations that are best described using complex quantities. In our current situation, however, we are working with quantities that can be described by real numbers and we expect that the functions we eventually get will be real functions as in Exercise V.8.16 above. We need some facts about the function e^z.

Euler's Formula

The following theorem states one of the more remarkable formulas in mathematics.

Theorem 24 (*Euler's Formula*)

$$e^{iy} = \cos y + i \sin y$$

Proof:

First recall that

$$\sin y = y - \frac{y^3}{3!} + \frac{y^5}{5!} + \cdots$$

$$\cos y = 1 - \frac{y^2}{2!} + \frac{y^4}{4!} + \cdots$$

Now

$$e^{iy} = 1 + iy + \frac{(iy)^2}{2!} + \frac{(iy)^3}{3!} + \frac{(iy)^4}{4!} + \frac{(iy)^5}{5!} + \cdots$$

$$= 1 - \frac{y^2}{2!} + \frac{y^4}{4!} + \cdots \quad \text{(the even terms)}$$

$$iy - i\frac{y^3}{3!} + i\frac{y^5}{5!} + \cdots \quad \text{(the odd terms)}$$

$$= 1 - \frac{y^2}{2!} + \frac{y^4}{4!} + \cdots +$$

$$i\left(y - \frac{y^3}{3!} + \frac{y^5}{5!} + \cdots\right)$$

$$= \cos y + i \sin y$$

The Property $e^{u+v} = e^u e^v$

You have known since high school days that

$$a^n \times a^k = \underbrace{(a \times a \times \cdots a)}_{n \text{ times}} \times \underbrace{(a \times a \times \cdots a)}_{k \text{ times}}$$

$$= \underbrace{(a \times a \times \cdots a)}_{n+k \text{ times}}$$

$$= a^{n+k}.$$

Thus it is not entirely surprising that for any two complex numbers u and v

$$e^{u+v} = e^u e^v.$$

In order to prove this we need to know that one of the results we discussed earlier for real functions applies also to complex functions, namely, that under reasonable conditions (but not always) an initial value problem has a *unique* solution.

Theorem 25 *For any complex numbers u and v, $e^{u+v} = e^u e^v$.*

Proof:

§8. Spring-and-Mass Systems

Consider the two functions

$$y_1(z) = e^{u+z}$$
$$y_2(z) = e^u e^z$$

Differentiating these two functions we see that

$$\left(\frac{d}{dz}\right) y_1(z) = e^{u+z} = y_1(z)$$

and

$$\left(\frac{d}{dz}\right) y_2(z) = e^u e^z = y_2(z).$$

Notice also that

$$y_1(0) = e^u = y_2(0),$$

so both these functions are solutions of the initial value problem

$$\left(\frac{d}{dz}\right) y = e^u y; \qquad y(0) = e^u.$$

Thus the two functions are identical and for every z

$$e^{u+z} = y_1(z) = y_2(z) = e^u e^z.$$

In particular, when $z = v$ we get

$$e^{u+v} = e^u e^v.$$

Complex Solutions of the Characteristic Equation

Now we return to the differential equation

$$ax'' + bx' + cx = 0$$

and its characteristic equation

$$ar^2 + br + c = 0.$$

If $b^2 - 4ac$ is negative then the characteristic equation has two complex solutions

$$\frac{-b \pm \sqrt{b^2 - 4ac}}{2a}.$$

We use the following notation

$$\omega = \frac{\sqrt{4ac - b^2}}{2a}$$

$$\lambda = \frac{-b}{2a}$$

With this notation the two solutions of the characteristic equation are

$$\lambda \pm \omega i$$

and the solutions of the differential equation are of the form

$$x(t) = c_1 e^{(\lambda + \omega i)t} + c_2 e^{(\lambda - \omega i)t}.$$

We can rewrite

$$\begin{aligned} e^{(\lambda + \omega i)t} &= e^{\lambda t} e^{\omega i t} \\ &= e^{\lambda}(\cos \omega t + i \sin \omega t) \end{aligned}$$

and

$$\begin{aligned} e^{(\lambda - \omega i)t} &= e^{\lambda t} e^{-\omega i t} \\ &= e^{\lambda t}(\cos(-\omega t) + i \sin(-\omega t)) \\ &= e^{\lambda t}(\cos \omega t - i \sin \omega t). \end{aligned}$$

Now we know that any function of the form

$$c_1 e^{(\lambda + \omega i)t} + c_2 e^{(\lambda - \omega i)t}$$

is a solution of the differential equation. If we let $c_1 = c_2 = 1/2$ we get the solution

$$\begin{aligned} y_1(t) &= \left(\frac{1}{2}\right) e^{(\lambda + \omega i)t} + \left(\frac{1}{2}\right) e^{(\lambda - \omega i)t} \\ &= \left(\frac{1}{2}\right) e^{\lambda t}(\cos \omega t + i \sin \omega t) + \left(\frac{1}{2}\right) e^{\lambda t}(\cos \omega t - i \sin \omega t) \\ &= e^{\lambda t} \cos \omega t. \end{aligned}$$

§8. Spring-and-Mass Systems

Similar calculations with $c_1 = 1/2i$ and $c_2 = -1/2i$ show that

$$y_2(t) = e^{\lambda t} \sin \omega t$$

is also a solution of the differential equation.

This gives us the following theorem.

Theorem 26 *If $b^2 - 4ac < 0$ then the solutions of the differential equation*

$$ax'' + bx' + cx = 0$$

are of the form

$$c_1 e^{\lambda t} \cos \omega t + c_2 e^{\lambda t} \sin \omega t$$

where

$$\omega = \frac{\sqrt{4ac - b^2}}{2a}$$

$$\lambda = \frac{-b}{2a}.$$

Proof:

The most direct way to prove this theorem is to substitute the function

$$x(t) = c_1 e^{\lambda t} \cos \omega t + c_2 e^{\lambda t} \sin \omega t$$

into the original differential equation.

Exercises

*Exercise VI.8.17: Find the solution of the initial value problem

$$x'' = -9x; \qquad x(0) = 1, \ x'(0) = 0.$$

Notice that this initial value problem might describe a spring-and-mass system with no friction. Explain how the lack of friction affects the longterm behavior of the system.

*Exercise VI.8.18: Find the solution of the initial value problem

$$x'' = -0.1x' - 9x; \qquad x(0) = 1, \ x'(0) = 0.$$

Notice that this initial value problem might describe a spring-and-mass system with very little friction. Explain how the friction affects the longterm behavior of the system.

Exercise VI.8.19: Find the solution of the initial value problem
$$x'' = -x - 9x; \qquad x(0) = 1, \ x'(0) = 0.$$
Notice that this initial value problem might describe a spring-and-mass system with more friction than the preceding problem. Explain how the friction affects the longterm behavior of the two systems.

Exercise VI.8.20: Find the solution of the initial value problem
$$x'' = -8x' - 9x; \qquad x(0) = 1, \ x'(0) = 0.$$
Notice that this initial value problem might describe a spring-and-mass system with more friction than the preceding problems. Explain how the friction affects the longterm behavior of the three systems.

Real solutions of the differential equation $ax'' + bx' + cx = 0$ look somewhat different depending on the quantity
$$\Delta = b^2 - 4ac.$$
This quantity is called the **discriminant**.

- If $\Delta > 0$ then the solutions are of the form
$$x(t) = c_1 e^{r_1 t} + c_2 e^{r_2 t}.$$

- If $\Delta = 0$ then the solutions are of the form
$$x(t) = c_1 e^{rt} + c_2 t e^{rt}.$$

- If $\Delta < 0$ then the real solutions are of the form
$$\begin{aligned} x(t) &= c_1 e^{\lambda t} \cos \omega t + c_2 e^{\lambda t} \sin \omega t \\ &= e^{\lambda t}(c_1 \cos \omega t + c_2 \sin \omega t) \end{aligned}$$

where
$$\lambda = -b/2a$$
and
$$\omega = \frac{\sqrt{4ac - b^2}}{2a}.$$

§8. Spring-and-Mass Systems

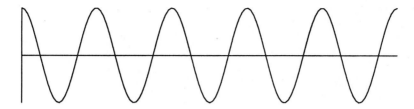

Figure VI.42: $x = c_1 \cos \omega t + c_2 \sin \omega t$

In all three cases the longterm behavior of the solutions is determined by the factors involving e. In particular, we have

$$\lim_{t \to \infty} x(t) = 0$$

if the real parts of the solutions to the characteristic equation

$$ax^2 + bx + c = 0$$

are negative. For example, in the last case

$$x(t) = e^{\lambda t}(c_1 \cos \omega t + c_2 \sin \omega t)$$

the last factor by itself oscillates back and forth like the function shown in Figure VI.42 but multiplying it by $e^{\lambda t}$ causes the oscillations to go back-and-forth within the "envelope" defined by $x = e^{\lambda t}$ and $x = -e^{\lambda t}$ as shown in Figure VI.43. If $\lambda < 0$ then the oscillations are "damped"— they get weaker and weaker. If $\lambda > 0$ the oscillations get stronger and stronger. If $\lambda = 0$ the oscillations continue forever with the same amplitude. The value of the constant ω determines the frequency with which the system oscillates.

Exercises

*Exercise VI.8.21: Describe how the strength of the friction affects the longterm behavior of a spring-and-mass system?

*Exercise VI.8.22: Describe how the strength of the spring affects the frequency with which a spring-and-mass system oscillates.

*Exercise VI.8.23: Describe how the strength of the friction affects the frequency with which a spring-and-mass system oscillates.

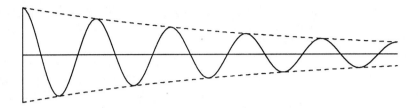

Figure VI.43: $x = e^{\lambda t}(c_1 \cos \omega t + c_2 \sin \omega t)$

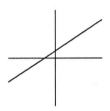

Figure VI.44: A typical linear function

*Exercise VI.8.24: Describe how the initial conditions affect the longterm behavior of a spring-and-mass system.

*Exercise VI.8.25: Describe how the initial conditions affect the frequency with which a spring-and-mass system oscillates.

§9. Linearity

Two of the hottest words in mathematics are "linear" and "nonlinear." The word "linear" originally comes from the word "line." A linear function is a function of the form $y = mx + b$ where m and b are constants. The graph of such a function is a straight line, hence, the name.

In this section we use the word "linear" in a slightly different but related way and we use it in many different situations. For example,

- Ordinary real-valued functions of a single real variable. These are the kinds of functions that we have been dealing with for

§9. Linearity

many years. Typical examples are the functions $y = 2x + 3$ and $y = \sin x$. These functions take as input a single real number and produce as output a single real number.

- Real-valued functions of several real variables. These are functions that take as input several real numbers and produce as output a single real number. For example, suppose that one is interested in the area of a rectangle whose height is h and whose width is w. The area is given by the function $A = wh$. This function takes as input the two real numbers h and w and produces as output the single real number A.

- Complex-valued functions of a single complex variable. For example, the function $w = e^z$ takes as input a single complex number z and produces as output a complex number w.

- More exotic processes like differentiation. Given a function $f(x)$ we can define a new function $f'(x)$. Think of this as a procedure $g = D(f)$ that takes as input a function $f(x)$ and produces as output another function $g(x)$. For example, $D(x^2) = 2x$ and $D(\sin x) = \cos x$.

- Even more exotic processes, called **differential operators**. For example, the differential operator $L(f) = 2f'' + 3f' + 4f$ takes as input a function $f(x)$ and produces as output a new function $g = L(f)$ defined by $g(x) = 2f''(x) + 3f'(x) + 4f(x)$.

We write the word *linear* in italics in this section because we are now using it in a different sense than we have been using it thus far.

Definition:

Suppose that f is a real-valued-function of a real variable. We say that f is *linear* if

1. For any two real numbers x and y,
$$f(x + y) = f(x) + f(y).$$

2. For any two real numbers k and x,
$$f(kx) = kf(x).$$

Example:

If m is a constant then the function $f(x) = mx$ is *linear*. The constant m is called the **slope** of the function f.
Proof:

$$f(x+y) = m(x+y) = mx + my = f(x) + f(y)$$
$$f(kx) = mkx = kmx = kf(x)$$

Example:

The function $f(x) = 2x - 1$ is not *linear*.
Proof:
All we need to to is find one example where the definition fails. There are many. Here is one

$$f(1+1) = f(2) = 3,$$

but

$$f(1) + f(1) = 1 + 1 = 2 \neq f(1+1).$$

Notice that a (unitalicized) linear function $f(x) = mx + b$ is (italicized) *linear* if and only if the constant b is zero.

Definition:

Suppose that f is a complex-valued-function of a complex variable. We say that f is *linear* if

1. For any two complex numbers u and v,

$$f(u+v) = f(u) + f(v).$$

2. For any two complex numbers k and z,

$$f(kz) = kf(z).$$

§9. Linearity

Example:

If m is a constant then the function $f(z) = mz$ is *linear*.

Proof:

$$\begin{aligned} f(u+v) &= m(u+v) = mu + mv = f(u) + f(v) \\ f(kz) &= mkz = kmz = kf(z) \end{aligned}$$

Definition:

Suppose that f is a real-valued function of n real variables. We say that f is *linear* if

1. For any real numbers $x_1, x_2, \ldots, x_n$ and $y_1, y_2, \ldots, y_n$

$$f(x_1 + y_1, x_2 + y_2, \ldots, x_n + y_n) = $$
$$f(x_1, x_2, \ldots, x_n) + f(y_1, y_2, \ldots, y_n)$$

2. For any real numbers $x_1, x_2, \ldots, x_n$ and any real number k,

$$f(kx_1, kx_2, \ldots kx_n) = kf(x_1, x_2, \ldots, x_n)$$

Example:

Suppose that $m_1, m_2, \ldots, m_n$ are real constants and let f be the function

$$f(x_1, x_2, \ldots, x_n) = m_1 x_1 + m_2 x_2 + \cdots + m_n x_n.$$

This function is *linear*. The constants $m_1, m_2, \ldots, m_n$ are called the **partial slopes** of the function f. In particular, for each i, m_i is called the **partial slope** of f with respect to the variable x_i.

Proof:

$$f(x_1 + y_1, x_2 + y_2, \ldots, x_n + y_n) =$$
$$m_1(x_1 + y_1) + m_2(x_2 + y_2) + \cdots + m_n(x_n + y_n) =$$
$$m_1 x_1 + m_1 y_1 + m_2 x_2 + m_2 y_2 + \cdots + m_n x_n + m_n y_n =$$
$$(m_1 x_1 + m_2 x_2 + \cdots + m_n x_n) + (m_1 y_1 + m_2 y_2 + \cdots + m_n y_n) =$$
$$f(x_1, x_2, \ldots, x_n) + f(y_1, y_2, \ldots, y_n)$$

Definition:

An **operator** is a procedure that takes as input a function and produces as output another function.

Example:

Differentiation is an operator because it starts with a function f and produces another function f'.

Example:

Suppose that a, b, and c are constants then the rule

$$L(y) = ay'' + by' + cy$$

defines an operator because starting with a function y it produces a new function $L(y)$. For example, if $L(y) = 2y'' + y' - y$ then $L(e^x) = 2e^x$.

Definition:

An operator L is said to be *linear* if

1. For any two functions f and g,

$$L(f + g) = L(f) + L(g).$$

§9. LINEARITY

2. For any function f and any real (if f is a real-valued function) or complex (if f is a complex-valued function) constant k,
$$L(kf) = kL(f).$$

Example:

If a, b and c are constants then the operator $L(y) = ay'' + by' + cy$ is *linear*.

Proof:

$$\begin{aligned} L(f+g) &= a(f+g)'' + b(f+g)' + c(f+g) \\ &= a(f'' + g'') + b(f' + g') + cf + cg \\ &= af'' + ag'' + bf' + bg' + cf + cg \\ &= (af'' + bf' + cf) + (ag'' + bg' + cg) \\ &= L(f) + L(g) \end{aligned}$$

$$\begin{aligned} L(kf) &= a(kf)'' + b(kf)' + c(kf) \\ &= akf'' + bkf' + ckf \\ &= k(af'' + bf' + cf) \\ &= kL(f) \end{aligned}$$

Exercises

*Exercise VI.9.1:

Prove that if f and g are *linear* functions of a single variable then so is the function $f + g$.

*Exercise VI.9.2:

Suppose that the operator H is defined by the rule: $H(f)$ is the function $g(t) = f(t-1)$. This operator is sometimes called a "shift operator" because it shifts the graph of f one unit to the right. See Figure VI.45. Is this operator *linear*?

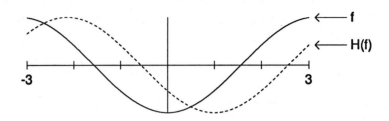

Figure VI.45: Comparison of f and $H(f)$

*Exercise VI.9.3:
Suppose the operator G is defined by
$$G(f)(x) = \int_0^x f(t)\,dt.$$
Is this operator *linear?*

*Exercise VI.9.4:
Is the function $f(x,y) = xy$ *linear?*

*Exercise VI.9.5:
Show that if f is any *linear* function then $f(0) = 0$.

*Exercise VI.9.6:
Suppose that $f(x_1, x_2, \ldots, x_n)$ is a *linear* function. Show that

- $f(0, 0, \ldots, 0) = 0$.

- If $f(x_1, x_2, \ldots, x_n) = 0$ and $f(y_1, y_2, \ldots, y_n) = 0$ then
$$f(x_1 + y_1, x_2 + y_2, \ldots, x_n + y_n) = 0.$$

*Exercise VI.9.7:
Consider the differential equation
$$ay'' + by' + cy = 0$$
and the *linear* operator
$$L(y) = ay'' + by' + cy.$$

§9. LINEARITY

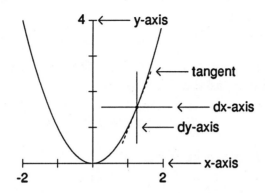

Figure VI.46: Local coordinates at the point $(1.25, 1.5625)$

Notice that y is a solution of the differential equation if and only if $L(y) = 0$.

Using just the properties of *linear* operators show that if c_1 and c_2 are two constants and y_1 and y_2 are two solutions of the differential equation then so is $c_1 y_1 + c_2 y_2$.

The Derivative Revisited

We introduced the concept of the derivative in order to study the behavior of a function near a particular point. Consider the function $y = x^2$ at the point $(1.25, 1.5625)$ as shown in Figure VI.46. In order to study this function near this point we introduce a new set of coordinates whose origin is at the point in question. These coordinates are called **local coordinates**. We call the new horizontal axis the dx-axis and the new vertical axis the dy-axis as shown in Figure VI.46.

Notice the slope of the tangent to the curve $y = x^2$ at the point $(1.25, 1.5625)$ is 2.5; so that the tangent line to the curve at this point has slope 2.5. Thus in local coordinates the tangent line has the equation

$$dy = 2.5 \, dx.$$

This is the best way to think of the derivative of a function $f(x)$ at a point $(x_0, f(x_0))$. It describes the tangent line in local coordinates by giving us its slope. The basic idea is that if we concentrate our attention on the part of a curve very close to the point $(x_0, f(x_0))$ then the graph looks very much like a straight line. Thus in local coordinates

close to the point $(x_0, f(x_0))$ the original function is very much like the *linear* function
$$dy = f'(x_0) \, dx.$$

Most mathematicians think of the derivative of the function $f(x)$ at the point x_0 as the *function*
$$dy = f'(x_0) \, dx$$
rather than as the *number*
$$f'(x_0).$$

We can convert between the original coordinates and local coordinates at the point $(x_0, f(x_0))$ as follows.

$$\begin{aligned} dx &= x - x_0 \\ dy &= y - f(x_0) \\ x &= dx + x_0 \\ y &= dy + f(x_0) \end{aligned}$$

Thus the equation of the tangent line in terms of the original coordinates is

$$\begin{aligned} y &= dy + f(x_0) \\ &= f'(x_0) \, dx + f(x_0) \\ &= f'(x_0)(x - x_0) + f(x_0) \end{aligned}$$

Exercises

*Exercise VI.9.8:
Find the equation of the tangent line to the curve $y = x^3 - 2$ at the point $(1, -1)$ in local coordinates.
Find the equation of the same tangent line in the original coordinates.

*Exercise VI.9.9:
Find the equation of the tangent line to the curve $y = \sqrt{25 - x^2}$ at the point $(3, 4)$ in local coordinates.

§9. LINEARITY

Find the equation of the same tangent line in the original coordinates.

*Exercise VI.9.10:
Find the equation of the tangent line to the curve $y = \cos x$ at the point $(0, 1)$ in local coordinates.

Find the equation of the same tangent line in the original coordinates.

*Exercise VI.9.11:
Find the equation of the tangent line to the curve $y = \sin x$ at the point $(0, 0)$ in local coordinates.

Find the equation of the same tangent line in the original coordinates.

Newton's Method Revisited

The derivative owes its usefulness to two facts.

- Linear functions are very simple. Hence we know a lot about them.

- Close to a point x_0 the derivative $dy = f'(x_0)dx$ (thought of as a linear function in local coordinates) is very close to the original function.

We exploited these ideas when we developed Newton's Method. Newton's Method is based on Figure VI.47. It is a method for producing a sequence of better and better estimates for a solution of an equation of the form
$$f(x) = 0.$$
The heart of Newton's Method is a procedure that starts with one estimate x_n for a solution and produces a better estimate x_{n+1}. This new estimate is obtained by looking at the tangent to the curve, i.e. the linear function that best approximates the curve, at the point $(x_n, f(x_n))$. The new estimate x_{n+1} is the point at which the tangent line intersects the x-axis. In local coordinates the tangent line is
$$dy = f'(x_n)\, dx$$

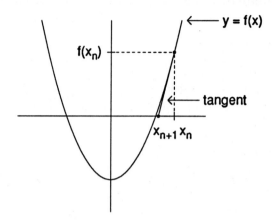

Figure VI.47: Newton's Method

and in the original coordinates it is

$$y = f'(x_n)(x - x_n) + f(x_n).$$

Thus the point at which this line intersects the x-axis can be obtained by solving the equation

$$0 = f'(x_n)(x - x_n) + f(x_n)$$

which leads to

$$x = x_n - \frac{f(x_n)}{f'(x_n)}$$

and the formula for x_{n+1} in terms of x_n is

$$x_{n+1} = x_n - \frac{f(x_n)}{f'(x_n)}.$$

The success of Newton's Method relies on the two facts about the derivative mentioned above, which imply that

- Because linear functions are so simple, it is easy to find a formula for x_{n+1} in terms of x_n.

- Because the linear function determined by the derivative at the point $(x_n, f(x_n))$ is close to the original curve, the point x_{n+1} at which this linear function intersects the x-axis is close to the point at which the curve intersects the x-axis.

§9. Linearity

It is important, however, to emphasize the local nature of this last observation. The tangent line is only close to the original curve close to the original point. This is the reason why Newton's method can fail to work without a good initial estimate for the solution.

Our original development of Newton's Method was motivated by the picture shown in Figure VI.47. We could, however, have developed the same method in a purely algebraic way using the two facts about the derivative mentioned above. In the next subsection we show how this approach allows us to use Newton's method for a complex-valued function of a complex variable.

The Derivative of A Complex Function

Suppose that $w = f(z)$ is a complex-valued function of a complex variable. One can define the derivative of the function f at a point z in exactly the same way as one does for a real-valued function of a real variable

$$f'(z) = \lim_{h \to 0} \frac{f(z+h) - f(z)}{h}.$$

Just as for a real-valued function of a real variable this limit may or may not exist. If the limit does exist for a particular value z_0 of z then the derivative gives us a linear function that approximates the original function near the point $(z_0, f(z_0))$ in exactly the same way as the derivative of a real-valued function of a real variable. In local coordinates

$$dw = f'(z_0)\, dz$$

or, in the original coordinates,

$$w = f'(z_0)(z - z_0) + f(z_0).$$

Again the best way to think about the derivative $f'(z_0)$ at the point $(z_0, f(z_0))$ is as the linear function that best approximates the original function f near the point $(z_0, f(z_0))$. It is very difficult to draw a picture of a complex-valued function of a complex variable because one would need four dimensions, two for the variable z and two for the variable w. Nonetheless we can work algebraically with such a function almost as easily as with a real-valued function of a real variable. For example, exactly the same computations that were used in developing Newton's Method work when the function is a complex-valued function of a complex variable. They lead to exactly the same method. The only difference is that traditionally one uses the letter z for a complex variable instead of the letter x. So we write Newton's Method in this case as

$$z_{n+1} = z_n - \frac{f(z_n)}{f'(z_n)}.$$

Of course, the arithmetic is a little harder because we are working with complex numbers.

Example:

Use Newton's Method to estimate $\sqrt{1+i}$. Start with the initial estimate $z_1 = (1+i)$.

The number $\sqrt{1+i}$ is the solution of the equation

$$z^2 - (1+i) = 0.$$

So we want to apply Newton's Method with the function

$$f(z) = z^2 - (1+i).$$

Thus Newton's Method becomes

$$z_{n+1} = z_n - \frac{z_n^2 - (1+i)}{2z_n}.$$

With the initial estimate $z_1 = (1+i)$ this leads to

$$\begin{aligned}
z_1 &= 1.00000 + 1.000000i \\
z_2 &= 1.00000 + 0.500000i \\
z_3 &= 1.10000 + 0.450000i \\
z_4 &= 1.09867 + 0.455088i \\
z_5 &= 1.09868 + 0.455090i
\end{aligned}$$

These results are very similar to the results we obtained earlier using Newton's Method with real numbers. The only difference is that the numbers are complex rather than real and that the arithmetic operations are complex.

---

§9. Linearity

Exercises

*Exercise VI.9.12: Use Newton's Method to estimate $\sqrt{2-i}$. Start with $z_1 = 2 - i$.

*Exercise VI.9.13: Use Newton's Method to estimate $\sqrt{2-i}$. Start with $z_1 = 1$.

Stability of an Equilibrium

One of the most important questions that we investigated last semester was the question of whether a given equilibrium point was attracting or repelling. For an (unitalicized) linear discrete dynamical system of the form

$$x_{n+1} = mx_n + b$$

the study of equilibrium points is very easy. If $m \neq 1$ there is always one equilibrium point that is found by solving the equation

$$\begin{aligned} x &= mx + b \\ x - mx &= b \\ (1-m)x &= b \\ x &= b/(1-m) \end{aligned}$$

If we start out with any initial point x_1 and consider the sequence

$$\begin{aligned} x_2 &= f(x_1) \\ x_3 &= f(x_2) \\ &\vdots \\ x_{n+1} &= f(x_n) \end{aligned}$$

then we see that if $|m| < 1$ then

$$\lim_{n \to \infty} x_n = b/(1-m).$$

Thus if $|m| < 1$ the equilibrium point is attracting in a very strong sense. It "attracts" the sequence regardless of the initial condition.

Last semester we saw that for a nonlinear discrete dynamical system

$$x_{n+1} = f(x_n)$$

there may be any number of equilibrium points. For each equilibrium point x_* the derivative $f'(x_*)$ gives us the same kind of information as the slope m did for a *linear* discrete dynamical system. If $|f'(x_*)| < 1$ then the equilibrium point is attracting. It is, however, attracting only in a weaker sense. *If the initial condition x_1 is sufficiently close to the equilibrium point x_** then

$$\lim_{n \to \infty} x_n = x_*.$$

If the initial condition is not sufficiently close to x_* then other behavior is possible.

Exactly the same observations can be made about a complex discrete dynamical system

$$z_{n+1} = f(z_n)$$

where f is a complex-valued function of a complex variable.

Exercises

*Exercise VI.9.14:

Consider the complex discrete dynamical system

$$z_{n+1} = z_n/2 - i$$

1. Find all the equilibrium points of this discrete dynamical system.

2. Determine whether each equilibrium point is attracting or repelling.

3. Experiment with several different initial conditions z_1 to test your answer to 2.

Exercise VI.9.15:

Consider the complex discrete dynamical system

$$z_{n+1} = 2(z_n + i)$$

1. Find all the equilibrium points of this discrete dynamical system.

2. Determine whether each equilibrium point is attracting or repelling.

3. Experiment with several different initial conditions z_1 to test your answer to 2.

*Exercise VI.9.16:
Consider the complex discrete dynamical system

$$z_{n+1} = z_n^2 + i$$

1. Find all the equilibrium points of this discrete dynamical system.

2. Determine whether each equilibrium point is attracting or repelling.

3. Experiment with several different initial conditions z_1 to test your answer to 2.

Exercise VI.9.17:
Consider the complex discrete dynamical system

$$z_{n+1} = z_n^3 - i$$

1. Find all the equilibrium points of this discrete dynamical system.

2. Determine whether each equilibrium point is attracting or repelling.

3. Experiment with several different initial conditions z_1 to test your answer to 2.

---

§10. Summary[+]

We began this chapter by looking at the plane from two new perspectives. The first perspective, polar coordinates, is particularly useful for describing phenomena, like the sensitivity of a microphone, that depend on angle and distance. The second perspective, complex numbers, is important because many real world phenomena, for example in physics, can be best described using complex numbers. The equation $z^2 = -1$ appears very frequently in a wide variety of settings and each time it does complex numbers are essential. Modern calculators and computers routinely can handle calculations involving complex numbers as easily as those involving the more familiar real numbers.

The main subject of this chapter was infinite series. We developed a framework for working with functions defined by power series of the form
$$p(z) = \sum_{n=0}^{\infty} a_n(z-a)^n.$$
This ability enables us to do many things.

- The solutions of many initial value problems can be expressed in this way.

- Any function that can be described by a power series can be estimated on the basis of the four elementary arithmetic operations—addition, subtraction, multiplication, and division. Hence, such a function can be approximated by a computer.

- Similarly any function that can be described by a power series is defined for complex numbers as well as real numbers.

By expressing three important functions—e^z, $\sin z$, and $\cos z$—as power series we discovered a surprising relationship among them, Euler's Formula,
$$e^{ix} = \cos x + i \sin x.$$
This formula was the key to finding solutions of the differential equation
$$ay'' + by' + cy = 0$$
and understanding the many phenomena, like spring-and-mass systems, that can be described by this differential equation.

The most important results about these differential equations involve complex numbers in a crucial way. The following theorem summarizes these results.

Theorem 27 *The long term behavior of solutions to the equation*
$$ay'' + by' + cy = 0$$

§10. Summary†

depends on the real parts of the solutions to the characteristic equation

$$ar^2 + br + c = 0.$$

- If the real parts of both solutions are negative then all solutions will approach zero.

- If the real parts of both solutions are positive then, unless the initial conditions are $y(t_0) = y'(t_0) = 0$, the solution will either approach $+\infty$, or $-\infty$, or will oscillate with increasing amplitude.

- If both solutions, r_1 and r_2, are real and one is positive and the other is negative then the solutions look like

$$y = c_1 e^{r_1 t} + c_2 e^{r_2 t}$$

and the long term behavior depends on the constants c_1 and c_2, which in turn depend on the initial conditions.

The following problems can help you review the tools developed in this chapter.

Exercises

*Exercise VI.10.1: Graph the function

$$r = \sin 5\theta$$

in Cartesian coordinates and in polar coordinates.

*Exercise VI.10.2: Find the area of one petal of the "flower" described by the equation

$$r = \sin 3\theta.$$

*Exercise VI.10.3: Let $u = 1 + i$ and $v = 1 - i$. Find $u + v$, $u - v$, uv and u/v. Express your answers in both the form $a + bi$ and the form $r \operatorname{cis} \theta$.

*Exercise VI.10.4: Find all the roots of the equation

$$z^8 = 16.$$

*Exercise VI.10.5: If you were offered a choice between four lottery prizes.

1. $50,000 per year for 20 years.

2. $100,000 per year for 10 years

3. $80,000 per year for 10 years.

4. $30,000 per year forever.

which would you take? In order to answer this question you will need to make some assumptions about interest rates or inflation. Explain your answer. How does your answer depend on the assumptions that you make about interest rates or inflation?

Exercise VI.10.6: Find the radius of convergence of the power series
$$p(z) = 1 + \frac{z}{2} + \frac{z^2}{3} + \frac{z^3}{4} + \frac{z^4}{5} + \cdots.$$

Exercise VI.10.7: Give examples of power series whose radius of convergence is 0, π, and ∞.

Exercise VI.10.8: Consider the initial value problem
$$y' = y^2, \quad y(0) = 1.$$

- Sketch the slope field for the differential equation $y' = y^2$. Based on this slope field describe the long term behavior of the solution to this initial value problem.

- Find the solution of this initial value problem using the method of separation of variables. What is
$$\lim_{t \to \infty} y(t) = ?$$

- Reconcile the two answers above.

- Find the power series solution to this initial value problem. What is the radius of convergence of this power series?

- There is a fundamental oddity about the solution to this initial value problem. What is it and how does it manifest itself in the two different ways in which we have expressed the solution?

§10. Summary†

Exercise VI.10.9:
Find the Taylor Series (centered at zero) for the function $y = e^{-z}$ three different ways. Your three answers should agree.

- Substitute $-z$ for z in the Taylor Series for e^z.

- Use the technique we developed for finding the Taylor Series of a function $f(z)$ based on the values $f(0), f'(0), f''(0), \ldots$.

- Find the power series solution of the initial value problem

$$y' = -y, \quad y(0) = 1.$$

*Exercise VI.10.10: The initial value problem

$$y'' = -4y - by', \quad y(0) = p$$

where b and p are nonnegative constants has several qualitatively different solutions. Describe them. How does the solution depend qualitatively on the values of b and p? Give specific examples. Discuss how the mathematics represents what is happening in the real world of spring-and-mass systems.

Chapter VII

Dynamical Systems

§1. One-Dimensional Dynamical Systems

In this section we study one-dimensional continuous dynamical systems. The adjective **one-dimensional** refers to the fact that we are interested in *one* quantity that is changing over time. For example, we might be studying the way in which the population of one species is changing or we might be studying the way in which the temperature of a cup of coffee is changing. The adjective **continuous** refers to the fact that, in contrast to the discrete dynamical systems we studied at the beginning of the first semester, the quantity in question is changing continuously rather than only at certain specific times. Here the changing quantity is modeled by a continuous function $p(t)$ rather than by a sequence and the way in which the quantity changes is described by a differential equation

$$p' = f(p)$$

or

$$p' = f(p,t)$$

rather than by an equation like

$$p_{n+1} = f(p_n).$$

Examples:

- As a hot object (for example, a cup of coffee) cools, its temperature changes continuously. Careful experiments have shown that the rate at which the object cools is proportional to the difference between its temperature and the ambient temperature. This is often called **Newton's Law of Cooling** and can be expressed by the differential equation

$$C' = -k(C - R)$$

where $C(t)$ denotes the temperature of the object at time t, the constant R is the ambient temperature, and k is a positive constant whose value depends on the particular situation. For example, a large well-insulated carafe of coffee will cool more slowly than a single cup of coffee and, thus, the value of the constant k for a large well-insulated carafe of coffee will be smaller than the value of the constant k for a small cup of coffee. This is an example of a one-dimensional continuous dynamical system because we are interested in a single quantity whose change is described by a differential equation.

- If we study a cup of coffee cooling in a large room then the model above is reasonable. If we have a large barrel of coffee in a small room, however, then as the coffee cools, the room will heat up. Thus there are two quantities, the temperature $C(t)$ of the coffee and the temperature $R(t)$ of the room, that change according to the pair of differential equations

$$\begin{aligned} C' &= -k_1(C - R) \\ R' &= k_2(C - R) \end{aligned}$$

where k_1 and k_2 are two positive constants depending on the particular object and the particular room. This is an example of a **two-dimensional** dynamical system because there are two quantities changing over time. We study two-dimensional dynamical systems later in this chapter.

- A typical bank account might pay interest at the rate of 6% per year. Usually banks compute the interest and add it to the account several times per year. This is called **compounding**. Many banks compound interest four times per year or **quarterly**. If you were to deposit $100 in an account earning interest at the rate of 6% per year compounded quarterly then the graph shown in

§1. One-Dimensional Dynamical Systems

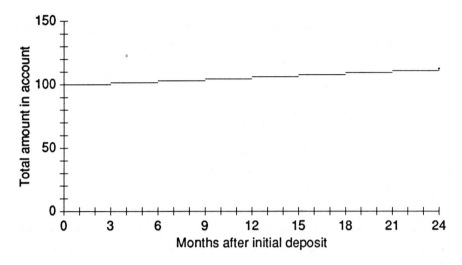

Figure VII.1: Record of bank account

Figure VII.1 would show the value of the account for the first 24 months. This situation cannot be described by a *continuous* dynamical system because the amount of money is changing only at the discrete times—three months, six months, nine months, etc.

Digression

Before looking at continuous dynamical systems we need to discuss some ambiguity in the English language. We often use the words "rate of growth" in two very different senses. For example, if we were looking at a particular town whose current population was 10,000 and rising at the rate of 200 people per year we might say "the rate of growth is 200 people per year" or we might say "the rate of growth is two percent per year." Both of these sentences are correct and seem to say the same thing but they capture two very different concepts. For example, compare the following two paragraphs.

- "The population of Westville is currently 10,000 and is growing at the rate of 200 people per year. We expect this rate of growth to continue for several years."

    This is the kind of population growth one would expect, for example, if the underlying cause of the population growth were immigration. It implies that in ten years the population of Westville will be 12,000.

- "The population of Westville is currently 10,000 and is growing at the rate of two percent per year. We expect this rate of growth to continue for several years."

This paragraph says that each year the population of Westville will increase by two percent. Thus in ten years the population of Westville will be 12,190. This is the kind of growth that one would expect if the reason for the population increase in Westville was that the per capita birth rate was higher than the per capita death rate.

It is important to keep the uses of the phrase "rate of growth" straight. We often use the term **relative rate of growth** for the second use. In English relative rates of growth are frequently expressed as percentages. We often use the term **absolute rate of growth** for the first use. If there is no danger of confusion we may omit the adjective and simply say "rate of growth."

End Digression

We illustrate our discussion in this section with a model of population change for a species that hunts in packs. We are interested in a particular species that has a natural growth rate of 20%. That is, if a good-sized pack were living in a habitat with an abundance of food, water, and shelter then its population would increase at the rate of 20% per year. This could be expressed by the differential equation

$$p' = 0.20p.$$

This 20% growth rate, however, applies only when the population is moderate—large enough so that the animals can form efficient hunting packs but not so large as to exceed the available food, water, and shelter. Thus the population change is expressed by a more complicated differential equation like

$$p' = R(p)p$$

where $R(p)$ is a function like the one shown in Figure VII.2.

Notice that the population grows at the rate of 20% per year if there are between 200 and 1000 individuals. If the population is below 200 it grows at a slower rate and if it is below 100 it declines. Similarly, if

§1. One-Dimensional Dynamical Systems

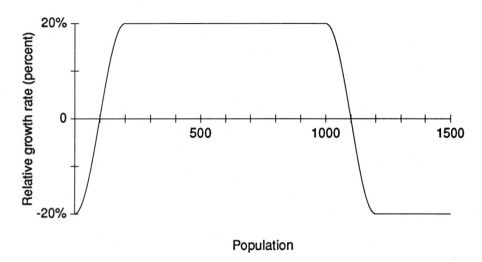

Figure VII.2: Relative growth rate, $R(p)$

the population is above 1000 it grows at a slower rate and if it is above 1100 it declines. We can rewrite our original differential equation as

$$p' = F(p)$$

where $F(p)$ is the function $F(p) = R(p)p$. Figure VII.3 shows the graph of the function $F(p)$. Notice that $F(p)$ is the *absolute* growth rate and that $R(p)$ is the *relative* growth rate.

The right side of this differential equation does not involve the time variable t, only the variable p whose change we are studying. This is a very common situation. Such equations are called **autonomous** and are particularly easy to visualize. Several computer programs can be used to visualize such equations. The display from one of them is shown in Figure VII.4. This program shows the solution of the initial value problem

$$p'(t) = f(p) \qquad p(a) = p_0.$$

As it runs the program constructs a graph of the function $p(t)$ on the right side of the screen and at the same time it shows the current value of p moving up and down a second "p-axis" on the left side of the screen. You will see two dots. One dot, which remains still, shows the initial value of p (i.e., p_0). The second dot, which moves, shows the current value of the quantity p. Figure VII.4 shows a snapshot of the computer screen while this program is running.

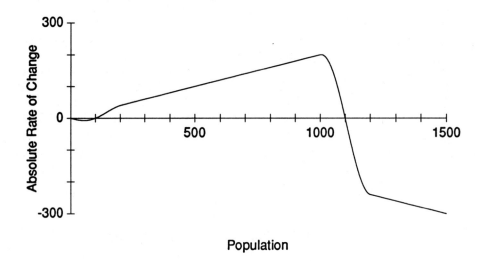

Figure VII.3: Absolute growth rate, $F(p)$

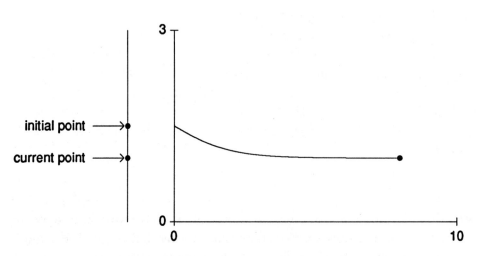

Figure VII.4: Program demonstration

§1. One-Dimensional Dynamical Systems

This program demonstrates two ways of visualizing the function $p(t)$. One way, on the right side of the screen, is the traditional way, a graph of the function $p = p(t)$. The second way, on the left side of the screen, is well-suited to computer-based animation. It shows a "movie" of the value of p as it changes.

This visualization of the function $p(t)$ leads to a particularly nice conceptualization. Think of the variable p as corresponding to possible points in a river. At each point p in the river the function $f(p)$ indicates the current, that is, how fast and in which direction the water is flowing at that point. We draw our metaphorical rivers as horizontal lines. If $f(p)$ is positive at a particular point p in the river then at that point the current is flowing to the right. If $f(p)$ is negative at a particular point p then at that point the current is flowing to the left. The size of the absolute value of $f(p)$ at a point p indicates how strong the current is or how rapidly the water is flowing at that point. If $f(p)$ is zero at a particular point p in the river then the current at that point is still. Solving an initial value problem

$$p' = f(p); \quad p(a) = p_0$$

is analogous to placing a cork into the river at time $t = a$ at the point p_0 and watching where the cork is carried by the currents.

Using this imagery we can analyze the differential equation

$$p' = F(p)$$

from the example involving a species that hunts in packs. Recall the function $F(p)$ shown in Figure VII.3. Figure VII.5 shows a map of the corresponding river with arrows pointing to the right where $F(p)$ is positive and to the left where $F(p)$ is negative. Notice that the river is flowing to the left (i.e., p is decreasing) when p is less than 100 because there are not enough individuals to form efficient hunting packs and when p is greater than 1100 because of shortages of food, water and shelter. When p is between 100 and 1100 the current is flowing to the right (i.e., p is increasing) because there are enough individuals to form efficient hunting packs and not enough to exceed the available food, water and shelter. The points at which the current is still—$p = 0, 100$, or 1100—are marked with big dots.

To embellish our imagery we might imagine an underwater spring feeding the river at the point marked "100" and underwater drains

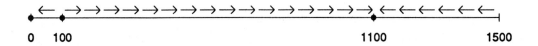

Figure VII.5: Map of the river and currents for $p' = F(p)$

draining the river at the points marked "0" and "1100." It is easy to visualize a cork placed in the river between $p = 0$ and $p = 100$ being carried down toward zero. This would be exactly the fate of a colony of fewer than 100 individuals of this species introduced into this habitat. Similarly, it is easy to imagine the travels of a cork placed into the river anyplace to the right of $p = 100$. Such a cork would be carried toward the point $p = 1100$. Again, this would be exactly the fate of a colony of more than 100 individuals of this particular species introduced into this habitat. In this situation one might say that 100 is the **critical mass** for this particular species in this particular habitat and that 1100 is the **carrying capacity** for this particular habitat for this particular species.

Example:

Consider the differential equation

$$p' = (p-1)(p-2)(p-3).$$

The function on the right side of this equation

$$f(p) = (p-1)(p-2)(p-3)$$

has three zeros, at $p = 1, p = 2$ and $p = 3$. These three points are marked by large dots in Figure VII.6. It is easy to see that $f(p)$ is positive when $p > 3$, negative when $2 < p < 3$, positive when $1 < p < 2$ and negative when $p < 1$. Thus the corresponding map of the river and its currents looks like Figure VII.6. A cork placed in the river between 1 and 3 would be carried toward the point 2; a cork placed in the river to the right of 3 would be carried off toward $+\infty$; a cork placed in the

§1. One-Dimensional Dynamical Systems

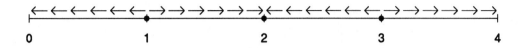

Figure VII.6: River and currents for $p' = (p-1)(p-2)(p-3)$

river to the left of 1 would be carried off toward $-\infty$; and a cork placed in the river exactly at 1, 2, or 3 would remain where it was placed.

Diagrams like the maps shown in Figures VII.5 and VII.6 are enormously useful. They have been studied by many people and have several different names. We shall call them **vector fields**. They are also called **phase diagrams**. Our diagrams are somewhat simplified. A true vector field has arrows at each point showing not only the direction of the current but also its strength. Stronger currents are indicated by longer arrows and weaker currents by shorter arrows. Because the strength of the currents can vary widely these diagrams can be difficult to draw effectively. The arrows frequently overlap each other. For this reason we use arrows that all have the same length. They indicate the direction of the current but not its strength. Our diagrams should properly be called **direction fields**.

Another program, MOVIE1, can be used to visualize one-dimensional autonomous dynamical systems. Imagine a river like the ones we have been considering. Suppose a large number of corks are placed in this river and then we make a movie showing the way that these corks are carried by the currents in the river. MOVIE1 shows this movie. Figure VII.7 gives a rough idea of what the program shows. This figure shows ten still "frames" from the movie for the differential equation

$$p' = (p-1)(p-2)(p-3)$$

with the corks placed in the river at evenly spaced points in the top frame. As you read this figure from top to bottom you see the story of the corks as it unfolds, as the corks are carried towards the "drain" at $p = 2$ and away from the "springs" at $p = 1$ and $p = 3$. A static

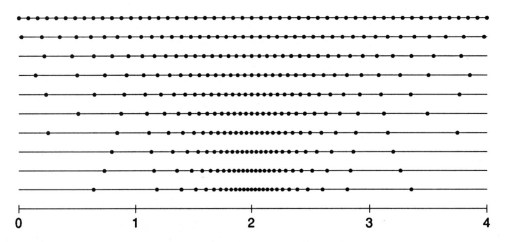

Figure VII.7: Animation of many corks

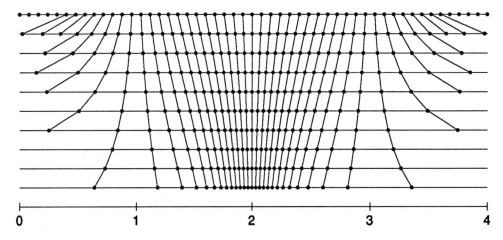

Figure VII.8: Animation of many corks

§1. ONE-DIMENSIONAL DYNAMICAL SYSTEMS

graph on a printed page doesn't do this display justice; you need to see the animated display. Figure VII.8 has some extra lines connecting the dots representing each cork from still frame to still frame. These lines may make it a little easier to see what is happening. As you might guess there is a two-dimensional version of the program, called MOVIE2, for two-dimensional dynamical systems.

We have observed several different kinds of behavior for one-dimensional autonomous dynamical systems. Now we would like to introduce some terminology and tools for discussing such behavior. The first thing to look for in a one-dimensional autonomous dynamical system is its equilibrium points.

Definition:

> An **equilibrium point** for a one-dimensional autonomous dynamical system
> $$p' = f(p)$$
> is a point p_* at which $p' = 0$, that is, a point p_* such that $f(p_*) = 0$. We find equilibrium points by solving the equation
> $$f(p) = 0.$$
> This equation may have any number of solutions and, hence, there may be any number of equilbrium points.
>
> The real significance of an equilibrium point p_* is that if the initial condition places $p(a) = p_0$ at an equilibrium point then $p(t)$ will remain at that same point for all time since at that point there is no metaphorical current to carry p away.

The next step in the analysis of a one-dimensional autonomous dynamical system is drawing its direction field. First mark the equilibrium points by dots on a line representing the p-axis. These points divide this line up into several intervals. We work only with differential equations in which $f(p)$ is continuous. Thus the only places at which the function $f(p)$ can change sign are the equilibrium points. Hence by evaluating $f(p)$ at one point in each interval we can determine where $f(p)$ is positive and where it is negative. Using this information sketch the direction field.

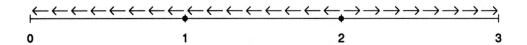

Figure VII.9: Direction field for $p' = (p-1)^2(p-2)$

Example:

Consider the system

$$p' = (p-1)^2(p-2)$$

This system has two equilibrium points, at $p = 1$ and $p = 2$, that break the "river" into three sections. The direction field shown in Figure VII.9 was determined by evaluating the sign of the function $f(p) = (p-1)^2(p-2)$ at one point in each of the sections.

We have already seen two different kinds of equilibrium points—equilibrium points corresponding to underwater drains that pull corks in—and equilibrium points corresponding to underwater springs that push corks away. This new example illustrates a third kind of equilibrium point. Notice that the river is flowing to the left *on both sides of the equilibrium point $p = 1$*. Thus if a cork is placed just to the right of $p = 1$ it will be carried toward $p = 1$ but if a cork is placed just to the left of $p = 1$ it will be carried away from $p = 1$.

We use the following terminology to describe the three different kinds of equilibrium points.

Definition:

- An equilibrium point p_* is called **stable** or **attracting** if the direction field immediately to the right of p_* is pointing to the left and the direction field immediately to the left of p_* is pointing to the right. That is, if on both sides of p_* the direction field is pointing towards p_*. A cork placed close to an attracting

§1. One-Dimensional Dynamical Systems

equilibrium point will be pulled or attracted toward it. If an equilibrium point is not stable then it is said to be **unstable**. We distinguish two kinds of unstable equilibrium points.

- An equilibrium point is called **repelling** if the direction field immediately to the right of p_* is pointing to the right and the direction field immediately to the left of p_* is pointing to the left. A cork placed close to a repelling equilibrium point will be pushed or repelled away from it.

- An equilibrium point is called **indeterminate** if it is neither attracting or repelling. The behavior of a cork placed close to an indeterminate equilibrium point is more difficult to predict and depends on exactly where it is placed.

Exercises

*Exercise VII.1.1: Consider the dynamical system

$$p' = (p-2)(p-4).$$

Sketch the direction field for this dynamical system. Find all the equilibrium points and classify them according to the previous definition.

Based on the direction field what would you expect the solution of the each of the initial value problems below to look like. Verify your predictions using the computer.

$$
\begin{aligned}
p' &= (p-2)(p-4); & p(0) &= 0.50 \\
p' &= (p-2)(p-4); & p(0) &= 1.50 \\
p' &= (p-2)(p-4); & p(0) &= 2.50 \\
p' &= (p-2)(p-4); & p(3) &= 2.50 \\
p' &= (p-2)(p-4); & p(0) &= 5.00
\end{aligned}
$$

Use the computer to get an overall picture of how solutions of this differential equation behave.

Exercise VII.1.2: Consider the dynamical system

$$p' = \sin p.$$

Sketch the direction field for this dynamical system. Find all the equilibrium points and classify them according to the previous definition.

Based on the direction field what would you expect the solution of each of the initial value problems below to look like. Verify your predictions using the computer.

$$\begin{aligned} p' &= \sin p; & p(0) &= 0 \\ p' &= \sin p; & p(0) &= \pi/2 \\ p' &= \sin p; & p(0) &= \pi \end{aligned}$$

***Exercise VII.1.3:** A one-dimensional autonomous dynamical system of the form

$$p' = mp + b$$

where m and b are constants is called **linear**. For this exercise you may assume that m is not zero. What can you say about linear one-dimensional autonomous dynamical systems? How many equilibrium points does such a system have? What can you say about whether the equilibrium point(s) are attracting, repelling or indeterminate? Suppose that $p(t)$ is a solution of an initial value problem

$$p' = mp + b; \qquad p(0) = p_0.$$

What can you say about the limit

$$\lim_{t \to +\infty} p(t).$$

Does the initial value p_0 make any difference for the limit? What does the slope m tell us about the dynamical system?

***Exercise VII.1.4:** Now consider the dynamical system

$$p' = f(p)$$

and suppose that p_* is an equilibrium point for this dynamical system. What does $f'(p_*)$ tell us about this equilibrium point? Make your

§1. ONE-DIMENSIONAL DYNAMICAL SYSTEMS

answer as complete as possible. Look at some examples. What can you say about the limit

$$\lim_{t \to +\infty} p(t).$$

Does the initial value p_0 make any difference for the limit?

*Exercise VII.1.5: In this exercise we would like to study what happens when an object is falling near the surface of the earth. We use the variable h to denote the height of the object above the surface of the earth and the variable $v = h'$ to denote its velocity. Since h denotes height v will be positive when the object is rising and v will be negative when the object is falling.

The force of gravity near the surface of the earth would produce an acceleration of -32 feet per second per second if it were unopposed. Notice that the acceleration is negative since it is in the downward direction. In this case the velocity of a falling body would be described by the one-dimensional autonomous dynamical system

$$v' = -32.$$

A falling body, however, encounters air resistance. The air resistance depends on the velocity; the higher the velocity the stronger the air resistance. The way in which the air resistance depends on the velocity depends on the shape of the falling body. Two very common models for the air resistance are

$$\text{air resistance} = -k_1 v$$

and

$$\text{air resistance} = -k_2 v^2 \text{sign}(v)$$

where k_1 and k_2 denote positive constants and $\text{sign}(v)$ is the sign of v. The minus sign represents the fact that air resistance is always trying to slow the object down. The first model is appropriate for a very streamlined object traveling at moderate velocities. The second model is part of the folklore of bicycling. The first model leads to a dynamical system for the velocity of a falling body of the form

$$v' = -32 - kv.$$

Analyze this dynamical system. What can you say about

$$\lim_{t \to +\infty} v(t)?$$

The second model leads to a dynamical system for the velocity of a falling body of the form

$$v' = -32 - kv^2 \text{sign}(v).$$

Analyze this dynamical system. What can you say about

$$\lim_{t \to +\infty} v(t)?$$

In fact, the way in which air resistance depends on velocity can be very complex. The force due to air resistance may be a very complicated function $g(v)$. It seems reasonable to make the following general assumptions about the function $g(v)$

- $g(0) = 0$, that is, a still body is not influenced at all by air resistance.

- $g(v)$ always has the opposite sign than v, that is, air resistance is always trying to slow the object down.

- If $|v_1| < |v_2|$ then $|g(v_1)| < |g(v_2)|$. This assumption says that the faster an object is moving the stronger the force due to air resistance.

- $g(-v) = -g(v)$, that is, the strength of air resistance depends only on the speed of an object not on the direction in which it is going.

- $\lim_{v \to +\infty} g(v) = +\infty$, that is, the force due to air resistance will grow without any bound as the velocity grows. This assumption is not very realistic *in theory* since there are limits to how fast an object can move. In practice, however, it is realistic since the force exerted due to air resistance will indeed be very large if the object is traveling very fast.

Based on these assumptions what can you say about the velocity of a falling object?

§2. Two-Dimensional Dynamical Systems

Last semester we studied a number of models for the population growth of a single species in a habitat all by itself. Such models are not very realistic. Most habitats outside the biology laboratory are inhabited by several species that interact in various ways, sometimes competing for the same food supply, for the same water and shelter, and sometimes even eating one another. In this section we investigate simple models of two species sharing a single habitat. We consider several kinds of interaction.

Examples:

- Two species that compete for some but usually not all of the same resources, for example, they might compete for the same food but not for the same shelter. We call this interaction **competitive** interaction.

- Two species, for example, the Montagues and the Capulets, or the Hatfields and the McCoys, that fight each other aggressively. We call this interaction **aggressive** interaction.

- Two species, one of which is biologically weaker than the other. We call this interaction **weak-strong** interaction.

- Two species, one of which eats the other as food. The eater is called the **predator** and the eatee is called the **prey**. This interaction is called **predator-prey** interaction.

The most naive population models, called **exponential** models, are based on the idea that a population grows at a constant *relative* or *per capita* rate. For example, a population might grow at a constant rate of 20% per year. Such a model would be described by the differential equation
$$p' = .20p$$
More generally, an exponential model is given by a differential equation of the form
$$p' = Rp$$

where R is some constant. It is easy to extend this idea to two species whose populations are denoted by p and q with a pair of differential equations

$$p' = R_1 p \qquad q' = R_2 q$$

where R_1 and R_2 are two constants corresponding to the relative rates of growth for each of the two species.

Such models, however, are unrealistic. The relative rate of growth of a species is not a constant but varies depending on many factors, the most important of which is the population itself. If there are only a few individuals in a habitat then there is plenty of food, water, and shelter, the animals are very healthy with low death rates and high birth rates, and the population increases at a rapid (relative) rate. Conversely, if there is a very large number of individuals in a habitat then there are shortages of food, water, and shelter, the animals are not very healthy, death rates are high and birth rates low, and the population grows more slowly, or may even decrease. The simplest single species model that captures this idea is the model

$$p' = (R - mp)p$$

where R and m are positive constants. We have replaced the constant relative rate R in an exponential model by a variable relative rate $(R - mp)$ that decreases as p increases. Figure VII.10 shows an example of such a variable relative rate, $R(p) = 1 - p/500,000$. This kind of model, in which the relative growth rate is a linear function of the population, is called a **logistic** model.

We want to modify this model for two species sharing the same habitat. The modified model will be expressed by a pair of differential equations of the form

$$\begin{aligned} p' &= (R_1 - c_{11}p - c_{12}q)p \\ q' &= (R_2 - c_{21}p - c_{22}q)q. \end{aligned}$$

The expression

$$R_1 - c_{11}p - c_{12}q$$

describes the relative growth rate for the species whose population is denoted by p. The constant c_{11} is a positive number so that as the population p rises the relative growth rate for p falls. This corresponds to the biological fact that with a higher population there is less food,

§2. Two-Dimensional Dynamical Systems

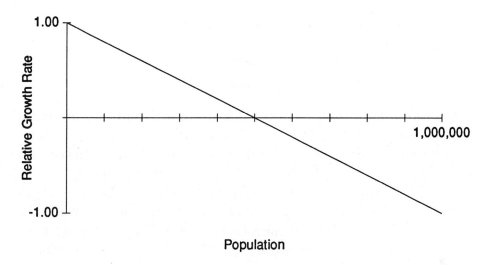

Figure VII.10: Relative growth rate as a function of population

water, and shelter for each individual. Normally the constant c_{12} is also positive because individuals of the species q compete with individuals of the species p for some things. Thus if there are more q's the relative growth rate for p will fall. Typical values of R_1, c_{11}, and c_{12} are shown in the following equation.

$$p' = (1.0 - .005p - .001q)p$$

Notice that the coefficient, $c_{11} = .005$, by which p is multiplied is larger then the coefficient, $c_{12} = .001$, by which q is multiplied. The reason for this is that each individual of species p competes with each other individual of p for everything. They eat the same food, require the same kind of shelter, and, in general, have all the same needs. On the other hand individuals of different species compete for some things but not for other things. For example, one species might eat nuts and berries while another species might eat grubs and berries. In the equation

$$p' = (1.0 - .005p - .001q)p$$

increasing p by one unit decreases the relative growth rate by .005 while increasing q by one unit decreases the relative growth rate by only .001. We say that the relative growth rate for p is **more sensitive** to changes in p than to changes to q. A complete model for two competing species might look like

$$p' = (1 - .005p - .001q)p$$
$$q' = (1 - .001p - .005q)q.$$

Notice that for each species the relative growth rate is more sensitive to changes in itself than to changes in the other species. This is the most common situation. We call such a model **competitive**. There are, however, other possibilities. For example, two highly aggressive species (e.g., the Montagues and Capulets) might always be ready to do battle; two Montagues may compete with each other for everything—food, water, shelter, the family business etc., but a Montague and a Capulet will actually get into a physical fight for no good reason. As a result the relative growth rate for each species would be more sensitive to the other species than to itself. The following pair of differential equations might model a situation of this kind.

$$p' = (2 - .001p - .01q)p$$
$$q' = (2 - .01p - .001q)q$$

Notice that the relative growth rate for p is more sensitive to changes in q than it is to changes in p. Similarly the relative growth rate for q is more sensitive to changes in p than it is to changes in p. This is an example of an **aggressive** model.

Another possibility is that one species is much more prolific than the other. Consider, for example, the model

$$p' = (2 - .01p - .001q)p$$
$$q' = (3 - .001p - .0005q)q.$$

We can compare the relative growth rate for p to the relative growth rate for q by noticing that

$$(3 - .001p - .0005q) - (2 - .01p - .001q) = 1 + .009p + .0005q$$

and that the right hand side is always greater than zero since p and q are never negative. Thus

$$(3 - .001p - .0005q) - (2 - .01p - .001q) > 0.$$

So $3 - .001p - .0005q > 2 - .01p - .001q$

§2. Two-Dimensional Dynamical Systems

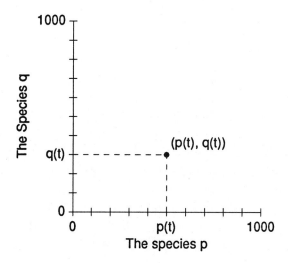

Figure VII.11: Population for two species at time t

and the species q always has a higher relative growth rate than the species p. This is an example of a **weak-strong** model.

There are even more possibilities. For example, suppose that the species p is a predator and that the species q is its prey. In this case more individuals of the species q will actually increase the food supply for the species p and, hence, the relative growth rate for the species p. The following pair of equations captures this general idea.

$$p' = (1.2 - .01p + .05q)p$$
$$q' = (2 - .01p - .001q)q$$

This is an example of a **predator-prey** model.

We would like to investigate all of these models. We use our first model, the one that describes a typical competitive situation as an example. The others appear in the homework.

We visualize this kind of model using a two-dimensional graph. The two axes represent the populations p (on the "x-axis") and q (on the "y-axis"). At any particular time, t, the current population level for both species is represented by a point on the graph whose "x-coordinate" is the population of the species represented by p and whose "y-coordinate" is the population of the species represented by q as shown in Figure VII.11.

We can think of this situation as analogous to the map of a sea rather than a river. We were able to use a river to visualize a one-

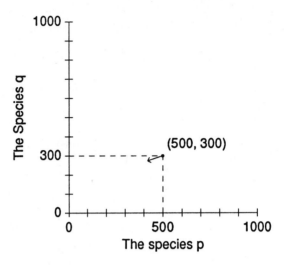

Figure VII.12: Current at the point (500, 300)

dimensional autonomous system because we only needed to keep track of one quantity. Now we need to keep track of two quantities and, hence, we need a two-dimensional analogy. The pair of differential equations

$$p' = (1 - .005p - .001q)p$$
$$q' = (1 - .001p - .005q)q$$

describes how both p and q are changing. For example, at the point $p = 500$, $q = 300$ we see that

$$p' = (1 - .005(500) - .001(300))(500) = -900$$
$$q' = (1 - .001(500) - .005(300))(300) = -300$$

so that both populations are declining very rapidly. We can indicate this on our map of the sea by an arrow as shown in Figure VII.12.

This arrow is pointing downward because the population of the species represented by q on the "y-axis" is declining and it is pointing to the left because the population represented by p on the "x-axis" is also declining. We are interested primarily in the direction of the current rather than its strength. One could indicate the strength by the length of the arrow. Such maps, however, frequently have so many arrows of such different lengths that the arrows begin to interfere with each other. Hence we will draw all our arrows with the same length indicating direction but not strength. This situation is analogous to

§2. Two-Dimensional Dynamical Systems

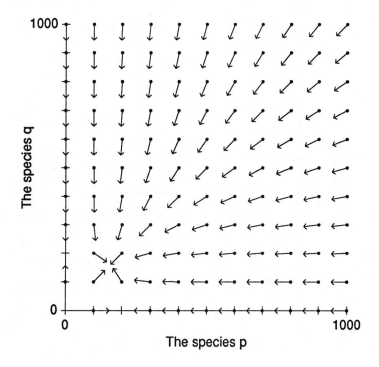

Figure VII.13: Direction field

the one-dimensional situation of section 1. In section 1 currents in a river indicated the way that one quantity was changing. Now currents in a sea represent the way that two quantities are changing. We can visualize the solution of an initial value problem like

$$p' = (1 - .005p - .001q)p; \qquad p(0) = 300$$
$$q' = (1 - .001p - .005q)q; \qquad q(0) = 400$$

by thinking about a cork that is placed in the sea at the point $(300, 400)$ and then carried along by the currents described by the pair of differential equations. Figure VII.13 shows a map of the sea with the direction of the current indicated at many points. Such a map is called a **direction field**.

Looking at Figure VII.13 it appears as if the currents are all converging on a point someplace near the point $(170, 170)$. We can investigate this point more closely by looking at the direction field shown in Figure VII.13 with a magnifying glass. Figure VII.14 shows this magnified view. As you can see, this magnified view seems to confirm our initial impression.

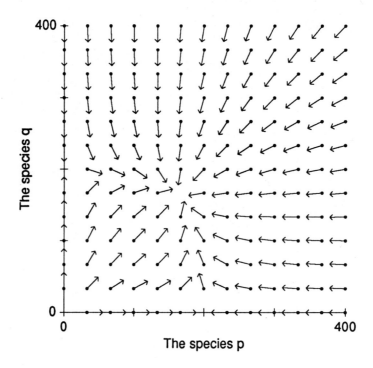

Figure VII.14: Magnified direction field

Figure VII.15 shows the solution of the initial value problem

$$p' = (1 - .005p - .001q)p; \qquad p(0) = 300$$
$$q' = (1 - .001p - .005q)q; \qquad q(0) = 400$$

superimposed on the direction field shown in Figure VII.14 with the initial point $(300, 400)$ marked by a large dot. Such a solution is sometimes called an **integral curve** or a **trajectory**.

Figure VII.16 shows several integral curves superimposed on the same graph.

The figures in this book were all drawn by computer programs. You will be using similar programs to produce similar pictures. You can also use a computer program that simulates the results of dropping many corks into the sea. This program, MOVIE2, is a two-dimensional version of the earlier one-dimensional animation program, MOVIE1.

Computers and computer graphics are powerful tools for studying dynamical systems. They complement more analytic tools. Now we would like to discuss some of these other tools. We begin by looking at

§2. Two-Dimensional Dynamical Systems

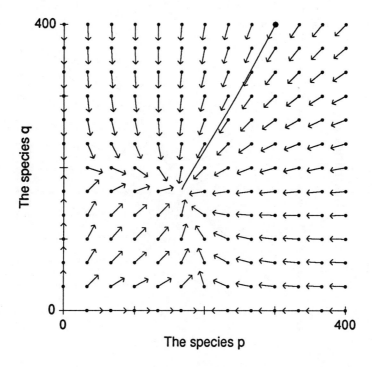

Figure VII.15: Solution of initial value problem

the notion of an **equilibrium point** for a two-dimensional autonomous dynamical system.

Definition:

An **equilibrium point** for a two-dimensional autonomous dynamical system given by a pair of equations

$$p' = F_1(p, q)$$
$$q' = F_2(p, q)$$

is a point (p_*, q_*) such that when $p = p_*$ and $q = q_*$ both p' and q' are zero. That is, such that

$$F_1(p_*, q_*) = 0$$
$$F_2(p_*, q_*) = 0$$

The real importance of this is that if at some time t there are p_* individuals of the species p and q_* individuals of the species q then both p and q will remain constant from that time on.

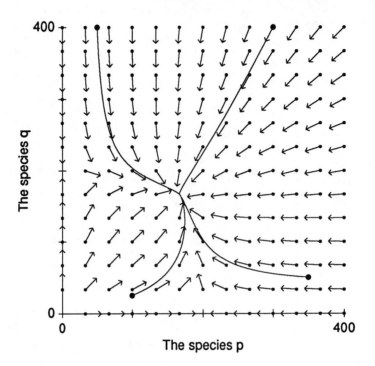

Figure VII.16: Solutions of several initial value problems

In order to find equilibrium points in this setting we need to solve two simultaneous equations. We illustrate this idea with our usual example, the dynamical system

$$p' = (1 - .005p - .001q)p$$
$$q' = (1 - .001p - .005q)q.$$

For this dynamical system we need to solve the pair of equations

$$(1 - .005p - .001q)p = 0$$
$$(1 - .001p - .005q)q = 0.$$

The first equation

$$(1 - .005p - .001q)p = 0$$

will be satisfied if either

$$p = 0$$

or if

$$1 - .005p - .001q = 0.$$

§2. Two-Dimensional Dynamical Systems

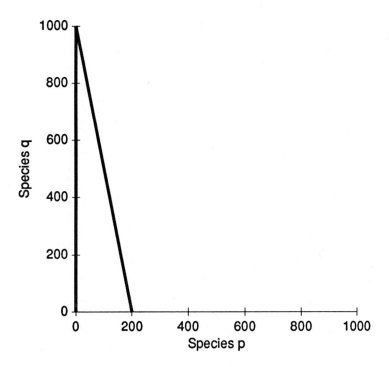

Figure VII.17: Points at which $p' = 0$

We want to draw a graph showing the points which satisfy either of these two equations. The set of points satisfying the first equation is simply the q-axis, $p = 0$. The points satisfying the second equation lie on a straight line. The easiest way to graph this straight line is by computing the points at which it intersects each of the axes. These two points are $(200, 0)$ and $(0, 1000)$. These two lines are indicated by heavy lines in Figure VII.17.

The second equation

$$(1 - .001p - .005q)q$$

will be satisfied if either

$$q = 0$$

or if

$$1 - .001p - .005q = 0.$$

We want to draw a graph indicating the points at which either of these equations is satisfied. The set of points satisfying the first equation is simply the p-axis, $q = 0$. The points satisfying the second equation

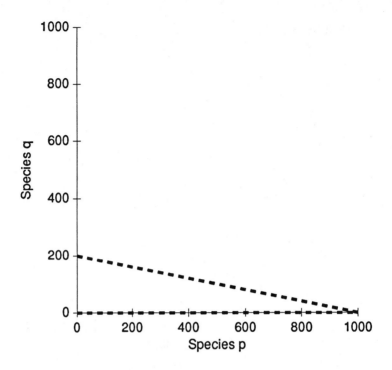

Figure VII.18: Points at which $q' = 0$

lie on a straight line. The easiest way to graph this straight line is by computing the points at which it intersects the two axes. These points are $(1000, 0)$ and $(0, 200)$. These two lines are indicated by heavy dashed lines in Figure VII.18.

Next we want to combine the information shown in both Figures VII.17 and VII.18 in one graph. Look at Figure VII.19. The solid heavy lines indicate points at which $p' = 0$. The dashed heavy lines indicate points at which $q' = 0$. There are four points at which a solid heavy line intersects a dashed heavy line. These points are circled in Figure VII.19.

These four points are the equilibrium points that we seek. They are the points at which *both* $p' = 0$ and $q' = 0$. Three of these points are

$$(0,0) \quad (200,0) \quad \text{and} \quad (0,200).$$

The fourth point is the intersection of the two lines

$$1 - .005p - .001q = 0$$

and

$$1 - .001p - .005q = 0.$$

§2. Two-Dimensional Dynamical Systems

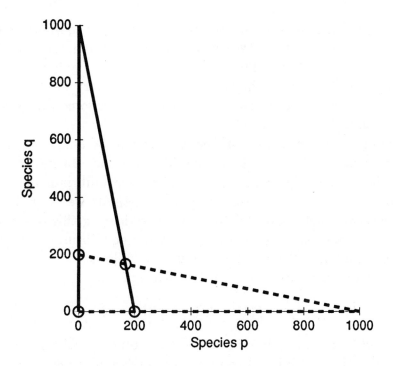

Figure VII.19: $p' = 0$ and $q' = 0$

We find the point of intersection by multiplying the second equation by 5 to obtain
$$5 - .005p - .025q = 0.$$
Subtracting this from the first equation, we obtain
$$-4 + .024q = 0.$$
So
$$q = \frac{4}{.024} = \frac{1}{.006} = 166.67.$$
Plugging this value back into either of the equations
$$1 - .005p - .001q = 0 \quad \text{or} \quad 1 - .001p - .005q = 0$$
we see that $q = 166.67$. Thus the last equilibrium point is the point $(166.67, 166.67)$.

Notice that all the solutions that we graphed in Figure VII.16 appear to be converging to this equilibrium point (See Figures VII.16 and VII.19).

Our analysis so far has allowed us to identify four equilibrium points for this dynamical system. We can see that one of them, $(166.67, 166.67)$, appears to be attracting. This fits very nicely with the biology. In this situation it appears as if the two species coexist very nicely. This is a reasonable result since the two species compete for some things but not for everything.

One of our main themes in this chapter will be analyzing the long term behavior of dynamical systems like this one. In particular, we would like to be able to determine when an equilibrium point is attracting. The computer-based graphical methods we have discussed so far are extremely powerful tools for investigating this question but we can also find out a great deal about the equilibrium points using graphical methods without the computer as follows.

First look again at Figure VII.17. The equation

$$p' = p(1 - .005p - .001q) = 0$$

is satisfied at all the points indicated by the heavy solid lines. These lines divide the region into two pieces—the triangular region to the left of the slanted heavy line and the other region to the right of the slanted heavy line. Using the intermediate value theorem one can show that in each of these two regions

$$p' = p(1 - .005p - .001q)$$

is either always positive or always negative. We can determine which of these two alternatives holds in each region by evaluating

$$p' = p(1 - .005p - .001q)$$

at any one point in each region. For example, in the triangular region to the left of the slanted heavy line we can pick the point $(100, 100)$. At this point we see that

$$p' = p(1 - .005p - .001q) = (100)(1 - .005(100) - .001(100)) = 40$$

is positive. Hence p' is positive in this whole triangular region. In the other region we can pick, for example, the point $(1000, 1000)$. At this point we see that

$$p' = p(1 - .005p - .001q) = (1000)(1 - .005(1000) - .001(1000)) = -5000$$

§2. Two-Dimensional Dynamical Systems

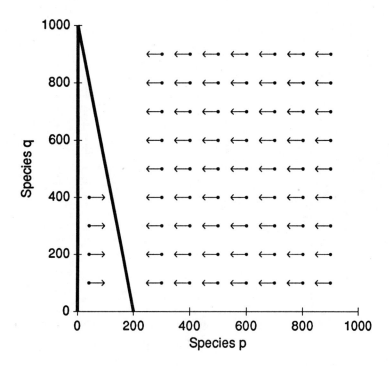

Figure VII.20: Left and right arrows according to p'

is negative. Hence p' is negative in this entire region.

This means that the currents are flowing to the right (because p' is positive) in the triangular region and are flowing to the left (because p' is negative) in the other region. We can represent this information by arrows pointing to the left and right in the appropriate regions as shown in Figure VII.20.

A similar analysis of the equation for q'

$$q' = q(1 - .005q - .001p)$$

leads to Figure VII.21. The arrows point up and down because q is graphed on the vertical (y-) axis.

Putting the information from Figures VII.20 and VII.21 together we see that the two equations

$$p' = p(1 - .005p - .001q) = 0$$

and

$$q' = q(1 - .001p - .005q) = 0$$

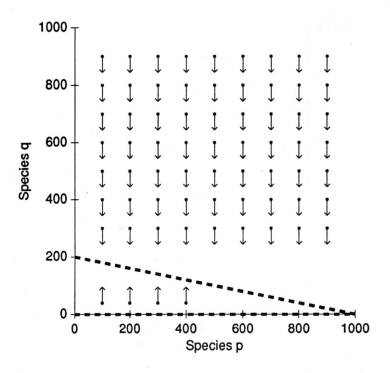

Figure VII.21: Up and down arrows according to q'

divide the region into four pieces and we can see *generally* which way the direction arrows point in each of the four regions. Figure VII.22 summarizes all this information in one graph.

We arrived at Figure VII.22 by putting together the information from Figures VII.20 and VII.21 as follows. First consider the small region in the lower left. This region is the intersection of the triangular regions from Figures VII.20 and VII.21. In Figure VII.20 we see that the arrows in the triangular region go to the right. In Figure VII.21 we see that the arrows in the triangular region go upward. Thus in the intersection of these two regions the arrows go upward and to the right as shown in Figure VII.22. The general directions of the arrows in the other three regions are determined in the same way.

This figure is not as precise as the computer-generated graphs we saw earlier but it does contain the same general information. In particular, one can see that the currents will carry corks toward the equilibrium point at $(166.67, 166.67)$.

§2. Two-Dimensional Dynamical Systems

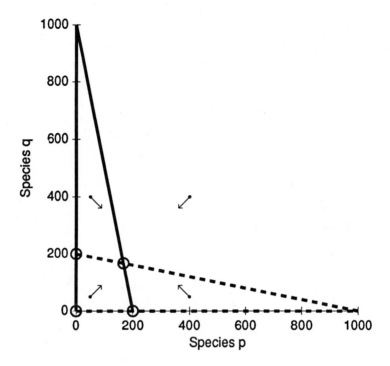

Figure VII.22: General direction field

Exercises

For each of the following models do all of the following without using the computer.

1. Find all the equilibrium points and draw a graph similar to Figure VII.19.

2. Find out generally which way the arrows in the direction field are pointing and draw a graph similar to Figure VII.22.

3. Discuss what you would expect the general behavior of solutions of the given differential equation to be with various different initial conditions.

4. Compare your discussion above with the "biology" of the models.

After you have done all the above without using the computer, then use the usual computer programs to investigate the same models and to check your conclusions.

*Exercise VII.2.1:
An aggressive model like

$$p' = (2 - .001p - .01q)p$$
$$q' = (2 - .01p - .001q)q.$$

Exercise VII.2.2:
A weak-strong model like

$$p' = (2 - .01p - .001q)p$$
$$q' = (3 - .001p - .0005q)q.$$

Exercise VII.2.3:
A predator-prey model like

$$p' = (1.2 - .01p + .05q)p$$
$$q' = (2 - .01p - .001q)q.$$

Exercise VII.2.4: In a small room in an unnamed building near the edge of an unnamed college campus there is a coffee urn. The temperature of coffee in the urn is denoted by $C(t)$ and the temperature of the room is denoted by $R(t)$. The two temperatures change according to the pair of differential equations

$$C' = -0.1(C - R)$$
$$R' = .001(C - R).$$

§3. Energy

One of the most important concepts in physics is the concept of energy, the capacity to do work. Perhaps the single most important principle of physics is the law of conservation of energy, you don't get something for nothing. Although energy can be converted from one form[1] to another, it cannot be created out of nothing nor may it be destroyed.

In section V.3 we discussed two forms of energy, the kinetic energy associated with a moving body and the potential energy associated with height, and used the law of conservation of energy to help us understand the travels of an object in a gravitational field. In this section we use the same law to examine the motion of a spring-and-mass system.

For a spring-and-mass system we need to consider two additional forms of energy, the heat energy created by friction and the potential energy stored in a compressed or stretched spring. Heat is often an important component of energy calculations. For example, when a car brakes, its kinetic energy is converted into heat energy by the friction of the brake shoes against the brake drums or pads. The law of conservation of energy requires that the kinetic energy cannot simply disappear; it must be converted into another form of energy.

We know from personal experience that a spring that has been either stretched or compressed contains a good deal of energy. We can determine how much energy such a spring contains by looking at the work required to stretch or compress it. As in section VI.8 we use the centimeter-gram-second system of measurement and measure force in dynes. Recall from section V.3 that in a simple situation involving a constant force, work is given by the equation

$$\text{work} = (\text{force})(\text{distance})$$

and in a more complicated situation in which force is not constant the multiplication becomes integration.

Consider the apparatus shown in Figure VII.23 with one end of a spring fixed in a solid wall and the other end free to move. We are interested in the amount of work required to stretch or compress this

---

[1] One of Einstein's contributions to our understanding of our physical world was the realization that matter is a form of energy. His famous equation $E = mc^2$ expresses the way in which energy in the form of matter can be converted to other forms of energy.

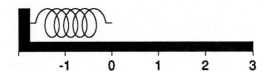

Figure VII.23: Spring at rest

spring. We let x denote the position of the free end of the spring measured in centimeters with coordinates chosen as shown in Figure VII.23 so that $x = 0$ is the equilibrium position in which the spring is neither stretched nor compressed. Let k denote the spring's spring constant measured as usual in dyne-centimeters so that when the free end of the spring is at x the spring exerts a force $-kx$ measured in dynes. Thus the work required to pull the spring from its rest position at $x = 0$ to a stretched position with $x = b$ is

$$\int_0^b kx \, dx = \frac{kb^2}{2}.$$

Exercises

*Exercise VII.3.1: Compute the amount of work required to stretch a spring whose spring constant is 3 dyne-centimeters 5 centimeters from its rest position at $x = 0$ to $x = 5$.

*Exercise VII.3.2: Compute the amount of work required to stretch a spring whose spring constant is 3 dyne-centimeters 5 centimeters from $x = 2$ to $x = 7$. Compare your answer to this problem with your answer to the preceding problem. They should be different. Why?

*Exercise VII.3.3: Compute the amount of work required to compress a spring whose spring constant is 3 dyne-centimeters from its rest position at $x = 0$ to $x = -2$.

*Exercise VII.3.4: Compute the amount of work required to move the end of a spring whose spring constant is k dyne-centimeters from $x = a$ to $x = b$.

§3. Energy

Figure VII.24: Spring and mass

*Exercise VII.3.5: Compute the amount of work required to move the end of a spring whose spring constant is k dyne-centimeters from $x = -2$ to $x = 1$. Explain why the answer is negative.

The preceding analysis shows that the amount of work required to move the free end of a spring with spring constant k from its rest position to the position x is $kx^2/2$. This work is stored in the spring as energy, called **spring energy**.

$$\text{spring energy} = \frac{kx^2}{2}.$$

Now we are ready to look at the spring-and-mass systems we studied in section VI.8 using energy considerations to give us another perspective. The basic situation looks like Figure VII.24 with a moving mass attached to one end of a spring and the other end fixed in a solid wall. At any given time t we want to keep track of two quantities, $x(t)$, the position of the mass and $v(t)$, its velocity. Since we need to consider two quantities this is a two-dimensional dynamical system and we indicate the state of the system at each time t as shown in Figure VII.25.

The way in which this system changes is described by a pair of differential equations.

$$\begin{aligned} x' &= F(x, v) \\ v' &= G(x, v) \end{aligned}$$

Notice that $x' = v$ giving us the first of these two differential equations. For the second differential equation notice that $v' = x''$ is just the acceleration of the mass. As in section VI.8 there are two forces acting on the mass.

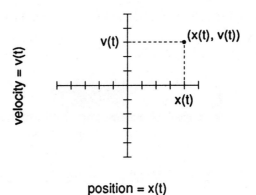

Figure VII.25: Position and velocity at time t

- A force $-kx$ due to the spring.
- A force $-px' = -pv$ due to friction.

So the acceleration of the mass is

$$\text{acceleration} = \left(\frac{1}{m}\right) \text{Force}$$

$$= \left(\frac{1}{m}\right)(-kx - pv)$$

leading to the second differential equation

$$v' = x''$$

$$= \left(\frac{1}{m}\right)(-kx - pv)$$

$$= -\left(\frac{k}{m}\right)x - \left(\frac{p}{m}\right)v.$$

Putting this all together we see that a spring-and-mass system can be described by the dynamical system

$$x' = v$$
$$v' = -\left(\frac{k}{m}\right)x - \left(\frac{p}{m}\right)v.$$

§3. ENERGY

Notice that by rewriting the original differential equation as a system of two equations we have replaced one equation that involved the second derivative by two equations that only involve first derivatives. This is an important trick that is often used to study second order differential equations using techniques like direction fields that were developed for first order differential equations.

As usual the first step in analyzing a dynamical system is finding any equilibrium points, points at which both x' and v' are zero. That is, we must solve the equations

$$v = 0$$
$$-\left(\frac{k}{m}\right)x - \left(\frac{p}{m}\right)v = 0$$

From the first equation we see that v must be zero. Substituting $v = 0$ into the second equation we obtain

$$-\left(\frac{k}{m}\right)x - \left(\frac{p}{m}\right)0 = 0$$
$$-\left(\frac{k}{m}\right)x = 0$$
$$x = 0$$

and see that the only equilibrium point is the point $(0,0)$. This is not surprising. Such a system is at equilibrium when the mass is still ($v = 0$) and the spring is neither stretched nor compressed ($x = 0$).

The next step is to sketch the direction field for this system using the method we developed in section 2. We begin with the variable x. Since $x' = v$, the x-axis (i.e., the line $v = 0$) is the set of points at which $x' = 0$. Above the x-axis the arrows are pointing right (since $x' = v$ is positive) and below the x-axis the arrows are pointing left. Figure VII.26 shows this information.

Next we look at the differential equation

$$v' = -\left(\frac{k}{m}\right)x - \left(\frac{p}{m}\right)v$$

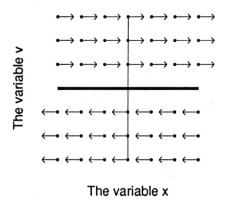

Figure VII.26: Left and right arrows according to x'

The right side of this equation is zero when

$$-\left(\frac{k}{m}\right)x - \left(\frac{p}{m}\right)v = 0$$
$$-kx - pv = 0$$
$$v = -\left(\frac{k}{p}\right)x.$$

This is the equation of a straight line passing through the origin. Its slope depends on the relative size of k and p. If the system runs smoothly so that there is relatively little friction then p will be small and the line will quite steep as shown in Figure VII.27. To the right of this line v' is negative and to its left v' is positive as indicated by the arrows pointing up and down in Figure VII.27.

Next we combine the information shown in Figures VII.26 and VII.27 into a single picture as we did in section 2 to produce the rough direction field shown in Figure VII.28.

Looking at Figure VII.28 we see that the currents are swirling around the equilibrium point $(0, 0)$ in a clockwise direction. Thus integral curves or trajectories would "spiral" or "circle" around the equilibrium point $(0, 0)$ in a clockwise direction. From this picture alone we cannot tell whether trajectories will spiral in, spiral out, or whether they will be circles or ellipses. Here is where the law of conservation of energy helps us to understand what will happen. At any given time t the system contains energy in two forms.

§3. Energy   751

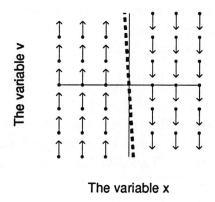

Figure VII.27: Up and down arrows according to v'

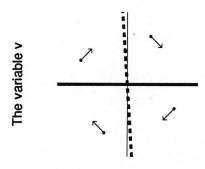

Figure VII.28: Rough direction field

- The kinetic energy $mv^2/2$.

- The spring energy $kx^2/2$.

First consider a system in which there is no friction. In this situation the law of conservation of energy implies that the total energy in the system will remain constant. That is,

$$\frac{mv^2}{2} + \frac{kx^2}{2} = c$$

where c is a constant. This is the equation of an ellipse. Thus for a system with no friction every trajectory will be an ellipse and the spring-and-mass system will oscillate back-and-forth forever.

Example:

Describe the longterm motion of a frictionless spring-and-mass system whose spring constant is 4 dyne-centimeters and whose mass is 9 grams. Suppose the mass is pulled to $x = 5$ and then released.

Answer:

When the mass is pulled to $x = 5$ and released it initially has no velocity and, hence, no kinetic energy. Its spring energy is $kx^2/2 = 4(5)^2/2 = 50$. Because the system has no friction it will always have the same energy. Thus at any time t

$$\frac{kx^2}{2} + \frac{mv^2}{2} = 50$$

$$\frac{4x^2}{2} + \frac{9v^2}{2} = 50$$

$$4x^2 + 9v^2 = 100$$

This is the equation of the ellipse shown in Figure VII.29. The mass will oscillate forever moving back-and-forth between $x = 5$ and $x = -5$ and its velocity will oscillate between $+10/3$ and $-10/3$.

Figures VII.30 and VII.31 show two examples of systems in which there is some friction. Now some of the energy in the system is converted into heat energy by friction. As a result trajectories will spiral

§3. Energy

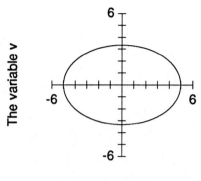

Figure VII.29: Trajectory of an initial value problem with no friction

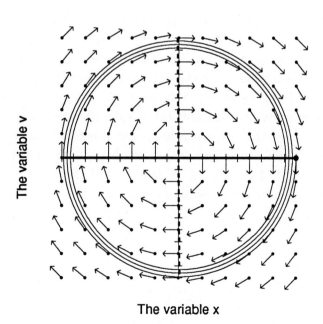

Figure VII.30: $x'' = -0.01x' - x$

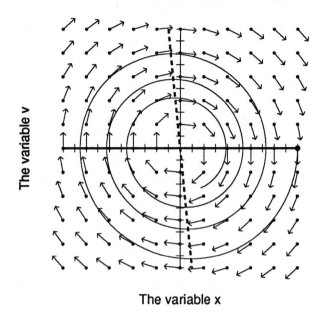

Figure VII.31: $x'' = -0.1x' - x$

in toward the equilibrium point $(0,0)$. The speed with which they spiral in is determined by the amount of friction. Figure VII.30 shows the differential equation $x'' = -0.01x' - x$ in which there is very little friction and the trajectory spirals in toward the equilibrium point very slowly. Figure VII.31 shows the differential equation $x'' = -0.1x' - x$ in which there is more friction and the trajectory spirals in more quickly. Both figures show the direction field, the lines at which $x' = 0$ (solid) and $v' = 0$ (dashed), and the trajectory beginning at the point $(10, 0)$ and continuing for 20 seconds. Notice that if there were no friction at all then the trajectory would be a circle of radius 10.

Exercises

*Exercise VII.3.6:
  Use the techniques of section VI.8 to solve the initial value problem

$$x'' = -0.01x' - x; \qquad x(0) = 10, \ x'(0) = 0.$$

Compare this solution to what you see in Figure VII.30.

Exercise VII.3.7:
Use the techniques of section VI.8 to solve the initial value problem

$$x'' = -0.1x' - x; \qquad x(0) = 10, \ x'(0) = 0.$$

Compare this solution to what you see in Figure VII.31.

---

§4. Classification of Equilibria, I

One of the main goals of this chapter is understanding the longterm behavior of dynamical systems. One of the key elements of this quest is understanding equilibrium points. In particular, we need to understand when an equilibrium point is attracting and when it is repelling. We have investigated a number of problems involving functions. A typical scenario is to look first at simple examples in which the function is linear, to solve that simple problem, and then to apply our knowledge about the linear situation to the nonlinear situation with the help of the derivative. We follow the same scenario in our study of equilibrium points. In this section we study equilibrium points for linear dynamical systems. In section 5 we look at some more examples of nonlinear dynamical systems and then, armed by our knowledge of linear systems and our experience with examples from sections 2, 5 and 6, we study nonlinear systems in section 7.

One-Dimensional Autonomous Systems

Our first job for this section is easy. We are interested in linear one-dimensional autonomous dynamical systems. Consider the dynamical system
$$y' = ky$$
where $k \neq 0$. The following theorem is easy to prove from our previous work.

Theorem 1 *If $k \neq 0$ the one-dimensional autonomous dynamical system*
$$y' = ky$$
has exactly one equilibrium point, $y_* = 0$.

- If $k < 0$ then this equilibrium point is *attracting*, and for every initial value problem
$$y' = ky; \qquad y(0) = y_0$$
we have
$$\lim_{t \to +\infty} y(t) = y_* = 0.$$

- If $k > 0$ then this equilibrium point is *repelling*, and for every initial value problem
$$y' = ky; \qquad y(0) = y_0$$
we have
$$\lim_{t \to +\infty} y(t) = \begin{cases} +\infty, & \text{if } y_0 > 0; \\ -\infty, & \text{if } y_0 < 0; \\ 0, & \text{if } y_0 = 0. \end{cases}$$

Proof:
This theorem follows immediately from the observation that the solution of the initial value problem is
$$y = y_0 e^{kt}$$
and the observation that
$$\lim_{t \to +\infty} e^{kt} = \begin{cases} 0, & \text{if } k < 0; \\ +\infty, & \text{if } k > 0. \end{cases}$$

Exercises

*Exercise VII.4.1: Suppose that $y(t)$ is the solution of the initial value problem
$$y' = 0.5y; \qquad y(0) = 2.$$
What can you say about
$$\lim_{t \to +\infty} y(t)?$$

*Exercise VII.4.2: Suppose that $y(t)$ is the solution of the initial value problem
$$y' = 0.5y; \qquad y(0) = -2.$$

§4. Classification of Equilibria, I

What can you say about
$$\lim_{t \to +\infty} y(t)?$$

Exercise VII.4.3: Suppose that $y(t)$ is the solution of the initial value problem
$$y' = -0.5y; \quad y(0) = 2.$$

What can you say about
$$\lim_{t \to +\infty} y(t)?$$

Exercise VII.4.4: Suppose that $y(t)$ is the solution of the initial value problem
$$y' = -0.5y; \quad y(0) = -2.$$

What can you say about
$$\lim_{t \to +\infty} y(t)?$$

*Exercise VII.4.5: Consider the linear one-dimensional autonomous dynamical system
$$y' = ky + b.$$
State and prove a theorem for this dynamical system similar to Theorem 1.

Exercise VII.4.6:

- Suppose $k < 0$ and sketch the direction field for the dynamical system $y' = ky + b$. What does this direction field tell you about the longterm behavior of solutions to initial value problems of the form
$$y' = ky + b; \quad y(0) = y_0?$$

- Suppose $k > 0$ and sketch the direction field for the dynamical system $y' = ky + b$. What does this direction field tell you about the longterm behavior of solutions to initial value problems of the form
$$y' = ky + b; \quad y(0) = y_0?$$

Compare your answers to these questions to the theorem you stated and proved in the preceding exercise.

Two-Dimensional Autonomous Systems

Now we would like to examine the linear two-dimensional autonomous dynamical system

$$x' = a_{11}x + a_{12}y$$
$$y' = a_{21}x + a_{22}y.$$

Our first observation is easy. This system has at least one equilibrium point, the point $(0,0)$. Notice, however, that the system may have more than one equilibrium point. For example, consider the system

$$x' = x + y$$
$$y' = 2x + 2y.$$

Any point (x, y) on the line $x + y = 0$ will be an equilibrium point since both x' and y' will be zero. Thus when we turn our first observation into a theorem we must be careful.

Theorem 2 *If* $(a_{11}a_{22} - a_{12}a_{21}) \neq 0$ *then the linear two-dimensional autonomous dynamical system*

$$x' = a_{11}x + a_{12}y$$
$$y' = a_{21}x + a_{22}y$$

has exactly one equilibrium point $(0,0)$.

Proof:
We find the equilibrium point(s) by solving the pair of equations

$$x' = 0$$
$$y' = 0.$$

That is

$$a_{11}x + a_{12}y = 0$$
$$a_{21}x + a_{22}y = 0.$$

§4. Classification of Equilibria, I

If we multiply the first equation by a_{22} and the second equation by a_{12} we get

$$a_{11}a_{22}x + a_{12}a_{22}y = 0$$
$$a_{12}a_{21}x + a_{12}a_{22}y = 0.$$

Subtracting the second equation from the first we obtain

$$(a_{11}a_{22} - a_{12}a_{21})x = 0.$$

Now, if $(a_{11}a_{22} - a_{12}a_{21}) \neq 0$, we can divide to get

$$x = 0$$

and, substituting $x = 0$ back into either of the original equations, we see $y = 0$. This completes the proof of the theorem.

For the remainder of this section we are interested only in dynamical systems that have exactly one equilibrium point, that is, for which $(a_{11}a_{22} - a_{12}a_{21}) \neq 0$. In the first part of this section we restricted our attention to one-dimensional linear dynamical systems $y' = ky$ for which $k \neq 0$. For two-dimensional linear dynamical systems the condition $(a_{11}a_{22} - a_{12}a_{21}) \neq 0$ is analogous to the condition $k \neq 0$ for one-dimensional linear dynamical systems.

One of our most powerful abilities as human beings is the ability to see commonalities among very different things. We can see past the obvious differences between a toy poodle and a German Shepherd to the common dogginess that both possess. According to legend, Isaac Newton was able to understand gravity by seeing the commonalities between a falling apple and the moon suspended in the sky. One of our most important tools for establishing commonalities, for drawing analogies, is mathematics. Often two apparently very different phenomena can be described by the same mathematics. This mathematics then allows us to use our knowledge about one phenomenon to help us understand the other.

In the last section we used our knowledge about first order dynamical systems to help us understand some second order dynamical systems. Now we are going to travel the same road in the opposite direction. We will use our knowledge about second order dynamical systems and spring-and-mass systems to help us in our present investigation. We start with the two equations

$$x' = a_{11}x + a_{12}y$$
$$y' = a_{21}x + a_{22}y.$$

Let us assume that a_{12} is not zero. Then we can rewrite the first equation as follows.

$$a_{12}y = x' - a_{11}x$$
$$y = \left(\frac{1}{a_{12}}\right)x' - \left(\frac{a_{11}}{a_{12}}\right)x$$

and, differentiating both sides,

(VII.1) $$y' = \left(\frac{1}{a_{12}}\right)x'' - \left(\frac{a_{11}}{a_{12}}\right)x'.$$

Now from the second of our original equations we obtain

$$y' = a_{21}x + a_{22}y$$
$$= a_{21}x + a_{22}\left[\left(\frac{1}{a_{12}}\right)x' - \left(\frac{a_{11}}{a_{12}}\right)x\right]$$
$$= a_{21}x + \left(\frac{a_{22}}{a_{12}}\right)x' - \left(\frac{a_{11}a_{22}}{a_{12}}\right)x.$$

Combining this result with equation VII.1 we see

$$\left(\frac{1}{a_{12}}\right)x'' - \left(\frac{a_{11}}{a_{12}}\right)x' = a_{21}x + \left(\frac{a_{22}}{a_{12}}\right)x' - \left(\frac{a_{11}a_{22}}{a_{12}}\right)x$$

$$x'' - a_{11}x' = a_{12}a_{21}x + a_{22}x' - a_{11}a_{22}x$$

$$x'' = (a_{11} + a_{22})x' - (a_{11}a_{22} - a_{12}a_{21})x.$$

Similar calculations show (if a_{21} is not zero) that

(VII.2) $$y'' = (a_{11} + a_{22})y' - (a_{11}a_{22} - a_{12}a_{21})y.$$

Thus both x and y are solutions of the same second order differential equation. We could use this fact to find actual solutions of initial value problems involving the linear autonomous dynamical system

§4. Classification of Equilibria, I

VII.1. This possibility is discussed in a brief appendix at the end of this section. We are currently interested, however, in investigating whether the equilibrium point $(0,0)$ is attracting or repelling and for this investigation we need only investigate the character of the solutions to equation VII.2. Now we need an old friend from section VI.8 reprinted below.

Reprise

The (real) solutions of the differential equation $ay'' + by' + cy = 0$ look somewhat different depending on the **discriminant**

$$\Delta = b^2 - 4ac.$$

- If $\Delta > 0$ then the solutions are of the form

$$y(t) = c_1 e^{r_1 t} + c_2 e^{r_2 t}$$

where r_1 and r_2 are solutions of the characteristic equation

$$ar^2 + br + c = 0.$$

- If $\Delta = 0$ then the solutions are of the form

$$y(t) = c_1 e^{rt} + c_2 t e^{rt}$$

where $r = -b/2a$ is the unique solution of the characteristic equation

$$ar^2 + br + c = 0.$$

- If $\Delta < 0$ then the real solutions are of the form

$$y(t) = e^{\lambda t}(c_1 \cos \omega t + c_2 \sin \omega t)$$

where

$$\lambda = -b/2a$$

is the real part and

$$\omega = \frac{\sqrt{4ac - b^2}}{2a}$$

is the imaginary part of the solution of the characteristic equation

$$ar^2 + br + c = 0,$$

$$r = \frac{-b + \sqrt{b^2 - 4ac}}{2a}.$$

In the present situation the discriminant is

$$\Delta = (a_{11} + a_{22})^2 - 4(a_{11}a_{22} - a_{12}a_{21}).$$

Our analysis of the character of the equilibrium point $(0,0)$ parallels the reprised result above.

Theorem 3 *Linear Classification Theorem*
Consider the linear two-dimensional autonomous dynamical system

$$\begin{aligned} x' &= a_{11}x + a_{12}y \\ y' &= a_{21}x + a_{22}y. \end{aligned}$$

Let

$$\begin{aligned} b &= -(a_{11} + a_{22}) \\ c &= a_{11}a_{22} - a_{12}a_{21} \\ \Delta &= b^2 - 4c \\ &= (a_{11} + a_{22})^2 - 4(a_{11}a_{22} - a_{12}a_{21}) \end{aligned}$$

- *If $\Delta > 0$ then solutions are of the form*

$$\begin{aligned} x(t) &= c_1 e^{r_1 t} + c_2 e^{r_2 t} \\ y(t) &= c_3 e^{r_1 t} + c_4 e^{r_2 t} \end{aligned}$$

where r_1 and r_2 are solutions of the characteristic equation

$$r^2 + br + c = 0$$

and the values of the constants c_1, c_2, c_3, and c_4 are determined by the initial conditions.

There are several possibilities for the longterm behavior of $x(t)$ and $y(t)$ depending on the values of r_1 and r_2.

§4. Classification of Equilibria, I

1. If both r_1 and r_2 are negative then

$$\lim_{t \to +\infty} x(t) = 0$$
$$\lim_{t \to +\infty} y(t) = 0$$

and $(0,0)$ is an attracting equilibrium point.

2. If both r_1 and r_2 are positive then usually

$$\lim_{t \to +\infty} x(t) = \pm\infty$$
$$\lim_{t \to +\infty} y(t) = \pm\infty$$

and $(0,0)$ is a repelling equilibrium point. If the initial condition is $x(t_0) = y(t_0) = 0$ the trajectory will stay at the equilibrium point $(0,0)$.

3. If one of the two roots r_1 and r_2 is positive and the other is negative then the longterm behavior of $x(t)$ and $y(t)$ depends on the initial conditions which determine the values of the constants $c_1, c_2, c_3,$ and c_4. Let r_1 denote the positive root and let r_2 denote the negative root. Then

$$\lim_{t \to +\infty} x(t) = \lim_{t \to +\infty} c_1 e^{r_1 t} + c_2 e^{r_2 t}$$
$$= \begin{cases} 0, & \text{if } c_1 = 0; \\ \pm\infty, & \text{if } c_1 \neq 0. \end{cases}$$

Similar comments apply to $y(t)$.

Thus $(0,0)$ is an indeterminate equilibrium point. Usually trajectories will fly off but sometimes they do go to $(0,0)$.

- If $\Delta = 0$ then the solutions are of the form

$$x(t) = c_1 e^{rt} + c_2 t e^{rt}$$
$$y(t) = c_3 e^{rt} + c_4 t e^{rt}$$

where $r = -b/2$ is the unique solution of the characteristic equation

$$r^2 + br + c = 0$$

and the values of the constants $c_1, c_2, c_3,$ and c_4 are determined by the initial conditions.

The longterm behavior of the solution depends on the sign of $r = -b/2 = (a_{11} + a_{22})/2$.

1. If $r < 0$ then
$$\lim_{t \to +\infty} x(t) = 0$$
$$\lim_{t \to +\infty} y(t) = 0$$

and the equilibrium point $(0,0)$ is attracting.

2. If $r > 0$ then usually
$$\lim_{t \to +\infty} x(t) = \pm\infty$$
$$\lim_{t \to +\infty} y(t) = \pm\infty.$$

Thus the equilibrium point $(0,0)$ is repelling. If the initial conditions are $x(t_0) = y(t_0) = 0$ then the trajectory will remain at the equilibrium point $(0,0)$.

- If $\Delta < 0$ then the real solutions are of the form
$$x(t) = e^{\lambda t}(c_1 \cos \omega t + c_2 \sin \omega t)$$
$$y(t) = e^{\lambda t}(c_3 \cos \omega t + c_4 \sin \omega t)$$

where
$$\lambda = -b/2$$
is the real part and
$$\omega = \frac{\sqrt{4c - b^2}}{2}$$
is the imaginary part the solution of the characteristic equation
$$r^2 + br + c = 0,$$
$$r = \frac{-b + \sqrt{b^2 - 4c}}{2},$$

§4. Classification of Equilibria, I

and the values of the constants $c_1, c_2, c_3,$ and c_4 are determined by the initial conditions. The longterm behavior of a solution depends on the sign of $\lambda = -b/2 = (a_{11} + a_{22})/2$.

1. If $\lambda < 0$ then
$$\lim_{t \to +\infty} x(t) = 0$$
$$\lim_{t \to +\infty} y(t) = 0$$

and the equilibrium point $(0,0)$ is attracting. These solutions correspond to spirals that spiral in toward the equilibrium point $(0,0)$.

2. If $\lambda = 0$ then
$$x(t) = c_1 \cos \omega t + c_2 \sin \omega t$$
$$y(t) = c_3 \cos \omega t + c_4 \sin \omega t.$$

These solutions correspond to ellipses, neither spiraling in nor spiraling out. Thus the equilibrium point $(0,0)$ is indeterminate. If the initial conditions are $x(t_0) = y(t_0) = 0$ then the trajectory will remain at the equilibrium point $(0,0)$.

3. If $\lambda > 0$ then the solutions will oscillate back-and-forth within an expanding "envelope." These solutions correspond to spirals that spiral outward in wider and wider sweeps. Thus the equilibrium point $(0,0)$ is repelling. If the initial conditions are $x(t_0) = y(t_0) = 0$ then the trajectory will remain at the equilibrium point $(0,0)$.

We can summarize the most important parts of this theorem as follows (See also Theorem 27 in section VI.10).

Theorem 4 *The longterm behavior of trajectories of the dynamical system*

$$x' = a_{11}x + a_{12}y$$
$$y' = a_{21}x + a_{22}y$$

depends on the real parts of the solutions to the equation

$$r^2 - (a_{11} + a_{22})r + (a_{11}a_{22} - a_{12}a_{21}).$$

- If the real parts of both solutions are negative then all trajectories will approach $(0,0)$.

- If the real parts of both solutions are positive then unless $x(0) = y(0) = 0$ a trajectory will "fly off to infinity."

- If the solutions are imaginary (i.e., if the real part of both solutions is zero) then the trajectories will be ellipses.

- If both solutions are real and one is positive and the other is negative, then the longterm behavior of a trajectory depends on its initial conditions. Usually trajectories "fly off to infinity." It is possible, however, for a trajectory to approach the equilibrium point $(0,0)$.

The best way to understand the Linear Classification Theorem is by applying it to some examples.

Example:

Consider the initial value problem

$$\begin{aligned} x' &= y \\ y' &= -x - 0.1y \\ x(0) &= 1 \\ y(0) &= 1 \end{aligned}$$

In order to apply the Linear Classification Theorem we need to calculate the following quantities

$$\begin{aligned} b &= -(a_{11} + a_{22}) = -(0 - 0.1) = 0.1 \\ c &= (a_{11}a_{22} - a_{12}a_{21}) = (0)(-0.1) - (1)(-1) = 1 \\ \Delta &= b^2 - 4a \\ &= (0.1)^2 - 4 = -3.99 \end{aligned}$$

§4. Classification of Equilibria, I

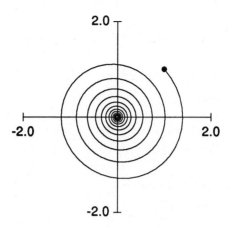

Figure VII.32: Numerical solution for first example

Since Δ is negative the character of the equilibrium point $(0,0)$ is determined by
$$\lambda = -\frac{b}{2} = -0.05$$
Since λ is negative the equilibrium point $(0,0)$ is attracting and regardless of the initial conditions we would expect
$$\lim_{t \to +\infty} x(t) = 0$$
$$\lim_{t \to +\infty} y(t) = 0.$$

Figure VII.32 shows a graph of the solution to this initial value problem that was obtained numerically. This graph confirms our expectations. Notice that the solution spirals in toward the equilibrium point $(0,0)$ because solutions of problems in which $\Delta < 0$ are either spirals or ellipses.

Example:

Consider the initial value problem
$$x' = y$$
$$y' = -x - 2y$$
$$x(0) = 1$$
$$y(0) = 1$$

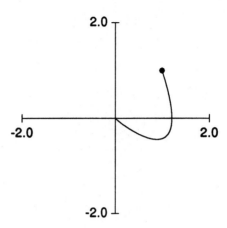

Figure VII.33: Numerical solution for second example

In order to apply the Linear Classification Theorem we need to calculuate the following quantities

$$\begin{aligned} b &= -(a_{11} + a_{22}) = -(0 - 2) = 2 \\ c &= (a_{11}a_{22} - a_{12}a_{21}) = (0)(-2) - (1)(-1) = 1 \\ \Delta &= b^2 - 4c \\ &= 2^2 - 4 = 0 \end{aligned}$$

Since Δ is zero the character of the equilibrium point $(0,0)$ is determined by

$$r = -\frac{b}{2} = -1.$$

Since r is negative the equilibrium point $(0,0)$ is attracting and regardless of the initial conditions we would expect

$$\lim_{t \to +\infty} x(t) = 0$$
$$\lim_{t \to +\infty} y(t) = 0.$$

Figure VII.33 shows a graph of the solution of this initial value problem that was obtained numerically. This graph confirms our expectations. Notice that this trajectory goes toward the equilibrium

§4. Classification of Equilibria, I

point $(0,0)$ without spiraling around. This behavior is characteristic of solutions when $\Delta = 0$ and $r < 0$.

Example:

Consider the dynamical system

$$x' = -y$$
$$y' = -x$$

In order to apply the Linear Classification Theorem we need to calculate the following quantities

$$b = -(a_{11} + a_{22}) = -(0+0) = 0$$

$$c = (a_{11}a_{22} - a_{12}a_{21}) = (0)(0) - (-1)(-1) = -1$$

$$\begin{aligned}\Delta &= b^2 - 4c \\ &= 0^2 - 4(-1) \\ &= 4\end{aligned}$$

Since Δ is positive the character of the equilibrium point $(0,0)$ is determined by

$$r_1 = \frac{-b + \sqrt{b^2 - 4c}}{2} = 1$$

and

$$r_2 = \frac{-b - \sqrt{b^2 - 4c}}{2} = -1.$$

Since one of these is positive and the other is negative the equilibrium point is indeterminate. The relevant sentence of the Linear Classification Theorem is

> If one of the two roots r_1 and r_2 is positive and the other is negative then the longterm behavior of $x(t)$ and $y(t)$ depends on the initial conditions.

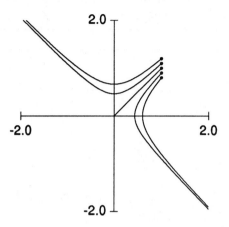

Figure VII.34: Trajectories for third example

Figure VII.34 shows a graph with five different trajectories for this dynamical system. The initial conditions for each of the trajectories are indicated by dots. Notice that four of these trajectories go flying off but one of them goes towards the equilibrium point $(0,0)$. One can show that for this particular dynamical system if the initial condition is on the line $y = x$ then the trajectory will approach the equilibrium point $(0,0)$ but if the initial condition is the slightest bit off this line then the solution will fly off. This is an example of what is meant by the phrase "*usually* trajectories will fly off but sometimes they do go to $(0,0)$.

Exercises

*Exercise VII.4.7: Consider the dynamical system

$$x' = x - 0.1y$$
$$y' = -0.1x + y.$$

1. What does our analysis above tell you about the equilibrium point $(0,0)$? Is it attracting, repelling or indeterminate?

2. What can you say on the basis of your analysis so far about the longterm behavior of the solution of the initial value problem

$$x' = x - 0.1y$$

§4. CLASSIFICATION OF EQUILIBRIA, I

$$y' = -0.1x + y$$
$$x(0) = 1$$
$$y(0) = 1?$$

3. What can you say on the basis of your analysis so far about the longterm behavior of the solution of the initial value problem

$$x' = x - 0.1y$$
$$y' = -0.1x + y$$
$$x(0) = 1$$
$$y(0) = -1?$$

4. What can you say on the basis of your analysis so far about the longterm behavior of the solution of the initial value problem

$$x' = x - 0.1y$$
$$y' = -0.1x + y$$
$$x(0) = 1$$
$$y(0) = 2?$$

5. Use the computer to investigate the solutions of the three initial value problems above.

Exercise VII.4.8: Consider the dynamical system

$$x' = 0.9x - 0.1y$$
$$y' = 0.1x + 0.9y.$$

1. What does our analysis above tell you about the equilibrium point $(0,0)$? Is it attracting, repelling or indeterminate?

2. What can you say on the basis of your analysis so far about the longterm behavior of the solution of the initial value problem

$$x' = 0.9x - 0.1y$$
$$y' = 0.1x + 0.9y$$
$$x(0) = 1$$
$$y(0) = 1?$$

3. What can you say on the basis of your analysis so far about the longterm behavior of the solution of the initial value problem

$$x' = 0.9x - 0.1y$$
$$y' = 0.1x + 0.9y$$
$$x(0) = 1$$
$$y(0) = -1?$$

4. What can you say on the basis of your analysis so far about the longterm behavior of the solution of the initial value problem

$$x' = 0.9x - 0.1y$$
$$y' = 0.1x + 0.9y$$
$$x(0) = 1$$
$$y(0) = 2?$$

5. Use the computer to investigate the solutions of the three initial value problems above.

Exercise VII.4.9: Consider the dynamical system

$$x' = 1.1x - 0.1y$$
$$y' = 0.1x + 1.1y.$$

1. What does our analysis above tell you about the equilibrium point $(0,0)$? Is it attracting, repelling or indeterminate?

2. What can you say on the basis of your analysis so far about the longterm behavior of the solution of the initial value problem

$$x' = 1.1x - 0.1y$$
$$y' = 0.1x + 1.1y$$
$$x(0) = 1$$
$$y(0) = 1?$$

§4. Classification of Equilibria, I

3. What can you say on the basis of your analysis so far about the longterm behavior of the solution of the initial value problem

$$x' = 1.1x - 0.1y$$
$$y' = 0.1x + 1.1y$$
$$x(0) = 1$$
$$y(0) = -1?$$

4. What can you say on the basis of your analysis so far about the longterm behavior of the solution of the initial value problem

$$x' = 1.1x - 0.1y$$
$$y' = 0.1x + 1.1y$$
$$x(0) = 1$$
$$y(0) = 2?$$

5. Use the computer to investigate the solutions of the three initial value problems above.

Exercise VII.4.10: Consider the dynamical system

$$x' = x + 0.2y$$
$$y' = y.$$

1. What does our analysis above tell you about the equilibrium point $(0,0)$? Is it attracting, repelling or indeterminate?

2. What can you say on the basis of your analysis so far about the longterm behavior of the solution of the initial value problem

$$x' = x + 0.2y$$
$$y' = y$$
$$x(0) = 0$$
$$y(0) = 1?$$

3. What can you say on the basis of your analysis so far about the longterm behavior of the solution of the initial value problem

$$\begin{aligned} x' &= x + 0.2y \\ y' &= y \\ x(0) &= 1 \\ y(0) &= 0? \end{aligned}$$

4. What can you say on the basis of your analysis so far about the longterm behavior of the solution of the initial value problem

$$\begin{aligned} x' &= x + 0.2y \\ y' &= y \\ x(0) &= 1 \\ y(0) &= 1? \end{aligned}$$

5. Use the computer to investigate the solutions of the three initial value problems above.

Exercise VII.4.11: Consider the dynamical system

$$\begin{aligned} x' &= a_{11}x + a_{12}y + b_1 \\ y' &= a_{21}x + a_{22}y + b_2. \end{aligned}$$

Analyze this dynamical system. What can you say about its equilibrium point(s)? Note: It is easiest to analyze this system by introducing new variables $X(t) = x(t) + c_1$ and $Y(t) = y(t) + c_2$.

Exercise VII.4.12: We analyzed our original dynamical system

$$\begin{aligned} x' &= a_{11}x + a_{12}y \\ y' &= a_{21}x + a_{22}y \end{aligned}$$

by converting it into two second order differential equations

$$\begin{aligned} x'' &= (a_{11} + a_{22})x' - (a_{11}a_{22} - a_{12}a_{21})x \\ y'' &= (a_{11} + a_{22})y' - (a_{11}a_{22} - a_{12}a_{21})y \end{aligned}$$

§4. Classification of Equilibria, I

In order to do this we assumed that a_{12} was nonzero (for the first equation) and that a_{21} was nonzero (for the second equation). What would happen if either or both of these quantities was zero?

Appendix: Extra Remarks

If you've studied matrices before you may recognize the quantity $(a_{11}a_{22} - a_{12}a_{21})$ that was so important in our analysis in this section. This quantity is called the **determinant** of the matrix

$$\begin{pmatrix} a_{11} & a_{12} \\ a_{21} & a_{22} \end{pmatrix}$$

The other important quantity $(a_{11} + a_{22})$ also has a name. It is called the **trace** of the matrix.

By expressing the original system

$$\begin{aligned} x' &= a_{11}x + a_{12}y \\ y' &= a_{21}x + a_{22}y \end{aligned}$$

as the second order differential equation

$$y'' = (a_{11} + a_{22})y' - (a_{11}a_{22} - a_{12}a_{21})y$$

we have seen how to determine the longterm behavior of the original dynamical system. We can actually do even more. We can find the exact solutions of initial value problems of the form

$$\begin{aligned} x' &= a_{11}x + a_{12}y \\ y' &= a_{21}x + a_{22}y \\ x(0) &= x_0 \\ y(0) &= y_0 \end{aligned}$$

To do this we need to look at one more aspect of the translation of our two-dimensional problem into a one-dimensional problem. So far we have only talked about the differential equations. For a system of two first order equations the initial conditions look like

$$x(0) = x_0, \ y(0) = y_0.$$

But, for a second order differential equation the initial conditions are typically
$$y(0) = y_0, \; y'(0) = y_1.$$

We translate the original first order initial conditions into second order initial conditions by
$$\begin{aligned} x'(0) &= a_{11}x(0) + a_{12}y(0) \\ y'(0) &= a_{21}x(0) + a_{22}y(0). \end{aligned}$$

This gives us two separate initial value problems—one for $x(t)$ and the other for $y(t)$—that can be solved using the methods of section VI.8.

---

§5. Optimization

Last semester we developed techniques for solving one-dimensional optimization problems, problems involving a function $f(x)$ of one variable. Many practical optimization problems, however, require the optimization of a function of two or more variables. In this section we show how our work on dynamical systems can be used to attack these higher dimensional optimization problems. We develop a numerical method called **the method of steepest descent** that is one of the most powerful techniques available for optimizing a function of two or even more variables. For simplicity we limit our work in this section to functions of two variables.

Suppose that we have a function $z = f(x, y)$ of two variables and want to find a local minimum of this function. We can visualize this problem by thinking of the function $z = f(x, y)$ as describing the landscape in a rather mountainous region. At any given point (x, y) the function $f(x, y)$ describes the height of the landscape at that point. A mountaintop corresponds to a local maximum of the function $f(x, y)$ and the bottom of a valley corresponds to a local minimum of the function $f(x, y)$. The idea behind the method of steepest descent is to search for the bottom of a valley by walking downhill. Standing at a particular point (x, y) we need to decide which direction is the downhill direction. We begin with a simple example.

§5. Optimization

Example:

Consider the function

$$z = 2x + 3y$$

and suppose that we are standing at the point $(0,0)$. Which way is down?

Answer:

In order to answer this question we need to build a picture of the function $z = 2x + 3y$. One of the most common tools for visualizing such a function is a topographic map, a map designed to show the three-dimensional shape of geographic features like mountains and valleys. One of the principal devices used in topographic maps is a set of **contour lines**. A contour line is a line connecting points that have the same height. For example, for the function $z = 2x + 3y$ we might draw a contour line connecting all the points whose height is 1. This would be the set of points

$$2x + 3y = 1$$

or

$$y = -\left(\frac{2}{3}\right)x + \left(\frac{1}{3}\right)$$

shown in Figure VII.35

We can get a very good picture of the topography of the mountain $z = 2x + 3y$ by a series of contour lines as shown in Figure VII.36.

Looking at Figure VII.36 we can deduce a great deal of information about the topography of the function $z = 2x + 3y$. For example, notice that if we were to walk north along the y-axis (the north-south axis) from the point $(0,0)$ to the point $(0,1)$ we would climb from $z = 0$ to $z = 3$. The underlying reason for this is the coefficient 3 in front of the y in the equation $z = 2x + 3y$. The coefficient in front of the y is sometimes called the **partial slope** of z with respect to y. This terminology comes from an analogy with a linear function $y = mx + b$ in which the coefficient m is called the *slope*. The slope m describes how fast y changes compared to x. The partial slope of z with respect to y tells how fast z changes compared to y. If x is held constant then as y changes, each unit change in y produces a change of 3 units in z.

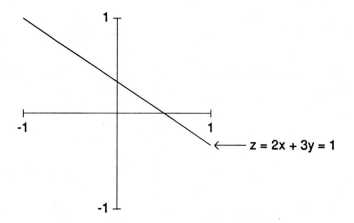

Figure VII.35: The single contour line $2x + 3y = 1$

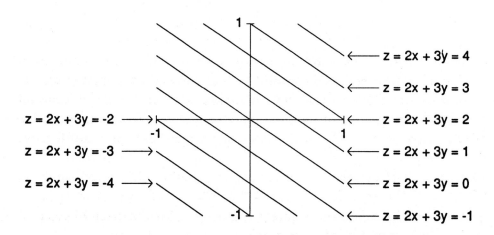

Figure VII.36: Several contour lines for $z = 2x + 3y$

§5. Optimization

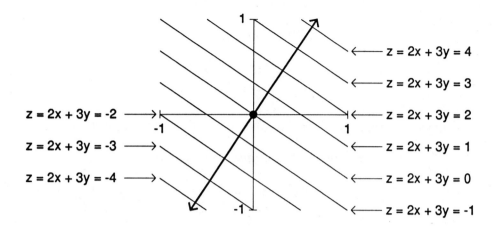

Figure VII.37: Steepest directions from $(0,0)$

Holding x constant and changing y can be visualized as walking along a north-south line in Figure VII.36.

We can make similar comments about the way z changes as x changes. Notice that as we walk along the x-axis from the point $(0,0)$ to the point $(1,0)$ we climb from $z = 0$ to $z = 2$. This climb of two units is caused by the coefficient 2 in front of x in the function $z = 2x + 3y$. In this case we say that the **partial slope** of z with respect to x is 2.

Standing at the point $(0,0)$ in Figure VII.36 we might contemplate walking in several different directions. Walking due east along the x-axis we see that every unit step produces a rise of two units in z. Walking due west along the x-axis we see that every unit step produces a fall of two units in z. Walking due north along the y-axis we see that every unit step produces a rise of three units in z. Walking due south along the y-axis we see that every unit step produces a fall of three units in z.

It is clear from this picture that if we want to climb as steeply as possible or to descend as steeply as possible we should walk at right angles to the contour lines. Figure VII.37 shows the direction of steepest ascent and the direction of steepest descent.

The previous example is typical of a linear function. The following theorem is motivated by this example. It tells us which way is down and which way is up for a linear function $z = ax + by + c$.

Theorem 5 *Consider a function $z = ax + by + c$.*

- *The direction field for the dynamical system*

$$x' = a$$
$$y' = b$$

always points uphill in the direction of steepest ascent.

- *The direction field for the dynamical system*

$$x' = -a$$
$$y' = -b$$

always points downhill in the direction of steepest descent.

Proof:
First notice that the slope of any contour line, $ax + by + c = k$, is $-a/b$ since, if

$$ax + by + c = k,$$

then

$$by = k - ax - c$$
$$y = -\left(\frac{a}{b}\right)x - \left(\frac{c}{b}\right) + \left(\frac{k}{b}\right).$$

Now the solution of an initial value problem

$$x' = a$$
$$y' = b$$
$$x(0) = x_0$$
$$y(0) = y_0$$

will be

$$x = at + x_0$$
$$y = bt + y_0.$$

Thus

$$t = \frac{x}{a} - \frac{x_0}{a}$$

§5. Optimization

and
$$y = bt + y_0 = \left(\frac{b}{a}\right) x - \left(\frac{b}{a}\right) x_0 + y_0.$$

Thus such a solution is a straight line with slope b/a and is perpendicular[2] to the contour lines. An identical argument works for the second dynamical system. These two dynamical systems both produce solutions that are straight lines perpendicular to the contour lines. Since the dynamical systems are pointing in opposite directions, one is pointing uphill in the direction of steepest ascent and the other is pointing downhill in the direction of steepest descent. To see which is which notice that when a and b are positive then northwest is the uphill direction. Solutions for the first dynamical system move to the northwest when a and b are positive. Thus the direction field for the first dynamical system is pointing uphill.

For a linear function $z = ax + by + c$ the two partial slopes give us the information that we need to find the direction of steepest ascent and the direction of steepest descent. The functions in which we are interested are not usually linear functions, so we need to find the directions of steepest ascent and steepest descent for a more general function $z = f(x, y)$. For functions of one variable the derivative serves the same purpose for a nonlinear function as the slope does for a linear function. Thus it is not surprising that something very much like the derivative, called the "partial derivative," serves the same purpose for a nonlinear function of two variables that the partial slope serves for a linear function of two variables.

For a linear function $z = ax + by + c$ the partial slope a tells us how fast z changes as x changes. We think of the variable y as being temporarily fixed, a constant, and examine what happens when just the variable x changes. For each unit that x changes the value of z will change by a units. For a nonlinear function we can do exactly the same thing. Consider the variable y as being temporarily fixed and ask how rapidly z changes as just the variable x changes. The derivative answers exactly this question. We only need to add one detail—we think of y as being a constant and take the derivative of z with respect to x exactly as if y were constant. This is called the **partial derivative of z with**

---

[2] Recall that two straight lines are perpendicular if the slope of one is the negative reciprocal of the slope of the other.

respect to x. We use the notation

$$\frac{\partial z}{\partial x}$$

for the partial derivative of z with respect to x. We can do exactly the same thing, finding the partial derivative of z with respect to y

$$\frac{\partial z}{\partial y}$$

by thinking of x as a constant.

Example:

The two partial derivatives of the function

$$z = x^2 y + xy + x \cos y$$

are

$$\frac{\partial z}{\partial x} = 2xy + y + \cos y$$

$$\frac{\partial z}{\partial y} = x^2 + x - x \sin y$$

The partial derivative $\partial z/\partial x$ was obtained by thinking of the variable y as a constant and taking the derivative of z with respect to x in the usual way. For example, the first term $x^2 y$ contributes $2xy$ to the derivative. This is entirely analogous, for example, to a term like $x^2(12.3456)$ whose derivative would be $2x(12.3456)$.

The partial derivative $\partial z/\partial y$ was obtained by thinking of x as a constant and taking the derivative of z with respect to y.

For nonlinear functions we have the following theorem.

Theorem 6 *Consider a function* $z = f(x,y)$.

- *The direction field for the dynamical system*

$$x' = \frac{\partial z}{\partial x}$$

$$y' = \frac{\partial z}{\partial y}$$

always points uphill in the direction of steepest ascent.

§5. Optimization

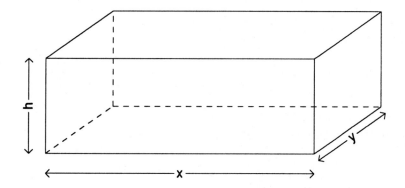

Figure VII.38: A box to hold sand

- *The direction field for the dynamical system*

$$x' = -\frac{\partial z}{\partial x}$$

$$y' = -\frac{\partial z}{\partial y}$$

always points downhill in the direction of steepest descent.

We will not prove this theorem here. This theorem is the basis of the method of steepest descent. The following example illustrates the method.

Example:

A manufacturer would like to build a box that will hold one cubic foot of sand. The box will be in the shape of a rectangular prism as shown in Figure VII.38. The manufacturer would like to minimize the cost of materials for the box by minimizing the total surface area of all six sides of the box. The total surface area is

$$S = 2xy + 2xh + 2yh$$

where each term corresponds to two opposite sides.

Notice that since the total volume must be one cubic foot, we have

$$xyh = 1$$

so that
$$h = \frac{1}{xy}$$
and we can rewrite our equation for the surface area as
$$S = 2xy + 2x\left(\frac{1}{xy}\right) + 2y\left(\frac{1}{xy}\right) = 2xy + \frac{2}{y} + \frac{2}{x}$$
giving us the function that we seek to minimize.

The next step is to calculate the partial derivatives.
$$\frac{\partial S}{\partial x} = 2y - \frac{2}{x^2}$$
$$\frac{\partial S}{\partial y} = 2x - \frac{2}{y^2}$$

Now we use the usual computer programs to examine trajectories of the dynamical system
$$x' = -2y + \frac{2}{x^2}$$
$$y' = -2x + \frac{2}{y^2}.$$

Figure VII.39 shows several such trajectories with their initial points indicated by dots. Notice that they all approach the point $(1, 1)$. This seems to indicate that the solution of this optimization problem is
$$x = 1, y = 1$$
and, since $h = 1/xy$,
$$h = 1.$$
Thus the manufacturer should make the box in the shape of a cube, not an entirely surprising result.

This problem was nice because there was only one local minimum and we did not need to worry about choosing an initial condition that would lead us to a local minimum that was not the global minimum.

§5. Optimization

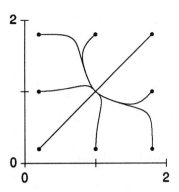

Figure VII.39: Trajectories for $x' = -2y + 2/x^2$; $y' = -2x + 2/y^2$

Many problems involve functions that have several different local minimums and depending on the initial conditions the method of steepest descent may find a local minimum that is not a global minimum.

Exercises

Exercise VII.5.1: Sketch the contour lines $z = -2$, $z = -1$, $z = 0$, $z = 1$ and $z = 2$ for the function

$$z = 4x - 2y.$$

Exercise VII.5.2: Sketch the contour lines $z = -2$, $z = -1$, $z = 0$, $z = 1$ and $z = 2$ for the function

$$z = 4x + 2y.$$

Exercise VII.5.3: Sketch the contour lines $z = -2$, $z = -1$, $z = 0$, $z = 1$ and $z = 2$ for the function

$$z = -4x + 2y.$$

Exercise VII.5.4: Sketch the contour lines $z = 0$, $z = 1$, $z = 2$, $z = 3$ and $z = 4$ for the function

$$z = x^2 + y^2.$$

Exercise VII.5.5: Sketch the contour lines $z = 0$, $z = 1$, $z = 2$, $z = 3$ and $z = 4$ for the function

$$z = \sqrt{x^2 + y^2}.$$

Exercise VII.5.6: Sketch the contour lines $z = -2$, $z = -1$, $z = 0$, $z = 1$ and $z = 2$ for the function
$$z = x^2 - y^2.$$

*Exercise VII.5.7: Find the partial derivatives of the function
$$z = xy + x^2y + 2x - 3y.$$

*Exercise VII.5.8: Find the partial derivatives of the function
$$w = x \cos y.$$

*Exercise VII.5.9: Find the partial derivatives of the function
$$f(x,y) = \sqrt{x^2 + y^2}.$$

*Exercise VII.5.10: A manufacturer wishes to build a box with an open top that will hold one cubic foot of sand. He wishes to minimize the cost of materials. What dimensions should the box have?

Exercise VII.5.11: A manufacturer wishes to build a box with a closed top that will hold one cubic foot of sand. She wishes to minimize the cost of materials. Because sand is so heavy the bottom of the box must be built with material whose cost is three times that of the material used for the sides and top. What dimensions should the box have?

*Exercise VII.5.12: Another way to look for an optimum is to look for points at which both partial derivatives are equal to zero. This is analogous to looking for critical points for a function of one variable. For a function of two variables a **critical point** is a point at which both partial derivatives are equal to zero (or possibly undefined). Show that you get the same answer for the box problems by looking for critical points directly. It can be very difficult to find the critical points directly. One of the reasons that the method of steepest descent is so important is that it can be used even when it is extraordinarily diffficult to find the critical points directly.

*Exercise VII.5.13: What would you change in our work above if you were looking for a maximum rather than a minimum? (Other than the name "steepest ascent" rather than "steepest descent.")

§5. OPTIMIZATION

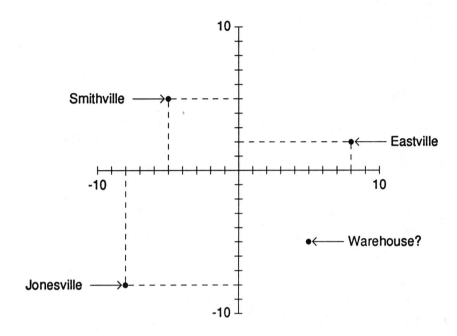

Figure VII.40: Three towns and a warehouse

*Exercise VII.5.14: Find a box like those that we have been discussing that has a total surface area of one square foot and that holds the largest possible volume.

Exercise VII.5.15: Figure VII.40 shows a map with three large cities. A major company would like to locate a warehouse near the three cities. The warehouse should be located at the place that minimizes the sum of the distances from the warehouse to each of the three cities. Where should the warehouse be located?

Exercise VII.5.16: Figure VII.41 shows a map with four large cities. A major company would like to locate a warehouse near the four cities. The warehouse should be located at the place that minimizes the sum of the distances from the warehouse to each of the four cities. Where should the warehouse be located?

Exercise VII.5.17: According to Fermat's Principle a light ray traveling from one point to another will follow the fastest path. Figure VII.42 shows two mirrors, an object and an eye. Find the path that a light ray will follow traveling from the object to one mirror, then the

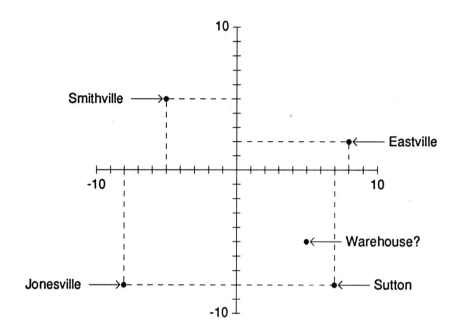

Figure VII.41: Four towns and a warehouse

second mirror and finally to the eye. For convenience the two mirrors are located along the x- and y-axes.

Exercise VII.5.18: One might guess that since one mirror "reverses" an image, two mirrors would reverse it twice, leaving it looking unreversed. Using ideas from this section and our work last semester investigate the apparent position of an object viewed using two mirrors as arranged in the previous exercise.

Exercise VII.5.19: All of the examples and exercises that we have looked at so far were chosen in part because they involved only one local minimum or local maximum. Hence, the method of steepest descent could be relied upon to produce the global minimum or the global maximum regardless of the initial condition chosen. The function below

$$z = x^4 - 2x^2 + y^2$$

has two local minimums. Investigate this function. What can you say about the equilibrium points of the dynamical system

$$x' = -\frac{\partial z}{\partial x}$$

§6. Peaks, Valleys, and Saddle Points

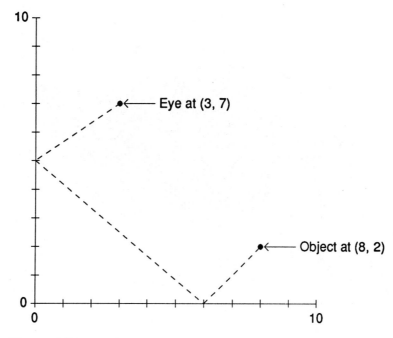

Figure VII.42: Two mirrors, an eye and an object

$$y' = -\frac{\partial z}{\partial y}?$$

Where are the two local minimums? Are there any other critical points? We say a point (x_0, y_0) is in the **basin of attraction** for a minimum (x_*, y_*) if the trajectory starting at (x_0, y_0) is pulled in to (x_*, y_*). What can you say about the basins of attraction for each of the two local minimums? Sketch some contour lines for the function $z = x^4 - 2x^2 + y^2$. They can help you answer the preceding questions.

§6. Peaks, Valleys, and Saddle Points

The topographic pictures of functions $z = f(x, y)$ that we used in the last section to develop the method of steepest descent also lead to some interesting insights about the character of the equilibrium points of a dynamical system. In this section we look at three examples that add a geometric perspective to the work we did in section 4.

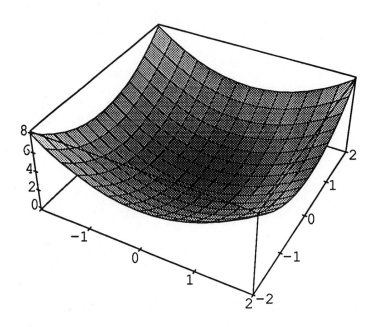

Figure VII.43: $z = x^2 + y^2$

Example 1:

Consider the function
$$z = x^2 + y^2.$$
The graph of this function looks like a bowl as shown in Figure VII.43. The contour lines
$$x^2 + y^2 = \text{constant}$$
for this function are circles centered at the origin as shown on the left side of Figure VII.44. The method of steepest descent applied to this function leads to the dynamical system

$$x' = -\left(\frac{\partial z}{\partial x}\right)(x^2 + y^2) = -2x$$
$$y' = -\left(\frac{\partial z}{\partial y}\right)(x^2 + y^2) = -2y.$$

Notice that this is a linear dynamical system. Its direction field is shown on the right side of Figure VII.44. At each point (x, y) the

§6. Peaks, Valleys, and Saddle Points

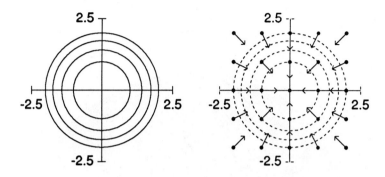

Figure VII.44: Contour lines $z = 1, 2, 3, 4$.

direction of steepest descent is back toward the minimum at the origin exactly as one would expect from Figure VII.43. Thus the one equilibrium point $(0,0)$ of this dynamical system is attracting.

We can gain some insight about the linear classification theorem from section 4 by applying it to the three examples in this section. In this case we compute

$$\begin{aligned} b &= -(a_{11} + a_{22}) = 4 \\ c &= (a_{11}a_{22} - a_{12}a_{21}) = 4 \\ \Delta &= b^2 - 4c = 0 \end{aligned}$$

and, since $\Delta = 0$, the character of the equilibrium point depends on the sign of $r = -b/2 = -2$. Since r is negative the equilibrium point is attracting exactly as we expected.

Example 2:

Next consider the function

$$z = -x^2 - y^2.$$

The graph of this function looks like an upside down bowl as shown in Figure VII.45. The contour lines

$$-x^2 - y^2 = \text{constant}$$

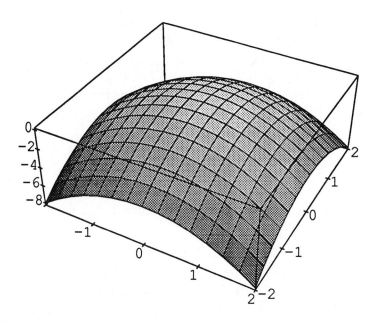

Figure VII.45: $z = -x^2 - y^2$

for this function are circles centered at the origin as shown in the left side of Figure VII.46. The method of steepest descent applied to to this function leads to the dynamical system

$$x' = -\left(\frac{\partial z}{\partial x}\right)(-x^2 - y^2) = 2x$$
$$y' = -\left(\frac{\partial z}{\partial y}\right)(-x^2 - y^2) = 2y.$$

Notice that this is a linear dynamical system. Its direction field is shown on the right side of Figure VII.46. At each point (x, y) the direction field points directly away from the maximum at the origin exactly as one would expect from Figure VII.45. Thus the one equilibrium point $(0, 0)$ of this dynamical system is repelling.

Applying the linear classification theorem to this example we compute

$$b = -(a_{11} + a_{22}) = -4$$

§6. Peaks, Valleys, and Saddle Points

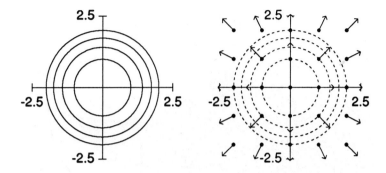

Figure VII.46: Contour lines $z = -1, -2, -3, -4$.

$$c = (a_{11}a_{22} - a_{12}a_{21}) = 4$$
$$\Delta = b^2 - 4c = 0$$

and, since $\Delta = 0$, the character of the equilibrium point depends on the sign of $r = -b/2 = 2$. Since r is positive this equilibrium point is repelling exactly as we expected.

Example 3:

Now consider the function

$$z = x^2 - y^2$$

whose graph is shown in Figure VII.47. This graph is more complicated to describe than the two previous graphs. It looks like a saddle. Imagine a horse with a saddle centered exactly at the origin so that a rider would be facing east along the x-axis. Thus the horse's head is on the x-axis with a positive coordinate and the horse's tail is also on the x-axis with a negative coordinate. A horsefly walking on the saddle above the horse's spine, conveniently located along the x-axis, starting at the back of the saddle will be walking downhill until he reaches the origin. As the horsefly continues past the origin he will begin walking uphill. If the horsefly had blinders on and could look neither to the right nor to the left he would think that the origin was a minimum. A second horsefly walking along the y-axis would get exactly the opposite

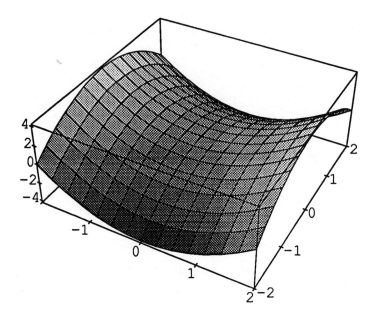

Figure VII.47: $z = x^2 - y^2$

impression. Walking along the the horse's right flank toward the origin this horsefly would be walking uphill until she reached the origin. As she continued past the origin she would begin to walk downhill. If she had blinders on and could look neither to her right nor to her left she would think that the origin was a maximum. Some of the contour lines for this function are shown in the left side of Figure VII.48.

The method of steepest descent applied to this function leads to the dynamical system

$$x' = -\left(\frac{\partial z}{\partial x}\right)(x^2 - y^2) = -2x$$
$$y' = -\left(\frac{\partial z}{\partial y}\right)(x^2 - y^2) = 2y.$$

Notice that this is a linear dynamical system. Its direction field is shown on the right side of Figure VII.48. Some trajectories for this dynamical system are shown in Figure VII.49. Notice that most of these trajectories fly off either toward $+\infty$ on the y-axis or $-\infty$ on the

§6. Peaks, Valleys, and Saddle Points

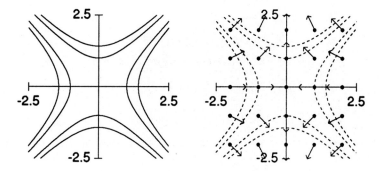

Figure VII.48: Contour lines $z = -2, -1, 1, 2$

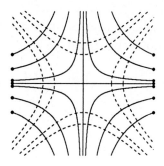

Figure VII.49: Trajectories on a saddle

y-axis. There are some exceptions. If its initial point is *exactly* on the x-axis then a trajectory will perform a delicate balancing act walking precariously along the horse's spine down to the "saddle point" at $(0,0)$.

Applying the linear classification theorem to this example we compute

$$b = -(a_{11} + a_{22}) = 0$$
$$c = (a_{11}a_{22} - a_{12}a_{21}) = -4$$
$$\Delta = b^2 - 4c = 16.$$

Since Δ is positive we must examine the roots of the characteristic equation

$$r^2 + br + c$$

to determine the character of the equilibrium point $(0,0)$. In this case the characteristic equation is

$$r^2 - 4$$

which has two roots $r = 2$ and $r = -2$. The relevant paragraph of the linear classification theorem is repeated below.

> If one of the two roots r_1 and r_2 is positive and the other is negative then the longterm behavior of $x(t)$ and $y(t)$ depends on the initial conditions which determine the values of the constants c_1, c_2, c_3, and c_4. Let r_1 denote the positive root and let r_2 denote the negative root. Then
>
> $$\lim_{t \to +\infty} x(t) = \lim_{t \to +\infty} c_1 e^{r_1 t} + c_2 e^{r_2 t}$$
> $$= \begin{cases} 0, & \text{if } c_1 = 0; \\ \pm\infty, & \text{if } c_1 \neq 0. \end{cases}$$
>
> Similar comments apply to $y(t)$.
>
> Thus $(0,0)$ is an indeterminate equilibrium point. Usually trajectories will fly off but sometimes they do go to $(0,0)$.

Now we can see exactly what is meant by "Usually trajectories fly off but sometimes they do go to $(0,0)$." The x-axis (or the spine of the horse) is like a continental divide. Trajectories that start on one side go flying off in one direction; trajectories the start on the other side go flying off in the other direction; and *trajectories that start exactly on the x-axis go to the equilibrium point* $(0,0)$.

Exercises

Exercise VII.6.1:
Look at the three examples we studied in this section but instead of applying the method of steepest descent apply the method of steepest ascent. What happens to each of these examples?

§7. Nonlinear Dynamical Systems

We now have the tools we need to solve the problem that first arose in section 2. In section 2 we looked at a number of different two-dimensional nonlinear dynamical systems, for example, the system

$$p' = (1 - .005p - .001q)p$$
$$q' = (1 - .001p - .005q)q.$$

This system might describe two competing species sharing the same habitat. In section 2 we saw that this dynamical system had four equlibrium points—$(0,0)$, $(200,0)$, $(0,200)$, and $(500/3, 500/3)$—which are circled in Figure VII.50.

By drawing the direction field for this dynamical system we were able to see visually that the equilibrium point in the middle seems to be attracting. See Figure VII.51.

Unfortunately it is not always immediately obvious by looking at a direction field whether a particular equilibrium point is attracting or not. We need a more powerful tool for determining the character of an equilibrium point—is it attracting? is it repelling? or is it indeterminate? Like many problems, this problem is easier for *linear* dynamical systems. In section 4 we looked at linear dynamical systems of the form

$$x' = a_{11}x + a_{12}y$$
$$y' = a_{21}x + a_{22}y.$$

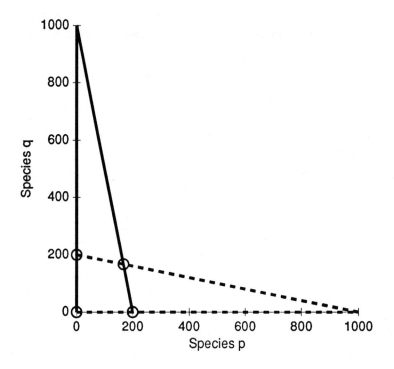

Figure VII.50: Four equilibrium points

These systems are particularly nice—there is always exactly one equilibrium point, the point $(0,0)$, and we developed formulas based on the partial slopes a_{11}, a_{12}, a_{21}, and a_{22} for classifying the equilibrium point. Our principal result in section 4 can be summarized as follows.

Theorem 7 *(Linear Classification Theorem) Consider the dynamical system*

$$x' = a_{11}x + a_{12}y$$
$$y' = a_{21}x + a_{22}y$$

Let

$$b = -(a_{11} + a_{22}) \quad \text{and} \quad c = (a_{11}a_{22} - a_{12}a_{21})$$

and

$$\Delta = b^2 - 4c$$

- *If Δ is negative and $\lambda = -b/2 = (a_{11} + a_{22})/2$ is positive then the equilibrium point $(0,0)$ is repelling.*

§7. Nonlinear Dynamical Systems

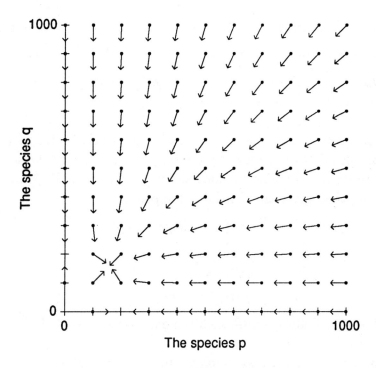

Figure VII.51: Direction field

- If Δ is negative and $\lambda = -b/2a = (a_{11} + a_{22})/2$ is negative then the equilibrium point $(0,0)$ is attracting.

- If Δ is positive let

$$r_1 = \frac{-b + \sqrt{b^2 - 4c}}{2} \quad \text{and} \quad r_2 = \frac{-b - \sqrt{b^2 - 4c}}{2}$$

If both r_1 and r_2 are negative then the equilibrium point $(0,0)$ is attracting.

If both r_1 and r_2 are positive then the equilibrium point $(0,0)$ is repelling.

If one is positive and the other is negative then the equilibrium point $(0,0)$ is indeterminate. We gained some additional insight into this kind of equilibrium point in section 6. This is the kind of equilibrium point that arises from a saddle point using the method of steepest descent.

- If Δ is zero then let $\lambda = -b/2 = (a_{11} + a_{22})$.

If λ is negative then the equilibrium point $(0,0)$ is attracting.

If λ is positive then the equilibrium point $(0,0)$ is repelling.

- In any situation not covered by the preceding list the equilibrium point $(0,0)$ is indeterminate.

Thus we have a recipe for determining the character of the single equilibrium point of a linear dynamical system. Linear dynamical systems are extremely nice in another way. If the equilibrium point is attracting then given any initial value problem

$$\begin{aligned} x' &= a_{11}x + a_{12}y \\ y' &= a_{21}x + a_{22}y \\ x(0) &= x_0 \\ y(0) &= y_0 \end{aligned}$$

the solution will always approach the equilibrium point $(0,0)$ *no matter what the initial point (x_0, y_0) is*.

Nonlinear dynamical systems are fundamentally much more complicated. First, there are usually several equilibrium points. Second, some of the equilibrium points may be attracting, others may be repelling and others may be indeterminate. Finally, the longterm behavior of the solution of an initial value problem depends on its initial value. We saw these complications in several contexts both for one-dimensional nonlinear dynamical systems in section 1 and for two-dimensional nonlinear dynamical systems in section 2.

For nonlinear dynamical systems we can analyze a particular equilibrium point in exactly the same way as we analyzed the equilibrium point for a linear dynamical system. There are two differences both of which you can probably guess on the basis of our work so far.

- For a linear dynamical system

$$\begin{aligned} x' &= F_1(x,y) = a_{11}x + a_{12}y \\ y' &= F_2(x,y) = a_{21}x + a_{22}y \end{aligned}$$

our analysis was based on the partial slopes a_{11}, a_{12}, a_{21}, and a_{22}. For a nonlinear dynamical system

$$\begin{aligned} x' &= F_1(x,y) \\ y' &= F_2(x,y) \end{aligned}$$

§7. Nonlinear Dynamical Systems

the analysis is based on the *partial derivatives*.

$$\frac{\partial F_1}{\partial x}, \frac{\partial F_1}{\partial y}, \frac{\partial F_2}{\partial x}, \frac{\partial F_2}{\partial y}.$$

This is exactly what we would expect from our work last semester. For a function $y = f(x)$ the derivative enabled us to use our knowledge about linear functions to study nonlinear functions. The underlying reason for this is that if one looks at a differentiable function near a particular point then the graph of the differentiable function looks very much like the tangent. Results about linear functions that depend on the slope often tell us something about nonlinear functions with the derivative providing information analogous to the information provided by the slope of a linear function. The same basic principle is at work for functions involving more than one variable. In this case the partial derivatives provide the same kind of information about a nonlinear function that the partial slopes provide for a linear function.

- For a nonlinear dynamical system if an equilibrium point is attracting then *if the initial values x_0 and y_0 of an initial value problem*

$$\begin{aligned} x' &= a_{11}x + a_{12}y \\ y' &= a_{21}x + a_{22}y \\ x(0) &= x_0 \\ y(0) &= y_0 \end{aligned}$$

are close to the equilibrium point then the solution will be attracted to the equilibrium point.

Again this is exactly what we would expect from earlier work. The derivative only gives us information about a function *close to* a point. In this situation it can only tell us that when the initial point of a trajectory is *close to* an attracting equilibrium point then the trajectory will be attracted to that equilibrium point.

The following theorem describes the main result. Notice that this theorem is virtually identical to theorem 7 except that it refers to the partial derivatives rather than the partial slopes.

Theorem 8 *(Nonlinear Classification Theorem)* Consider the dynamical system

$$x' = F_1(x, y)$$
$$y' = F_2(x, y)$$

Suppose that (x_*, y_*) is an equilibrium point for this dynamical system (that is, $F_1(x_*, y_*) = 0$ and $F_2(x_*, y_*) = 0$).
Let

$$a_{11} = \left(\frac{\partial F_1}{\partial x}\right)(x_*, y_*)$$

$$a_{12} = \left(\frac{\partial F_1}{\partial y}\right)(x_*, y_*)$$

$$a_{21} = \left(\frac{\partial F_2}{\partial x}\right)(x_*, y_*)$$

$$a_{22} = \left(\frac{\partial F_2}{\partial y}\right)(x_*, y_*)$$

and

$$b = -(a_{11} + a_{22}) \quad \text{and} \quad c = (a_{11}a_{22} - a_{12}a_{21}).$$

Let

$$\Delta = b^2 - 4c$$

- If Δ is negative and $\lambda = -b/2 = (a_{11} + a_{22})/2$ is positive then the equilibrium point (x_*, y_*) is repelling.

- If Δ is negative and $\lambda = -b/2 = (a_{11} + a_{22})/2$ is negative then the equilibrium point (x_*, y_*) is attracting.

- If Δ is positive let

$$r_1 = \frac{-b + \sqrt{b^2 - 4c}}{2} \quad \text{and} \quad r_2 = \frac{-b - \sqrt{b^2 - 4c}}{2}$$

If both r_1 and r_2 are negative then the equilibrium point (x_*, y_*) is attracting.

§7. Nonlinear Dynamical Systems

If both r_1 and r_2 are positive then the equilibrium point (x_*, y_*) is repelling.

If one is positive and the other is negative then the equilibrium point is **indeterminate**. *This particular kind of indeterminate equilibrium point is called a* **saddle** *point because it behaves very much like an equilibrium point corresponding to a saddle point for the method of steepest descent.*

- *If Δ is zero then let*

$$\lambda = -\frac{b}{2}.$$

If λ is negative then the equilibrium point (x_*, y_*) is attracting.

If λ is positive then the equilibrium point (x_*, y_*) is repelling.

- *In any situation not covered by the preceding list the equilibrium point (x_*, y_*) is indeterminate. Note that a saddle point (see above) is only one of several different kinds of indeterminate equilibrium points.*

We will not give a proof of this theorem. As you might guess the proof relies on the fact that the partial derivatives provide information that allows a nonlinear function to be approximated by a linear function.

We illustrate the way this theorem is used by looking at the same example we cited earlier.

Example:

Consider the dynamical system

$$\begin{aligned} p' &= (1 - .005p - .001q)p \\ q' &= (1 - .001p - .005q)q \end{aligned}$$

Because this dynamical system arose as a model for population growth we used the letters p and q to denote the variables. Of course, it doesn't make any difference what letters we use to denote our variables. The letter p is analogous to the letter x in theorem 8. The letter

q is analogous to the letter y. In order to avoid any potential confusion we will rewrite this dynamical system using the same letters as in theorem 8.

$$x' = (1 - .005x - .001y)x$$
$$y' = (1 - .001x - .005y)y$$

Thus, in the terminology of theorem 8,

$$\begin{aligned} F_1(x,y) &= (1 - .005x - .001y)x \\ &= x - .005x^2 - .001xy \\ F_2(x,y) &= (1 - .001x - .005y)y \\ &= y - .001xy - .005y^2 \end{aligned}$$

and we can calculate the partial derivatives

$$\frac{\partial F_1}{\partial x} = 1 - .01x - .001y$$

$$\frac{\partial F_1}{\partial y} = -.001x$$

$$\frac{\partial F_2}{\partial x} = -.001y$$

$$\frac{\partial F_2}{\partial y} = 1 - .001x - .01y.$$

With these calculations done we are ready to examine each of the four equilibrium points. We found these four equilibrium points back in section 2—$(0,0), (200,0), (0,200)$, and $(500/3, 500/3)$.

- The equilibrium point $(0,0)$. At this point we see that

$$a_{11} = \left(\frac{\partial F_1}{\partial x}\right)(0,0) = 1 - .01(0) - .001(0) = 1$$

$$a_{12} = \left(\frac{\partial F_1}{\partial y}\right)(0,0) = -.001(0) = 0$$

§7. Nonlinear Dynamical Systems

$$a_{21} = \left(\frac{\partial F_2}{\partial x}\right)(0,0) = -.001(0) = 0$$

$$a_{22} = \left(\frac{\partial F_2}{\partial y}\right)(0,0) = 1 - .001(0) - .01(0) = 1$$

Thus

$$b = -(a_{11} + a_{22}) = -2 \quad \text{and} \quad c = (a_{11}a_{22} - a_{12}a_{21}) = 1$$

So

$$\Delta = b^2 - 4c = 0.$$

Now

$$\lambda = -\frac{b}{2} = 1$$

and according to theorem 8 the equilibrium point $(0,0)$ is repelling. This result matches our intuition about population models. The point $(0,0)$ is almost always an equilibrium point for such models (the exceptions involve models with immigration). If there are even a few individuals, however, then there will be lots of food, water and shelter to go around and one would expect the population to rise.

- The equilibrium point $(200, 0)$. At this point we see that

$$a_{11} = \left(\frac{\partial F_1}{\partial x}\right)(200,0) = 1 - .01(200) - .001(0) = -1$$

$$a_{12} = \left(\frac{\partial F_1}{\partial y}\right)(200,0) = -.001(200) = -0.2$$

$$a_{21} = \left(\frac{\partial F_2}{\partial x}\right)(200,0) = -.001(0) = 0$$

$$a_{22} = \left(\frac{\partial F_2}{\partial y}\right)(200,0) = 1 - .001(200) - .01(0) = 0.8$$

Thus

$$b = -(a_{11} + a_{22}) = 0.2 \quad \text{and} \quad c = (a_{11}a_{22} - a_{12}a_{21}) = -0.8$$

So
$$\Delta = b^2 - 4c = 3.24.$$

Now

$$r_1 = \frac{-b + \sqrt{b^2 - 4c}}{2} = \frac{-0.2 + 1.8}{2} = 0.8$$

$$r_2 = \frac{-b - \sqrt{b^2 - 4c}}{2} = \frac{-0.2 - 1.8}{2} = -1.0$$

and by theorem 8 the equilibrium point $(200, 0)$ is indeterminate. This result is in agreement with our intuition. This equilbrium point corresponds to a situation with 200 individuals of species x (né species p) and none of species y. If one starts out close to this equilibrium point at some point on the x-axis (i.e., if one starts with no individuals of species y) then the population will be attracted towards the equilibrium point $(200, 0)$. If there are even a few individuals, however, of species y (i.e. if one starts out close to the point $(200, 0)$ but ever so slightly above the x-axis) then the population will be repelled away from the equilibrium point $(200, 0)$. Biologically, this happens because the population of the species y will grow to fill its ecological "niche." There are many well-known examples of situations in which a few individuals of a new species have been introduced into a habitat with the result that the new species grows to a sizable population and in the process "crowds out" some of the species already present. Notice that with the population of the species y equal to zero, the habitat we have been studying supports an x population of 200. But, with the species y present, the habitat can only support an x population of 500/3. You should verify these observations by some computer experiments.

- The equilibrium point $(0, 200)$. This equilibrium point is very similar to the equilibrium point $(200, 0)$. The calculations are left to the reader.

- The equilibrium point $(500/3, 500/3)$.

  At this point we see that

§7. Nonlinear Dynamical Systems

$$a_{11} = \left(\frac{\partial F_1}{\partial x}\right)\left(\frac{500}{3}, \frac{500}{3}\right) = -\frac{5}{6}$$

$$a_{12} = \left(\frac{\partial F_1}{\partial y}\right)\left(\frac{500}{3}, \frac{500}{3}\right) = -\frac{1}{6}$$

$$a_{21} = \left(\frac{\partial F_2}{\partial x}\right)\left(\frac{500}{3}, \frac{500}{3}\right) = -\frac{1}{6}$$

$$a_{22} = \left(\frac{\partial F_2}{\partial y}\right)\left(\frac{500}{3}, \frac{500}{3}\right) = -\frac{5}{6}$$

So

$$b = -(a_{11} + a_{22}) = \frac{5}{3} \quad \text{and} \quad c = (a_{11}a_{22} - a_{12}a_{21}) = \frac{2}{3}$$

So

$$\Delta = b^2 - 4c = 1/9.$$

Now

$$r_1 = \frac{-b + \sqrt{b^2 - 4c}}{2} = \frac{-\frac{5}{3} + \sqrt{\frac{1}{9}}}{2} = -\frac{2}{3}$$

$$r_2 = \frac{-b - \sqrt{b^2 - 4c}}{2} = \frac{-\frac{5}{3} - \sqrt{\frac{1}{9}}}{2} = -1.$$

Since both r_1 and r_2 are negative the equilibrium point $(500/3, 500/3)$ is attracting. Again, this matches our earlier work.

Exercises

The first three exercises below ask you to investigate the same models at which you looked at the end of section 2. For each of the following models recall your results from section 2. Then use theorem 8 to analyze each of the equilibrium points. Compare your results here with your results from section 2.

Exercise VII.7.1: An aggressive model like

$$p' = (2 - .001p - .01q)p$$
$$q' = (2 - .01p - .001q)q$$

Exercise VII.7.2: A weak-strong model like

$$p' = (2 - .01p - .001q)p$$
$$q' = (3 - .001p - .0005q)q$$

Exercise VII.7.3: A predator-prey model like

$$p' = (1.2 - .01p + .05q)p$$
$$q' = (2 - .01p - .001q)$$

Exercise VII.7.4: Analyze the dynamical system

$$x' = (x-y)^2 - 1$$
$$y' = y^2 - 4$$

Find all of its equilibrium points and determine whether each one is attracting, repelling or indeterminate.

Exercise VII.7.5: Consider the function

$$z = y^2 + x^4 - x^2$$

which is shown in Figure VII.52. This function has two local minimums and one saddle point. Suppose that one applied the method of steepest descent to this function. Find the dynamical system involved and analyze its equilibrium points using theorem 8.

§8. Summary[+]

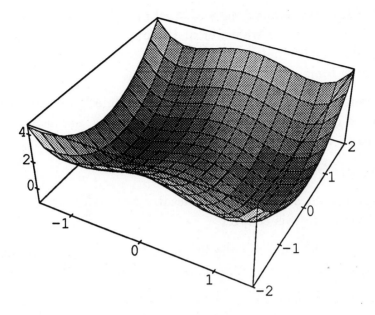

Figure VII.52: $z = y^2 + x^4 - x^2$

§8. Summary[+]

This is the summary[+] section for the last chapter of this book and thus, in a way, for the book as a whole. In this chapter we come back to the point at which we began. Above all calculus studies change. We began this book studying change in a situation that conceptually was very simple but turned out to be surprisingly complex—the changing population for a species like temperate zone insects for which a discrete dynamical system is a good model. We concluded by returning to population growth, but this time we studied two species interacting. Moreover, the change of the population of these species is best described by continuous dynamical systems. These systems are conceptually more difficult because they are described by systems of differential equations. But, in many ways they are simpler. For example, one-dimensional continuous dynamical systems do not exhibit the rich repertoire of behavior—2-cycles, 4-cycles, chaos, and so forth—that one-dimensional discrete dynamical systems exhibit.

During this year we have built up a powerful framework for studying change. We can model such phenomena with discrete dynamical systems or with continuous dynamical systems. The choice depends on the physics or biology or whatever of the phenomena involved. Once we have built a mathematical model we can study that model using a broad array of tools.

- Computer-based numerical methods like Euler's Method for continuous dynamical systems or the programs we introduced in chapter I for discrete dynamical systems.

- Computer-based graphics.

- Rough graphical analysis like the direction fields we constructed in section 2 in this chapter.

- Analytic techniques like the linear classification theorem and the nonlinear classification theorem.

We have become sophisticated mathematicians and, like carpenters, we have a workroom full of tools—saws and computers, hammers and integration, chisels and direction fields. To build a cabinet or to build understanding, we use a combination of tools, each for its part of the whole project.

The main concepts and tools involved in autonomous continuous dynamical systems show up for one-dimensional dynamical systems of the form
$$p' = f(p).$$

- An **equilibrium point** is a point p_* such that $f(p_*) = 0$. This implies that if $p(t_0) = p_*$ then for every later time $t \geq t_0$,
$$p(t) = p_*.$$
Actually, it also implies that for earlier times $t \leq t_0$,
$$p(t) = p_*$$
but we haven't discussed this particular consequence.

- There are three kinds of equilibrium points—**attracting equilibrium points** that pull in trajectories that start nearby, **repelling equilibrium points** that push nearby trajectories away, and **indeterminate equilibrium points** whose behavior is somewhat more complicated (in particular they may pull in some nearby trajectories and push away others).

§8. Summary[+]

- We can study the general behavior of these systems by sketching direction fields, either by hand or using a computer.

- We can determine the character of an equilibrium point p_* either from a direction field or from the sign of $f'(p_*)$.

- We can often find exact solutions by using an algebraic technique like, for example, separation of variables.

- We can estimate solutions numerically using a computer-based method like Euler's Method.

The same array of concepts and tools are important for two-dimensional autonomous continuous dynamical systems, but since the systems are more complicated the tools and concepts are more complicated. For example, instead of looking at the sign of the derivative we have the more complicated nonlinear classification theorem.

We have come full circle this year. We started with population models and returned to the same place but with a difference. We are different. We now have tools and, more importantly, we are mathematicians able to use those tools. Now we can study much more complicated models, in which two species interact, and we can learn a great deal about those models. Although this is the end of this course. There is a great deal more that can be done, that you can do. I hope that this year has answered some of your questions, has posed new questions and, most importantly, has given you a sense of adventure about mathematics and the way in which it can be used to help us understand the world in which we live.

It is fitting that we close this year not with a review of what we have done, but rather by looking at a new problem, one that we haven't looked at before but which can be studied using the ideas from this semester.

The SIR Model

In this section we consider one model for the way in which an epidemic spreads. We are interested in a relatively benign disease, one from which people recover and which confers immunity on its victims. At any given time we must keep track of three quantities.

- Those people who have not yet been infected by the disease and, hence, are susceptible to it. This group of people is called the **susceptibles**. The number of susceptibles is given by the function $S(t)$.

- Those people who are infected by the disease and are capable of transmitting it to other. This group of people is called the **Infecteds**. The number of infected people is given by the function $I(t)$.

- Those people who have had the disease and have recovered. These people can no longer transmit the disease and they cannot catch it again. They are called the **recovereds**. The number of recovered people is given by the function $R(t)$.

People who are infected eventually recover. In the SIR model we assume that the infected people recover at a fixed relative rate. Thus

$$R' = bI$$

where b is a fixed constant whose value depends on how long it typically takes for an infected person to recover.

Susceptible people become infected on the basis of contacts with infected people. Suppose that each susceptible person comes in contact with an average of c people per day. If T is the total number of people involved then I/T of the people are infected and on the average a susceptible person will come into contact with cI/T infected people. In the SIR model we assume that a fixed proportion k of these contacts result in transmission of the infection from an infected person to a susceptible person. For a highly contagious disease the constant k may be equal to 1 because every contact between an infected person and a susceptible person results in transmission of the infection. For a less contagious disease the constant k may be close to 0. The probability of a single susceptible person becoming infected is thus kcI/T. Let a denote the constant kc/T. Thus the probability of a single susceptible person becoming infected is aI and the rate at which susceptible people become infected is aIS yielding the differential equation

$$S' = -aIS.$$

§8. Summary†

People move into the infected group from the susceptible group by becoming infected and move out of the infected group into the recovered group by recovering. This give us the differential equation

$$I' = aIS - bI.$$

These three equations describe the SIR model,

$$S' = -aIS$$
$$I' = aIS - bI$$
$$R' = bI.$$

This is a three-dimensional continuous autonomous dynamical system, but if we concentrate on just the two variables S and I we can learn a lot from the two-dimensional dynamical system

$$S' = -aIS$$
$$I' = aIS - bI$$

This smaller model is called the SI model. Since we are assuming that no one dies from this disease the total number T of people stays constant and $R = T - S - I$. So the SI model gives us all the information we need to completely understand the SIR model.

Exercises

*Exercise VII.8.1: Find all the equilibrium points for the SI model.

*Exercise VII.8.2: Sketch the direction field for SI model.

Exercise VII.8.3: Do the first batch of SI and SIR experiments in your lab manual.

In the exercises above you discovered that there is an entire line of equilibrium points, somewhat like the coffee urn in a small room model of Exercise VI.2.4. In this situation the equilibrium points are all indeterminate because for any equilibrium point there are other equilibrium

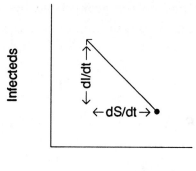

Figure VII.53: One arrow in the SI vector field

points close by and thus no equilibrium point is either attracting or repelling. If we look at a particular point on the SI vector field (note now we are looking at the vector field, not just the direction field) we see an arrow pointing in the direction of change. Look at Figure VII.53. The slope of this arrow is

$$\begin{aligned} \text{slope} &= I'/S' \\ &= \frac{aIS - bI}{-aIS} \\ &= \frac{b}{aS} - 1 \end{aligned}$$

Thus, if we think of a particular trajectory as a curve $I = f(S)$ we see that

$$\frac{dI}{dS} = \frac{b}{aS} - 1.$$

Exercises

The following exercises refer to the model.

$$\begin{aligned} S' &= -0.001 IS \\ I' &= 0.001 IS - 0.5 I \end{aligned}$$

*Exercise VII.8.4: Find the equation of the trajectory for the initial values

$$S(0) = 990, \quad I(0) = 10.$$

§8. Summary†

Describe the course of the epidemic described by this model.

Exercise VII.8.5: Find the equation of the trajectory for the initial values
$$S(0) = 980, \quad I(0) = 20.$$
Describe the course of the epidemic described by this model.

Exercise VII.8.6: Find the equation of the trajectory for the initial values
$$S(0) = 960, \quad I(0) = 40.$$
Describe the course of the epidemic described by this model.

Exercise VII.8.7: Find the equation of the trajectory for the initial values
$$S(0) = 920, \quad I(0) = 80.$$
Describe the course of the epidemic described by this model.

The following exercises refer to the model.

$$S' = -0.001IS$$
$$I' = 0.001IS - 0.1I$$

*Exercise VII.8.8: Find the equation of the trajectory for the initial values
$$S(0) = 990, \quad I(0) = 10.$$
Describe the course of the epidemic described by this model.

Exercise VII.8.9: Find the equation of the trajectory for the initial values
$$S(0) = 980, \quad I(0) = 20.$$
Describe the course of the epidemic described by this model.

Exercise VII.8.10: Find the equation of the trajectory for the initial values
$$S(0) = 960, \quad I(0) = 40.$$
Describe the course of the epidemic described by this model.

Exercise VII.8.11: Find the equation of the trajectory for the initial values
$$S(0) = 920, \quad I(0) = 80.$$

Describe the course of the epidemic described by this model.

Exercise VII.8.12: What can you say about the course of an epidemic and the way in which this course depends on the initial conditions, and the values of the constants a and b?

Appendix A

Selected Solutions

Exercise I.1.1:

Year	Population	Year	Population
1	100.00	11	457.05
2	180.00	12	465.64
3	244.00	13	472.51
4	295.20	14	478.01
5	336.16	15	482.41
6	368.93	16	485.93
7	395.14	17	488.74
8	416.11	18	490.99
9	432.89	19	492.79
10	446.31	20	494.24

The population is rising and seems to be leveling off at about 500.

Exercise I.1.5: The thin line in Figure A.1 is the function

$$R_1(p) = 2.3 - .0001p$$

and the thick line is the function

$$R_2(p) = 2.3 - .00005p.$$

Notice that the thick line is higher than the thin line. Thus, the population multiplier R_2 is greater than the population multiplier R_1 and corresponds to a habitat that is more favorable.

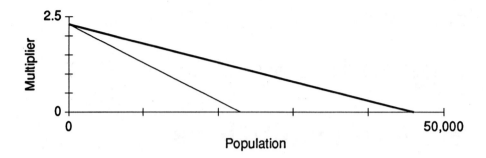

Figure A.1: $R_1(p) = 2.3 - .0001p$ and $R_2(p) = 2.3 - .00005p$.

Exercise I.1.6:

Comparing Figures A.2 which shows $R_1(p)$ and A.3 which shows $R_2(p)$ we see that $R_2(p)$ is low when p is low. This behavior is typical of the population multiplier for a species that hunts in packs.

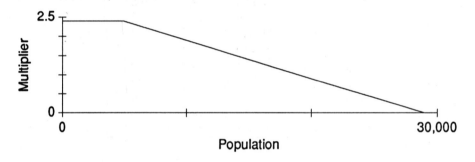

Figure A.2: $R_1(p)$

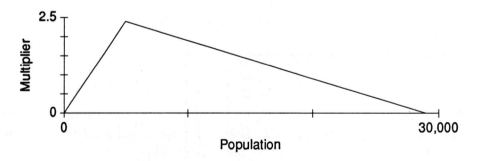

Figure A.3: $R_2(p)$

Exercise I.1.7:

Year	Population	Year	Population
1	10.00	11	440.98
2	17.82	12	443.73
3	31.50	13	444.30
4	54.92	14	444.42
5	93.43	15	444.44
6	152.46	16	444.44
7	232.59	17	444.44
8	321.28	18	444.44
9	392.51	19	444.44
10	429.20	20	444.44

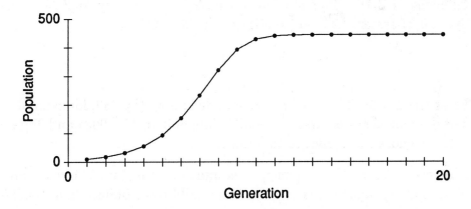

Figure A.4: $p_{n+1} = 1.8(1 - .001 p_n) p_n \quad p_1 = 10$

Exercise I.1.9:

Year	Population	Year	Population
1	10.00	11	843.07
2	33.66	12	449.83
3	110.59	13	841.44
4	334.43	14	453.62
5	756.79	15	842.69
6	625.80	16	450.72
7	796.20	17	841.74
8	551.71	18	452.92
9	840.91	19	842.46
10	454.86	20	451.25

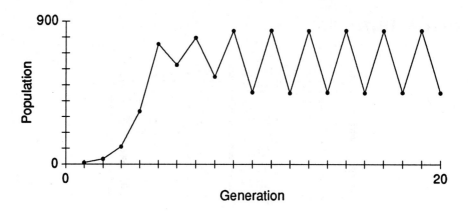

Figure A.5: $p_{n+1} = 3.4(1 - .001p_n)p_n \quad p_1 = 10$

Exercise I.2.1: The sequence is bounded above by 1000/3. The dynamical system has one equilibrium point, 1000/3.

$$\lim_{n \to \infty} p_n = 1000/3.$$

Exercise I.2.5: The sequence is bounded above (by 850, for example). The dynamical system has two equilibrium points, 2.4/.0034 and 0. The sequence does not converge to a limit.

Exercise I.2.9: The sequence is bounded above (by 1000, for example). The dynamical system has two equilibrium points, 2.96/.00396 and 0. The sequence does not converge to a limit.

Exercise I.2.11: You will have tried your own experiments. Here are two that I tried. First, for $a = 2$

Year	Population	Year	Population
1	50.00	11	626.35
2	100.00	12	624.09
3	200.00	13	625.60
4	400.00	14	624.60
5	640.00	15	625.27
6	614.40	16	624.82
7	631.77	17	625.12
8	620.37	18	624.92
9	628.03	19	625.05
10	622.95	20	624.96

Second, for $a = 2.5$

Year	Population	Year	Population
1	50.00	11	574.23
2	125.00	12	814.97
3	312.50	13	502.66
4	716.15	14	833.31
5	677.60	15	463.02
6	728.19	16	828.77
7	659.76	17	473.03
8	748.25	18	830.91
9	627.90	19	468.33
10	778.80	20	829.99

In the first example the population is converging to a limit. In the second example the population appears to be settling down into a pattern but not converging to a limit. This is an important exercise because it shows that the kinds of unexpected behavior we have already seen for logistic models are not unusual. This family of models exhibits the same kinds of behavior.

Exercise I.3.1: The year 2004.

Exercise I.3.3:

Tolerance	N
1.0	9
0.5	11
0.1	17

Exercise I.3.5: The limit is 400 and $N = 10$.

Exercise 1.3.9: The equilibrium point is $-1,000$ but the sequence is not converging to a limit.

Exercise I.3.11: 3.304518.

Exercise I.3.12: At least 401 rectangles.

Exercise I.3.13: The exact answer (correct to eight digits) is 0.84147098. Your answer should be within 0.1 of this value.

Exercise I.4.1: The exact solution (correct to eight digits) is 0.72449196. Your answer should be within 0.1 of this value.

Exercise I.4.2: This equation has three solutions. They are (correct to eight digits) 1.53208889, -1.87938524, and 0.34729636. Your answer should be within 0.1 of one of these values.

Exercise I.5.1:

Exact output	Exact input	Output tolerance	Input tolerance
48	7.2	8	1.20
48	7.2	6	0.90
48	7.2	4	0.60
48	7.2	1	0.15
32	4.8	8	1.20
32	4.8	6	0.90
32	4.8	4	0.60
32	4.8	2	0.30

Exercise I.5.2:

Exact output	Exact input	Output tolerance	Input tolerance
3.0	5.4772	0.25	0.2237
3.0	5.4772	0.20	0.1796
3.0	5.4772	0.15	0.1353
4.0	6.3246	0.25	0.1946
4.0	6.3246	0.20	0.1562
4.0	6.3246	0.15	0.1175

Exercise I.5.3:
This function (Figure A.6) is continuous at every point except $x = 2$. At $x = 2$ it is not continuous.

Exercise I.5.5:
This function (Figure A.7) is continuous at every point.

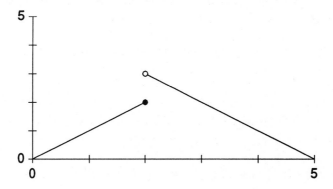

Figure A.6: Graph for Exercise I.5.3

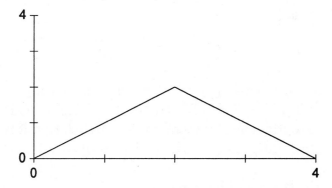

Figure A.7: Graph for Exercise I.5.5

Exercise I.5.6:

Given an output tolerance $\epsilon > 0$ we must find an input tolerance $\delta > 0$ such that if $|x - a| < \delta$ then $|3f(x) - 3f(a)| < \epsilon$. Since $f(x)$ is continuous at a there is a δ such that if $|x - a| < \delta$ then $|f(x) - f(a)| < \epsilon/3$. This δ will do the trick, since, if $|x - a| < \delta$ then

$$|3f(x) - 3f(a)| = 3|f(x) - f(a)| < 3\epsilon/3 = \epsilon.$$

Exercise I.5.7:

Given an output tolerance $\epsilon > 0$ we must find an input tolerance $\delta > 0$ such that if $|x - a| < \delta$ then $|(f(x) + g(x)) - (f(a) + g(a))| < \epsilon$. Since $f(x)$ and $g(x)$ are continuous at a we know there is a $\delta_1 > 0$ such that if $|x - a| < \delta_1$ then $|f(x) - f(a)| < \epsilon/2$ and there is a $\delta_2 > 0$ such that if $|x-a|<\delta_2$ then $|g(x)-g(a)|<\epsilon/2$. Now let δ be the smaller of δ_1 and δ_2. If $|x-a|<\delta$ then

$$|f(x)-f(a)|<\epsilon/2$$

and

$$|g(x)-g(a)|<\epsilon/2.$$

So

$$\begin{aligned}|(f(x)+g(x))-(f(a)+g(a))| &= |(f(x)-f(a))+(g(x)-g(a))| \\ &\leq |f(x)-f(a)|+|g(x)-g(a)| \\ &< \epsilon/2+\epsilon/2 \\ &= \epsilon.\end{aligned}$$

Exercise I.5.8:

Given an output tolerance $\epsilon>0$ we must find an input tolerance $\delta>0$ such that if $|x-a|<\delta$ then $|f(x)g(x)-f(a)g(a)|<\epsilon$.

Since $g(x)$ is continuous at a there is a $\delta_1>0$ such that if $|x-a|<\delta_1$ then $|g(x)-g(a)|<1$. In particular, if $|x-a|<\delta_1$ then $|g(x)|<|g(a)|+1$.

Let M be the larger of $|f(a)|$ and $|g(a)|+1$.

Since $f(x)$ is continuous at a there is a $\delta_2>0$ such that if $|x-a|<\delta_3$ then $|f(x)-f(a)|<\epsilon/2M$.

Similarly, since $g(x)$ is continuous at a there is a $\delta_3>0$ such that if $|x-a|<\delta_3$ then $|g(x)-g(a)|<\epsilon/2M$.

Now let δ be the smallest of δ_1, δ_2, and δ_3. This is the desired input tolerance, since if $|x-a|<\delta$ then

$$\begin{aligned}|f(x)g(x)-f(a)g(a)| &= |f(x)g(x)-f(a)g(x)+f(a)g(x)-f(a)g(a)| \\ &= |g(x)(f(x)-f(a))+f(a)(g(x)-g(a))| \\ &\leq |g(x)||f(x)-f(a)|+|f(a)||g(x)-g(a)| \\ &\leq M\epsilon/2M+M\epsilon/2M \\ &= \epsilon.\end{aligned}$$

Exercise I.6.1:

Generation	Population
1	50.000
2	133.000
3	322.871
4	612.151
5	664.782

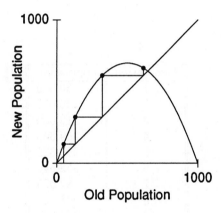

Figure A.8: Exercise I.6.1

Exercise I.6.2:

Generation	Population
1	50.000
2	152.000
3	412.467
4	775.482
5	557.152

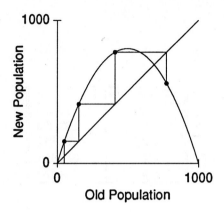

Figure A.9: Exercise I.6.2

Exercise I.6.3:

Generation	Population
1	50.000
2	140.000
3	212.000
4	269.600
5	315.680
6	352.544
7	382.035
8	405.628
9	424.503
10	439.602

Exercise I.6.4:

Generation	Population
1	50.000
2	−40.000
3	−148.000
4	−277.600
5	−433.120
6	−619.744
7	−843.693
8	−1,112.431
9	−1,434.918
10	−1,821.901

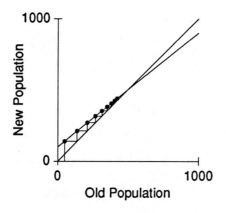

Figure A.10: Exercise I.6.3

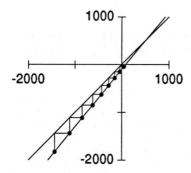

Figure A.11: Exercise I.6.4

Exercise I.6.5:

Generation	Population
1	50.000
2	60.000
3	52.000
4	58.400
5	53.280
6	57.376
7	54.099
8	56.721
9	54.623
10	56.301

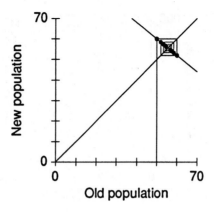

Figure A.12: Exercise I.6.5

Exercise I.7.1:

Based on the calculations in the following table it looks as if the equilibrium point (500) is attracting. This exercise supports the conjecture since $|m| = 0.8$ is less than 1.

Year	Population	Year	Population
1	50.00	11	451.68
2	140.00	12	461.35
3	212.00	13	469.08
4	269.60	14	475.26
5	315.68	15	480.21
6	352.54	16	484.17
7	382.04	17	487.33
8	405.63	18	489.87
9	424.50	19	491.89
10	439.60	20	493.51

Exercise I.7.2:

Based on the calculations in the following table it looks as if the equilibrium point (55.56) is attracting. This exercise supports the conjecture since $|m| = 0.8$ is less than 1.

Year	Population	Year	Population
1	50.00	11	54.96
2	60.00	12	56.03
3	52.00	13	55.17
4	58.40	14	55.86
5	53.28	15	55.31
6	57.38	16	55.75
7	54.10	17	55.40
8	56.72	18	55.68
9	54.62	19	55.46
10	56.30	20	55.64

Exercise I.7.7:

Since $|m| = 1.02$ is bigger than 1 our conjecture implies that the equilibrium point $(-5,000)$ is not attracting. The following calculations support the conjecture.

Year	Population	Year	Population
1	50.00	11	1,155.92
2	151.00	12	1,279.04
3	254.02	13	1,404.62
4	359.10	14	1,532.71
5	466.28	15	1,663.37
6	575.61	16	1,796.64
7	687.12	17	1,932.57
8	800.86	18	2,071.22
9	916.88	19	2,212.64
10	1,035.22	20	2,356.90

Exercise I.7.8:

Since $|m| = 1.02$ is bigger than 1 our conjecture implies that the equilibrium point $(100/2.02)$ is not attracting. The following calculations support the conjecture.

Year	Population	Year	Population
1	50.00	11	50.11
2	49.00	12	48.89
3	50.02	13	50.13
4	48.98	14	48.86
5	50.04	15	50.16
6	48.96	16	48.84
7	50.06	17	50.18
8	48.94	18	48.81
9	50.08	19	50.21
10	48.91	20	48.78

Exercise I.7.9: The dynamical system

$$p_{n+1} = -p_n + b$$

has one equilibrium point, $-b/2$. If $p_1 \neq -b/2$ then the sequence $p_1, p_2, p_3 \ldots$ bounces back-and-forth between p_1 and $-p_1 + b$.

Exercise I.7.10:

Based on the following calculuations it looks as if

$$\lim_{n \to \infty} p_n = 14/15.$$

Day	Price	Day	Price
1	.5000	6	.8605
2	.6300	7	.8824
3	.7210	8	.8976
4	.7847	9	.9084
5	.8293	10	.9158

Exercise I.7.13:

Based on the following calculuations it looks as if

$$\lim_{n \to \infty} p_n = 14/15.$$

Day	Price	Day	Price
1	.5000	6	1.0753
2	1.2800	7	.8197
3	.6560	8	1.0242
4	1.1552	9	.8606
5	.7558	10	.9915

Exercise I.7.14:

Notice that

$$\begin{aligned} p_{n+1} &= p_n + 1500k\left[\left(\frac{14}{15}\right) - p_n\right] \\ &= p_n + 1400k - 1500kp_n \\ &= (1 - 1500k)p_n + 1400k \end{aligned}$$

This is a linear dynamical system and the equilibrium point will be attracting if $|1 - 1500k| < 1$, that is, if

$$\begin{aligned} -1 < \; & 1 - 1500k \; < 1 \\ -2 < \; & -1500k \; < 0 \\ 0 < \; & 1500k \; < 2 \\ 0 < \; & k \; < 2/1500 = 0.001333 \end{aligned}$$

Exercise I.7.15:

The equilibrium price is 0.933333 and based on the following calculations we see that for $n \geq 15$, p_n is within 0.05 of this price.

Day	Price	Day	Price
1	.5000	11	.8480
2	.5650	12	.8608
3	.6203	13	.8717
4	.6672	14	.8809
5	.7071	15	.8888
6	.7411	16	.8955
7	.7699	17	.9012
8	.7944	18	.9060
9	.8153	19	.9101
10	.8330	20	.9136

Exercise I.7.19:

The new equilibrium price is $1550/1500 = 1.0333$.

Exercise I.7.20:

Based on the following calculations the price will be within .01 of the new equilibrium when $n \geq 5$.

Day	Price	Day	Price
1	.9333	11	1.0331
2	.9783	12	1.0332
3	1.0031	13	1.0333
4	1.0167	14	1.0333
5	1.0242	15	1.0333
6	1.0283	16	1.0333
7	1.0306	17	1.0333
8	1.0318	18	1.0333
9	1.0325	19	1.0333
10	1.0329	20	1.0333

Exercise I.7.21:

We can use either the demand function or the supply function to compute the total number of Byties sold at the two equilibrium prices. At the old equilibrium price:

$$D(14/15) = 533.33.$$

At the new equilibrium price:

$$D(1550/1500) = 483.33.$$

The difference is 50 Byties, so 2.5 bakers will go out of business.

Exercise I.7.22:

Before the rise in the cost of ingredients each baker made a profit of $14/15 - 0.40 = 0.533333$ on each Bytie, for a total profit of \$10.67. After the rise in the cost of ingredients each baker makes a profit of $1550/1500 - 0.55 = 0.483333$ on each Byties, for a total profit of \$9.67. Thus each baker's daily profit has dropped by \$1.00.

Exercise I.7.23:

Both bakers and customers are paying. The price of each Bytie rose by $0.10 and the cost of ingredients rose by $0.15, so the bakers were able to pass on only two-thirds of the rise to their customers. Even so they lost some customers.

Exercise I.8.1: -0.858572.

Exercise I.8.4: -0.703293.

Exercise I.8.6: -0.442962.

Exercise I.8.8: The exact slope of the tangent to the curve at the point $(0.6, 0.8)$ is -0.75. The exact slope of the tangent to the curve at the point $(0.4, ?)$ is -0.436436. Notice that the estimates are close to these exact values.

Exercise I.8.9: $f'(x) = 3x^2$ and $f'(2) = 12$.

Exercise I.8.10: $f'(x) = 4x^3$ and $f'(3) = 108$.

Exercise I.8.11: $f'(x) = 3x^2 + 4x^3$. Notice that the derivative of $x^3 + x^4$ is just the sum of the derivatives of x^3 and x^4.

Exercise I.8.12: $f'(x) = 5x^4$. Notice that the derivative of $x^2 x^3$ is *not* the product of the derivatives of x^2 and x^3.

Exercise I.8.13: $y = 7x - 4$.

Exercise I.8.14:

(a) $p_* = 1.9/.0029 = 655.1724$.

(b) $f'(p) = 2.9 - .0058p$.

(c) $f'(p_*) = -0.9$.

(d) Since $|f'(p_*)| < 1$ we would expect the equilibrium point to be attracting.

(e) The following calculations agree with this expectation.

Year	Population	Year	Population
1	50.00	11	655.34
2	137.75	12	655.02
3	344.45	13	655.31
4	654.83	14	655.05
5	655.48	15	655.28
6	654.89	16	655.08
7	655.42	17	655.26
8	654.95	18	655.09
9	655.37	19	655.24
10	654.99	20	655.11

Exercise I.8.15:

(a) $p_* = 2.1/.0031 = 677.4194$.

(b) $f'(p) = 3.1 - .0062p$.

(c) $f'(p_*) = -1.1$.

(d) Since $|f'(p_*)| > 1$ we would expect this equilibrium point not to be attracting.

(e) The following calculations agree with this expectation.

Year	Population	Year	Population
1	50.00	11	575.19
2	147.25	12	757.47
3	389.26	13	569.49
4	736.98	14	760.03
5	600.90	15	565.39
6	743.44	16	761.74
7	591.29	17	562.62
8	749.17	18	762.84
9	582.54	19	560.83
10	753.88	20	763.53

Exercise I.8.16: The curve (a) is the function and (b) is its derivative.

Exercise I.8.17: The curve (a) is the function and (b) is its derivative.

Exercise I.8.18: The curve (a) is the function and (b) is its derivative.

Exercise I.9.1:
 (a) 28 feet per second.
 (b) −4 feet per second.
 (c) The ball will be rising as long as $h'(t)$ is positive. Since $h'(t) = 60 - 32t$ the ball will be rising while

$$60 - 32t > 0$$
$$60 > 32t$$
$$60/32 > t.$$

So the ball will be rising until $t = 60/32$.
 (d) It will hit the ground when $h(t) = 0$. That is,

$$60t - 16t^2 = 0$$
$$t(60 - 16t) = 0$$

This equation has two solutions: $t = 0$ and $t = 60/16$. The first solution is the time at which the ball was thrown. The second is the time when it hits the ground.
 (e) $h'(60/16) = -60$ feet per second.

Exercise I.9.2:
 (a) The obvious estimate is

$$\frac{195,000,000 - 200,000,000}{1} = -5,000,000 \text{ miles per day.}$$

 (b) The obvious estimate is

$$\frac{190,100,000 - 95,000,000}{1} = -4,900,000 \text{ miles per day.}$$

 (c) The obvious estimate is

$$\frac{185,300,000 - 190,100,000}{1} = -4,800,000 \text{ miles per day.}$$

 (d) The obvious estimate is

$$\frac{180,600,000 - 185,300,000}{1} = -4,700,000 \text{ miles per day.}$$

(e) The obvious estimate is

$$\frac{176{,}000{,}000 - 180{,}600{,}000}{1} = -4{,}600{,}000 \text{ miles per day.}$$

(f) There are several different ways that you might answer this, all quite reasonable. The most obvious way doesn't work since we have no data for September 7. One possibility would be to assume that the obvious pattern above continues and that the velocity on September 6 would be $-4{,}500{,}000$ miles per day.

(g) Notice that the velocity seems to be increasing. (The absolute value of the velocity, sometimes called the speed, is decreasing, however.) If we estimate velocity at time t by

$$\frac{f(t+h) - f(t)}{h}$$

with h positive then we will get an overestimate. If we estimate velocity at time t using h negative then we will get an underestimate. We can get a better estimate by looking at the fraction

$$\frac{f(x+h) - f(x-h)}{2h}.$$

For example, at noon on September 2 we can estimate the velocity by

$$\frac{190{,}100{,}000 - 200{,}000{,}000}{2} = -4{,}950{,}000 \text{ miles per day.}$$

Exercise I.9.4: -372 gallons per hour.

Exercise I.9.6:

(a) 1600. Since the sensitivity is positive increasing the price will increase total profit. Thus I would recommend that the company raise its price.

(b) 600. Since the sensitivity is positive increasing the price will increase total profit. Thus I would recommend that the company raise its price.

Exercise I.10.1:

(a) The following table shows $p_1, p_2 \ldots p_{10}$.

Year	Population	Year	Population
1	100.00	6	1,160.58
2	560.00	7	1,196.35
3	836.00	8	1,217.81
4	1,001.60	9	1,230.68
5	1,100.96	10	1,238.41

(b) The population seems to be getting close to the equilibrium population 1,250.

(c)
$$\lim_{n \to \infty} p_n = 1,250.$$

(d) Because this is a linear dynamical system we will see the same longterm behavior for any p_1.

(e) The population would converge to the limit point 1,500 if the 500 were changed to 600. If it were changed to another value the population would converge the new equilibrium point.

(f) If the 0.6 were changed to 0.95 the population would converge but more slowly. If it were changed to 1.2 it would not converge. For a linear dynamical system

$$p_{n+1} = mp_n + b$$

the sequence converges if $|m| < 1$.

Exercise I.10.2:

(a) The following table shows $p_1, p_2 \ldots p_{10}$.

Year	Population	Year	Population
1	800.00	6	1,246.50
2	860.00	7	1,395.80
3	932.00	8	1,574.95
4	1,018.40	9	1,789.95
5	1,122.08	10	2,047.93

(b) The population is getting farther away from the equilibrium population 500.

(c) The sequence does not converge.

(d) Yes, if $p_1 > 500$ then the population will skyrocket getting larger and larger without any bound. If $p_1 < 500$ then the population will eventually become very large (without any bound) and negative! (Thus if $p_1 < 500$ this is not a realistic model.) If p_1 is exactly equal to 500 it will remain so.

(e) If the 100 is changed to 200 then the population will eventually become negative!! The key thing is whether the initial population is above or below the equilibrium population. If it is above the equilibrium population then the population gets larger and larger without any bound and, for that reason, is unrealistic. If it is below the equilibrium population then the model predicts negative population and is also unrealistic.

(f) As long as the 1.2 is changed to a number greater than 1 the same general remarks apply. If it is changed to a number less than 1 then the equilibrium point will be attracting but since it will also be negative the model will predict negative population and be unrealistic.

One does not see populations behaving like these models in the natural world. These models can, however, be used as a basis for modeling artificial situations, for example, situations with "harvesting." Mathematically, harvesting is like emmigration.

Exercise I.10.8:

(a) $f(-10) = -899, f(-2) = 13, f(2) = -11, f(10) = 901$.

(b) There are solutions to this equation in each of the intervals $[-10, -2], [-2, 2]$, and $[2, 10]$.

(c) The solutions (correct to eight digits) are: $-3.21113935, 0.10010030$, and 3.11103905.

Exercise I.10.9:

(a) 2.63705.

(b) It is an underestimate. It cannot be off by more then 0.8.

(c) 3.03705. This estimate is much better than either of the earlier estimates based on five strips.

Exercise I.10.10:

(a) 3.17199.

(b) and (c) This is the best estimate yet based on five strips.

(d) The other two methods either consistently underestimate or consistently overestimate. The midpoint method does a little bit of both and the under and overestimates tend to cancel out.

Exercise I.10.11:
(a) Every two seconds the velocity increases by 50, then 25, then 12.5, then 6.25, etc.
(b) After 14 seconds 99.2188, after 16 seconds 99.6094.
(c)
$$100 - \frac{100}{2^{\frac{n}{2}}}$$

(d) 100 feet per second.
(e) 14 seconds.
(f) 20 seconds.
(g) 27 seconds. Your answer to this problem may be 28 seconds. That is fine.

Exercise I.10.12: $4x^3$.

Exercise I.10.13: $-1/(x+2)^2$.

Exercise I.10.14: $4x^3 - 1/(x+2)^2$.

Exercise I.10.15: (a) is the original function and (b) is its derivative.

Exercise II.1.1: $6x^5 - 12x^3 + 54x$.

Exercise II.1.2: $161x^6 - 20x^4$.

Exercise II.1.3: $6x + 5$.

Exercise II.1.4: $-6t + 30t^4$.

Exercise II.1.5: 50/9 degrees Celsius per hour.

Exercise II.1.7: $y = 2x$.

Exercise II.1.10:
We give two proofs. The first is algebraic.

$$f'(-x) = \lim_{h \to 0} \left(\frac{f(-x+h) - f(-x)}{h} \right)$$

$$= \lim_{h \to 0} \left(\frac{-f(x-h) + f(x)}{h} \right)$$
$$= \lim_{h \to 0} \left(\frac{f(x-h) - f(x)}{-h} \right)$$
$$= \lim_{h \to 0} \left(\frac{f(x+(-h)) - f(x)}{-h} \right)$$
$$= f'(x)$$

where the final line comes from noticing that when h is very small so is $-h$, so
$$\frac{f(x+(-h)) - f(x)}{-h}$$
is very close to $f'(x)$.

We've shown that $f'(-x) = f'(x)$, in other words, that $f'(x)$ is symmetric about the y-axis.

The second proof is geometric and is based on Figure A.13 below. Notice that when this picture is rotated 180° about the origin the tangent at the point $(x, f(x))$ is rotated 180° and placed exactly on top of the tangent at the point $(-x, f(-x))$. Thus, the tangent at $(-x, f(-x))$ has the same slope as the tangent at $(x, f(x))$ rotated 180°. But rotating a line 180° does not change its slope. So that the tangents at $(x, f(x))$ and $(-x, f(-x))$ have the same slope, that is, $f'(-x) = f'(x)$.

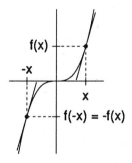

Figure A.13: Tangents at x and $-x$

Exercise II.1.12:

$$\begin{aligned} f(-x) &= p(-x) + q(-x) \\ &= p(x) + q(x) \\ &= f(x) \end{aligned}$$

Exercise II.1.18:
If $f(x)$ is symmetric about the y-axis then for every x,

$$f(-x) = f(x).$$

If $f(x)$ is symmetric about the origin then for every x,

$$f(-x) = -f(x).$$

Hence, for every x,

$$\begin{aligned} f(x) &= -f(x) \\ 2f(x) &= 0 \\ f(x) &= 0 \end{aligned}$$

Thus there is exactly one function, the function $f(x) = 0$, that is both symmetric about the origin and symmetric about the y-axis.

Exercise II.1.19:
We would expect that

$$\lim_{n \to \infty} p_n = 0.$$

The following calculations agree with this expectation.

Year	Population	Year	Population	Year	Population	Year	Population
1	50.00	11	21.14	21	10.80	31	5.94
2	45.12	12	19.66	22	10.15	32	5.61
3	40.93	13	18.31	23	9.54	33	5.30
4	37.30	14	17.08	24	8.98	34	5.01
5	34.11	15	15.94	25	8.45	35	4.74
6	31.30	16	14.91	26	7.96	36	4.48
7	28.80	17	13.95	27	7.50	37	4.24
8	26.57	18	13.07	28	7.07	38	4.01
9	24.58	19	12.25	29	6.67	39	3.79
10	22.77	20	11.50	30	6.30	40	3.59

Exercise II.1.20:

We would expect that

$$\lim_{n\to\infty} p_n = .05/.00105 = 47.62.$$

The following calculations agree with this expectation.

Year	Population	Year	Population	Year	Population	Year	Population
1	50.00	11	49.02	21	48.44	31	48.11
2	49.88	12	48.94	22	48.40	32	48.08
3	49.76	13	48.88	23	48.36	33	48.06
4	49.65	14	48.81	24	48.33	34	48.04
5	49.54	15	48.75	25	48.29	35	48.02
6	49.44	16	48.69	26	48.26	36	48.00
7	49.35	17	48.64	27	48.22	37	47.98
8	49.26	18	48.58	28	48.19	38	47.96
9	49.17	19	48.54	29	48.16	39	47.94
10	49.09	20	48.49	30	48.14	40	47.93

Exercise II.1.23:

Our conjecture does not tell us anything about the longterm behavior of this model. Looking at the following calculations we see that it does look as if the population *might* be approaching zero, but it is doing so very slowly.

Year	Population	Year	Population	Year	Population	Year	Population
1	50.00	11	32.86	21	24.56	31	19.63
2	47.50	12	31.78	22	23.95	32	19.24
3	45.24	13	30.77	23	23.38	33	18.87
4	43.20	14	29.83	24	22.83	34	18.52
5	41.33	15	28.94	25	22.31	35	18.17
6	39.62	16	28.10	26	21.81	36	17.84
7	38.05	17	27.31	27	21.34	37	17.52
8	36.60	18	26.56	28	20.88	38	17.22
9	35.26	19	25.86	29	20.45	39	16.92
10	34.02	20	25.19	30	20.03	40	16.63

Exercise II.1.25:

The constant b affects the value of the equilibrium point but it does not affect the nature of the longterm behavior. Biologically, the constant b reflects the size and abundance of resources in the habitat. It appears as if the habitat controls the size of the equilibrium population but not the nature of the longterm behavior. The constant a has a very dramatic effect on the longterm behavior. This constant reflects the natural population multiplier in a habitat of abundance. If it is very high then the population can skyrocket so rapidly that in a single generation the population can overshoot the equilibrium population. It is this possibility that allows longterm behavior like 2-cycles or worse.

Exercise II.1.26:

The point $p = 0$ is always an equilibrium point. At $p = 0$, $f(p) = .009ap^2$. So $f'(p) = .018ap$ and $4f'(0) = 0$. Thus, the equilibrium point $p = 0$ is always attracting.

The first clause of the definition of $R(p)$ gives us

$$p_{n+1} = .009ap^2, \quad \text{if } p \leq 100$$

which gives us a second equilibrium point $p = 1000/9a$. But this clause only applies when $p \leq 100$. So we only see this equilibrium point if

$$1000/9a \leq 100,$$

that is, if $1/9 \leq a$.

In this case[1] $f'(p) = .018ap$, so that $f'(1000/9a) = 2$ and this equilibrium point is not attracting.

The second clause of the defintion of $R(p)$ gives us a third equilibrium point,

$$p_* = \frac{a-1}{.001a}.$$

Since this clause only applies when $100 < p$ it turns out that this equilibrium point only shows up when $a > 1/9$ the same range of values of a for which the second equilibrium point shows up. For this clause

$$f(p) = a(1 - .001p)p,$$

so

$$f'(p) = a - .002ap$$

and

$$f'(p_*) = 2 - a.$$

Thus, this equilibrium point is attracting if $1 < a < 3$ and not attracting otherwise.

Exercise II.2.1: $f'(x) = 4x^3$.

Exercise II.2.2: $p'(x) = -8t^{-3} + 6t - 3t^2 = -8/t^3 + 6t - 3t^2$.

Exercise II.2.3: $-48x^7 + 7x^6 - 468x^2 + 52x$.

Exercise II.2.13:

$$\frac{(x^2 - 12)(9x^2 - 4x) - 2x(3x^3 - 2x^2 + 4)}{(x^2 - 12)^2}$$

Exercise II.2.15:

$$3x^2 + \frac{2x(x-3) - (x^2 - 4)}{(x-3)^2} = 3x^2 + \frac{x^2 - 6x + 4}{(x-3)^2}$$

---

[1] This formula for the derivative $f'(p)$ only applies at the equilibrium point when $1/9 < a$. If $a = 1/9$ then the function $f(p)$ is not differentiable at this equilibrium point.

Exercise II.2.16:

$$\frac{(2x(x^3-1)+3x^2(x^2+1))(x^2-7)-2x(x^2+1)(x^3-1)}{(x^2-7)^2} = \frac{3x^6-34x^4-21x^2+16x}{(x^2-7)^2}$$

Exercise II.2.27:
The current total yearly energy expenditure is $\$15 \times 10^9$. It is growing at the rate of $\$13 \times 10^7$ per year.

Exercise II.2.28:
The current area of the iceberg is 10,000 square feet. It is shrinking at the rate of 400 square feet per day. In five days it will be shrinking at the rate of 360 square feet per day.

Exercise II.2.29:
The iceberg is currently shrinking at the rate of 60,000 cubic feet per day. In five days it will be shrinking at the rate of 48,600 cubic feet per day.

Exercise II.2.30: Per capita personal income is rising at the rate of $\$16.58$ per year.

Exercise II.2.31:

$$f^*(x) = \frac{f'(x)}{f(x)}$$

$$= \frac{u(x)v'(x)+v(x)u'(x)}{v(x)^2} \frac{v(x)}{u(x)}$$

$$= \frac{v'(x)}{v(x)} + \frac{u'(x)}{u'(x)}$$

$$= v^*(x) + u^*(x)$$

Exercise II.2.32:

$$f^*(x) = u^*(x) - v^*(x).$$

Exercise II.2.33:

$$f^*(x) = \frac{u^*(x)}{1 + \frac{v(x)}{u(x)}} + \frac{v^*(x)}{1 + \frac{u(x)}{v(x)}}.$$

Notice that this formula is not as neat as the formulas obtained in Exercises II.2.31 and II.3.32.

Exercise II.3.1: 40.92 feet.

Exercise II.3.2: The sine of the angle is $12/20 = 3/5$. So we need to solve the equation

$$\sin\theta = 0.6.$$

The answer can be obtained by a numerical method like the Bisection Method or, if you've seen some trigonometry before, by using the arcsine function. In either case the answer is 0.6435 radians. The end of the rope is 16 feet from the base of the pole.

Exercise II.3.3: The cosine and the secant are symmetric about the y-axis.

Exercise II.3.5:

$$\begin{aligned}
\sin(\alpha - \beta) &= \sin(\alpha + (-\beta)) \\
&= \sin\alpha\cos(-\beta) + \sin(-\beta)\cos\alpha \\
&= \sin\alpha\cos\beta - \sin\beta\cos\alpha
\end{aligned}$$

Exercise II.3.8:

$$\begin{aligned}
\tan x &= \frac{\sin x}{\cos x} \\
(\tan x)' &= \frac{(\cos x)(\cos x) - (\sin x)(-\sin x)}{(\cos x)^2} \\
&= \frac{(\cos x)^2 + (\sin x)^2}{(\cos x)^2} \\
&= \frac{1}{(\cos x)^2}
\end{aligned}$$

$$= \sec^2 x$$

Exercise II.3.12: $f'(x) = 2\cos x - 2x \sin x$.

Exercise II.3.16: $2(\cos^2 \theta - \sin^2 \theta)$.

Exercise II.3.17: $2\cos 2\theta$. Note this exercise and the preceding exercise are the same since $\sin 2\theta = 2\sin\theta \cos\theta$. Use an appropriate trigonometric identity to show that the answers agree.

Exercise II.3.18: There are two ways to do this. First, $y = \sin^2 \theta + \cos^2 \theta = 1$, so $y' = 0$. Second,

$$\begin{aligned} y &= \sin^2 \theta + \cos^2 \theta \\ y' &= 2(\sin\theta)(\cos\theta) + 2(\cos\theta)(-\sin\theta) \\ &= 2(\sin\theta)(\cos\theta) - 2(\cos\theta)(\sin\theta) \\ &= 0 \end{aligned}$$

Exercise II.3.19: $f'(t) = 2\sin t \cos t$.

Exercise II.3.20: $g'(t) = 2\cos t \sin t$. These two functions have the same derivative because they differ by a constant. That is,

$$(\sin t)^2 - -(\cos t)^2 = 1.$$

Exercise II.4.1:
The best price now is $1.15. Of the $0.10 rise in the unit manufacturing cost, $0.05 was passed on to the customers.

Exercise II.4.2:
When the unit manufacturing cost is $0.30 the best price is $2.15. When the unit manufacturing cost is $0.50 the best price is $2.25. Of the $0.20 rise in unit manufacturing cost, $0.10 was passed on to the consumers.

Exercise II.4.4: $(1+\sqrt{5})/2 = 1.618034$.

After the first price rise the best price is $1+\sqrt{2} = 2.414214$. Notice that when the unit manufacturing cost rose from \$0.50 to \$1.00 the price paid by consumers rose by almost \$0.80. More than 100% of the rise in manufacturing cost was passed on to the consumers. The reason this happens is that those customers for whom price is most important have been priced out of the market. Thus the manufacturer is free to raise prices because the remaining customers are willing to pay a lot for the product. Was the rise in the manufacturing cost good for the manufacturer? Why?

After the second price rise the best price is $(3+\sqrt{13})/2 = 3.302776$. Once again more than 100% of the rise in manufacturing cost was pased on to the consumers.

Exercise II.4.5: $2x - 64x + 27 = -62x + 27$.

Exercise II.4.8: $6s^5 + 5s^4 - 28s^3 - 9s^2 + 8s + 12$.

Exercise II.4.10:

$$\frac{t^4 - 6t^3 + 36t^2 + 2t - 3}{(1-t^3)^2}$$

Exercise II.4.16: The global maximum is at $x = 2$ and the global minimum is at $x = -2$.

Exercise II.4.17: The global maximum is at $x = 3$ and the global minimum is at $x = -3$.

Exercise II.4.20: The global maximum is at $x = 1$ and the global minimum is at $x = -1$.

Exercise II.4.21: The height of the gutter should be 2.5 inches.

Exercise II.4.22: 6.158403 square units. The rectangle's width will be $2\sqrt{4/3}$ and its height will be 8/3.

Exercise II.4.23: The field should be a square 25 feet by 25 feet.

Exercise II.4.24: The field should be 50 feet (along the river) by 25 feet.

Exercise II.4.25: The field should be 25 feet by 50/3 feet. The fence down the middle should be parallel to the shorter sides.

Exercise II.4.26: The warehouse should be located in the same place as the middle retail store.

Exercise II.4.27: The warehouse should be located anyplace between the two middle retail stores.

Exercise II.4.29: If there are an odd number of retail stores then the warehouse should be located at the same place as the middle retail store. If there are an even number of retail stores then the warehouse should be located anyplace between the two middle retail stores.

Exercise II.5.1: See Figure A.14.

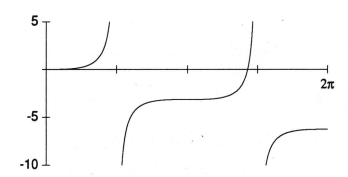

Figure A.14: $y = \tan x - x$

Exercise II.5.2: See Figure A.15

Exercise II.5.4: See Figure A.16

Exercise II.6.1: He must add at least 7,000,000/3 gallons of uncontaminated water.

Exercise II.6.2: He must add at least 9,000,000 gallons of uncontaminated water.

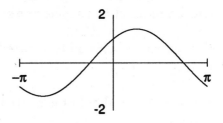

Figure A.15: $y = \sin x + \cos x$

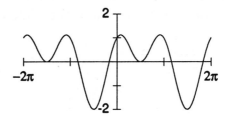

Figure A.16: $y = \sin t + \cos 2t$

Exercise II.6.3:

$$G(x) = \frac{10,000,000 + 0.5x}{1,000,000 + x}.$$

$$\lim_{x \to +\infty} G(x) = 0.5$$

To meet the 5 mg per gallon standard the bottler must add at least 5,000,000/4.5 gallons from his normal water supply.

To meet the 3 mg per gallon standard the bottler must add at least 7,000,000/2.5 gallons from his normal water supply.

To meet the 1 mg per gallon standard the bottler must add at least 18,000,000 gallons from his normal water supply.

The bottler could not meet the 0.3 mg per gallon standard.

Exercise II.6.4: $t > 23.3333$.

Exercise II.6.6:

There is one critical point at $x = 1$. This is the global maximum. There are no other local maxima or minima.

$$\lim_{x \to +\infty} \frac{1}{x^2 + (2-x)^2} = 0.$$

$$\lim_{x \to -\infty} \frac{1}{x^2 + (2-x)^2} = 0.$$

Figure A.17 shows a sketch of this graph.

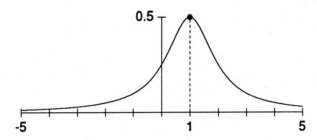

Figure A.17: $y = 1/(x^2 + (2-x)^2)$

Exercise II.6.7:

There are three critical points, at $x = -\sqrt{1/2}$, $x = 0$, and $x = \sqrt{1/2}$.

$x =$	$y =$
$-\sqrt{2}$	$-1/4$
0	0
$\sqrt{2}$	$-1/4$

There is one local maximum, at $x = 0$, and two global minimums, at $x = -\sqrt{1/2}$ and $x = \sqrt{1/2}$.

$$\lim_{x \to +\infty} x^4 - x^2 = +\infty$$

$$\lim_{x \to -\infty} x^4 - x^2 = +\infty$$

Figure A.18 shows a sketch of this graph.

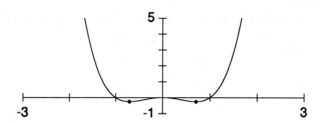

Figure A.18: $y = x^4 - x^2$

Exercise II.6.8:

There are two critical points, at $x = -2$ and $x = 2$.

$x =$	$y =$
-2	$-1/4$
2	$1/4$

$$\lim_{x \to +\infty} \frac{x}{x^2 + 4} = 0$$

$$\lim_{x \to -\infty} \frac{x}{x^2 + 4} = 0$$

The point $x = -2$ is the global minimum and the point $x = 2$ is the global maximum. Figure A.19 shows a sketch of this graph.

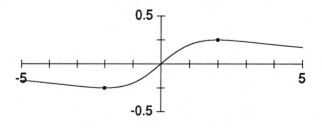

Figure A.19: $y = x/(x^2 + 4)$

Exercise II.6.9:
There is one critical point, $x = 0$, which is the global minimum.

$$\lim_{x \to +\infty} \frac{x^2}{x^2 + 4} = 1$$

$$\lim_{x \to -\infty} \frac{x^2}{x^2 + 4} = 1$$

Figure A.20 shows a sketch of this graph.

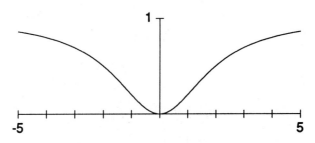

Figure A.20: $y = x^2/(x^2 + 4)$

Exercise II.6.10:
There is one critical point, at $x = 0$, but the derivative is positive on both sides of the critical point and the function is increasing.

$$\lim_{x \to +\infty} \frac{x^3}{x^2 + 4} = +\infty$$

$$\lim_{x \to -\infty} \frac{x^3}{x^2 + 4} = -\infty$$

In fact, since

$$\frac{x^3}{x^2 + 4} = x - \frac{4x}{x^2 + 4},$$

as x gets very large (either positive or negative) y gets very close to x.
Figure A.21 shows a sketch of this graph.

Exercise II.6.11: $x = -1.324718$.

Exercise II.6.12: $x = -0.682328$.

Exercise II.6.13: $x = -0.0668657$.

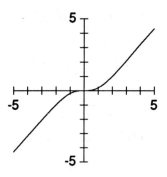

Figure A.21: $y = x^3/(x^2 + 4)$

Exercise II.6.14: $x = 0.100507$.

Exercise II.6.15: $x = 0, \pm 1.89549$.

Exercise II.6.16: $x = 0, \pm 2.47458$.

Exercise II.6.17: $x = 0, \pm 7.95732$.

Exercise II.6.18: $x = 12.3$.

Exercise II.6.19: The median of a_1, a_2, and a_3.

Exercise II.6.20: Any number between the two middle values.

Exercise II.6.21: If n is odd the middle value minimizes the sum. If n is even any number between the two middle values will minimize the sum.

Exercise II.6.22: $x = 12.266667$.

Exercise II.6.23: The average $(a_1 + a_2 + a_3)/3$.

Exercise II.6.24: The average $(a_1 + a_2 + \cdots + a_n)/n$.

Exercise II.7.1:
The function $f(x) = x^3 + x + 1$ is one-to-one. Figure A.22 shows f and its inverse. The domain of f^{-1} is the whole real line.

Exercise II.7.2:
The function $f(x) = x^3 - x + 1$ is not one-to-one. It has no inverse.

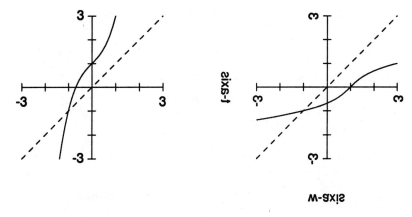

Figure A.22: $f(x) = x^3 + x + 1$ and its inverse

Figure A.23 shows the function $f(x) = x^3 - x + 1$.

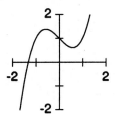

Figure A.23: $f(x) = x^3 - x + 1$

Exercise II.7.3:
The function $f(x) = 2^x$ is one-to-one. The domain of its inverse is the set $(0, +\infty)$. Figure A.24 shows $f(x)$ and its inverse.

Exercise II.7.4:
Let
$$x = y^3$$
So
$$\frac{dx}{dy} = 3y^2$$
and
$$\frac{dy}{dx} = \frac{1}{dx/dy}$$

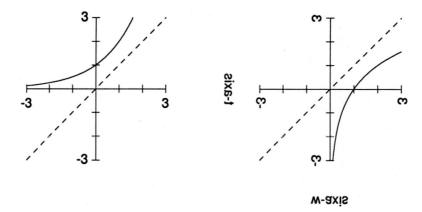

Figure A.24: $f(x) = 2^x$ and its inverse

$$= \frac{1}{3y^2}$$

$$= \frac{1}{3}y^{-2}$$

$$= \frac{1}{3}(x^{\frac{1}{3}})^{-2}$$

$$= \frac{1}{3}x^{-\frac{2}{3}}$$

Exercise II.7.6:

Let
$$x = y^q$$

So
$$\frac{dx}{dy} = qy^{q-1}$$

and
$$\frac{dy}{dx} = \frac{1}{dx/dy}$$

$$= \frac{1}{qy^{q-1}}$$

$$= \frac{1}{q}y^{1-q}$$

$$= \frac{1}{q}(x^{\frac{1}{q}})^{1-q}$$

$$= \frac{1}{q}x^{(\frac{1}{q}-1)}$$

Exercise II.7.7:

$$f' = \frac{5}{9}$$

$$f^{-1}(t) = 32 + \frac{9}{5}t$$

$$(f^{-1})' = \frac{9}{5}$$

Exercise II.7.9:

$$-\frac{1}{|x|\sqrt{x^2-1}}.$$

Exercise II.7.10:

$$x = 0.8481 + 2k\pi,\ 2.2935 + 2k\pi, \quad k = 0, \pm 1, \pm 2, \ldots.$$

Exercise II.7.12:

$$x = 0.896055 + k\pi, \quad k = 0, \pm 1, \pm 2, \ldots.$$

Exercise II.7.13:

$$x = \pi/4 + k\pi, \quad k = 0, \pm 1, \pm 2, \ldots.$$

Exercise II.7.15: $y' = 0.$

Exercise II.7.16:

$$y' = \frac{2}{\sqrt{1-x^2}}.$$

Exercise II.8.1: $(g \circ f)(4) = 17$, $(g \circ f)(9) = 82$, $(f \circ g)(4) = 25$.

Exercise II.8.2: $(c \circ f)(t)$ or $c(f(t))$.

Exercise II.8.3: $p \circ (c^{-1})$.

Exercise II.8.4: $z = \sqrt{y}$ where $y = x^2 + 2x + 4$.

Exercise II.8.6: $g(f(x)) = (x^2 - 4)^3$.

Exercise II.8.7: $g(f(x)) = x$ for any x in the domain of the function f.

Exercise II.8.8: $dz/dx = 8(12 - 3x^7)^7(-21x^6)$.

Exercise II.8.9: $dy/dx = 8(12 - 3x^7)^7(-21x^6)$.

Exercise II.8.11:

$$\frac{dy}{dx} = -\frac{x}{\sqrt{1-x^2}} + \frac{x}{\sqrt{1+x^2}}$$

Exercise II.8.12:

$$\frac{dy}{dx} = \frac{1}{2}\sqrt{\frac{1-x}{1+x^2}}\left(\frac{2x - x^2 + 1}{(1-x)^2}\right)$$

Exercise II.8.14: $3\cos(3x + 1)$.

Exercise II.8.16: $2x \tan(1 - 2x^3) - 6x^4 \sec^2(1 - 2x^3)$.

Exercise II.8.18: $(-\sin x)(\cos(\cos x))$.

Exercise II.8.20: $-3x^2(\sin x^3)\cos(\cos(x^3))$.

Exercise II.8.21:

$$\frac{dy}{dx} = \frac{2}{3}x^{-\frac{1}{3}}$$

Exercise II.8.22:

$$\frac{dy}{dx} = -\frac{3}{2}(1-2x)^{-\frac{1}{4}}$$

Exercise II.8.25: The water treatment plant should be located at the point that is four miles from the point on the Green River which is even with Elmville.

Exercise II.8.26: The water treatment plant should be located in Oakville.

Exercise II.8.27: The water treatment plant should be located at the point that is $2/\sqrt{3}$ miles from the point on the Green River which is even with Elmville and Mapleville.

Exercise II.8.28: As long as A is at least $2/\sqrt{3}$ miles then the answer is the same as in the preceding problem. If a is less than $2/\sqrt{3}$ then the water treatment plant should be located in Oakville.

Exercise II.8.29:

From Figure A.25 we see that there is no 2-cycle. The table below shows the population converging to the equilibrium point 600.

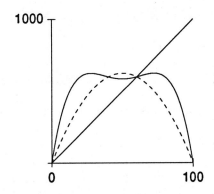

Figure A.25: Looking for 2-cycles of $p_{n+1} = 2.5p_n(1 - .001p_n)$

Year	Population	Year	Population
1	50.00	11	600.40
2	118.75	12	599.80
3	261.62	13	600.10
4	482.94	14	599.95
5	624.27	15	600.03
6	586.39	16	599.99
7	606.34	17	600.01
8	596.73	18	600.00
9	601.61	19	600.00
10	599.19	20	600.00

Exercise II.8.30:

From Figure A.26 we see that there is a 2-cycle. The table below shows the population "converging" to this 2-cycle.

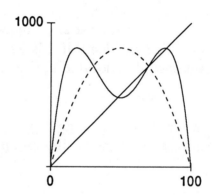

Figure A.26: Looking for 2-cycles of $p_{n+1} = 3.3 p_n (1 - .001 p_n)$

Year	Population	Year	Population
1	50.00	11	479.18
2	156.75	12	823.57
3	436.19	13	479.50
4	811.56	14	823.61
5	504.66	15	479.41
6	824.93	16	823.60
7	476.59	17	479.43
8	823.19	18	823.60
9	480.31	19	479.43
10	823.72	20	823.60

Exercise II.8.31:

From Figure A.27 we see that there is a 2-cycle. The table below shows the population "converging" to this 2-cycle.

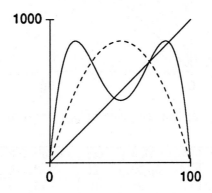

Figure A.27: Looking for 2-cycles of $p_{n+1} = 3.4p_n(1 - .001p_n)$

Year	Population	Year	Population	Year	Population	Year	Population
1	50.00	11	454.66	21	451.29	31	452.13
2	161.50	12	843.01	22	841.93	32	842.21
3	460.42	13	449.97	23	452.47	33	451.83
4	844.67	14	841.49	24	842.32	34	842.11
5	446.08	15	453.51	25	451.58	35	452.06
6	840.11	16	842.65	26	842.03	36	842.19
7	456.69	17	450.81	27	452.26	37	451.89
8	843.62	18	841.77	28	842.25	38	842.13
9	448.54	19	452.85	29	451.74	39	452.02
10	841.00	20	842.44	30	842.08	40	842.17

Exercise II.8.32:

From Figure A.28 we see that there is a 2-cycle. The table below, however, shows the population converging to a 4-cycle.

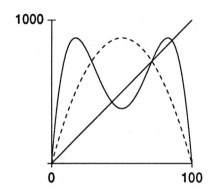

Figure A.28: Looking for 2-cycles of $p_{n+1} = 3.5p_n(1 - .001p_n)$

Year	Population	Year	Population
1	50.00	11	500.75
2	166.25	12	875.00
3	485.14	13	382.82
4	874.23	14	826.94
5	384.84	15	500.89
6	828.58	16	875.00
7	497.12	17	382.82
8	874.97	18	826.94
9	382.89	19	500.88
10	827.00	20	875.00

Exercise II.8.33:

From Figure A.29 we see that there is a 2-cycle. It is very hard to discern any longterm pattern on the basis of the table below.

Year	Population	Year	Population	Year	Population	Year	Population
1	50.00	11	391.11	21	598.35	31	522.99
2	171.00	12	857.31	22	865.18	32	898.10
3	510.33	13	440.38	23	419.92	33	329.47
4	899.62	14	887.20	24	876.91	34	795.31
5	325.11	15	360.27	25	388.57	35	586.06
6	789.88	16	829.71	26	855.30	36	873.34
7	597.48	17	508.66	27	445.55	37	398.22
8	865.79	18	899.73	28	889.33	38	862.71
9	418.31	19	324.78	29	354.33	39	426.39
10	875.98	20	789.47	30	823.61	40	880.49

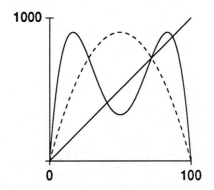

Figure A.29: Looking for 2-cycles of $p_{n+1} = 3.6p_n(1 - .001p_n)$

Exercise II.9.1: $-4/3$.

Exercise II.9.8: The slope of the outgoing line will be negative if $0 < m < 1$ or $m < -1$. The slope of the outgoing line will be positive if $1 < m$ or $-1 < m < 0$.

Exercise II.9.9: $y = 1.8074x + 3.8688$

Exercise II.9.13: $y = .75x + .25$

Exercise II.9.20: The y-intercept for every outgoing ray is 0.25.

Proof:

For the function $f(x) = x^2$ the slope of the outgoing ray that bounces off the mirror at the point (x_0, x_0^2) is

$$\begin{aligned} m &= \frac{f'(x)^2 - 1}{2f'(x)} \\ &= \frac{(2x_0)^2 - 1}{2(2x_0)} \\ &= \frac{4x_0^2 - 1}{4x_0}. \end{aligned}$$

Thus the equation describing the outgoing line is

$$y = \left(\frac{1 - 4x_0^2}{4x_0}\right)x + b$$

We determine the value of b by using the fact that the outgoing ray passes through the point (x_0, x_0^2).

$$x_0^2 = \left(\frac{4x_0^2 - 1}{4x_0}\right)x_0 + b$$

$$x_0^2 = \frac{4x_0^2 - 1}{4} + b$$

$$x_0^2 = x_0^2 - \frac{1}{4} + b$$

$$b = \frac{1}{4}$$

This says that all the light that comes in to the mirror from a distant source directly overhead will be focused on one point, the point $(0, 0.25)$. This is a special property of this mirror shape. The shape is sometimes called a **parabola** and a mirror with this shape is called a **parabolic mirror**. Such mirrors can be used to focus the sun's rays to make use of solar energy. They are also used in astronomical mirrors.

Exercise II.9.21: 0.7219 feet.

Exercise II.9.23: 0.2894 feet.

Exercise II.9.25: 0.75 feet.

Exercise II.9.27: $(82.645, -15.05)$.

Exercise II.9.28: $(86.49, -18.078)$.

Exercise II.9.29:
The total distance traveled by a light ray following the path shown in Figure II.106 is

$$T(x) = \sqrt{a^2 + x^2} + \sqrt{(L-x)^2 + b^2}$$

To find the minimum we proceed as follows.

$$T' = \frac{x}{\sqrt{a^2 + x^2}} - \frac{L-x}{\sqrt{(L-x)^2 + b^2}}$$

$$0 = \frac{x}{\sqrt{a^2 + x^2}} - \frac{L-x}{\sqrt{(L-x)^2 + b^2}}$$

$$\frac{x}{\sqrt{a^2 + x^2}} = \frac{L-x}{\sqrt{(L-x)^2 + b^2}}$$

which shows that the two triangles are similar, so the angle of coincidence is equal to the angle of incidence.

Exercise II.9.30: You would see two reflections, at the two points where the walls intersect the x-axis.

Exercise II.9.31: Suppose that a light ray is traveling from one medium in which its velocity is v_1 to another in which its velocity is v_2 as shown in Figure A.30. The Snell's Law says

$$\frac{\sin \alpha}{v_1} = \frac{\sin \beta}{v_2}.$$

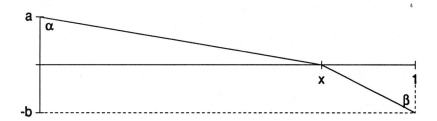

Figure A.30: Snell's Law

The total time required for a light ray to traverse the path shown in Figure A.30 is

$$T(x) = \frac{\sqrt{a^2 + x^2}}{v_1} + \frac{\sqrt{b^2 + (1-x)^2}}{v_2}.$$

A light ray will follow a path for which $T(x)$ is a minimum or maximum, hence, $T'(x) = 0$. This leads to the equation

$$T'(x) = \frac{x}{v_1\sqrt{a^2+x^2}} - \frac{1-x}{v_2\sqrt{b^2+(1-x)^2}} = 0.$$

leading to

$$\frac{\sin\alpha}{v_1} = \frac{\sin\beta}{v_2}$$

which is Snell's Law.

Exercise II.10.1: 0.4070042475.

Exercise II.10.2: 1.324717957.

Exercise II.10.3: 0.6823278038.

Exercise II.10.4:
Let $f(x) = x^2 - a$. We want to solve $f(x) = 0$

$$\begin{aligned}
x_{n+1} &= x_n - \frac{f(x_n)}{f'(x_n)} \\
&= x_n - \frac{x_n^2 - a}{2x_n} \\
&= x_n - \frac{x_n}{2} + \frac{a}{2x_n} \\
&= \frac{x_n}{2} + \frac{a}{2x_n} \\
&= \frac{x_n + \frac{a}{x_n}}{2}
\end{aligned}$$

Exercise II.10.7:
To find equilibrium points of this dynamical system we solve the equation $x = F(x)$ leading to

$$x = x - \frac{f(x)}{f'(x)}$$

$$\frac{f(x)}{f'(x)} = 0$$
$$f(x) = 0.$$

So equilibrium points of this dynamical system are solutions of the original equation.

To check whether an equilibirum point x_* is attracting we examine $F'(x_*)$.

$$F'(x) = 1 - \frac{f'(x)f'(x) - f''(x)f(x)}{(f'(x))^2}$$

and at an equilibrium point x_* we get, since $f(x_*) = 0$, $F'(x_*) = 0$. Thus not only is the equilibrium point attracting since $|F(x_*)| < 1$ but it is attracting very fast since $|F(x_*)| = 0$.

Exercise II.11.1: The equilibrium point 0 is stable. In all the following models the limit is 0.

Year	Pop	Pop	Pop	Pop
1	10.00	70.00	130.00	190.00
2	8.55	40.95	40.95	8.55
3	7.37	29.31	29.31	7.37
4	6.39	22.51	22.51	6.39
5	5.56	17.98	17.98	5.56
6	4.87	14.73	14.73	4.87
7	4.27	12.28	12.28	4.27
8	3.76	10.37	10.37	3.76
9	3.32	8.85	8.85	3.32
10	2.94	7.61	7.61	2.94
11	2.61	6.59	6.59	2.61
12	2.32	5.74	5.74	2.32
13	2.06	5.01	5.01	2.06
14	1.84	4.40	4.40	1.84
15	1.64	3.87	3.87	1.64
16	1.46	3.42	3.42	1.46
17	1.31	3.02	3.02	1.31
18	1.17	2.68	2.68	1.17
19	1.04	2.38	2.38	1.04
20	.94	2.12	2.12	.94

Exercise II.11.2: The equilibrium point 0 is unstable. In all the following models the population is not approaching 0.

Year	Pop	Pop	Pop	Pop
1	10.00	70.00	130.00	190.00
2	10.45	50.05	50.05	10.45
3	10.89	41.28	41.28	10.89
4	11.33	36.03	36.03	11.33
5	11.76	32.50	32.50	11.76
6	12.17	29.94	29.94	12.17
7	12.58	28.00	28.00	12.58
8	12.96	26.49	26.49	12.96
9	13.34	25.28	25.28	13.34
10	13.69	24.29	24.29	13.69
11	14.03	23.48	23.48	14.03
12	14.35	22.79	22.79	14.35
13	14.65	22.21	22.21	14.65
14	14.94	21.72	21.72	14.94
15	15.20	21.30	21.30	15.20
16	15.45	20.93	20.93	15.45
17	15.68	20.62	20.62	15.68
18	15.90	20.34	20.34	15.90
19	16.10	20.10	20.10	16.10
20	16.28	19.89	19.89	16.28

Exercise II.11.3:

This model is of the form

$$p_{n+1} = f(p_n)$$

where

$$f(p) = R(p)p.$$

Using the Product Rule to compute $f'(p)$ we see that

$$\begin{aligned} f'(p) &= R(p) + R'(p)p \\ f'(0) &= R(0) + R'(0)(0) \\ f'(0) &= R(0). \end{aligned}$$

So if $R(0) < 1$ then 0 is a stable equilibrium. This is to be expected since near zero, R will be less than 1 and the population will be decreasing toward zero.

Exercise II.11.4: The equilibrium point $1.9/.0145$ is attracting. In all the following examples the sequence is converging to this point. Notice that *in this case* the initial population doesn't seem to make any difference.

Year	Pop	Pop	Pop	Pop
1	10.00	70.00	130.00	190.00
2	27.55	131.95	131.95	27.55
3	68.89	130.20	130.20	68.89
4	130.97	131.78	131.78	130.97
5	131.10	130.36	130.36	131.10
6	130.98	131.64	131.64	130.98
7	131.08	130.49	130.49	131.08
8	130.99	131.52	131.52	130.99
9	131.07	130.59	130.59	131.07
10	131.00	131.43	131.43	131.00
11	131.07	130.68	130.68	131.07
12	131.00	131.36	131.36	131.00
13	131.06	130.74	130.74	131.06
14	131.01	131.29	131.29	131.01
15	131.06	130.80	130.80	131.06
16	131.02	131.24	131.24	131.02
17	131.05	130.84	130.84	131.05
18	131.02	131.21	131.21	131.02
19	131.05	130.88	130.88	131.05
20	131.02	131.17	131.17	131.02

Exercise II.11.5: The equilibrium point $2.1/.0155$ is unstable. Thus the behavior shown in the following examples is not surprising.

Year	Pop	Pop	Pop	Pop
1	10.00	70.00	130.00	190.00
2	29.45	141.05	141.05	29.45
3	77.85	128.88	128.88	77.85
4	147.40	142.07	142.07	147.40
5	120.18	127.57	127.57	120.18
6	148.69	143.22	143.22	148.69
7	118.26	126.04	126.04	118.26
8	149.83	144.49	144.49	149.83
9	116.51	124.32	124.32	116.51
10	150.78	145.83	145.83	150.78
11	115.04	122.44	122.44	115.04
12	151.49	147.19	147.19	151.49
13	113.90	120.48	120.48	113.90
14	152.01	148.50	148.50	152.01
15	113.08	118.54	118.54	113.08
16	152.35	149.67	149.67	152.35
17	112.52	116.76	116.76	112.52
18	152.57	150.65	150.65	152.57
19	112.17	115.24	115.24	112.17
20	152.71	151.40	151.40	152.71

Exercise II.11.6:

The nonzero equilibrium point is

$$1000\sqrt{\frac{a-1}{a}}$$

which positive and, hence, of interest if $a > 1$. This equilibrium point is stable if $1 < a < 2$. The following calculations for $a = 1.9$ reflect this.

Year	Pop	Pop	Pop	Pop
1	50.00	200.00	800.00	950.00
2	94.76	364.80	547.20	175.99
3	178.43	600.88	728.37	324.02
4	328.23	729.46	649.71	551.00
5	556.45	648.48	713.36	729.06
6	729.89	713.98	665.65	648.93
7	647.99	665.03	704.34	713.75
8	714.22	704.73	674.35	665.26
9	664.79	673.99	698.61	704.59
10	704.88	698.86	679.53	674.12
11	673.85	679.31	694.92	698.77
12	698.96	695.08	682.73	679.39
13	679.23	682.59	692.54	695.03
14	695.15	692.65	684.74	682.64
15	682.54	684.65	691.01	692.61
16	692.69	691.07	686.01	684.68
17	684.62	685.96	690.02	691.05
18	691.10	690.06	686.82	685.98
19	685.93	686.78	689.38	690.04
20	690.08	689.41	687.33	686.80

The following calculations for $a = 2.1$ reflect the fact that the nonzero equilibrium point is unstable.

Year	Pop	Pop	Pop	Pop
1	50.00	200.00	800.00	950.00
2	104.74	403.20	604.80	194.51
3	217.54	709.07	805.51	393.02
4	435.21	740.39	594.01	697.86
5	740.83	702.51	807.27	751.79
6	701.90	747.20	590.48	686.46
7	747.80	693.07	807.66	762.26
8	692.21	756.33	589.71	670.64
9	757.12	679.74	807.73	774.93
10	678.54	767.90	589.56	650.10
11	768.87	661.69	807.74	788.23
12	660.12	781.16	589.54	626.85
13	782.18	639.43	807.75	799.12
14	637.64	793.77	589.53	606.49
15	794.61	616.64	807.75	805.15
16	615.07	802.55	589.53	594.72
17	803.00	599.84	807.75	807.18
18	598.95	806.43	589.53	590.66
19	806.57	592.18	807.75	807.64
20	591.88	807.48	589.53	589.74

Exercise II.11.7:

In this model
$$p_{n+1} = f(p_n) = a p_n R(p_n)$$
where
$$R(p) = \begin{cases} .0243p, & \text{if } p \leq 100; \\ 2.7(1 - .001p), & \text{otherwise.} \end{cases}$$

Zero is an equilibrium point and
$$f'(0) = 0.$$

So 0 is always attracting.

Looking for nonzero equilibrium points we begin with the first clause of the defintion of $R(p)$ and solve

$$\begin{aligned} p &= ap.0243p \\ 1 &= .0243ap \\ p &= 1/.0243a \end{aligned}$$

We will denote this equilibrium point p_u. That is, $p_u = 1/.0243a$. This equilibrium point only makes sense if it is less than 100 so that we are in the range of the first clause in the definition of $R(p)$.

$$\begin{aligned} 1/.0243a &< 100 \\ 1/a &< 2.43 \\ a &> 1/2.43 \end{aligned}$$

Now if a is in this range we compute $f'(p_u)$ as follows

$$\begin{aligned} f(p) &= .0243ap^2 \\ f'(p) &= .0486ap \\ f'(p_u) &= .0486a(1/.0243a) \\ &= 2. \end{aligned}$$

So this equilibrium point is always unstable.

To find the other nonzero equilibrium point we look at the second clause in the definition of $R(p)$ and solve

$$\begin{aligned} p &= ap(2.7 - .001p) \\ p &= 2.7ap - .0027ap^2 \\ 1 &= 2.7a - .0027ap \\ .0027ap &= 2.7a - 1 \\ p &= (2.7a - 1)/.0027a \end{aligned}$$

This equilibrium point

$$p_* = \frac{2.7a - 1}{.0027a}$$

only makes sense if it is greater than 100, so that we are in the range of the second clause in the definition of $R(p)$.

$$\frac{2.7a - 1}{.0027a} > 100$$

$$2.7a - 1 > .27a$$
$$2.43a > 1$$
$$a > 1/2.43$$

Now in this range

$$\begin{aligned} f(p) &= 2.7ap(1 - .001p) \\ f(p) &= 2.7ap - .0027ap^2 \\ f'(p) &= 2.7a - .0054ap \\ f'(p_*) &= 2.7a - .0054a\left(\frac{2.7a - 1}{.0027a}\right) \\ &= 2.7a - 2(2.7a - 1) \\ &= 2 - 2.7a \end{aligned}$$

So p_* is stable if

$$\begin{aligned} -1 &< 2 - 2.7a < 1 \\ -3 &< -2.7a < 1 \\ 3 &> 2.7a > 1 \\ 3/2.7 &> a > 1/2.7 \end{aligned}$$

But, since this equilibrium point only comes into play when $a > 1/2.43$ we see that this equilibrium point exists and is stable if

$$1/2.43 < a < 3/2.7$$

Summarizing all this work we see

- The equilibrium point 0 is always stable.
- If $a > 1/2.43$ there are two nonzero equilibrium points

$$p_u = 1/.0243a$$

and

$$p_* = \frac{2.7a - 1}{.0027a}$$

- The nonzero equilibrium point p_u is always unstable.

- The nonzero equilibrium point p_* is stable if $1/2.43 < a < 3/2.7$ and unstable if $3/2.7 < a$.

The following calculations illustrate what happens when $a = 1$.

Year	Pop	Pop	Pop	Pop
1	20.00	40.00	60.00	80.00
2	9.72	38.88	87.48	155.52
3	2.30	36.73	185.96	354.60
4	.13	32.79	408.73	617.92
5	.00	26.12	652.51	637.46
6	.00	16.58	612.20	623.99
7	.00	6.68	641.01	633.49
8	.00	1.09	621.31	626.88
9	.00	.03	635.26	631.53
10	.00	.00	625.60	628.29
11	.00	.00	632.41	630.56
12	.00	.00	627.67	628.97
13	.00	.00	630.99	630.09
14	.00	.00	628.67	629.31
15	.00	.00	630.30	629.85
16	.00	.00	629.16	629.47
17	.00	.00	629.96	629.74
18	.00	.00	629.40	629.55
19	.00	.00	629.79	629.68
20	.00	.00	629.52	629.59

The following calculations illustrate what happens when $a = 1.2$.

Year	Pop	Pop	Pop	Pop
1	20.00	40.00	60.00	80.00
2	11.66	46.66	104.98	186.62
3	3.97	63.47	304.42	491.82
4	.46	117.49	686.06	809.78
5	.01	335.94	697.83	499.07
6	.00	722.79	683.19	810.00
7	.00	649.18	701.27	498.64
8	.00	737.89	678.75	809.99
9	.00	626.64	706.48	498.65
10	.00	758.04	671.87	809.99
11	.00	594.26	714.29	498.65
12	.00	781.21	661.22	809.99
13	.00	553.78	725.79	498.65
14	.00	800.63	644.82	809.99
15	.00	517.18	742.05	498.65
16	.00	809.04	620.18	809.99
17	.00	500.55	763.20	498.65
18	.00	810.00	585.55	809.99
19	.00	498.64	786.29	498.65
20	.00	809.99	544.44	809.99

Exercise II.12.1: $y' = 15x^2 + \frac{1}{2\sqrt{x}}$

Exercise II.12.2: $A' = 2\pi r$

Exercise II.12.3: $f'(t) = \frac{8}{3}t^{-\frac{1}{3}} - 2 - 2t^{-3}$

Exercise II.12.4:

$$D'(p) = \frac{-2p}{(1+p^2)^2}$$

Exercise II.12.5:

$$y' = 4(x - \sqrt{x})^3 \left(1 - \frac{1}{2\sqrt{x}}\right)$$

Exercise II.12.6:

$$\frac{dz}{dx} = -\frac{1}{2t^2}\sqrt{\frac{t}{t+1}}$$

Exercise II.12.7:
$$\frac{dy}{dx} = \frac{1}{2}\sqrt{\frac{2x^3 - 12x + 3}{1 - x^2}} \left(\frac{2x^4 + 6x^2 - 6x + 12}{(2x^3 - 12x + 3)^2} \right)$$

Exercise II.12.8:
$$\frac{dz}{dx} = \frac{2}{5} \left(\frac{1 - 2x^5}{x + x^2} \right)^{-\frac{3}{5}} \left(\frac{-6x^6 - 8x^5 - 2x - 1}{(x + x^2)^2} \right)$$

Exercise II.12.9:
$$f'(t) = \frac{4x - 10}{2\sqrt{2x^2 - 10x + 17}} + \frac{1}{2}\sqrt{\frac{1 + x}{1 - x}} \left(\frac{-2}{(1 + x)^2} \right)$$

Exercise II.12.10:
$$\frac{-3}{2\sqrt{x} - 8x + 8x^{\frac{3}{2}}}$$

Exercise II.12.11: $f'(\theta) = -2\theta \sin(\theta^2 + 1)$.

Exercise II.12.12: $y' = -(\sin x)(\cos(\cos x))$.

Exercise II.12.13: $y' = 3\sec^2 x(4 + \tan x)^2$.

Exercise II.12.14:
$$y' = \frac{\cos x}{1 + \sin^2 x}.$$

Exercise II.12.15:
$$w' = \frac{4}{\sqrt{1 - (4x + 1)^2}}.$$

Exercise II.12.16: 0.

Exercise II.12.17:
The derivative is always positive. Hence, the function is increasing, the minimum is at the left endpoint, and the maximum is at the right endpoint. Figure A.31 shows a sketch of the graph.

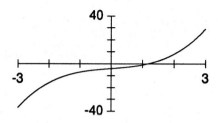

Figure A.31: $y = x^3 + 2x - 4$

Exercise II.12.18:

This problem is similar to the preceding one except that we add the information that
$$\lim_{x \to +\infty} x^3 + 2x - 4 = +\infty$$
and
$$\lim_{x \to -\infty} x^3 + 2x - 4 = -\infty.$$
There is no global maximum or global minimum.

Exercise II.12.19:

There is a local minimum at $\sqrt{2/3}$ and a local maximum at $-\sqrt{2/3}$. The global minimum is at the left endpoint and the global maximum is at the right endpoint. Figure A.32 shows a sketch of the graph.

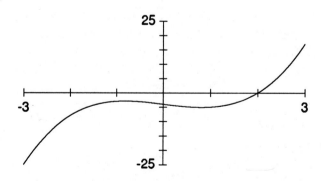

Figure A.32: $y = x^3 - 2x - 4$

Exercise II.12.20:
This problem is similar to the preceding one except that we add the information that
$$\lim_{x \to +\infty} x^3 + 2x - 4 = +\infty$$
and
$$\lim_{x \to -\infty} x^3 + 2x - 4 = -\infty.$$
There is no global maximum or global minimum.

Exercise II.12.21:
There is one critical point, at $x = 0$, but the derivative is negative everywhere else. hence, the function is decreasing. The global minimum is at the left endpoint and the global maximum is at the right endpoint. Figure A.33 shows a sketch of this graph.

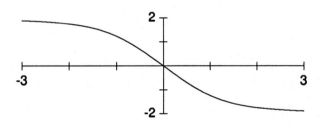

Figure A.33: Graph for Exercise II.10.15

Exercise II.12.22:
This problem is similar to the preceding one except there is no global minimum or global maximum. It is possible to show that
$$\lim_{x \to +\infty} \sqrt{(x-1)^2 + 1} - \sqrt{(x+1)^2 + 1} = -2$$
and
$$\lim_{x \to -\infty} \sqrt{(x-1)^2 + 1} - \sqrt{(x+1)^2 + 1} = 2$$
by multiplying by
$$\frac{\sqrt{(x-1)^2 + 1} + \sqrt{(x+1)^2 + 1}}{\sqrt{(x-1)^2 + 1} + \sqrt{(x+1)^2 + 1}}$$
and doing some algebra.

Exercise II.12.23:
There is a global minimum at $x = -1$ and a global maximum at $x = 1$.
$$\lim_{x \to +\infty} \frac{2x}{x^2 + 1} = 0$$
and
$$\lim_{x \to -\infty} \frac{2x}{x^2 + 1} = 0.$$
Figure A.34 shows a sketch of this graph.

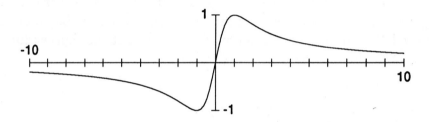

Figure A.34: $y = 2x/(x^2 + 1)$

Exercise II.12.24:
There is a global minimum at $x = 0$. There is no global maximum.
$$\lim_{x \to +\infty} \frac{2x^2}{x^2 + 1} = 2$$
and
$$\lim_{x \to -\infty} \frac{2x^2}{x^2 + 1} = 2.$$
Figure A.35 shows a sketch of this graph.

Exercise II.12.25:
There is one critical point, at $x = 0$, but the derivative is positive everywhere else and the function is increasing. There is no global maximum or global minimum.
$$\lim_{x \to +\infty} \frac{2x^3}{x^2 + 1} = +\infty$$
and
$$\lim_{x \to -\infty} \frac{2x^3}{x^2 + 1} = -\infty.$$

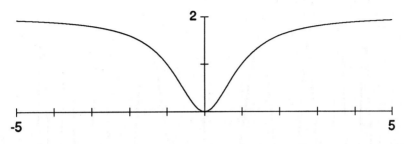

Figure A.35: $y = 2x/(x^2+1)$

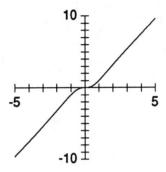

Figure A.36: $y = 2x^3/(x^2+1)$

Figure A.36 shows a sketch of this graph.

Exercise II.12.26: See Figure A.37

Exercise II.12.27: See Figure A.38

Exercise II.12.28:
She should run on the sand to the point (80,0) on the shoreline.

Exercise II.12.29:
It is a square the length of whose sides is $R/\sqrt{2}$.

Exercise II.12.30:
The point on the curve $y = 2x - 1$ closest to the origin is $(0.4, -0.2)$.

Exercise II.12.31:
The point on the curve $y = mx + b$ closest to the origin is given by

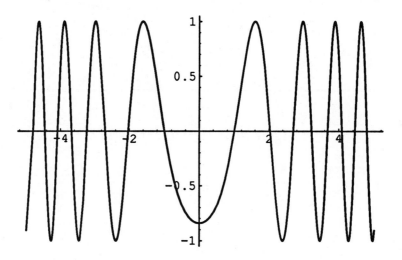

Figure A.37: $y = \sin(x^2 - 1)$

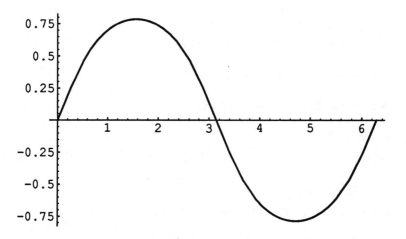

Figure A.38: $y = \arctan(\sin x))$

$x = -mb/(1+m^2)$.
Now
$$\frac{y}{x} = \frac{mx+b}{x}$$
$$= m + \frac{b}{x}$$
$$= m + \frac{b}{-mb/(1+m^2)}$$
$$= m - \frac{1+m^2}{m}$$
$$= -\frac{1}{m}.$$

So the line from the origin to this point is perpendicular to the original line.

Exercise II.12.32:

We find the point on the curve closest to the origin by minimizing
$$T = \sqrt{x^2 + f(x)^2}$$
or, equivalently, by minimizing
$$S = x^2 + f(x)^2$$
Now
$$S'(x) = 2x + 2f(x)f'(x)$$
and setting this equal to zero we get
$$x = -f(x)f'(x).$$
So at this point
$$\frac{f(x)}{x} = -\frac{1}{f'(x)}$$
which shows that the line from the origin to the closest point on the curve is perpendicular to the curve.

Exercise II.12.33: The base should be $\sqrt[3]{2} \times \sqrt[3]{2}$ and the height should be $1/\sqrt[3]{4}$.

Exercise II.12.34: The box should be a cube 1 foot by 1 foot by 1 foot.

Exercise II.12.35: -0.8631233415

Exercise II.12.36: 1.24905

Exercise II.12.37: 40 feet.

Exercise II.12.38:
This problem requires lots of algebra. First find points in the 2-cycle by solving the equation
$$p = f(f(p))$$
where
$$f(p) = ap(1 - bp).$$
This is a fourth degree equation.
$$a^3b^3p^4 - 2a^3b^2p^3 + (a^2b + a^3b)p^2 + (1 - a^2)p = 0.$$
Two of the roots are known because they are equilibrium points, namely, 0 and $(a-1)/ab$. Thus, the factors p and $p - (a-1)/b$ can be factored out giving us
$$a^2b^2p^2 - (ab + a^2b)p + (a + 1) = 0$$
which has two solutions:
$$p = \frac{1 + a \pm \sqrt{(a+1)(a-3)}}{2ab}.$$
If $a > 3$ this equation has two distinct real roots.

Notice, parenthetically, that if $a = 3$ then there is one root which is identical to the nonzero equilibrium point $p_* = (a-1)/ab = 2/3b$.

Next evaluate the derivative of $f \circ f$ at either of the points in the 2-cycle. We will use the notation
$$p_{*1} = \frac{1 + a + \sqrt{(a+1)(a-3)}}{2ab}$$
$$p_{*2} = \frac{1 + a - \sqrt{(a+1)(a-3)}}{2ab}$$

Now
$$(f \circ f)'(p) = f'(f(p))f'(p)$$

So
$$\begin{aligned}(f \circ f)')(p_{*1}) &= f'(f(p_{*1}))f'(p_{*1}) \\ &= f'(p_{*2})f'(p_{*1})\end{aligned}$$

where the last line follows from the fact that $f(p_{*1}) = p_{*2}$.

Finally, a straightforward calculation shows that

$$f'(p_{*1})f'(p_{*2}) = -a^2 + 2a + 4$$

which is less than one in absolute value if $3 < a < 3.44949$.

Thus, the 2-cycle is attracting if $3 < a < 3.44949$.

Exercise III.1.1: The weight has returned to its rest position. It is heading downward at its maximum downward (negative) velocity. The forces exerted by gravity and the spring are in balance, so the acceleration is 0.

Exercise III.1.2: The weight is at the bottom of its travels. Its velocity is zero. The spring is stretched very tightly and exerts a force stronger than gravity, so the acceleration is positive.

Exercise III.1.3: The graph is shown in Figure A.39. The curve is concave upward when $x > 1$ and concave downward when $x < 1$.

Exercise III.1.7: The graph is shown in Figure A.40. The curve is concave downward everywhere.

Exercise III.2.1: The motion of the ball is described by the function

$$y(t) = -16t^2 + 120t + 6.$$

The ball will hit the ground at time $t = 7.5497$. Just before it hits the ground its velocity will be -121.5895 feet per second. The highest point in its travels will be 231 feet. The initial velocity in this exercise is twice the initial velocity in the example. Yet the ball goes roughly *four times* as high. If you subtract the height at which the ball was originally thrown the ball goes exactly four times as high.

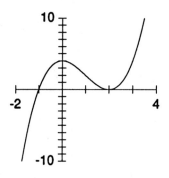

Figure A.39: $y = x^3 - 3x^2 + 4$

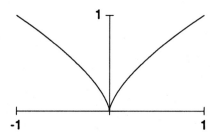

Figure A.40: $y = x^{2/3}$

Exercise III.2.2: The ball will take 2.5 seconds to reach the ground. Just before it hits the ground it will be traveling at 80 feet per second.

Exercise III.2.4: The kinetic energy is proportional to the square of the velocity. This has important implications for the relative safety and energy efficiency of traveling at 55 miles per hour or 65 miles per hour. The amount of damage a car can do depends in part on its kinetic energy. Comparing the kinetic for the two speeds we get a factor of $(65/55)^2 \approx 1.40$. So a car traveling at 65 miles per hour can do 40% more damage than a car traveling at 55 miles per hour. The cost of accelerating a car to a given speed depends in part on the kinetic energy at that speed. Thus the cost of accelerating to 65 miles per hour is roughly 40% higher than the cost of accelerating to 55 miles per hour. These rough comparisons are only a small part of the story. For example, a driver traveling faster has less time to react and will get into more accidents than a slower driver.

Exercise III.2.5: The graph is shown in Figure A.41. The ball will hit the ground 78.58 feet from the point at which it was thrown.

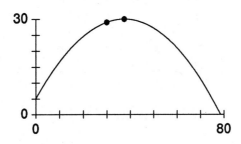

Figure A.41: Exercise III.2.5

Exercise III.2.7: The initial horizontal velocity is 31.8198 feet per second. The initial vertical velocity is also 31.8198 feet per second.

Exercise III.2.8: The projectile will remain in the air for 3.125 seconds. It will hit the ground 270.63 feet from the point at which it was fired.

Exercise III.2.11: The angle of elevation should be $\pi/4$ radians. The projectile will hit the ground 312.50 feet from the point at which it was fired.

Exercise III.2.12: Range $= v_0^2/32$ where v_0 is the initial speed.

Exercise III.2.13: $\pi/2$.

Exercise III.3.1: The slope field is shown in Figure A.42. This differential equation has no equilibrium points. Its solutions are straight lines with a slope of 1.

Exercise III.3.3: The slope field is shown in Figure A.43. This differential equation has three equilibrium points, at $p = 0, 1$, and 2. The equilibrium point $p = 1$ is attracting, the others are repelling. If the initial value of p is between 0 and 2 then $p(t)$ will approach 1. If the initial value of p is 2, $p(t)$ will stay there forever. If the initial value of p is 0, $p(t)$ will stay there. If the initial value of p is greater than 2

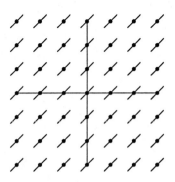

Figure A.42: $p' = 1$

then $p(t)$ will shoot off to $+\infty$. If the initial value of p is less than 0 then $p(t)$ will shoot off to $-\infty$.

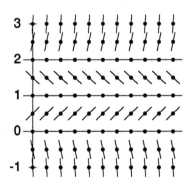

Figure A.43: $p' = p(p-1)(p-2)$

Exercise III.3.4: The slope field is shown in Figure A.44. This differential equation has one equilibrium point, at $p = 1$. It is neither repelling nor attracting. If the initial value of p is less than 1 then $p(t)$ will approach 1. If the initial value of p is greater than 1 then $p(t)$ will shoot off to $+\infty$. If the initial value of p is 1 then $p(t)$ will stay there.

Exercise III.3.6: The slope field is shown in Figure A.45. Any integral multiple of π is an equilibrium point. The odd multiplies of π are attracting and the even multiples of π are repelling. If the initial value of p is on an equilibrium point then p will stay there. Otherwise

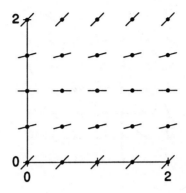

Figure A.44: $p' = (p-1)^2$

$p(t)$ will be attracted to the nearest odd multiple of π.

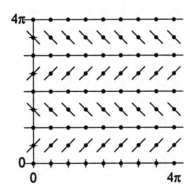

Figure A.45: $p' = \sin p$

Exercise III.3.7: The slope field is shown in Figure A.46

Exercise III.3.9: The slope field is shown in Figure A.47

Exercise III.4.1:

i	t_{i-1}	w_{i-1}	w_i
1	.0000	1.0000	1.2500
2	.2500	1.2500	1.5625
3	.5000	1.5625	1.9531
4	.7500	1.9531	2.4414

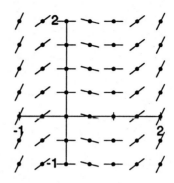

Figure A.46: $p' = t(t-1)$

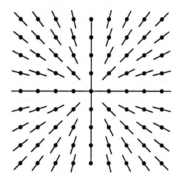

Figure A.47: $p' = p/t$

Exercise III.4.2:

i	t_{i-1}	w_{i-1}	$f(t_{i-1}, w_{i-1})$	w_i
1	1.0	1.0000	1.5000	1.3000
2	1.2	1.3000	1.9000	1.6800
3	1.4	1.6800	2.3800	2.1560
4	1.6	2.1560	2.9560	2.7472
5	1.8	2.7472	3.6472	3.4766

Exercise III.5.1: 19.56%.

Exercise III.5.2: 19.72%

Exercise III.5.3: 19.72%.

Exercise III.5.4: 20.02%.

Exercise III.5.6: You can estimate $y'(0)$ by
$$\frac{y(h) - y(0)}{h}$$
for small h. If $h = 0.0001$ for example, we get 0.6932 which is nowhere near $2^0 = 1$. $y'(1)$ is approximately 1.3863 which is nowhere near $2^1 = 2$.

Exercise III.5.7: 56,953,125

Exercise III.5.9: The effective rate for 6% compounded quarterly is 6.1364%. The effective rate for 5.5% compounded continuously is only 5.6540%, so the first account is better.

Exercise III.6.1: $y' = e^{\sin x} \cos x$

Exercise III.6.7:
$$y' = \frac{2x - 2}{x^2 - 2x}$$

Exercise III.6.13:
$$y' = \frac{1}{1+x} + \frac{1}{1-x}$$

Exercise III.6.14:
$$y' = (\sin x)^x \left(\frac{x \cos x}{\sin x} + \ln \sin x \right)$$

Exercise III.6.17: The graph is shown in Figure A.48. There is a minimum at $x = -1$.

Exercise III.6.19:
$$A(t) = 100(0.9441)^t$$
or
$$A(t) = 100 \left(\frac{1}{2} \right)^{\frac{t}{12.05}}$$

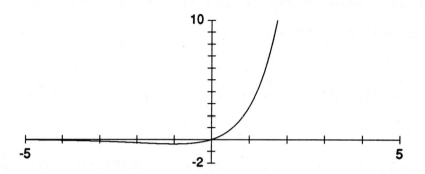

Figure A.48: $y = xe^x$

Exercise III.6.21:

$$b^{\log_b x} = \exp\left((\ln b)\frac{\ln x}{\ln b}\right) = \exp \ln x = x$$

Exercise III.6.22: 3.

Exercise III.6.23: 2.

Exercise III.6.24: 3.

Exercise III.6.25: 2.7268.

Exercise III.6.26: −3.

Exercise III.6.27:

$$y' = \frac{1}{x \ln b}$$

Exercise III.7.1:

For Figure III.44, the graph of $y = \tan \theta$, we have

$$\lim_{\theta \to \frac{\pi}{2}^-} \tan \theta = +\infty$$

and

$$\lim_{\theta \to \frac{\pi}{2}^+} \tan \theta = -\infty.$$

Since the tangent function is periodic of period π, we see the same behavior at
$$\frac{\pi}{2}+\pi, \frac{\pi}{2}-\pi, \frac{\pi}{2}+2\pi, \frac{\pi}{2}-2\pi$$
and so forth.

For Figure III.45, the graph of $y = \ln x$, we have
$$\lim_{x \to 0^+} \ln x = -\infty.$$

Exercise III.7.2: The function
$$y = \frac{1}{x(x-3)}$$
has the following vertical asymptotes.
$$\lim_{x \to 3^+} \left(\frac{1}{x(x-3)} \right) = +\infty$$

$$\lim_{x \to 3^-} \left(\frac{1}{x(x-3)} \right) = -\infty$$

$$\lim_{x \to 0^+} \left(\frac{1}{x(x-3)} \right) = -\infty$$

$$\lim_{x \to 0^-} \left(\frac{1}{x(x-3)} \right) = +\infty$$

and the following horizontal asymptotes
$$\lim_{x \to +\infty} \left(\frac{1}{x(x-3)} \right) = 0$$

$$\lim_{x \to -\infty} \left(\frac{1}{x(x-3)} \right) = 0.$$

There is a local maximum at $(3/2, -4/9)$. Its graph is sketched in Figure A.49.

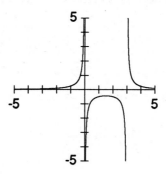

Figure A.49: $y = 1/x(x-3)$

Exercise III.7.7:
The function
$$y = \frac{1}{x^2(x-1)(x-2)}$$
has the following vertical asymptotes.

$$\lim_{x \to 0^-} \left(\frac{1}{x^2(x-1)(x-2)} \right) = +\infty$$

$$\lim_{x \to 0^+} \left(\frac{1}{x^2(x-1)(x-2)} \right) = +\infty$$

$$\lim_{x \to 1^-} \left(\frac{1}{x^2(x-1)(x-2)} \right) = +\infty$$

$$\lim_{x \to 1^+} \left(\frac{1}{x^2(x-1)(x-2)} \right) = -\infty$$

$$\lim_{x \to 2^-} \left(\frac{1}{x^2(x-1)(x-2)} \right) = -\infty$$

$$\lim_{x \to 2^+} \left(\frac{1}{x^2(x-1)(x-2)} \right) = +\infty$$

and the following horizontal asymptotes

$$\lim_{x \to +\infty} \left(\frac{1}{x^2(x-1)(x-2)} \right) = 0$$

$$\lim_{x \to -\infty} \left(\frac{1}{x^2(x-1)(x-2)} \right) = 0.$$

It has a local maximum at $x = 1.64039$ and a local minimum at $x = 0.609612$. Its graph is sketched in Figure A.50

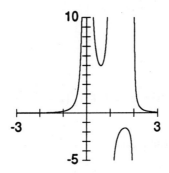

Figure A.50: $y = 1/x^2(x-1)(x-2)$

Exercise III.7.8:

The function
$$z = t + \frac{1}{t-1} + \frac{2}{t-2}$$
has the following vertical asymptotes.

$$\lim_{t \to 1^-} \left(t + \frac{1}{t-1} + \frac{2}{t-2} \right) = -\infty$$

$$\lim_{t \to 1^+} \left(t + \frac{1}{t-1} + \frac{2}{t-2} \right) = +\infty$$

$$\lim_{t \to 2^-} \left(t + \frac{1}{t-1} + \frac{2}{t-2} \right) = -\infty$$

$$\lim_{t \to 2^+} \left(t + \frac{1}{t-1} + \frac{2}{t-2} \right) = +\infty.$$

It has a local maximum at $t = -0.2762$ and a local minimum at $t = 3.5386$. Its graph is sketched in Figure A.51

Exercise III.7.10: 10 feet × 10 feet.

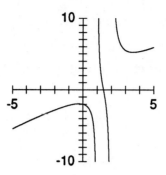

Figure A.51: $z = t + 1/(t-1) + 1/(t-2)$

Exercise III.7.11: The radius should be

$$\sqrt[3]{\frac{500}{\pi}}$$

and its height should be

$$2\sqrt[3]{\frac{500}{\pi}}.$$

Exercise III.8.1: The limit is 0.

Exercise III.8.3: The limit is 1/2.

Exercise III.8.5: The limit is 0.

Exercise III.8.6: See Figure A.52. Notice that this function is undefined at $x = 0$ but that

$$\lim_{x \to 0} \frac{e^x - 1}{x} = 1.$$

Exercise III.8.7: See Figure A.53.

Exercise III.8.10:

$$\lim_{x \to +\infty} \frac{x^n}{e^x} = 0.$$

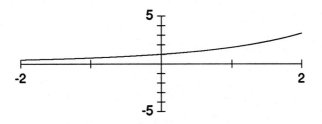

Figure A.52: $y = (e^x - 1)/x$

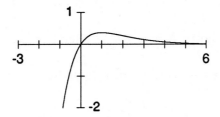

Figure A.53: $y = xe^{-x}$

Exercise III.8.11:

$$\lim_{x \to +\infty} \frac{x^c}{e^x} = 0.$$

Exercise III.8.12:

$$\lim_{x \to +\infty} \frac{x^c}{\ln x} = +\infty.$$

Thus the function e^x grows faster than the functions x^c, which in turn grow faster than the function $\ln x$.

Exercise IV.1.1:

The radius of a slice of the cone x feet from its apex is $x/5$. Thus, the area of this slice is $\pi x^2/25$ and the exact volume of the cone is

$$\int_0^5 \pi \frac{x^2}{25} \, dx.$$

Estimating this integral using the midpoint method and five subintervals we obtain 5.1836279.

Exercise IV.1.2: See Figure A.54.

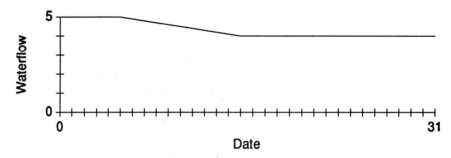

Figure A.54: Water over the dam

The exact total expressed as an integral is

$$\int_0^{31} r(t)\, dt.$$

Estimating this integral using the midpoint method and 31 subintervals we obtain 134.

Drawing a graph of the function $r(t)$ we obtain Figure A.54.

Because this region can be broken into rectangles and triangles its exact area can be found by elementary geometry. The area is 134, so the exact total waterflow in May is 134 million gallons.

Exercise IV.1.3:

The exact change in altitude (elevation) is

$$\int_0^{10} s(d)\, dd.$$

which is the same as the area of the region under the curve $y = s(d)$ shown in Figure A.55. Because this region is made up of triangles and rectangles it is easy to find its exact area by elementary geometry. The area is 0.4, so the exact change in elevation is 0.4 miles. Your estimate using the midpoint method will also be 0.4 miles. Can you see why the midpoint method gave the exact answer for this exercise and the preceding exercise? Would you get the exact answer if you used the midpoint method with any number of subintervals?

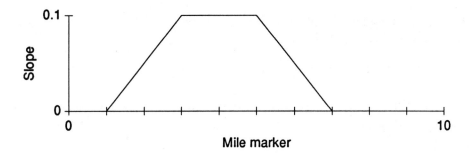

Figure A.55: Slope of trail

Exercise IV.1.4:
The exact distance the object will fall in the first two seconds is

$$\int_0^2 -32t \, dt.$$

which is the same as the area of the triangle shown in Figure A.56. Notice the area is negative because it is below the x-axis. This corresponds to the fact that the object is falling. The area of the triangle is easily seen to be -64, so the object will fall 64 feet in the first two seconds after it is dropped. Your estimate using the midpoint method will be the same as the exact answer. Would you get the exact answer for this exercise using the midpoint method and any number of subintervals?

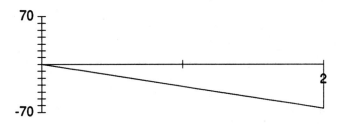

Figure A.56: $v = -32t$

Exercise IV.1.5:
The net waterflow from time $t = 0$ to time $t = 25$ is

$$\int_0^{25} 2 + 3\sin\left(\frac{2\pi t}{12.5}\right) dt.$$

The total unsigned waterflow during the same period is

$$\int_0^{25} \left|2 + 3\sin\left(\frac{2\pi t}{12.5}\right)\right| dt.$$

Using the midpoint method and ten subintervals we obtain the following estimates.

$$\text{net water flow} \approx 50$$
$$\text{total water flow} \approx 58.531695$$

Exercise IV.2.1: 8.9375.

Exercise IV.2.2: 2.0.

Exercise IV.2.3:

$$\int_0^{3\pi/2} \sin x \, dx = 1 + \frac{1}{\sqrt{2}}$$
$$\int_0^{\pi/2} \sin x \, dx = 3$$
$$\int_{\pi/2}^0 4\sin x \, dx = -4$$

Exercise IV.2.4:

$$\int_0^a f(x) \, dx = \text{the area under the curve } y = f(x) \text{ between } x = 0 \text{ and } x = b.$$

$$\int_{-a}^0 f(x) \, dx = \text{the area under the curve } y = f(x) \text{ between } x = -a \text{ and } x = 0.$$

Because the curve $y = f(x)$ is symmetric about the y-axis these two regions have the same area.

Exercise IV.2.5:

$$\int_0^a f(x)\,dx = \text{the area under the curve } y = f(x) \text{ between } x = 0 \text{ and } x = b.$$

$$\int_{-a}^0 f(x)\,dx = \text{the area under the curve } y = f(x) \text{ between } x = -a \text{ and } x = 0.$$

Because the curve $y = f(x)$ is symmetric about the origin these two regions have the same unsigned area and the signed area of one is the −the signed area of the other.

Exercise IV.3.1: 1.5.

Exercise IV.3.2:
$$y = \frac{x^2}{2} - 2x + 4.$$

Exercise IV.3.3:
$$y = \frac{x^2}{2} - 2x + 5.5$$

Exercise IV.3.4: 1/6.

Exercise IV.3.5: $\sin t + c$.

Exercise IV.3.6:
$$\frac{1}{2}\sin 2t + c.$$

Exercise IV.3.9: $y = t^3/3 + 4$.

Exercise IV.4.1: $4 + \ln 3$.

Exercise IV.4.2: 6.

Exercise IV.4.3: 764/9.

Exercise IV.4.5: $201 - \cos 10$.

Exercise IV.4.7: $3 + \ln 5 - \ln 2$.

Exercise IV.4.9: $4\sqrt{6} - 4\sqrt{2}/3$.

Exercise IV.4.11: $16/3$.

Exercise IV.4.14: For x negative

$$\ln|x| = \ln(-x)$$
$$\left(\frac{d}{dx}\right)\ln|x| = \frac{1}{-x}(-1)$$
$$= \frac{1}{x}$$

Exercise IV.4.15: See Figure A.57.

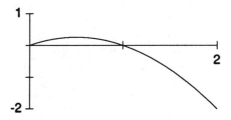

Figure A.57: $f(x) = x(1-x)$ between $x = 0$ and $x = 2$.

- The signed area of the piece above the x-axis is $1/6$.

- The signed area of the piece below the x-axis is $-5/6$.

- The signed area of the whole region is $-2/3$.

- The unsigned area of the whole region is 1.

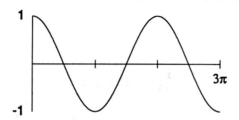

Figure A.58: $f(x) = \cos x$ between $x = 0$ and $x = 3\pi$.

Exercise IV.4.16: See Figure A.58.

- The signed area of the piece above the x-axis is 3.
- The signed area of the piece below the x-axis is -3.
- The signed area of the whole region is 0.
- The unsigned area of the whole region is 6.

Exercise IV.4.17: See Figure A.59.

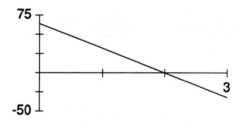

Figure A.59: $v(t) = 64 - 32t$ between $t = 0$ and $t = 3$.

- The signed distance traveled while the ball is traveling upward is 64 feet.
- The signed distance traveled while the ball is traveling downward is -16 feet.
- The total signed distance traveled is 48 feet.

- The total unsigned distance traveled is 80 feet.

Exercise IV.5.1:
$$\frac{2}{3}(x+1)^{\frac{3}{2}} + c.$$

Exercise IV.5.3:
$$\left[\frac{\sin^4 x}{4}\right]_0^\pi = 0.$$

Exercise IV.5.4:
$$\left[\frac{1}{2}\arctan(2x)\right]_1^2 = \frac{\arctan(4)}{2} - \frac{\arctan(2)}{2} = 0.109334.$$

Exercise IV.5.6:
$$\frac{\arctan(\frac{x}{k})}{k} + c.$$

Exercise IV.5.9:
$$\frac{\arctan(x/4)}{2} + \frac{3\log(16+x^2)}{2} + c.$$

Exercise IV.5.10:
$$\arcsin(x/2) + c.$$

Exercise IV.5.11:
$$-\sqrt{4-x^2} + c.$$

Exercise IV.5.12:
$$2\arcsin(x/2) - 3\sqrt{4-x^2} + c.$$

Exercise IV.5.17: The two different-looking answers are

$$\frac{\sin^2 x}{2} + c$$

and

$$\frac{-\cos^2 x}{2} + c.$$

The difference between these two answers is a constant.

$$\frac{\sin^2 x}{2} - \frac{-\cos^2 x}{2} = \frac{1}{2}.$$

Exercise IV.5.18:

$$\arctan(3) + \frac{\ln(10)}{2} - \frac{\ln(5)}{2} - \arctan(2) = 0.488471.$$

Exercise IV.5.19: $\frac{\ln 9}{2} - \frac{\ln 5}{2} = 0.293893.$

Exercise IV.5.21:

$$-\frac{1}{3} + \frac{2^{3/2}}{3} = 0.609476.$$

Exercise IV.5.24: The trick here is to notice that $x^2 + 4x + 5 = (x+2)^2 + 1$ and then make the substitution $u = (x+2)$. The result is

$$\arctan 4 - \arctan 3 = 0.0767719.$$

Exercise IV.5.25:

$$\frac{\pi}{12} - \frac{\arctan(\frac{3}{4})}{3} = 0.047299.$$

Exercise IV.5.27: See Figure A.60.

$$\int_{1-\sqrt{2}}^{1+\sqrt{2}} (2x+1) - x^2 \, dx = 2\sqrt{2} - \left(1 - \sqrt{2}\right)^2 + \frac{\left(1 - \sqrt{2}\right)^3}{3} + \left(1 + \sqrt{2}\right)^2 - \frac{\left(1 + \sqrt{2}\right)^3}{3} = 3.77124.$$

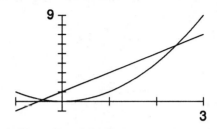

Figure A.60: $y = 2x + 1$ and $y = x^2$.

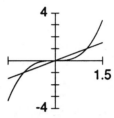

Figure A.61: $y = x$ and $y = x^3$.

Exercise IV.5.29: See Figure A.61.

Notice that there are two pieces to this region. Their total area is is 1/2.

Exercise IV.5.31: see Figure A.62.

The total food needs will be

$$\int_0^{10} Ce^{.05t} \, dt = 12.9744C.$$

The total food supply will be

$$\int_0^{10} Ce^{.02t} \, dt == 11.0701C.$$

The necessary food imports will be

$$\int_0^{10} (Ce^{.05t} - Ce^{.052t}) \, dt = 1.9043C.$$

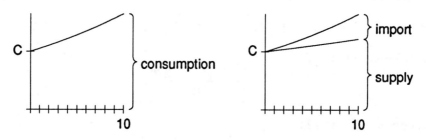

Figure A.62: Food supply and consumption

Exercise IV.6.8: The formula for the trapezoidal rule is

$$\sum_{i=1}^{n} \frac{h}{2}(f(x_i) + f(x_{i-1})) = \frac{h}{2}(f(x_0) + f(x_1)) + \frac{h}{2}(f(x_1) + f(x_2)) + \cdots + \frac{h}{2}(f(x_{n-1}) + f(x_n))$$

$$= \frac{h}{2}f(x_0) + hf(x_1) + \cdots + hf(x_{n-1}) + \frac{h}{2}f(x_n)$$

$$= \frac{h}{2}f(a) + hf(x_1) + \cdots + hf(x_{n-1}) + \frac{h}{2}f(b)$$

$$= \frac{h}{2}f(a) + \frac{h}{2}f(b) + \sum_{i=1}^{n} f(x_i)h.$$

Exercise IV.9.1: 2/3.

Exercise IV.9.2: This integral doesn't make sense now since the integrand is undefined when $x = 0$. We will talk about this kind of integral later in section V.3.

Exercise IV.9.3: 73/6.

Exercise IV.9.4: This integral must be estimated numerically. The answer correct to six decimal places is 0.828116.

Exercise IV.9.5: $\frac{1-\cos(\pi^2/4)}{2}$.

Exercise IV.9.6: This integral must be estimated numerically. The answer is approximately 1.03813.

Exercise IV.9.7: 2.

Exercise IV.9.8: 14/3.

Exercise IV.9.9: 116/15.

Exercise IV.9.10:
$$\frac{\log 13 - \log 9}{8}.$$

Exercise IV.9.11:
$$\frac{\arctan(2/3)}{6}.$$

Exercise IV.9.12:
$$\frac{5\arctan(2/3)}{6} - \frac{3\log 9}{8} + \frac{3\log 13}{8}.$$

Exercise V.1.1:
$$\frac{x 2^x}{\ln 2} - \frac{2^x}{(\ln 2)^2} + c.$$

Exercise V.1.2:
$$x \arcsin x + \sqrt{1-x^2} + c.$$

Exercise V.1.3:
$$x \arctan x - \frac{\ln(1+x^2)}{2} + c.$$

Exercise V.1.5:
$$-x^2 e^{-x} - 2x e^{-x} - 2 e^{-x} + c.$$

Exercise V.1.6:
$$\int x^n e^{-x}\, dx = -x^n e^{-x} + n \int x^{n-1} e^{-x}\, dx.$$

Exercise V.1.8:
$$\frac{x^2}{2} \ln x - \frac{x^2}{4} + c.$$

Exercise V.1.10:
$$\frac{x}{2} - \frac{\sin x \cos x}{2}.$$

Exercise V.1.11:

$$\int \sin^n x \, dx = \int \sin^{(n-1)} x \sin x \, dx$$
$$= -\sin^{(n-1)} x \cos x + \int (n-1) \sin^{(n-2)} x \cos^2 x \, dx$$
$$= -\sin^{(n-1)} x \cos x + \int (n-1) \sin^{(n-2)} x (1 - \sin^2 x) \, dx$$
$$= -\sin^{(n-1)} x \cos x + (n-1) \int \sin^{(n-2)} x \, dx - (n-1) \int \sin^n x \, dx$$
$$n \int \sin^n x \, dx = -\sin^{(n-1)} x \cos x + (n-1) \int \sin^{(n-2)} x \, dx$$
$$\int \sin^n x \, dx = \frac{-\sin^{(n-1)} x \cos x}{n} + \frac{n-1}{n} \int \sin^{(n-2)} x \, dx$$

Exercise V.1.12:
$$\int x^n \sin x \, dx = -x^n \cos x + n \int x^{n-1} \cos x \, dx.$$

Exercise V.1.15:
```
MyIntegral[x Sin[x]] := -x Cos[x] - Sin[x]
MyIntegral[x Cos[x]] :=  x Sin[x] - Cos[x]
MyIntegral[x^n_ Sin[x]] :=
    -x^n Cos[x] + n MyIntegral[x^(n-1) Cos[x]]
MyIntegral[x^n_ Cos[x]] :=
    x^n Sin[x] - n MyIntegral[x^(n-1) Sin[x]]
```

Exercise V.1.16:
If n is odd then we can write $n = 2k+1$ and
$$\int \sin^m x \cos^n x \, dx = \int \sin^n x \cos^{(2k+1)} x \, dx$$
$$= \int \sin^n x (\cos^2 x)^k \cos x \, dx$$
$$= \int \sin^n x (1 - \sin^2 x)^k \cos x \, dx$$

which can be evaluated by the substitution $u = \sin x$.

If m is odd then similar calculations lead to the substitution $u = \cos x$.

If both m and n are even then we can write $n = 2k$ and

$$\int \sin^m x \cos^n x \, dx = \int \sin^n x \cos^{(2k)} x \, dx$$
$$= \int \sin^n x (\cos^2 x)^k \, dx$$
$$= \int \sin^n x (1 - \sin^2 x)^k \, dx$$

which can be evaluated using the reduction formula found in Exercise V.1.11.

Exercise V.1.17:
$$\frac{\beta \sin \alpha x \sin \beta x + \alpha \cos \alpha x \cos \beta x}{\beta^2 - \alpha^2} + c.$$

Exercise V.1.18:
$$\frac{x}{2} - \frac{1}{4\alpha} \sin 2\alpha x + c.$$

Exercise V.2.1: $16\pi/3$.

Exercise V.2.2: $\pi R^2 H/3$.

Exercise V.2.4: $4\pi ab^2/3$.

Exercise V.2.6: $2\pi/15$.

Exercise V.2.7: $250,000/3$.

Exercise V.2.8: $x^2 h/3$.

Exercise V.2.9: $4\pi\sqrt{3}$.

Exercise V.3.2: 56.5685 feet per second.

Exercise V.3.4: 11.4796 meters.

Exercise V.3.6: 25 feet

Exercise V.3.8: $9.86m$.

Exercise V.3.9: $1.634m$ or roughly one-sixth that near the surface of the earth.

Exercise V.3.11: 23.1 feet per second.

Exercise V.3.13: 68.85 meters.

Exercise V.3.15: 150 feet.

Exercise V.3.17: 5,463 meters per second.

Exercise V.3.19:

$$\sqrt{7.98 \times 10^{14} \left(\frac{1}{6,380,000} - \frac{1}{6,380,000 + 1,000H} \right)}$$

where H is measured in kilometers.

Exercise V.3.20: 11,200 meters per second.

Exercise V.3.21: 1.

Exercise V.3.22: ∞.

Exercise V.3.23: π.

Exercise V.3.24: ∞.

Exercise V.3.25: ∞.

Exercise V.3.26: 2.

Exercise V.3.27: ∞.

Exercise V.4.1: No.

Exercise V.4.3:

$$\frac{dy}{dx} = -\frac{y}{x}.$$

Exercise V.4.6: $y = 3$.

Exercise V.4.8:
$$\frac{dy}{dx} = \frac{x(2x^2 - 21)}{y(2y^2 - 29)}.$$

If, in addition,
$$x^2 + y^2 - 25 = 0$$

then $y^2 = 25 - x^2$ and
$$\begin{aligned}
\frac{dy}{dx} &= \frac{x(2x^2 - 21)}{y(2y^2 - 29)} \\
&= \frac{x(2x^2 - 21)}{y(2(25 - x^2) - 29)} \\
&= \frac{x(2x^2 - 21)}{y(21 - 2x^2)} \\
&= -\frac{x}{y}
\end{aligned}$$

as we wanted to show.

The calculation when (x, y) lies on the curve $y^2 - x^2 - 4 = 0$ is very similar.

Exercise V.4.9:

We will use the notation $f'(x, y)$ for the result of differentiating the factor $f(x, y)$ with respect to x and $g'(x, y)$ for the result of differentiating $g(x, y)$ with respect to x. For example, if
$$f(x, y) = x^2 + y^2$$

then
$$f'(x, y) = 2x + 2yy'.$$

Applying implicit differentiation to
$$f(x, y)g(x, y) = 0$$

we get
$$f'(x, y)g(x, y) + f(x, y)g'(x, y) = 0$$

and if (x, y) is on the curve $f(x, y) = 0$ this becomes
$$f'(x, y)g(x, y) = 0.$$

As long as (x, y) does not lie on the curve $g(x, y) = 0$ we can divide by $g(x, y)$ to get
$$f'(x, y) = 0$$
which is exactly what we get if we apply implicit differentiation to the equation
$$f(x, y) = 0.$$

Exercise V.4.10:

If a point (x, y) lies on both the curve $f(x, y) = 0$ and $g(x, y) = 0$ then the equation
$$f'(x, y)g(x, y) + f(x, y)g'(x, y) = 0$$
obtained in IV.10.5 becomes
$$0 = 0$$
and dy/dx is undefined, as might be expected since at these crossing points there are usually two possible tangents.

Exercise V.5.2: $y = 5x/3$.

Exercise V.5.5:
$$\begin{aligned} T &= 70 + 142e^{-0.0255347t} \\ T(25) &= 145 \\ \lim_{t \to \infty} T(t) &= 70. \end{aligned}$$

This is what we would expect. The temperature of the coffee would approach room temperature.

Exercise V.6.1:
$$p = 500 - 400e^{-0.2t}.$$

$$\lim_{t \to \infty} p(t) = 500.$$

Exercise V.6.4:

$$\frac{x-2}{x^2+3x+2} = -\frac{3}{x+1} + \frac{4}{x+2}.$$

Exercise V.6.5:

$$\frac{x^2+2x+3}{x^3+x} = \frac{3}{x} + \frac{-2x+2}{x^2+1}.$$

Exercise V.6.7:

$$\frac{1}{3}\ln|x-1| + \frac{5}{3}\ln|x-4| + c.$$

Exercise V.6.10:

$$p(t) = \frac{1000e^{2t}}{9+e^{2t}}.$$

$$\lim_{t\to+\infty} p(t) = 1000.$$

Exercise V.6.16:

$$p(t) = \frac{Ce^{at}}{1+Ce^{at}}.$$

$$\lim_{t\to+\infty} p(t) = 1.$$

The value of the constant a affects the speed with which $p(t)$ approaches the limit 1. In contrast to discrete logistic models, however, the longterm behavior is qualitatively the same regardless of the value of a. We do not, for example, ever see 2-cycles as we do for discrete logistic models.

Exercise V.7.1: 7.6404.

Exercise V.7.2:
Using the formula we get

$$\int_0^1 \sqrt{1+4}\, dx = \sqrt{5}.$$

This curve is a straight line from $(0,0)$ to $(1,2)$. By the Pythagorean Theorem its length is

$$\sqrt{1+2^2} = \sqrt{5}.$$

Exercise V.7.3:
Using the formula we get

$$\int_a^b \sqrt{1+m^2}\, dx = (b-a)\sqrt{1+m^2}.$$

This curve is a straight line from (a, ma) to (b, mb). By the Pythagorean Theorem its length is

$$\sqrt{(b-a)^2 + (mb-ma)^2} = (b-a)\sqrt{1+m^2}.$$

Exercise V.7.4: 9.68845.

Exercise V.7.7: The arc described by the linear function $y = mx + k$ between $x = a$ and $x = b$ is a straight line from the point $(a, ma + k)$ to the point $(b, mb + k)$. By the Pythagorean Theorem the length of this line is

$$\sqrt{(b-a)^2 + (mb+k-ma-k)^2} = (b-a)\sqrt{1+m^2}.$$

Exercise V.7.8: The formula obtained in the preceding exercise computes the arclength for a curve whose slope is the constant m by simple multiplication. For a curve whose slope is not constant, multiplication is replaced by integration and the slope by the derivative yielding

$$\int_a^b \sqrt{1+f'(x)^2}\, dx.$$

Exercise V.7.9:
The arclength of the curve described by the pair of functions $x(f(s))$ and $y(f(s))$ for s between p and q is

$$\int_p^q \sqrt{\left(\frac{dx}{ds}\right)^2 + \left(\frac{dy}{ds}\right)^2}\, ds.$$

or

$$\int_p^q \sqrt{\left(\frac{dx}{dt}\frac{dt}{ds}\right)^2 + \left(\frac{dy}{dt}\frac{dt}{ds}\right)^2}\, ds$$

Recall that $t = f(s)$. This integral is

$$\int_p^q \sqrt{\left(\frac{dx}{dt}\right)^2 + \left(\frac{dy}{dt}\right)^2}\, \frac{dt}{ds}\, ds.$$

Making the substitution $t = f(s)$ and noting that $a = f(p)$ and $b = f(q)$ this becomes

$$\int_a^b \sqrt{\left(\frac{dx}{dt}\right)^2 + \left(\frac{dy}{dt}\right)^2}\, dt$$

as we wanted to show.

Exercise V.8.1: $1/12$.

Exercise V.8.3: $\pi/16$.

Exercise V.8.4:

$$\frac{\log\left(\frac{1+\sin(x)}{\cos(x)}\right)}{2} + \frac{\sin(x)}{2\cos^2(x)}$$

Exercise V.8.6: $25\pi/4$.

Exercise V.8.9:

$$\frac{\arctan(200)}{2}.$$

Exercise V.8.12:
$$\frac{3\arctan(\frac{-4+2x}{8})}{4} + \frac{\ln(20-4x+x^2)}{2} + c.$$

Exercise V.9.1: The exact answer is
$$\frac{1}{2} - \frac{\cos(1)\sin(1)}{2}.$$

Exercise V.9.2: An approximate answer found by numerical methods is 0.310268.

Exercise V.9.3: The exact answer is $\sin 1 - \cos 1$.

Exercise V.9.4: The exact answer is $-2 + \cos 1 + 2\sin 1$.

Exercise V.9.5: The exact answer is
$$\frac{3}{8} - \frac{\cos 2}{8} - \frac{\sin 2}{4}.$$

Exercise V.9.6: The exact answer is
$$\frac{1}{2} - \frac{\cos 1}{2}.$$

Exercise V.9.7: An approximate answer found by numerical methods is 0.182111.

Exercise V.9.8: The exact answer is
$$\frac{\sin(1)}{2} - \frac{\cos(1)}{2}.$$

Exercise V.9.9: The exact answer is
$$\frac{\ln(3)}{2} - \frac{\ln(1)}{4}.$$

Exercise V.9.10: The exact answer is $\pi/3$.

Exercise V.9.11: The exact answer is

$$-\frac{2}{3} - \frac{8}{3\left(-1 + \frac{\sin(1)^2}{(1+\cos(1))^2}\right)^3} - \frac{4}{\left(-1 + \frac{\sin(1)^2}{(1+\cos(1))^2}\right)^2} - \frac{2}{-1 + \frac{\sin(1)^2}{(1+\cos(1))^2}}.$$

Exercise V.9.12: The exact answer is

$$\frac{\sqrt{5}}{2} + \frac{\ln(2 + \sqrt{5})}{4}.$$

Exercise V.9.13: The exact answer is

$$-\frac{1}{12} + \frac{5^{\frac{3}{2}}}{12}.$$

Exercise V.9.14: The exact answer is

$$\frac{-\sqrt{5}}{32} + \frac{5^{\frac{3}{2}}}{16} - \frac{\ln(2 + \sqrt{5})}{64}.$$

Exercise V.9.15: The exact answer is π.

Exercise V.9.16: The exact answer is $8/3$.

Exercise V.9.17: The exact answer is π.

Exercise V.9.18: An approximate answer found by numerical methods is 57.9534.

Exercise V.9.19: The exact answer is $1/4$.

Exercise V.9.20: The exact answer is $2/15$.

Exercise V.9.21: The exact answer is $\pi/16$.

Exercise V.9.22: The value of this improper integral is $+\infty$.

Exercise V.9.23: The value of this improper integral is $+\infty$.

Exercise V.9.24: The value of this improper integral is -1.

Exercise V.9.25: The exact answer is
$$-\frac{3}{10} + \frac{3e^3 \cos(1)}{10} + \frac{e^3 \sin(1)}{10}.$$

Exercise V.9.26: The exact answer is
$$\frac{-\arctan(\frac{1}{2})}{2} + \frac{\arctan(2)}{2}.$$

Exercise V.9.27: The exact answer is
$$\frac{\arctan(\frac{1}{2})}{2} - \frac{\arctan(2)}{2} - \frac{\ln(5)}{2} + \frac{\ln(20)}{2}.$$

Exercise V.9.28: The exact answer is
$$\frac{-5\arctan(\frac{1}{2})}{2} + \frac{5\arctan(2)}{2} - \ln(5) + \ln(20).$$

Exercise V.9.29: The exact answer is
$$2 - \frac{3\pi}{4} + 3\arctan(3) + \ln(2) + \ln(3) - \ln(10).$$

Exercise V.9.30:
$$F(x) = \int_0^x \cos t \, dt = \sin x - \sin 0 = \sin x.$$
So the graph of $F(x)$ is as shown in Figure A.63.

Exercise V.9.31:
$$F(x) = \int_1^x \cos t \, dt = \sin x - \sin 1.$$
So the graph of $F(x)$ is as shown in Figure A.64.

Exercise V.9.32:
$$F(x) = \int_x^0 \cos t \, dt = \sin 0 - \sin t = -\sin t.$$
So the graph of $F(x)$ is as shown in Figure A.65.

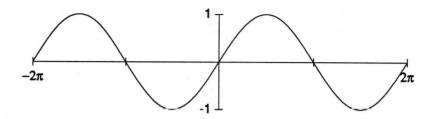

Figure A.63: $F(x) = \sin x$

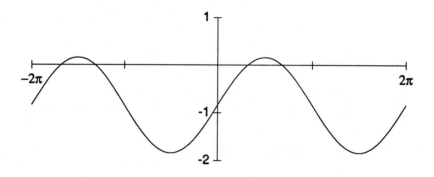

Figure A.64: $F(x) = \sin x - \sin 1$

Exercise V.9.33:

$$F(x) = \int_x^1 \cos t \, dt = \sin 1 - \sin x.$$

So the graph of $F(x)$ is as shown in Figure A.66.

Exercise V.9.34: This function

$$F(x) = \int_0^x e^{-t^2/2} \, dt$$

is much more difficult to graph than the preceding functions because none of our antidifferentiation techniques enable us to find a simple antiderivative. Notice, however, that

$$F'(x) = e^{-t^2/2}$$

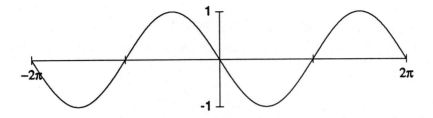

Figure A.65: $F(x) = -\sin x$

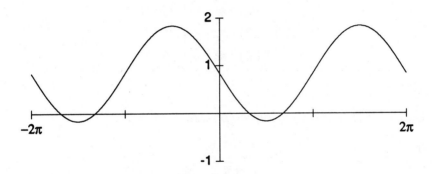

Figure A.66: $F(x) = \sin 1 - \sin x$

is always positive, so that we know $F(x)$ is always increasing. In addition, notice that

$$F(0) = \int_0^0 e^{-t^2/2}\, dt = 0.$$

We can estimate some additional values of $F(x)$ using numerical methods. For example,

$$\begin{aligned}
F(1) &\approx 0.855624 \\
F(2) &\approx 1.19629 \\
F(3) &\approx 1.24993 \\
F(4) &\approx 1.25323 \\
F(5) &\approx 1.25331 \\
F(10) &\approx 1.25331 \\
F(100) &\approx 1.25331
\end{aligned}$$

One can show that the function $F(x)$ is symmetric about the origin (This is a good exercise.) or estimate the additional values of $F(x)$

$$F(-1) \approx -0.855624$$
$$F(-2) \approx -1.19629$$
$$F(-3) \approx -1.24993$$
$$F(-4) \approx -1.25323$$
$$F(-5) \approx -1.25331$$
$$F(-10) \approx -1.25331$$
$$F(-100) \approx -1.25331$$

Using all this information we obtain Figure A.67.

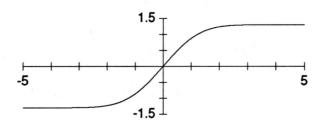

Figure A.67: $F(x) = \int_0^x e^{-t^2/2}\, dt$

Exercise V.9.35: This seemingly bizarre function

$$F(x) = \frac{1}{\sqrt{2\pi}} \int_{-\infty}^{x} e^{-t^2/2}\, dt$$

is actually one of the most important functions in probability and statistics. It is closely related to the function studied in Exercise V.9.34. The first observation,

$$F'(x) = \frac{1}{\sqrt{2\pi}} e^{-t^2/2},$$

shows us that the function $F(x)$ is always increasing since $F'(x)$ is always positive. To estimate $F(0)$ we estimate numerically

$$\frac{1}{\sqrt{2\pi}} \int_{-1}^{x} e^{-t^2/2} \, dt \approx 0.341345$$

$$\frac{1}{\sqrt{2\pi}} \int_{-10}^{x} e^{-t^2/2} \, dt \approx 0.500000$$

$$\frac{1}{\sqrt{2\pi}} \int_{-20}^{x} e^{-t^2/2} \, dt \approx 0.500000$$

$$\frac{1}{\sqrt{2\pi}} \int_{-100}^{x} e^{-t^2/2} \, dt \approx 0.500000$$

Thus,
$$F(0) \approx 0.500000$$

Working out some additional numerical estimates we see that

$$F(-5) \approx 0.000000$$
$$F(-4) \approx 0.000032$$
$$F(-3) \approx 0.001350$$
$$F(-2) \approx 0.022750$$
$$F(-1) \approx 0.158655$$
$$F(0) \approx 0.500000$$
$$F(1) \approx 0.841345$$
$$F(2) \approx 0.977250$$
$$F(3) \approx 0.998650$$
$$F(4) \approx 0.999968$$
$$F(5) \approx 1.000000$$

leading to Figure A.68.

Exercise V.9.36: $y = t^2 + 2t + 1$. (Note: You may choose a different letter for the independent variable, for example, $y = x^2 + 2x + 1$ is another possible answer.)

Exercise V.9.37: $y = t^2$.

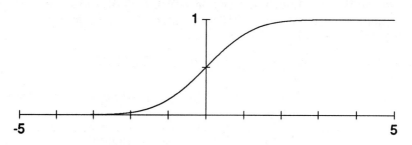

Figure A.68: $F(x) = (2\pi)^{-1/2} \int_{-\infty}^{x} e^{-t^2/2} \, dt$

Exercise V.9.38: $y = t^2 + 1$.

Exercise V.9.39:
$$p = \frac{2e^{4t}}{1 + e^{4t}}.$$

Exercise V.9.40: $y = 2x^2$.

Exercise V.9.41:
$$y = \sqrt{3x^2 + 1}.$$

Exercise V.9.42: -60. The seesaw will accelerate in the counter-clockwise direction.

Exercise V.9.43: -170. The seesaw will accelerate in the counter-clockwise direction.

Exercise V.9.44: If the fulcrum is placed at the point x then the total torque is

$$T(x) = 50(9-x) + 30(5-x) + 20(-6-x) + 10(-9-x) = 390 - 110x.$$

To find the balance point we solve the equation
$$\begin{aligned} T(x) &= 0 \\ 390 - 110x &= 0 \\ x &= 39/11 \end{aligned}$$

So the seesaw will be balanced if the fulcrum is placed at $x = 39/11$.

Exercise V.9.45: The balance point is $x = -1006/117$.

Exercise V.9.46: The torque is -1.

Exercise V.9.47: The torque is $0.1(-4 - 6x_0)$.

Exercise V.9.48: The rod should be placed along the line $x = -2/3$.

Exercise V.9.49: First we must consider torque around the x-axis. This leads to the integral

$$\int_{-1}^{2} 0.1 \left(\frac{8}{3} - \frac{4y}{3}\right) y \, dy.$$

More generally, torque around the line $y = y_0$ is given by

$$\int_{-1}^{2} 0.1 \left(\frac{8}{3} - \frac{4y}{3}\right) (y - y_0) \, dy = -0.6 y_0$$

which leads to the equation

$$-0.6 y_0 = 0$$
$$y_0 = 0.$$

So if the piece of metal is supported at the point $(-2/3, 0)$ it will be in balance.

Exercise V.9.50:

$$\text{Torque} = \int_a^b k(f(x) - g(x)) x \, dx.$$

Exercise V.9.51:

$$\text{Torque} = \int_a^b k(f(x) - g(x))(x - x_0) \, dx.$$

Exercise V.9.52: $1{,}000 \cos(\pi/4)$ units per hour.

Exercise V.9.53: $1{,}000 \cos(\pi/4)$ units per hour.

Exercise V.9.54: 0 units per hour.

Exercise V.9.55:

$$\int_{-6}^{6} 1,000 \cos\left(\frac{\pi t}{12}\right) dt = \frac{24,000}{\pi}.$$

Exercise VI.1.1:

$$\begin{aligned}
(1,1) \text{ Cartesian} &= (\sqrt{2}, \pi/4) \text{ polar} \\
(1,-1) \text{ Cartesian} &= (\sqrt{2}, -\pi/4) \text{ polar} \\
(-1,1) \text{ Cartesian} &= (\sqrt{2}, 3\pi/4) \text{ polar} \\
(-1,-1) \text{ Cartesian} &= (\sqrt{2}, -3\pi/4) \text{ polar}
\end{aligned}$$

Exercise VI.1.3: The Cartesian coordinates of the point whose polar coordinates are $(4, \pi)$ are $(-4, 0)$.

Exercise VI.1.5: See Figure A.69.

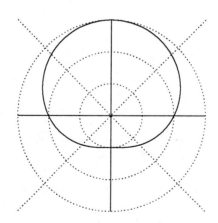

Figure A.69: $r = 2 + \sin \theta$

Exercise VI.1.7: See Figure A.70.

Exercise VI.1.11: $\pi^3/6$.

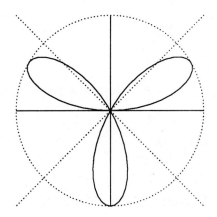

Figure A.70: $r = \sin 3\theta$

Exercise VI.2.1:

$$\begin{aligned} |z| &= 5 \\ |w| &= 13 \\ \bar{z} &= 3 - 4i \\ \bar{w} &= 12 - 5i \\ z + w &= 15 + 9i \\ z - w &= -9 - i \\ zw &= 16 + 63i \\ \frac{z}{w} &= \left(\frac{56}{169}\right) + \left(\frac{33}{169}\right)i \end{aligned}$$

Exercise VI.2.6: The number 1 has three cube roots

$$1, \; -\frac{1}{2} + i\frac{\sqrt{3}}{2}, \; \text{and} \; -\frac{1}{2} - i\frac{\sqrt{3}}{2}.$$

Exercise VI.3.1: $530,180.

Exercise VI.3.6: $4,089.

Exercise VI.3.8: 1,145,500 houses.

Exercise VI.3.10: $675,000.

Exercise VI.4.1: It diverges by the integral test.

Exercise VI.4.2: It diverges by the integral test.

Exercise VI.4.3: The series diverges if $p \geq -1$ and converges if $p < -1$.

Exercise VI.4.6: Nothing.

Exercise VI.4.7: Yes.

Exercise VI.5.1: 1, 2, 6, 24, 120, 720, 5040, 40,320, 362,880.

Exercise VI.5.3: Yes, by the ratio test.

Exercise VI.5.4:

$y(0)$	1
$y'(0)$	-1
$y''(0)$	1
$\vdots$	$\vdots$
$y^{(n)}(0)$	$(-1)^n$
$\vdots$	$\vdots$

Exercise VI.6.1: Only the first term is nonzero. $e^0 = 1$.

Exercise VI.6.3:

$$\sum_{n=0}^{5} \frac{(-1)^n}{n!} = 0.3666666667$$

$$\sum_{n=0}^{10} \frac{(-1)^n}{n!} = 0.3678794643$$

$$\sum_{n=0}^{15} \frac{(-1)^n}{n!} = 0.3678794412$$

$$e^{-1} = 0.3678794412$$

Exercise VI.6.6:
$$\cos x = 1 - \frac{x^2}{2} + \frac{x^4}{4!} - \frac{x^6}{6!} + \cdots$$

The radius of convergence is ∞.

Exercise VI.6.9:
$$y(x) = 1 - x + \frac{x^2}{2} - \frac{x^3}{3!} + \frac{x^4}{4!} - \frac{x^5}{5!} + \cdots$$

Exercise VI.6.10:
$$y(x) = 2 + 2(x-1) + 2\left(\frac{(x-1)^2}{2}\right) + 2\left(\frac{(x-1)^3}{3!}\right) + \cdots$$

Exercise VI.6.11:
$$y(x) = 1 + 2x - \frac{x^2}{2} - \frac{2x^3}{3!} + \frac{x^4}{4!} + \frac{2x^5}{5!} - \frac{x^6}{6!} - \frac{2x^7}{7!} + \cdots$$

Exercise VI.7.1:
$$p_{14}(x) = x - \frac{x^3}{3!} + \frac{x^5}{5!} - \cdots + \frac{x^{13}}{13!}$$

Exercise VI.7.5:
$$p_9(x) = 1 + x + \frac{x^2}{2!} + \cdots + \frac{x^9}{9!}$$

Exercise VI.7.6:
$$p_4(x) = (x-1) - \frac{(x-1)^2}{2} + \frac{(x-1)^3}{3} - \frac{(x-1)^4}{4} + \cdots - \frac{(x-1)^{100}}{100}.$$

Exercise VI.7.8:
$$p_2(x) = (x-1) - \frac{x-1)^2}{2}.$$

Exercise VI.7.9:

The righthand estimate is

$$\frac{\sin 1.01 - \sin 1}{0.01} = 0.5361.$$

The lefthand estimate is

$$\frac{\sin 0.99 - \sin 1}{-0.01} = 0.5445.$$

The symmetric difference estimate is

$$\frac{\sin 1.01 - \sin 0.99}{0.02} = 0.5403.$$

The actual derivative is

$$f'(1) = \cos 1 = 0.5403.$$

According to the theorem the errors for the two one-sided estimates should be "in the same ballpark" as h or 0.01 and they are. The error for the symmetric difference estimate should be in the same ballpark as h^2 or 0.0001 and, in fact, this error is less than 0.0001.

Exercise VI.8.1: Yes.

Exercise VI.8.4:

$$\begin{aligned} x &= c_1 \cos t + c_2 \sin t \\ x' &= -c_1 \sin t + c_2 \cos t \\ x'' &= -c_1 \cos t - c_2 \sin t \\ &= -x \end{aligned}$$

Exercise VI.8.5: $x = \cos t.$

Exercise VI.8.8:

$$\begin{aligned} x &= c_1 \cos \omega t + c_2 \sin \omega t \\ x' &= -\omega c_1 \sin \omega t + \omega c_2 \cos \omega t \\ x'' &= -\omega^2 c_1 \cos \omega t - \omega^2 c_2 \sin \omega t \\ &= -\omega^2 x \\ &= -\left(\frac{k}{m}\right) x \end{aligned}$$

Exercise VI.8.9: $x = \cos 3t$.

Exercise VI.8.12:
The second statement is proved as follows.

$$\begin{aligned} x &= c_1 e^{r_1 t} + c_2 e^{r_2 t} \\ x' &= r_1 c_1 e^{r_1 t} + r_2 c_2 e^{r_2 t} \\ x'' &= r_1^2 c_1 e^{r_1 t} + r_2^2 c_2 e^{r_2 t} \end{aligned}$$

So

$$ax'' + bx' + cx =$$
$$a(r_1^2 c_1 e^{r_1 t} + r_2^2 c_2 e^{r_2 t}) + b(r_1 c_1 e^{r_1 t} + r_2 c_2 e^{r_2 t}) + c(c_1 e^{r_1 t} + c_2 e^{r_2 t}) =$$
$$(ar_1^2 + br_1 + c)c_1 e^{r_1 t} + (ar_2^2 + br_2 + c)c_2 e^{r_2 t} =$$
$$0$$

The third statement is proved as follows.

$$\begin{aligned} x &= c_1 e^{rt} + c_2 t e^{rt} \\ x' &= r c_1 e^{rt} + c_2 e^{rt} + r c_2 t e^{rt} \\ x'' &= r^2 c_1 e^{rt} + r c_2 e^{rt} + r c_2 e^{rt} + r^2 c_2 t e^{rt} \\ &= r^2 c_1 e^{rt} + 2 r c_2 e^{rt} + r^2 c_2 t e^{rt} \end{aligned}$$

$$
\begin{aligned}
ax'' + bx' + cx &= \\
&a(r^2 c_1 e^{rt} + 2rc_2 e^{rt} + r^2 c_2 t e^{rt}) + \\
&b(rc_1 e^{rt} + c_2 e^{rt} + rc_2 t e^{rt}) + \\
&c(c_1 e^{rt} + c_2 t e^{rt}) = \\
&(ar^2 + br + c)c_1 e^{rt} + (a2r + b)c_2 e^{rt} + (ar^2 + br + c)c_2 t e^{rt} = 0
\end{aligned}
$$

Exercise VI.8.13: $x = e^{2t}$

Exercise VI.8.16:

By the theorem preceding the exercises the solution is of the form

$$x(t) = c_1 e^{it} + c_2 e^{-it}.$$

Thus

$$x'(t) = ic_1 e^{it} - ic_2 e^{-it}$$

Using the initial conditions to determine c_1 and c_2 we obtain

$$
\begin{aligned}
c_1 + c_2 &= 1 \\
ic_1 - ic_2 &= 0
\end{aligned}
$$

So

$$c_1 = c_2$$

and

$$c_1 = c_2 = \frac{1}{2}$$

giving us the solution

$$
\begin{aligned}
x(t) &= = \frac{1}{2}e^{it} + \frac{1}{2}e^{-it} \\
&= \left(\frac{1}{2}\right)(\cos t + i \sin t) + \left(\frac{1}{2}\right)(\cos(-t) + i\sin(-t)) \\
&= \left(\frac{1}{2}\right)(\cos t + i \sin t) + \left(\frac{1}{2}\right)(\cos t - i \sin t) \\
&= \cos t
\end{aligned}
$$

using the fact proved in Exercise V.6.8 that $e^{it} = \cos t + i \sin t$.

Exercise VI.8.17:

$$x(t) = \cos 3t$$

Since there is no friction this system will continue to oscillate back-and-forth between -1 and 1 forever.

Exercise VI.8.18:

$$x(t) = e^{-.05t}(\cos 2.999583t + .0167 \sin 2.999583t)$$

The small amount of friction will cause the oscillations to have smaller and smaller amplitude.
$$\lim_{t \to \infty} x(t) = 0.$$

Exercise VI.8.21: If there is no friction at all then the system oscillates forever with the same amplitude. If there is a moderate amount of friction then the system oscillates with decreasing amplitude. If there is a great deal of friction the system does not oscillate; it approaches the rest position.

Exercise VI.8.22: The stronger the spring, the faster the system oscillates, i.e., the higher the frequency.

Exercise VI.8.23: Friction decreases the frequency of oscillation.

Exercise VI.8.24: Unless $x(0) = x'(0) = 0$ the initial conditions do not affect the long term behavior of a spring-and-mass system.

Exercise VI.8.25: Unless $x(0) = x'(0) = 0$ the initial conditions do not affect the frequency with which a spring-and-mass system oscillates.

Exercise VI.9.1:

$$\begin{aligned}
(f+g)(x+y) &= f(x+y) + g(x+y) \\
&= f(x) + f(y) + g(x) + g(y) \\
&= f(x) + g(x) + f(y) + g(y) \\
&= (f+g)(x) + (f+g)(y)
\end{aligned}$$

$$
\begin{aligned}
(f+g)(kx) &= f(kx)+g(kx) \\
&= kf(x)+kg(x) \\
&= k(f(x)+g(x)) \\
&= k(f+g)(x)
\end{aligned}
$$

Exercise VI.9.2: Yes, it is *linear*.

$$
\begin{aligned}
H(f+g))(t) &= (f+g)(t-1) \\
&= f(t-1)+g(t-1) \\
&= H(f)(t)+H(g)(t)
\end{aligned}
$$

$$
\begin{aligned}
H(kf)(t) &= (kf)(t-1) \\
&= kf(t-1) \\
&= kH(f)(t)
\end{aligned}
$$

Exercise VI.9.3: The operator G is *linear*.

Exercise VI.9.4: The function $f(x,y) = xy$ is not *linear*.

Exercise VI.9.5:

$$
\begin{aligned}
f(0) &= f(0+0) \\
f(0) &= f(0)+f(0) \\
0 &= f(0)
\end{aligned}
$$

Exercise VI.9.6:

$$\begin{aligned}
f(0,0,\ldots,0) &= f(0+0, 0+0, \ldots, 0+0) \\
&= f(0,0,\ldots,0) + f(0,0,\ldots,0) \\
0 &= f(0,0,\ldots,0)
\end{aligned}$$

If $f(x_1, x_2, \ldots, x_n) = 0$ and $f(y_1, y_2, \ldots, y_n) = 0$ then

$$\begin{aligned}
f(x_1+y_1, x_2+y_2, \ldots, x_n+y_n) &= f(x_1, x_2, \ldots, x_n) + f(y_1, y_2, \ldots, y_n) \\
&= 0 + 0 \\
&= 0
\end{aligned}$$

Exercise VI.9.7:
Let $y = c_1 y_1 + c_2 y_2$. Then

$$\begin{aligned}
ay'' + by' + cy &= L(y) \\
&= L(c_1 y_1 + c_2 y_2) \\
&= L(c_1 y_1) + L(c_2 y_2) \\
&= c_1 L(y_1) + c_2 L(y_2) \\
&= c_1(0) + c_2(0) \\
&= 0
\end{aligned}$$

Exercise VI.9.8: In local coordinates $dy = 3\ dx$. In the original coordinates $y = 3x - 4$.

Exercise VI.9.9: In local coordinates $dy = -3dx/4$. In the original coordinates $y = -3x/4 + 25/4$.

Exercise VI.9.10: In local coordinates $dy = 0$. In the original coordinates $y = 1$.

Exercise VI.9.11: In local coordinates $dy = dx$. In the original coordinates $y = x$. Notice that at the point $(0,0)$ the local and the original coordinates happen to be the same.

Exercise VI.9.12:

$$\begin{aligned} z_1 &= 2 - i \\ z_2 &= 1.5 - 0.5i \\ z_3 &= 1.45 - 0.35i \\ z_4 &= 1.45534 - 0.343539i \end{aligned}$$

Exercise VI.9.13:

$$\begin{aligned} z_1 &= 1 \\ z_2 &= 1.5 - 0.5i \\ z_3 &= 1.45 - 0.35i \\ z_4 &= 1.45534 - 0.343539i \end{aligned}$$

Exercise VI.9.14: There is one equilibrium point $-2i$. This equilibrium point is attracting and since the dynamical system is linear for any initial z_1

$$\lim_{n \to \infty} z_n = -2i.$$

Exercise VI.9.16: There are two equilibrium points,

$$\frac{1 \pm \sqrt{1 - 4i}}{2}$$

or

$$\begin{aligned} z_1 &= 1.300 - 0.625i \\ z_2 &= -0.300 + 0.625i \end{aligned}$$

Now, letting

$$f(z) = z^2 + i$$

we see that

$$\begin{aligned} f'(z) &= 2z \\ f'(z_1) &= 2.600 - 1.250i \\ |f'(z_1)| &= 2.885 \\ f'(z_2) &= -0.600 + 1.250i \\ |f'(z_2)| &= 1.386 \end{aligned}$$

and neither equilibrium point is attracting.

Exercise VI.10.1: See Figures A.71 and A.72.

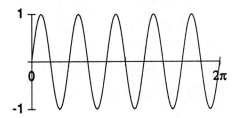

Figure A.71: $y = \sin 5x$

Figure A.72: $r = \sin 5\theta$

Exercise VI.10.2: $\pi/12$.

Exercise VI.10.3:

$$u + v = 2$$
$$u - v = 2i$$
$$uv = 2$$
$$u/v = i$$

Exercise VI.10.4:

$$\sqrt{2} \text{ cis } 0 = \sqrt{2}$$
$$\sqrt{2} \text{ cis } \pi/4 = 1 + i$$
$$\sqrt{2} \text{ cis } \pi/2 = i\sqrt{2}$$
$$\sqrt{2} \text{ cis } 3\pi/4 = -1 + i$$
$$\sqrt{2} \text{ cis } \pi = -\sqrt{2}$$
$$\sqrt{2} \text{ cis } 5\pi/4 = -1 - i$$
$$\sqrt{2} \text{ cis } 3\pi/2 = -i\sqrt{2}$$
$$\sqrt{2} \text{ cis } 7\pi/4 = 1 - i$$

Exercise VI.10.5: Your answer to this question depends on how you compare the value of future money with present money, that is, with your assumptions about interest rates or inflation. The table below was computed using the indicated interest rates.

Prize	2%	4%	6%	8%	10%	12%
$50,000 per year for 20 years	833,923	706,697	607,906	530,180	468,246	418,289
$100,000 per year for 10 years	916,224	843,533	780,169	724,689	675,902	632,825
$80,000 per year for 10 years	732,979	674,827	624,135	579,751	540,722	506,260
$30,000 per year forever	1,530,000	780,000	530,000	405,000	330,000	280,000

In most of the examples in the table above the best prize is $100,000 per year for 10 years. When the interest rate is very low (2%), however, the $30,000 per year forever prize is the best. When the interest rate is very high (12%) this prize is the worst.

Exercise VI.10.6: The radius of convergence is 1.

Exercise VI.10.7: There are many possible answers.
The power series
$$p(z) = \sum_{n=0}^{\infty} n! z^n$$
has radius of convergence 0.
The power series
$$p(z) = \sum_{n=0}^{\infty} \frac{z^n}{\pi^n}$$
has radius of convergence π.
The power series
$$p(z) = \sum_{n=0}^{\infty} \frac{z^n}{n!}$$
has radius of convergence ∞.

Exercise VI.10.8: See Figure A.73.

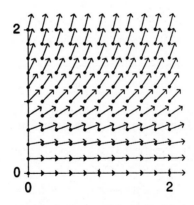

Figure A.73: Slope field for $y' = y^2$

Based on this slope field one would expect that the solution would go to $+\infty$. However, the exact solution is $y = 1/(1-t)$ and
$$\lim_{t \to \infty} \frac{1}{1-t} = 0!!!!$$
The solution to this apparent paradox is that the solution does go to ∞. However, it does so so quickly that
$$\lim_{t \to 1^-} y(t) = +\infty.$$

The power series solution is $1 + t + t^2 + t^3 + \cdots$. The radius of convergence of this power series is 1.

The fundamental oddity of this initial value problem is that it goes haywire (i.e., goes off to infinity) as t approaches 1. This behavior is manifested by the denominator $(1-t)$ in the exact solution and by the radius of convergence 1 in the power series solution.

Exercise VI.10.10: If b is zero the solution oscillates forever with the same amplitude. If b is moderate then the solution oscillates with decreasing amplitude. If b is large then the solution doesn't oscillate and as t approaches infinity the solution approaches zero. The behavior described above is independent of p. In the real-world of spring-and-mass systems the constant b represents friction.

Exercise VII.1.1: The diagram below shows the direction field for this function. There are two equilibrium points, $p = 2$ and $p = 4$. The equilibrium point $p = 2$ is attracting and the equilibrium point $p = 4$ is repelling.

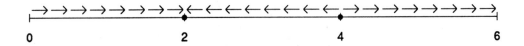

Direction field for exercise 1

The solution of the initial value problem:

$$p' = (p-2)(p-4) \qquad p(0) = 0.50$$

will approach the equilibrium point $p = 2$.

The solution of the initial value problem:

$$p' = (p-2)(p-4) \qquad p(0) = 1.50$$

will approach the equilibrium point $p = 2$.

The solution of the initial value problem:

$$p' = (p-2)(p-4) \qquad p(0) = 2.50$$

will approach the equilibrium point $p = 2$.

The solution of the initial value problem:
$$p' = (p-2)(p-4) \qquad p(3) = 2.50$$
will approach the equilibrium point $p = 2$. The only difference between this initial value problem and the preceding one is the time at which "the cork is placed in the river."

The solution of the initial value problem:
$$p' = (p-2)(p-4) \qquad p(0) = 5.00$$
will shoot off towards $+\infty$.

Exercise VII.1.3: A linear one-dimensional autonomous dynamical system of the form:
$$p' = mp + b$$
has exactly one equilibrium point $p_* = -b/m$. This equilibrium point is attracting if $m < 0$ and is repelling if $m > 0$.

If the equilibrium point is attracting then
$$\lim_{t \to +\infty} p(t) = p_*$$
regardless of the initial condition p_0.

If the equilibrium point is repelling then

- If $p_0 < p_*$ then
$$\lim_{t \to +\infty} p(t) = -\infty$$

- If $p_* < p_0$ then
$$\lim_{t \to +\infty} p(t) = +\infty$$

Exercise VII.1.4: If $f'(p_*)$ is positive then p_* is repelling. If $f'(p_*)$ is negative then p_* is attracting. If p_* is attracting then if the initial condition p_0 is *close* to p_* then
$$\lim_{t \to +\infty} p(t) = p_*$$
If p_* is repelling then there are many different possibilities for the longterm behavior of $p(t)$. For example, $p(t)$ might be attracted to another equilibrium point as with the repelling equilibrium point 0 for our model of a species that hunts in packs, or $p(t)$ might shoot off to plus or minus infinity.

The initial value p_0 makes a great deal of difference.

Exercise VII.1.5: For the first dynamical system:

$$v' = -32 - kv$$

there is exactly one equilibrium point, $v_* = -32/k$. This equilibrium point is attracting since this is a linear dynamical system with negative slope $(-k)$. The velocity v_* is frequently called the *terminal velocity*. Because this is a linear dynamical system

$$\lim_{t \to +\infty} p(t) = v_*$$

regardless of the initial velocity.

For the second dynamical system it helps to graph the function

$$-32 - kv^2 \text{sign}(v)$$

See the figure below. In order to graph this function notice that

$$-32 - k\text{sign}(v)v^2 = \begin{cases} -32 + kv^2, & \text{if } v \leq 0; \\ -32 - kv^2, & \text{if } 0 \leq v. \end{cases}$$

This function has one zero, at the point

$$v_* = -\sqrt{\frac{32}{k}}$$

and is positive to the left of v_* and negative to the right of v_*. Hence, the equilibrium point v_* is attracting. Thus on both sides of v_* the direction field is pointing towards v_* and for any initial condition v_0 we will have:

$$\lim_{t \to +\infty} p(t) = v_*$$

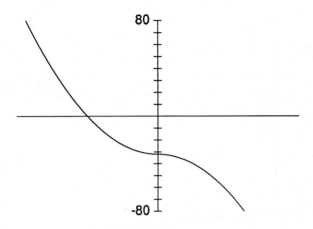

Graph of $-32 - kv^2\text{sign}(v)$

If the force due to air resistance is given by $g(v)$ then the velocity will be described by the dynamical system:

$$v' = -32 + g(v)$$

The conditions postulated for the function $g(v)$ imply that the function $-32 + g(v)$ will always have one zero, v_*, and will be positive to the left of that zero and negative to the right of that zero. This, in turn, implies that v_* will be an attracting equilibrium and, moreover, that regardless of the initial velocity v_0.

$$\lim_{t \to +\infty} v(t) = v_*$$

Exercise VII.2.1: There are four equilibrium points; $(0,0), (2000,0), (0,2000)$, and $(181.82, 181.82)$. These points are circled in Figure A.74. Left and right arrows indicating the way in which p is changing are shown in Figure A.75 and up and down arrows indicating the way in which q is changing are shown in Figure A.76. The information from these two pictures is combined to give a rough picture of the direction field in Figure A.77. Notice that there are two arrows pointing toward the equilibrium at $(181.82, 181.82)$ and two pointing away from this equilibrium. We can gain more understanding of this dynamical system

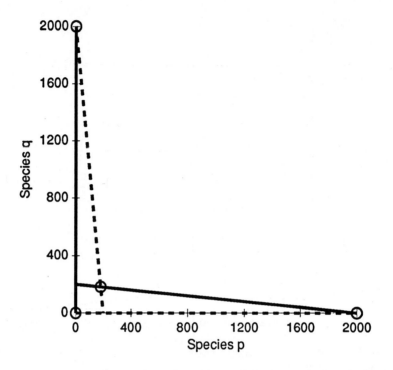

Figure A.74: $p' = 0$ and $q' = 0$

by looking at a few trajectories. Figure A.78 shows six trajectories. Most of them go to either the equilibrium point $(2000, 0)$ or to the equilibrium point $(0, 2000)$. Experimentation shows that if the initial condition (p_0, q_0) has exactly the same population for p and q (i.e., if $p_0 = q_0$) then the trajectory will head towards the equilibrium point in the middle at $(181.82, 181.82)$. However, if either species has even the slightest advantage then the one with the advantage will survive and the other one will be wiped out. Thus, for this particular model of an aggressive interaction we see that it is extremely unlikely that both species will survive. Usually one species wipes the other one out.

Exercise VII.3.1: $75/2$ dyne-centimeters.

Exercise VII.3.2: $135/2$ dyne-centimeters. The two answers are different because although the spring is stretched the same distance in both exercises, in the second exercise it is stretched from $x = 2$ to $x = 7$ and is under higher tension then when it is stretched from $x = 0$ to $x = 5$ in the first exercise.

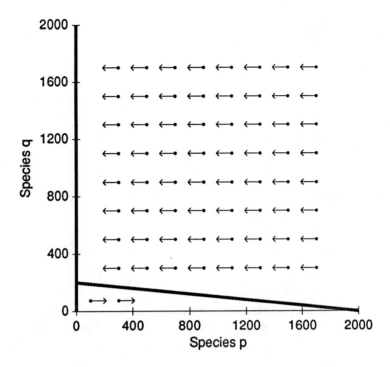

Figure A.75: Left and right arrows according to p'

Exercise VII.3.3: 6 dyne-centimeters.

Exercise VII.3.4: $kb^2/2 - ka^2/2$.

Exercise VII.3.5: $-3k/2$. The answer is negative because less energy is stored in the spring when $x = 1$ then when $x = -2$.

Exercise VII.3.6:

$$x(t) = e^{-0.005t}\left(10\cos(\sqrt{3.9999}\,t/2) + \frac{0.1}{\sqrt{3.9999}}\sin(\sqrt{3.9999}\,t/2)\right).$$

The factor $e^{-0.005t}$ causes the solution to approach zero as t goes to $+\infty$.

Exercise VII.4.1:

$$\lim_{t \to +\infty} y(t) = +\infty$$

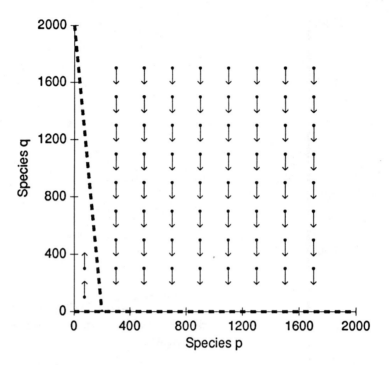

Figure A.76: Up and down arrows according to q'

Exercise VII.4.2:
$$\lim_{t \to +\infty} y(t) = -\infty$$

Exercise VII.4.5:

$$\begin{aligned} ky + b &= 0 \\ ky &= -b \\ y &= -\frac{b}{k} \end{aligned}$$

Thus this dynamical system has exactly one equilibrium point $y = -b/k$ if $k \neq 0$.

$$\begin{aligned} \frac{dy}{dt} &= ky + b \\ \frac{dy}{ky+b} &= dt \end{aligned}$$

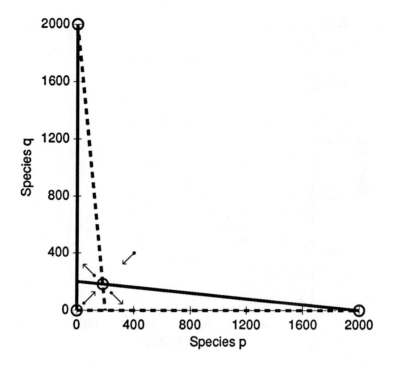

Figure A.77: General direction field

$$\left(\frac{1}{k}\right)\left(\frac{dy}{y+\frac{b}{k}}\right) = dt$$

$$\left(\frac{1}{k}\right)\ln(y+b/k) = t + c_1$$

$$\ln(y+b/k) = kt + kc_1 = kt + c$$

$$y + b/k = e^{kt+c}$$

$$y = Ce^{kt} - \frac{b}{k}$$

Thus the equilibrium point is attracting is $k < 0$ and repelling if $k > 0$. If $k > 0$ and the initial condition $y_0 > -k/a$ then

$$\lim_{t \to +\infty} y(t) = +\infty$$

and if $y_0 < -k/a$ then

$$\lim_{t \to +\infty} y(t) = -\infty.$$

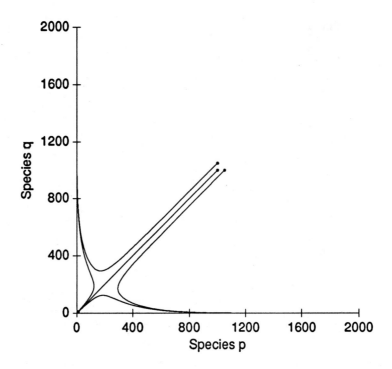

Figure A.78: Six trajectories for aggressive competition

Exercise VII.4.7: The equilibrium point $(0,0)$ is repelling. In this case the solutions of initial value problems will fly off to plus or minus infinity.

Exercise VII.5.1: See Figure A.79

Exercise VII.5.5: See Figure A.80

Exercise VII.5.7:

$$\frac{\partial z}{\partial x} = y + 2xy + 2$$

$$\frac{\partial z}{\partial y} = x + x^2 - 3$$

Exercise VII.5.8:

$$\frac{\partial z}{\partial x} = \cos y$$

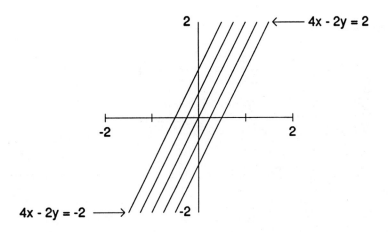

Figure A.79: Contour lines for $z = 4x - 2y$

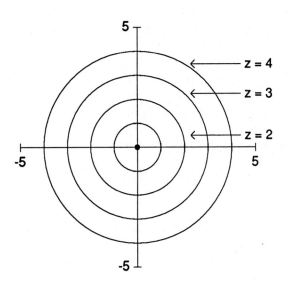

Figure A.80: Contour lines for exercise VI.5.5

$$\frac{\partial z}{\partial y} = -x \sin y$$

Exercise VII.5.9:

$$\frac{\partial z}{\partial x} = \frac{x}{\sqrt{x^2 + y^2}}$$

$$\frac{\partial z}{\partial y} = \frac{y}{\sqrt{x^2 + y^2}}$$

Exercise VII.5.10: The manufacturer must find the minimum of the function:

$$S = 2xh + 2yh + xy = \frac{2}{y} + \frac{2}{x} + xy$$

The partial derivatives are

$$\frac{\partial S}{\partial x} = -\frac{2}{x^2} + y$$

$$\frac{\partial S}{\partial y} = -\frac{2}{y^2} + x$$

So one needs to look at solutions of the dynamical system:

$$x' = \frac{2}{x^2} - y$$

$$y' = \frac{2}{y^2} - x$$

Every solution of this dynamical system converges to the equilibrium point $(1.2599, 1.2599)$.

Exercise VII.5.12: The following discussion is for the problem we looked at in exercise VI.5.10. We need to solve the equations:

$$-\frac{2}{x^2} + y = 0$$

$$-\frac{2}{y^2} + x = 0$$

or
$$y = \frac{2}{x^2}$$
$$x = \frac{2}{y^2}$$

or
$$x^2 y = 2$$
$$xy^2 = 2$$

or
$$xy = \frac{2}{x}$$
$$xy = \frac{2}{y}$$

So,
$$\frac{2}{x} = \frac{2}{y}$$

and
$$x = y$$

Substituting $x = y$ back into the equation $x^2 y = 2$ we get
$$y^3 = 2$$

So
$$y = x = \sqrt[3]{2}$$

Exercise VII.5.13: To find a local maximum rather than a local minimum of the function $f(x, y)$ look at the dynamical sytem:

$$x' = \frac{\partial f}{\partial x}$$
$$y' = \frac{\partial f}{\partial y}$$

(i.e., the direction of steepest ascent) rather than
$$x' = -\frac{\partial f}{\partial x}$$
$$y' = -\frac{\partial f}{\partial y}$$
(i.e., the direction of steepest descent).

Exercise VII.5.14: $x = y = h = 0.4082$

Exercise VII.7.1: We are looking at the model
$$p' = F_1(p, q) = (2 - .001p - .01q)p$$
$$q' = F_2(p, q) = (2 - .01p - .001q)q$$
Thus,
$$\frac{\partial F_1}{\partial p} = 2 - .002p - .01q$$
$$\frac{\partial F_1}{\partial q} = -.01p$$
$$\frac{\partial F_2}{\partial p} = -.01q$$
$$\frac{\partial F_2}{\partial q} = 2 - .01p - .002q$$
There are four equilibrium points

1. The point $(0,0)$. At this point we have
$$a_{11} = \left(\frac{\partial F_1}{\partial p}\right)(0,0) = 2$$
$$a_{21} = \left(\frac{\partial F_1}{\partial q}\right)(0,0) = 0$$
$$a_{21} = \left(\frac{\partial F_2}{\partial p}\right)(0,0) = 0$$
$$a_{22} = \left(\frac{\partial F_2}{\partial q}\right)(0,0) = 2$$

So
$$b = -(a_{11} + a_{22}) = -4 \quad \text{and} \quad c = (a_{11}a_{22} - a_{12}a_{21}) = 4$$
and
$$\Delta = b^2 - 4c = 0$$
Since Δ is zero we need to examine
$$r = -\frac{b}{2} = 2.$$
Since r is positive this equilbrium point is repelling. This is exactly what we would expect form our earlier work. The equilibrium point at $(0,0)$ is always repelling for these kinds of models.

2. The point $(2000, 0)$. At this point we have

$$a_{11} = \left(\frac{\partial F_1}{\partial p}\right)(2000, 0) = -2$$

$$a_{21} = \left(\frac{\partial F_1}{\partial q}\right)(2000, 0) = -20$$

$$a_{21} = \left(\frac{\partial F_2}{\partial p}\right)(2000, 0) = 0$$

$$a_{22} = \left(\frac{\partial F_2}{\partial q}\right)(2000, 0) = -18$$

So
$$b = -(a_{11} + a_{22}) = 20 \quad \text{and} \quad c = (a_{11}a_{22} - a_{12}a_{21}) = 36$$
and
$$\Delta = b^2 - 4c = 256$$
Since Δ is positive we need to examine
$$r_1 = -2 \quad \text{and} \quad r_2 = -18$$
Since both are negative the equilibrium point $(2000, 0)$ is attracting. This is exactly what we would expect from our earlier work.

3. The analysis of the equilibrium point $(0, 2000)$ is very similar.

4. The point $(181.82, 181.82)$. At this point we have

$$a_{11} = \left(\frac{\partial F_1}{\partial p}\right)(181.82, 181.82) = -0.1818$$

$$a_{21} = \left(\frac{\partial F_1}{\partial q}\right)(181.82, 181.82) = -1.8182$$

$$a_{21} = \left(\frac{\partial F_2}{\partial p}\right)(181.82, 181.82) = -1.8182$$

$$a_{22} = \left(\frac{\partial F_2}{\partial q}\right)(181.82, 181.82) = -0.1818$$

So

$$b = -(a_{11} + a_{22}) = 0.3636 \quad \text{and} \quad c = (a_{11}a_{22} - a_{12}a_{21}) = -3.2727$$

and

$$\Delta = b^2 - 4c = 13.22$$

Since Δ is positive we need to examine

$$r_1 = 1.64 \quad \text{and} \quad r_2 = -2.00$$

Since one of these is positive and the other is negative this point is indeterminate. In this case usually solutions will move away from the point $(181.82, 181.82)$ but sometimes they can move towards the point $(181.82, 181.82)$. This matches our earlier observations. It is possible for two aggressive species to coexist in a fragile state of equilibrium but usually one or the other species will die out.

Exercise VII.8.1: Any point on the S-axis (i.e. $I = 0$) is an equilibrium point.

Exercise VII.8.2: See Figure A.81

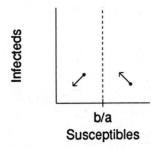

Figure A.81: The direction field for the SIR model

Exercise VII.8.4:

$$\frac{dI}{dS} = \frac{dI/dt}{dS/dt}$$
$$= \frac{0.001IS - 0.5I}{-0.001IS}$$
$$= \frac{500}{S} - 1$$

So looking at the trajectory as a function $I = f(S)$ we see that

$$I = \int \frac{500}{S} - 1 \, dS$$
$$= 500 \log S - S + c$$

Using the initial conditions $S(0) = 990$ and $I(0) = 10$ we see that

$$10 = 500 \log 990 - 990 + c.$$

So

$$c = 1000 - 500 \log 990$$

and

$$I = 500 \log S - S + 1000 - 500 \log 990.$$

We find where this trajectory hits the line $I = 0$ by solving the equation

$$500 \log S - S + 1000 - 500 \log 990 = 0$$

numerically. The result is $S \approx 199.796$. Thus, the epidemic will approach an equilibrium with approximately 199.796 people left who never catch the disease.

Exercise VII.8.8:

$$\frac{dI}{dS} = \frac{dI/dt}{dS/dt}$$
$$= \frac{0.001IS - 0.1I}{-0.001IS}$$
$$= \frac{100}{S} - 1$$

So looking at the trajectory as a function $I = f(S)$ we see that

$$I = \int \frac{100}{S} - 1 \, dS$$
$$= 100 \log S - S + c$$

Using the initial conditions $S(0) = 990$ and $I(0) = 10$ we see that

$$10 = 100 \log 990 - 990 + c.$$

So
$$c = 1000 - 100 \log 990$$

and
$$I = 100 \log S - S + 1000 - 100 \log 990.$$

We find where this trajectory hits the line $I = 0$ by solving the equation
$$100 \log S - S + 1000 - 100 \log 990 = 0$$
numerically. The result is $S \approx 0.0449661$. Thus, the epidemic will approach an equilibrium with approximately 0.0449661 people left who never catch the disease, that is, virtually everyone will eventually catch the disease.

Exercise VII.8.12: Some epidemics continue until virtually everyone has caught and eventually recovered from the disease. The most important factor is the ratio b/a. The number of infected people will continue to rise until the remaining susceptible population is b/a and then the number of infected people will begin to decline. Depending on how big b/a is, some fraction of the population will never catch the disease.

Index

2-cycles 246-248, 300-301, 313, 315-317
4-cycles 316-317
absolute convergence 623
absolute rate of growth 714
absolute value 95-98
acceleration 304-306, 308
additivity 422
aggressive interaction 727, 730, 744, 808
air pressure 361
air resistance 725
angle of coincidence 256-257, 259-260, 272
angle of elevation 326-327
angle of incidence 256-257, 259-260, 272
angles 148-149
animation 717, 719
antiderivative 433, 435
antidifferentiation 437, 440
apparent image 261, 263-270, 272
apparent position 261, 263-270, 272
approaches 20-21, 29
approximation 33, 114-115
arccosecant 234
arccosecant, derivative of 234
arccosine 230-231
arccosine, derivative of 230-231
arccotangent 234
arccotangent, derivative of 234
arclength 546, 549
arcsecant 232, 234
arcsecant, derivative of 232, 234
arcsin, integration of 485
arcsine 228-231
arcsine, derivative of 229
arctan 231
arctan, derivative of 231
arctan, integration of 485
arctangent 231-232
arctangent, derivative of 231-232
area 34-36, 403, 406-409, 433, 435, 455-458
area in polar coordinates 594-595
area, signed 442-444, 456-458
area, unsigned 442-444, 456-458
Aristotle 317
Asimov, Isaac 507
attracted 806
attracting 722, 740, 755, 797, 799-800, 802-803, 807
attracting cycle 301
attracting equilibrium point 59, 71, 73, 76, 98-99, 110-111, 130-132, 280-281, 286-289, 298, 338, 756, 763-765, 767-768
autonomous 331, 715, 732
average 214

average rate of change 282
average velocity 282
basin of attraction 789
Binary Chop 41
binocular vision 261, 263-270, 272
Binomial Formula 124
biomass 5
Bisection Method 38, 41-44, 46-48, 50, 59, 114, 274, 278
black holes 495, 506-507, 509
bound 6
bounded 19
bounded above 19-21
bounded below 19-21
bounded intervals 192
bounded series 618
Byties 77, 79-82
calculators 23
Capulets 730
cardioid 589
carrying capacity 343, 529, 718
caustics 249, 256-257, 259-260
centroids 575
Chain Rule 238-241, 292, 445-446
characteristic equation 681, 683, 685
chord 83, 85-89, 91-93
circle, area of 556
circle, circumference of 550
coexist 740
coffee 711-712, 744
Comparison Test 621-622, 627
competitive interaction 727, 730, 797
completing the square 564
complex numbers 596, 598-599, 601, 603, 637, 639, 641, 692, 702
composing functions 236-237
compound 712
compounding 362-363
computer 734
computer graphics 45, 734
computer programs 715
computers 23
concave down 308-311
concave up 308-311
cone 488, 491, 493
cone, volume of 491, 493
conjugate 600
conservation laws 497
conservation of energy 497, 745, 750, 752
Constant Multiple Rule 125, 126, 291
continuity 50, 52-54, 57-59
continuous 711, 713
continuous dynamical system 525
continuous dynamical systems 112, 298
continuous exponential model 525
continuous logistic model 528-529
contour lines 777
converges 21, 612
cooling 711
cork 717, 733
critical mass 718
critical point 170, 295, 786
currents 717, 733
curve sketching 188, 190-191, 206-208, 210-212, 295, 387-391
cycles 246-248, 298, 300-301
definite integral 411, 416, 418-423, 425, 427-433, 435,

INDEX

440
degrees 148
demand 77, 79-82
dependent variable 179, 219, 401
derivative 93-100, 102, 104-106, 108, 110, 118, 225, 642, 667, 697-701, 704
derivative of a complex function 701
derivative, partial 781
derivatives, higher 629
derivatives, numerical estimates 664-665, 667
determinant 775
differential equations 317-319, 321-322
differential operators 691
differentiation formulas 291-292
direction field 719, 721, 733, 797
discrete 711
discrete dynamical systems 18, 110-112, 298, 525
discriminant 688, 761
displacement 445
distance, signed 445
distance, unsigned 445
diverges 612
domain 191, 383, 401
domain of a power series 637
domain of power series 638-640
Doppler Effect 429
dynamical systems 110-111, 525, 703-704
dyne 675
e 369
economics 77, 79-82, 112, 166-168
Einstein, Albert 497, 745

elastic 225
ellipse 551
ellipsoid 493
ellipsoid, volume of 493
emmigration 113
endpoints 164, 295
energy 323, 495-496, 745
energy, conservation of 497, 745
epidemic 811
equilibrium point 14, 18, 21, 31-33, 60-63, 98-99, 110-111, 130-132, 338-339, 721, 735, 738, 755, 797
equilibrium population 17
equilibrium position 676
equilibrium price 77, 79-82
escape velocity 495, 500, 506
estimation 33-36, 41, 114-115, 117
Euler's Formula 683
Euler's Method 345-359, 366-367, 402
Euler's Theorem 359-360
even 123
excess demand 79
experimentation 10-14, 22
exponential function 360, 365-371, 373
exponential function, derivative of 367
exponential function, properties of 367, 369, 371
exponential growth 727
exponential model with emmigration 113
exponential model with immigration 7, 113
exponential models 4-6, 8, 110-111, 525

exponentiation 374, 376-378
extended initial value table 629-630
Extreme Value theorem 285
factorial 628
falling bodies 317-319, 321-322, 672
Fermat's Principle 249, 251, 261, 263-270, 272, 787
first order differential equations 749
food 458
force 495-496, 745
free lunch 497
friction 675-676, 680, 745, 752
function 401
Fundamental Theorem of Calculus 425, 427-433, 435, 437
Galileo 317
geometric series 605-606, 622, 638-639
global maximum 164
global minimum 164
graphical computation 64, 66-67, 69
graphical iteration 64, 66-67, 69
graphical methods 740
gravitational constant 500
gravity 317-319, 321-322, 383, 500, 725, 759
growth rate 713
habitats 9
half-life 381
handle 604
harmonic series 613-615, 617-618, 628, 652
heat energy 745
higher derivatives 629

Hooke's Law 676
horizontal symptote 194-197
Hugo, Victor 494
image 219
imaginary part 596
immigration 7, 31, 526, 805
implicit differentiation 513-516
improper integral 505, 508
indefinite integral 433, 435
independent variable 179, 219, 401
indeterminate 723, 797, 800, 803, 806
indeterminate equilibrium point 763, 765, 769
infecteds 812
infinite series 610
inflection point 311
initial conditions 319
initial value problem 319, 401, 717
integers 597
integral 411, 416, 418-423, 425, 427-433, 435
integral curve 734
Integral Test 618-619, 621
integration 595
integration by parts 477-482
integration formulas 437
integration, numerical 460, 464
interest 361-364, 712
Intermediate Value Theorem 48, 50, 59, 206, 740
intervals 191-192
inverse 217-219, 221-222, 225
Inverse Function Rule 222, 292
inverse trigonometric functions 228, 230-232, 234

inverting 217-219
kinetic energy 323, 497-498, 745
L'Hôpital's Rule 394-397, 399
large 194
levers 575
light caustics 256-257, 259-260
light ray 787
limit 20-21, 24-26, 28-33, 36, 89, 91-93, 117, 196-198, 200-202, 385-386, 411
linear 419-420, 690-694, 755, 797
linear classification theorem 762
linear demand function 166-168
linear dynamical systems 70-71, 73, 75-76, 110-111, 724, 757
linear equation 38, 83
linear functions 699
linear operator 694
linearity 437
local coordinates 697
local maximum 164
local minimum 164
location problems 176-177, 183, 243-245
Log, Taylor Series for 649
logarithm 372-374, 382
logarithm, integration of 480
logarithmic differentiation 378-379
logistic equation 525, 529, 538-539
logistic models 10-14, 21-23, 59, 98-99, 110-111, 113-114, 130-132, 528-529, 728
long division for polynomials 562
lottery 604, 607
lower bound 19-20

lunch, free 497
magnitude 599
mass 495-496
Mathematica 23, 45, 183, 229
maximum 164, 789
mean 214-216
Mean Value Theorem 282-285, 548
median 214-216
method of steepest ascent 786
method of steepest descent 776
microphone 589
microscope 100, 102, 104
Midpoint Method 116, 460
minimum 164, 789
mirror 249, 251, 261, 263-270, 272, 787
mirror image 121
models 1-2, 18
Montagues 730
movie 717, 719
MOVIE1 719
MOVIE2 721, 734
multiplication 403
nanosecond 261
natural logarithm 372-374
newton 496
Newton's Method 274-275, 277-279, 300-301, 699
Newton's Method, Complex Case 701-702
Newton, Isaac 759
niche 806
nonautonomous 331
nonlinear 690, 755, 800
nonlinear classification theorem 802
nonlinear dynamical systems 111

nonnegative series 618
notation 110
Notre Dame de Paris 494
numerical integration 437, 460, 464
odd 123
omnidirectional 589
one-dimensional 711, 755
one-sided estimate of the derivative 667
one-to-one 218, 221
One-way Convergence Test 615, 628
operator 694
optics 249, 251, 261, 263-270, 272, 300
optimization 162-164, 166-168, 170, 173-174, 176-177, 179-180, 182-184, 187, 191, 193-194, 203-212, 243-245, 297-298, 328, 393
order of a differential equation 671
pack-hunters 9, 15, 17, 714
partial derivative 359, 781, 801
partial fractions 525, 531, 562
partial slope 693, 777, 798, 800
partial sum 610
payoff 604
peaks 789
per capita 714
period 157
period-doubling 313, 315-317
periodic 157
phase diagrams 719
pi 34-36
piecewise linear 546
point of inflection 311

polar coordinates 585-587, 589-592, 594-595, 598-600, 603
polar coordinates, area 594
polynomial derivative of 128
polynomials 121, 128
population 458
population growth 797, 803
population models 1, 4-6, 110-111, 525
population multiplier 4, 8-9, 15
positivity 421
potential energy 497-498, 745
power functions 121
Power Rule 124, 143-144, 222, 242, 291
power series 635-636
power series centered at a 639
power series centered at zero 637
power series, derivative of 641-642
predator-prey interaction 727, 731, 744, 808
present value 605-609
product of two functions 134
Product Rule 134-136, 138-139, 291, 477
programs 715
proof 73
pyramid 493-494
pyramid, volume of 493
quadrants 154
Quadratic Formula 39, 204, 564
Quotient Rule 139-140, 142-143, 291
radians 148-149
radioactive decay 360, 371-372, 379, 381
radius of convergence 640-642,

INDEX

648-649, 652
range 217, 219
rate of change 105-106, 108, 110
rate of growth 713
Ratio Test 626, 628, 648, 652
rational functions 562
real numbers 597
real part 596
recovereds 812
refraction 261, 263-270, 272
relative derivative 146
relative growth rate 525, 714
Relativity Theory 506
repelled 806
repelling 723, 755, 797-798, 800, 802-803, 805
repelling equilibrium point 339, 756, 763-765
Riemann Sums 409-411, 414-416, 437, 548
river 717
saddle point 789, 793, 799
sample points 407, 410
scientific method 73
sea 731
secant 83
secant, integral of 553
second derivative 304-306, 308-311
second order differential equations 749
sensitivity 108, 110, 729
separation of variables 519, 522-523, 526
sequence 3
sigma notation 410-411
signed area 442-444, 456-458
signed distance 445

Simpson's Rule 463-464
sine, derivative of 158, 160
Sine, Taylor Series for 646
SIR model 811
slope 83, 85-89, 91-98, 118, 692
slope field 336-339
slope, partial 777
small 194
smooth 104
solar power 582
solids of revolution 487-488, 491
solution of a differential equation 333
solution of an initial value problem 332-333
speed limit 323
speeding ticket 283-284
sphere 488, 492
sphere, volume of 492
spiral 551, 590
spring 306
spring constant 677, 746
spring energy 747
spring-and-mass system 671, 675-678, 683
stable 722
stable equilibrium 287-289, 298
steepest ascent, method of 786
steepest descent, method of 776
substitution 445-449
Sum Rule 127, 291
summation notation 410-411
supply 77, 79-82
susceptibles 812
symmetric difference estimate of the derivative 667
symmetry 123, 129
tangent 83, 85-89, 91-98, 104

tangent field 336-339
tangent, integral of 449
taxes 603
Taylor Polynomial 659-661, 663
Taylor Series 645-647, 649-652, 656, 658-660
Taylor's Theorem 655, 658, 660, 664, 667-668
temperate zone insects 2
temperature 711, 744
term 610
Theory of Relativity 506
threshold 343
tides 501
time-dependent 331
tolerance 28-33, 36, 44, 46, 52-54, 57-58, 197-198, 411
topographic maps 777
torque 575
totally nonautonomous 331
trace 775
trajectory 324-326, 734
trapezoid method 461
trigomometric functions 157
trigonometric functions 149-151, 153-154, 157
trigonometric functions, derivatives of 160
trigonometric functions, integration of 486-487, 553
trigonometric identities 157
trigonometric substitution 556-557
trigonometry 147
two-dimensional 712, 732, 758
two-dimensional motion 324-326
unbounded intervals 192
unsigned area 442-444, 456-458

unsigned distance 445
unstable 723
unstable equilibrium 287-289, 298
upper bound 19-21
valleys 789
vector fields 719
velocity 104-105
vertical asymptotes 383, 385-391
volume 403-404, 417, 487-488, 491
weak-strong interaction 727, 731, 744, 808
work 323, 495-496, 745